Cell Adhesion

Frontiers in Molecular Biology

SERIES EDITORS

B. D. Hames

*Department of Biochemistry
and Molecular Biology
University of Leeds, Leeds LS2 9JT, UK*

D. M. Glover

*Department of Genetics,
University of Cambridge, UK*

TITLES IN THE SERIES

Preface

It has long been recognized that cell adhesion is essential for the development and co-ordinated function of multicellular organisms. Cells can bind selectively to components present in the extracellular matrix or on the surfaces of other cells. These recognition and adhesion events are now known to be mediated by a number of different classes of cell surface receptors. In recent years, it has also become clear that engagement of cell surface adhesion molecules does much more for cells than simply tethering them to some location within an organism. Rather, it is now evident that cell adhesion events can provide critical information about environmental conditions and can affect many central aspects of cell structure and function, including cytoskeletal organization, cell polarity, cell proliferation, and gene expression.

In the past two decades, there have been rapid advances in our understanding of the molecular mechanisms by which cells recognize adhesive ligands in the extracellular environment and on the surfaces of other cells. The first part of this book provides an up-to-date discussion of the structure and function of the seven major classes of eukaryotic cell adhesion molecules. The first chapter, contributed by Hansjürgen Volkmer, describes the immunoglobulin superfamily of cell adhesion molecules, which plays a central role in development. Selectins, which represent a group of lectins that mediate physiologically important cell adhesion events in the vascular system, are described by Rod McEver. An important group of cell–cell adhesion molecules required for development and implicated in tumorigenesis, the cadherin family, is discussed by Glenn Radice and Masatoshi Takeichi. Integrins, which represent a major class of cell surface receptors for extracellular matrix, are described by Doug DeSimone and colleagues. In the past few years, three other groups of molecules that are involved in cell adhesion have been identified and characterized. These include the cell surface heparan sulfate proteoglycans that are described in the chapter by Mert Bernfield and colleagues, the ADAMs family discussed in the chapter by Judy White and colleagues, and the cell surface protein tyrosine phosphatases described in the chapter by Susann Brady-Kalnay. All of the chapters in the first section of the book include a discussion of the molecular features of the protein family, an analysis of their functional properties and patterns of expression, as well as a discussion of their roles in development and disease. Many exciting new insights derived from genetic analysis of cell adhesion molecule function are considered.

In the second part of the book, we discuss the structure and function of adhesive junctions. Junctional complexes represent specialized regions of the cell surface that mediate adhesion. These junctions have unique morphologies, molecular compositions, and physiological roles. At these specialized regions of the cell membrane, specific adhesion receptors are linked to cytoskeletal elements and signal transduction machineries. The biochemical organization of these adhesive membranes and the molecular mechanisms by which they transmit information to the cell interior are addressed. The features of adherens junctions, a major class of cell–cell junctions, are discussed in the chapter by Barry Gumbiner and colleagues. Focal adhesions and focal complexes, which mediate interactions between cells and the extracellular matrix, are described by Lynda Peterson and Keith Burridge. Desmosomes and hemidesmosomes, which play a critical role in tissue integrity, are described in the chapter by Kathy Green and colleagues. Finally, the tight junction is discussed by Shoichiro Tsukita and colleagues.

Our goal in writing this book was to provide a series of current, in-depth reviews of the field of cell adhesion that would be useful for students, instructors, and researchers. The authors, who are leaders in their fields, have worked very hard to provide a broad perspective and sufficient background information to make the information easily accessible to students and scientists outside the area of cell adhesion. Yet, each section also includes the most recent developments and a discussion of emerging trends in the area in an effort to ensure that the contents will also be stimulating reading for workers in the field. I want to thank all of the contributing authors for their hard work and enthusiastic commitment to this project and Jana Brubaker for skilled administrative assistance. It is my fondest hope that this book will serve as a valuable primer on cell adhesion for both students and researchers and will convey the tremendous importance of cell adhesion for cell function, development, and disease.

Salt Lake City   M. C. B.
September 2001

Contents

4 The integrin family of cell adhesion molecules   100

BETTE J. DZAMBA, MARGARET A. BOLTON, AND DOUGLAS W. DESIMONE

5 Cell surface heparan sulfate proteoglycans   155

OFER REIZES, PYONG WOO PARK, AND MERTON BERNFIELD

6 ADAMs

JUDITH M. WHITE, DORA BIGLER, MICHELLEE CHEN, YUJI TAKAHASHI,
AND TYRA G. WOLFSBERG

7 Protein tyrosine phosphatases

SUSANN M. BRADY-KALNAY

8 The adherens junction   259

CARA J. GOTTARDI, CARIEN M. NIESSEN, AND BARRY M. GUMBINER

9 Focal adhesions and focal complexes

LYNDA PETERSON AND KEITH BURRIDGE

10 Desmosomes and hemidesmosomes

LESLIE J. BANNON, LAWRENCE E. GOLDFINGER, JONATHAN C. R. JONES, AND KATHLEEN J. GREEN

11 The tight junction   369

SHOICHIRO TSUKITA, MIKIO FURUSE, AND MASAHIKO ITOH

Index   397

Contributors

LESLIE J. BANNON
Department of Pathology , Northwestern University, 303 East Chicago Avenue, Chicago, IL 60611, USA.

MARY C. BECKERLE
Huntsman Cancer Institute and Department of Biology, University of Utah, 2000 Circle of Hope, Room 3270, Salt Lake City, UT 84112, USA.

MERTON BERNFIELD
The Children's Hospital and Harvard Medical School, 300 Longwood Avenue, Enders 950, Boston, MA 02115–5737, USA.

DORA BIGLER
Department of Cell Biology, University of Virginia Health System, School of Medicine, PO Box 800732, Charlottesville, VA 22908 0732, USA.

MARGARET A. BOLTON
Department of Cell Biology, University of Virginia Health System, School of Medicine, PO Box 800732, Charlottesville, VA 22908–0732, USA.

SUSANN M. BRADY-KALNAY
Department of Molecular Biology and Microbiology, Case Western Reserve University, 10900 Euclid Avenue, Cleveland, OH 44106–4960, USA.

KEITH BURRIDGE
Department of Cell and Developmental Biology and Lineberger Comprehensive Cancer Center, 108 Taylor Hall, CB 7090, University of North Carolina at Chapel Hill, NC 27599–7090, USA.

MICHELLEE CHEN
Department of Cell Biology, University of Virginia Health System, School of Medicine, PO Box 800732, Charlottesville, VA 22908–0732, USA.

DOUGLAS W. DESIMONE
Department of Cell Biology, University of Virginia Health System, School of Medicine, PO Box 800732, Charlottesville, VA 22908–0732, USA.

BETTE J. DZAMBA
Department of Cell Biology, University of Virginia Health System, School of Medicine, PO Box 800732, Charlottesville, VA 22908–0732, USA.

MIKIO FURUSE
Department of Cell Biology, Faculty of Medicine, Kyoto University, Sakyo-ku, Kyoto 606, Japan.

LAWRENCE E. GOLDFINGER
Department of Vascular Biology, The Scripps Research Institute, 10550 North Torrey Pines Road, La Jolla, CA 92037, USA.

CARA J. GOTTARDI
Cellular Biochemistry & Biophysics Program, Memorial Sloan-Kettering Cancer Center, 1275 York Avenue, Box 564, New York, NY 10021, USA.

KATHLEEN J. GREEN
Departments of Pathology and Dermatology and the Robert H. Lurie Cancer Center, Northwestern University Medical School, 303 East Chicago Avenue, Chicago, IL 60611, USA.

BARRY M. GUMBINER
Cellular Biochemistry & Biophysics Program, Memorial Sloan-Kettering Cancer Center, 1275 York Avenue, Box 564, New York, NY 10021, USA.

MASAHIKO ITOH
Department of Cell Biology, Faculty of Medicine, Kyoto University, Sakyo-ku, Kyoto 606, Japan.

JONATHAN C. R. JONES
Departments of Cell and Molecular Biology and the Robert H. Lurie Cancer Center, Northwestern University Medical School, 303 East Chicago Avenue, Chicago, IL 60611, USA.

RODGER P. MCEVER
W. K. Warren Medical Research Institute, University of Oklahoma Health Sciences Center, 825 N. E. 13th Street, Oklahoma City, OK 73104, USA.

CARIEN M. NIESSEN
Cellular Biochemistry & Biophysics Program, Memorial Sloan-Kettering Cancer Center, 1275 York Avenue, Box 564, New York, NY 10021, USA.

PYONG WOO PARK
The Children's Hospital and Harvard Medical School, 300 Longwood Avenue, Enders 950, Boston, MA 02115–5737, USA.

LYNDA PETERSON
Department of Cell and Developmental Biology and Lineberger Comprehensive Cancer Center, 108 Taylor Hall, CB 7090, University of North Carolina at Chapel Hill, NC 27599–7090, USA.

GLENN L. RADICE
Center for Research on Reproduction and Women's Health, Department of Obstetrics and Gynecology, University of Pennsylvania School of Medicine, Philadelphia, PA 19104–6142, USA.

OFER REIZES
The Children's Hospital and Harvard Medical School, 300 Longwood Avenue, Enders 950, Boston, MA 02115–5737, USA.

YUJI TAKAHASHI
Department of Cell Biology, University of Virginia Health System, School of Medicine, PO Box 800732, Charlottesville, VA 22908–0732, USA.

MASATOSHI TAKEICHI
Department of Cell and Developmental Biology, Graduate School of Biostudies, Kyoto University, Kitashirakawa, Sakyo-ku, Kyoto 606–8502, Japan.

SHOICHIRO TSUKITA
Department of Cell Biology, Faculty of Medicine, Kyoto University, Sakyo-ku, Kyoto 606, Japan.

HANSJÜRGEN VOLKMER
Naturwissenschaftliches und Medizinisches Institut (NMI), Markwiesenstr. 55, 72770 Reutlingen, Germany.

JUDITH M. WHITE
Department of Cell Biology, University of Virginia Health System, School of Medicine, PO Box 800732, Charlottesville, VA 22908–0732, USA.

TYRA G. WOLFSBERG
National Center for Biotechnology Information (NIH/NLM), Building 38A, Room 8N805, 8600 Rockville Pike, Bethesda, MD, 20894, USA.

Abbreviations

ADAM	a disintegrin and a metalloprotease
AJ	adherens junction
ASIP	atypical PKC isotype-specific interacting protein
Bnl	*branchless*
BP180	bullous pemphigoid antigen II
BP230	bullous pemphigoid antigen I
Btl	*breathless*
CAM	cell adhesion molecule
CASPR	contactin-associated protein
CCBD	central cell binding domain (of fibronectin)
C-CPE	C-terminal half of *Clostridium perfringens* enterotoxin
CEPU-1	cerebellar Purkinje cell-1
CHO	Chinese hamster ovary
CKII	casein kinase II
CLIBS	cation- and ligand-induced binding site
CMT	Charcot–Marie–Tooth syndrome
CNR	cadherin-related neuronal receptor
CNS	central nervous system
Col	collagen
CPE	*Clostridium perfringens* enterotoxin
CPE-R	*Clostridium perfringens* enterotoxin receptor
CSAT	cell substrate attachment antigen
C-terminus	carboxy terminus
DCC	deleted in colorectal cancer
dlg	*discs large*
DP	desmoplakin
dpp	*decapentaplegic*
DS	desmosome
Dsc	desmocollin
Dsg	desmoglein
DSS	Dejerine–Sottas syndrome
E	embryonic day
EC	extracellular domain
E-cadherin	epithelial cadherin
ECM	extracellular matrix
EGF	epidermal growth factor
ERK	extracellular signal-regulated kinase
ERM	ezrin–radixin–moesin

ES cell	embryonic stem cell
ESL-1	E-selectin ligand-1
FAK	focal adhesion kinase
FBG	fibrinogen
FGF-2	fibroblast growth factor-2
FGFR	fibroblast growth factor receptor
FN	fibronectin
FNIII repeat	fibronectin type III repeat
FNIII-like	fibronectin type III-like
GAG	glycosaminoglycan
GAP	GTPase activating protein
GDI	guanine nucleotide dissociation inhibitor
GEF	guanine nucleotide exchange factor
GlcNAc	*N*-acetyl glucosamine
GlcUA	glucuronic acid
GlyCAM-1	glycosylated cell adhesion molecule-1
GPI	glycosyl phosphatidylinositol
Grb2	Grb2 adaptor protein
GUK domain	guanylate kinase-like domain
Hh	*hedgehog*
HIP	HS/heparin interacting protein
HS	heparan sulfate
HSDs	highly sulfated domains
HSPG	heparan sulfate proteoglycan
Htl	*heartless*
IAP	integrin-associated protein
IdUA	iduronic acid
Ig	immunoglobulin domain
IgCAM	immunoglobulin-like domain containing cell adhesion molecules
IGF-2	insulin-like growth factor-2
Ig-like	immunoglobulin-like
IL-1	interleukin-1
ISDs	intermediate sulfated domains
JAM	junction-associated membrane protein
LAMP	limbic system-associated protein
LAP	latency-associated peptide
LFA	lymphocyte function-associated antigen
LIBS	ligand-induced binding site
LM	laminin
LPA	lysophosphatidic acid
LPS	lipopolysaccharide
mAb	monoclonal antibody
mADAM	mouse ADAM
MAdCAM-1	mucosal addressin cell adhesion molecule-1

MAGUK	membrane-associated guanylate kinase
MAM domain	a sequence motif found in meprins, A5 (neuropilin), PTPMu
MAM	meprin, A5, μ
MAPK	mitogen-activated protein kinase
MDC	metalloprotease/disintegrin/cysteine-rich
MDCK cell	Madin–Darby canine kidney cell
MIDAS	metal ion-dependent adhesion site
MLC	myosin light chain
MLCK	myosin light chain kinase
MMP	matrix metalloprotease
N-cadherin	neural cadherin
NCAM	neural cell adhesion molecule
NDST	*N*-deacetylase *N*-sulfotransferase
NgCAM	neuron–glia cell adhesion molecule
NrCAM	NgCAM-related cell adhesion molecule
N-terminus	amino terminus
OBCAM	opioid binding cell adhesion molecule
OPN	osteopontin
p120	phosphoprotein 120 kDa
pAb	polyclonal antibody
PAK	p21-activated kinase
P-cadherin	placental cadherin
PCR	polymerase chain reaction
PDGF	platelet-derived growth factor
PDZ	PSD-95, discs-large, and zonula occludens-1
Pg	plakoglobin
PG	proteoglycan
PI3-K	phosphoinositide 3-kinase
PIP2	phosphatidylinositol 4,5-bisphosphate
PKC	protein kinase C
PKP	plakophilin
PMA	phorbol 12-myristate 13-acetate
PS	position-specific
PSGL-1	P-selectin glycoprotein ligand-1
PTK	protein tyrosine kinase
PTP	protein tyrosine phosphatase
PTP-PEST	protein tyrosine phosphatase with PEST repeats
R-cadherin	retinal cadherin
RGCs	retinal ganglion cells
RGD(E)S	arginine–glycine–aspartic acid–(glutamic acid)–serine
RNAi	RNA interference
RPTP	receptor protein tyrosine phosphatase
SDS–PAGE	sodium dodecyl sulfate–polyacrylamide gel electrophoresis
sfl	*sulfateless*

SGB	Simpson–Golabi–Behmel
sgl	*sugarless*
sLe^x	sialyl Lewis x
SOS	son of sevenless gene product
SPPK	striate palmoplantar keratoderma
Src	Src kinase
SVMP	snake venom metalloprotease
TER	*trans*-epithelial resistance
TJ	tight junction
TN	tenascin
TNF	tumour necrosis factor
TSP	thrombospondin
ttv	*tout velu*
Tyr	tyrosine
UMDs	unmodified domains
VASP	vasodilator-stimulated phosphoprotein
VCAM-1	vascular cell adhesion molecule-1
VE-cadherin	vascular endothelial cadherin
VLA	very late antigen
VN	vitronectin
VWF	von Willebrand factor
wg	*wingless*
ZO	*Zonula occludens*
ZOT	*Zonula occludens* toxin

1 | The immunoglobulin superfamily of cell adhesion molecules

HANSJÜRGEN VOLKMER

1. Introduction

IgCAMs (immunoglobulin-like domain containing cell adhesion molecules) are part of the immunoglobulin superfamily, a widespread and complex protein family of more than 100 members that is associated with many different functions (see ref. 1 for a comprehensive review). IgCAMs function as cell adhesion and signalling receptors that transduce extracellular signals from neighbouring cells or the extracellular matrix (ECM) to the intracellular signalling machinery.

Cell–cell and cell–ECM interactions during embryonal development and in the adult organism may be regulated by IgCAMs. For instance, the development of the nervous system requires many highly specific molecular interactions for neural differentiation to occur. IgCAMs are involved in crucial steps of neural wiring like migration, neurite induction, pathfinding, and plasticity (2). To build up complex structures, complementary molecular cues are needed. Such cues may be supplied by IgCAM molecules that are highly diverse in their structure, expression pattern, and functional interactions.

IgCAMs contain one or an array of Ig-like domains that can be combined with other protein motifs (1). Ig-like domains that have been extensively studied in immunoglobulins usually contain two conserved cysteine residues that are linked by intrachain disulfide bonds. A growing list of IgCAMs has been described and some vertebrate members are depicted in Fig. 1. Structural diversity is achieved by post-translational modification of extracellular domains by glycosylation, or by modification of intracellular domains through serine, threonine, or tyrosine phosphorylation (3). Additional variants of some IgCAMs are created by alternative splicing that changes the domain organization of IgCAMs.

Some of the IgCAMs are expressed in a wide variety of tissues suggesting a general biological role (1). Other IgCAMs show more restricted expression patterns consistent with tissue-specific functions. Expression is often regulated during

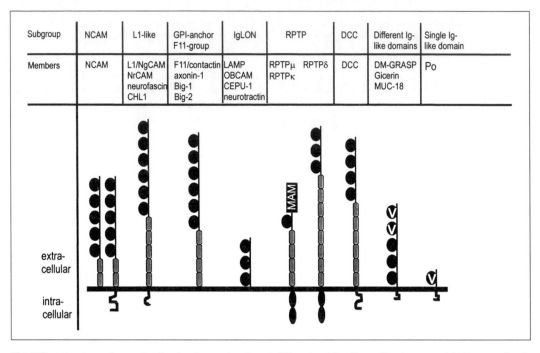

Subgroup	NCAM	L1-like	GPI-anchor F11-group	IgLON	RPTP		DCC	Different Ig-like domains	Single Ig-like domain
Members	NCAM	L1/NgCAM NrCAM neurofascin CHL1	F11/contactin axonin-1 Big-1 Big-2	LAMP OBCAM CEPU-1 neurotractin	RPTPμ RPTPκ	RPTPδ	DCC	DM-GRASP Gicerin MUC-18	Po

Fig. 1 The structure of neural cell adhesion molecules of different subfamilies of the immunoglobulin superfamily (IgCAM) are represented schematically. Members of the respective subfamilies are depicted. A black bar represents the membrane plane, Ig-like domains are shown in dark grey circles, and the FNIII-like repeats are light grey shaded boxes. Phosphatase-like domains are shown as black elipses and 'MAM' refers to MAM domains (see text).

development. The diversity of IgCAMs created by post-translational modification or alternative splicing may also be subject to developmental or tissue-specific regulation (2, 4).

IgCAMs interact with many other cell surface molecules including other IgCAMs as well as molecules belonging to other protein families. The capacity of IgCAMs to interact with a large variety of other molecules suggests that IgCAMs participate in a sophisticated network of molecular interactions. Interaction patterns can be subject to regulation and may thereby influence the biological function of IgCAMs. Examples will be given for different functional complexes formed by IgCAMs with respect to axon–axon, axon–glia, or axon–ECM interactions.

In addition to the many different and complex extracellular interactions, the cytoplasmic domains of transmembrane IgCAMs also participate in intracellular interactions. Thus IgCAMs may play an important role in aspects of cell motility. For example, the cytoplasmic domains of IgCAMs are able to bind to components of the cortical cytoskeleton like ankyrins (3).

Signal transduction is also modulated by the action of IgCAMs. Evidence has been provided that IgCAMs are associated with many different signalling molecules like receptor and non-receptor tyrosine kinases, receptor protein tyrosine phosphatases

(RPTP), and various serine/threonine kinases. In the nervous system, IgCAM signalling has been shown to be important for neurite extension and fasciculation. Recently, evidence has been provided that IgCAMs may also control neuronal survival.

Investigations of IgCAM functions *in vivo* have demonstrated a pivotal role for these molecules in the development of the nervous system. Studies of commissural axon pathfinding at the midline of chick embryos have elucidated a role of NrCAM and axonin-1 in axon guidance. Analysis of mutated mice demonstrated the importance of several IgCAMs, including L1/NgCAM, NCAM, P_0, and DCC, for cell migration, axon guidance, myelination, learning and memory, as well as motor functions.

IgCAMs are also implicated in neurodegenerative disorders (5). For example mutations of L1 are associated with X-linked hereditary hydrocephalus. IgCAMs have also been shown to be involved in the migration of malignant glioma and melanoma cells. Interestingly, DCC has been shown to play a dual role as a tumour suppressor gene and a signalling receptor for axon guidance. These aspects of IgCAM structure and function will be discussed in detail below.

2. Structure of IgCAMs

A main feature for the classification of IgCAMs is the presence of at least one extracellular Ig-like domain as in the case of P_0. Ig-like domains can be found either as the only extracellular sequence motif or in combination with other sequence motifs (Fig. 1).

2.1 Structural features of Ig-like domains

Knowledge of the structural features of immunoglobulins helped to deduce the structure of related IgCAMs (6). Individual Ig-like domains are 70 to 110 amino acid residues in length. The primary sequence contains characteristic, highly conserved amino acid residues. Two cysteine residues are separated by a stretch of 55 to 75 amino acid residues. Ig-like domains consist of two opposed β-sheets stabilized by a disulfide bridge formed by the two conserved cysteine residues (6). The β-sheets are formed by a series of antiparallel β-strands connected by loops of different length (1).

Inspection of the sequence of Ig-like domains led to the suggestion to categorize Ig-like domains in groups that were termed 'sets'. Sets differ in the number and size of the β-strands as well as the size and conformation of the loops. Three categories were originally defined including the C1-set, the C2-set, and the V-set (6). V and C1 refer to variable and constant domains of immunoglobulins. V domains are generally larger than C1-type domains. The C2-set was found to combine features of the C1-set and V-set since the fold is C-type but amino acid sequences of the β-strands near the C-terminal conserved cysteine residue are similar to the V-set. Later, a further set of Ig-like domains, termed I-set, was defined. Members of the I-set are more closely related to the V-set based on re-examination and modelling of Ig-like domains of

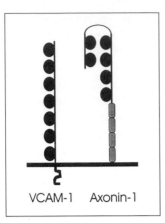

VCAM-1   Axonin-1

Fig. 2 Structures proposed for extracellular domains of IgCAM. VCAM may have a rod-like structure. The two NH$_2$-terminal Ig-like domains axonin-1 may fold back to generate a horseshoe-like structure.

IgCAMs against turkey telokin, the C-terminal domain of smooth muscle myosin light chain kinase, which has a known Ig-like structure resolved by X-ray analysis (7, 8). So far, only few Ig-like domains or arrays of Ig-like domains have been analysed by X-ray or NMR analysis. Therefore, the precise classification of Ig-like domains remains to be established in the future.

Ig-like domains in IgCAMs may be arranged in an open chain to create a rod-like structure as in the case of VCAM-1 (Fig. 2) (9). In contrast, evidence has been provided that an array of Ig-like domains may also fold back with a sharp bend resembling a horseshoe. Evidence for such a horseshoe structure has been shown for axonin-1 as shown by negative staining electron microscopy (10). Furthermore, indirect evidence has been provided that Ig-like domains of L1-like IgCAMs may also fold back in a horseshoe-like structure (11). The crystal structure of silk moth haemolin, a protein that contains Ig-like domains similar to Ig-like domains of L1, revealed that a bend is located between the second and third of the four Ig-like domains and allows intimate contact between the first and fourth as well as second and third Ig-like domain. Accordingly, linkage of the Ig-like domains 2 and 3 of L1 may be more flexible than linkage of all other neighbouring Ig-like domains that are fixed to each other by inter-domain β-sheets (12).

2.2   FNIII-like repeats

Fibronectin type III-like repeats (FNIII-like repeats) are protein domains that are frequently combined with Ig-like domains. This motif is repeated 16 times in the extracellular matrix component fibronectin. It has a length of about 90 amino acid residues and contains conserved tryptophan and tyrosine residues (1). These conserved residues are also found in FNIII-like repeats of IgCAMs. The overall sequence of these repeats is distantly related to the FNIII repeats of fibronectin and they are therefore termed FNIII-like repeats. FNIII-like repeats may be found in tandem downstream of the Ig-like domains. Crystallographic analysis revealed a striking structural similarity of FNIII-like repeats and Ig-like domains. FNIII-like repeats consist of two β-sheets composed of a series of antiparallel β-strands. However, in contrast to Ig-like domains, the FNIII-like repeats are usually not stabilized by disulfide bridges.

2.3 Membrane linkage of IgCAMs

Most IgCAMs are type I transmembrane proteins that contain a membrane spanning α-helical hydrophobic sequence of about 25 amino acid residues in length. A cytoplasmic domain is located at the carboxy terminus of the transmembrane domain and may participate in intracellular signal transduction or cytoskeletal association. Alternatively, IgCAMs may be anchored in the cytoplasmic membrane by a glycosyl phosphatidylinositol (GPI) anchor. In the case of P_0, neurofascin, and NCAM, the polypeptide chains may be additionally linked to the membrane by palmitoylation (13–15). This modification may localize IgCAMs to specific membrane microdomains.

2.4 IgCAM subgroups

IgCAMs are grouped into several subfamilies according to sequence similarities and structural composition (Fig. 1) (1). Sections 2.4.1 to 2.4.8 briefly summarize the structural features of different IgCAM subgroups.

2.4.1 NCAM

Different subfamilies of IgCAMs share a combination of Ig-like domains and FNIII-like repeats. Neural cell adhesion molecule (NCAM) and related molecules from *Drosophila melanogaster*, grasshopper (fasciclin II), and *Aplysia* (ApCAM) are found in many different species from nematodes to mammals and consist of five Ig-like and two FNIII-like repeats. Three major isoforms of mammalian NCAM are known including NCAM-120, -140, and -180. They are generated by alternative splicing. NCAM-120 is linked to the membrane via a GPI anchor and NCAM-140 and -180 contain transmembrane domains followed by cytoplasmic domains of different length at the COOH-terminus. A small alternatively spliced exon, 30 bp in length, termed VASE or π, may be integrated into the fourth Ig-like domain. Further small miniexons located between the two FNIII-like repeats may be spliced in different combinations. They code for peptides rich in serine and threonine residues and may be O-glycosylated. O-linked glycosylation may serve to stabilize an extended stretch of amino acid residues, thereby raising NCAM above the membrane plane (16). The expression of many variants in different tissues and at different stages of development as described in Section 3.1 and 3.2 underlines the importance of NCAM for multiple cell–cell interactions.

2.4.2 L1-like IgCAMs

L1/NgCAM, neurofascin, NrCAM, and CHL1 are grouped into the L1-subfamily of IgCAMs. L1 is found in mammalian species whereas NgCAM is considered as the chick species homologue based on common features like structure, expression, and function (1). Therefore 'L1/NgCAM' will be used in the text except for the presentation of data that refer to L1 or NgCAM exclusively.

An L1-related molecule termed neuroglian was also described in *Drosophila*. All these molecules contain six Ig-like and five FNIII-like domains, a transmembrane segment, and a cytoplasmic domain. The sequence of the cytoplasmic domain is most conserved among the different members of this subgroup. Neurofascin differs from other L1-like molecules because it contains a domain of 80 amino acid residues in length termed PAT-domain that mainly consists of proline, alanine, and threonine residues. This domain is located between the fourth and fifth FNIII-like repeat and may be a target for *O*-linked glycosylation causing an extended structure (16).

Different alternatively spliced sequences have been identified at conserved positions within the polypeptides of L1-like IgCAMs, thus enhancing structural diversity. The highest number of different isoforms generated by alternative splicing has been identified for neurofascin (4). At least 50 isoforms can be expressed from 10 alternatively spliced exons. Four small exons code for sequences between 4 and 17 amino acid residues in length and are located at the NH_2-terminus, between the second and third Ig-like domain, between the Ig-like and FNIII-like repeats, and in the cytoplasmic domain. Three larger domains including the third FNIII-like repeat, the PAT-domain, and the fifth FNIII-like repeat are encoded by pairs of adjacent exons.

L1-subgroup members are important for neuron–neuron and neuron–glia interactions which are key determinants for the development of the nervous system. For example mutations of the L1 gene may result in severe and complex disorders as described in Section 8.

2.4.3 GPI-anchored IgCAMs of the F11/contactin subgroup

Representatives of the F11/contactin subgroup of GPI-anchored IgCAMs include chick F11, also termed contactin, axonin-1, and BIG-1 and -2 (1). They consist of six Ig-like and four FNIII-like domains. As they lack a cytoplasmic domain, signalling into the cell presumably requires association with other transmembrane receptors.

2.4.4 GPI-anchored IgCAMs of the IgLON subgroup

The IgLON subfamily includes LAMP, OBCAM, and neurotrimin as well as CEPU-1 and neurotractin (17). They lack other extracellular sequence motifs beside Ig-like domains and contain three Ig-like domains that are linked to the membrane by a GPI anchor. In the case of CEPU-1, a soluble isoform lacking the GPI anchor has been described (18). Some members like CEPU-1 or LAMP, show a restricted expression pattern arguing for a very specific role in the function of the nervous system. CEPU-1 is primarily expressed in Purkinje cells of the cerebellum and LAMP was found in the limbic system.

2.4.5   Receptor protein tyrosine phosphatases

Several receptor protein tyrosine phosphatases (RPTP) including RPTPδ, RPTPκ, and RPTPμ are IgCAMs that are composed of Ig-like domains and FNIII-like repeats (19–21). All display two phosphatase domains in the cytoplasmic region. RPTPμ and RPTPκ contain a MAM domain at the NH_2-terminus, one Ig-like domain, and four FNIII-like repeats. MAM refers to a putative adhesive motif found in zinc-metalloprotease meprin, neural cell surface molecule A5, and RPTPμ (22). RPTPδ contains three Ig-like and eight FNIII-like repeats in the extracellular domain (23). RPTPs are highly interesting as they may directly couple cell adhesion to intracellular signalling. RPTPs are also discussed in detail in Chapter 7.

2.4.6   DCC

DCC (Deleted in Colorectal Cancer) was originally described as a tumour suppressor gene (24). DCC and the chick homologue neogenin contain four Ig-like and six FNIII-like domains followed by a transmembrane domain and a cytoplasmic domain at the COOH-terminus (25). DCC was found on neural cells and plays an important role in the guidance of commissural axons during development (see Section 7.1).

2.4.7   IgCAMs with different types of Ig-like domains

The IgCAM subgroups described above are believed to contain Ig-like domains belonging to the same set. Other IgCAMs are composed of Ig-like domains of different sets. For example, MUC18, DM-GRASP, and Gicerin have two V-set Ig-like domains at the NH_2-terminus and membrane proximal Ig-like domains of the C2-set (1).

2.4.8   Single Ig-like domain (P_0)

P_0 contains a single V-set Ig-like domain, a transmembrane domain, and a COOH-terminal cytoplasmic domain (1). It is an example for a small IgCAM that is important for myelination. Mutations in humans may contribute to peripheral neuropathy as detailed in Section 7.5.

3.   Expression of IgCAMs

Expression patterns of IgCAMs may give a first hint of the function of IgCAMs. The structural diversity of IgCAMs caused by alternative splicing and/or post-translational modifications is accompanied by variable expression patterns. IgCAMs are found in many different tissues of neural and non-neural origin (1). For example NCAM, NrCAM, and L1/NgCAM are all expressed by neurons or glial cells in many different regions of the nervous system. On the other hand, some IgCAMs are expressed in restricted areas. For example, neurofascin as well as the GPI-linked molecules axonin-1/Tag-1, F11/contactin, and Big-1, -2 are mainly expressed in fibre-rich areas of the nervous system. Furthermore, expression can be regulated such that specific IgCAMs or IgCAM isoforms are confined to particular developmental stages, to certain tissues and cells, or even specific membrane regions of a single cell as described in Section 3.3.

3.1 Tissue-specific IgCAM expression

NCAM expression is an example for regulation of different isoforms in different tissues. 180 kDa and 140 kDa isoforms of NCAM are expressed in neurons, whereas glial cells mainly express the 140 kDa isoform. Myotubes were shown to express the 120 kDa GPI-anchored isoform of NCAM. Interestingly, NCAM expression in muscle is concentrated at the neuromuscular junction of myofibres in the adult, suggesting a role for synapse function (26).

Although L1 is found in different tissues, the cytoplasmic alternatively spliced exon of L1 is regulated by tissue-specific factors. An isoform lacking this exon was found in non-neural cells, myoblasts, and B cells but not in brain (27–29).

3.2 Regulation of IgCAM variants in development

Several examples demonstrate regulated expression of IgCAMs in the course of development. In the case of NCAM, the VASE exon is integrated into NCAM isoforms synthesized late in development, but is missing in brain regions that maintain plasticity and cellular growth in adult stages including hippocampus and

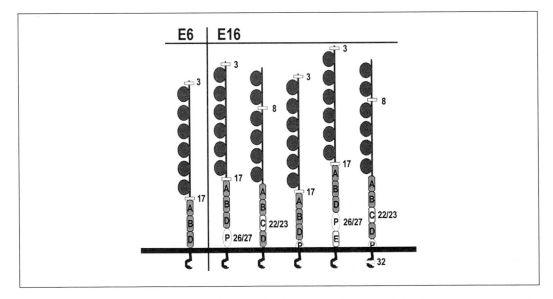

Fig. 3 Alternatively spliced isoforms of chick neurofascin. Early in development (embryonal day 6) one major isoform of neurofascin is expressed in brain. Later in development (embryonal day 16) the 'early' isoform is replaced by five different major 'late' isoforms of neurofascin. Ig-like domains are shown in dark grey circles and the FNIII-like repeats are light grey shaded boxes. FNIII-like repeats are marked with capital letters A to E and the PAT domain with P. Alternatively spliced exons are marked with exon numbers of the *neurofascin gene* (4).

the olfactory system (2). Developmental regulation was also shown for polysialic acid modifications (PSA) (2). PSA-NCAM is expressed in embryonal brain on extending axons or migrating cells. Later in development, PSA modification is down-regulated but is still maintained in regions of high plasticity. The expression of PSA suggests a role in plastic events either in the development of the nervous system or in learning processes in the adult brain. In the hippocampus, for instance, PSA is important for long-term potentiation (LTP), a mechanism that may be involved in learning and memory processes (30).

NrCAM is found on axon fibres, but it is also expressed on peripheral glial cells (31). In chick, transient expression of NrCAM was found in floor plate cells of the ventral prospective spinal cord, at a time-point when commissural axons are crossing the ventral midline (32). Later in development, NrCAM expression is down-regulated, implicating a specific function of NrCAM for axon guidance across the floor plate.

Alternative splicing of neurofascin is regulated during development (4). Early in development, when neurites grow out to form nerve tracts, one major isoform of neurofascin is expressed (Fig. 3). Later in development, when a considerable fraction of neural cells will go into terminal differentiation, the early isoform of neurofascin is replaced by five major late isoforms that are different from the early isoform. Oligodendrocyte-derived neurofascin is only transiently expressed during early myelinogenesis, but it is down-regulated later (33).

3.3 Expression of neurofascin in specific membrane domains

A neurofascin isoform lacking the third FNIII-like repeat is expressed in specific membrane compartments. It is confined to the nodes of Ranvier of sciatic nerves and to axon initial segments of cerebellar Purkinje cells (34). Interaction of neurofascin with ankyrin may be responsible for the restricted location. Neurofascin co-localizes with the cytoskeletal component ankyrin$_G$ at the nodal segments that bind to the intracellular region of neurofascin (34, for details see Section 5.1).

4. Extracellular interactions of IgCAMs

IgCAMs are transmembrane receptors that bind to molecules in the surrounding environment via their extracellular domains. Thus, signals may be transduced to the cellular signalling machinery to regulate cellular functions. Some interaction partners of IgCAMs are membrane receptors themselves, which raises the possibility of bidirectional intercellular signalling. Different modes of partner recognition have been described. Some IgCAMs, including the L1-like molecules, NCAM, and axonin-1, display homophilic binding; that is, the interaction occurs between two identical molecules. However, most IgCAMs have been shown to interact heterophilically either with non-identical IgCAMs or with structurally unrelated partners. IgCAMs expressed by one cell can bind to interaction partners on membranes of opposing cells in *trans*, but there are also lateral interactions with molecules within the same membrane plane in *cis*. Interactions with components of the extracellular matrix have also been described.

IgCAMs can interact with many different interaction partners as summarized in Fig. 4. Some IgCAMs have been shown to participate in multiprotein complexes that are important for neurite induction. Examples will be given to illustrate the importance of IgCAMs for neuron–neuron, neuron–glia, and neuron–ECM interactions.

4.1 Neuron–neuron interactions

L1/NgCAM and axonin-1 are both expressed on extending axons suggesting a role in axon–axon interactions and neuronal fasciculation. The homophilic interactions of L1/NgCAM in *trans* have been shown to induce neurite extension when expressed in transfected cells or supplied as a immobilized substrate protein (35, 36). In addition to homophilic interactions of L1/NgCAM, further heterophilic interactions of L1/NgCAM are important for neurite outgrowth. For example, inhibition of axonin-1 expression by antisense oligonucleotides revealed that axonal axonin-1 is required

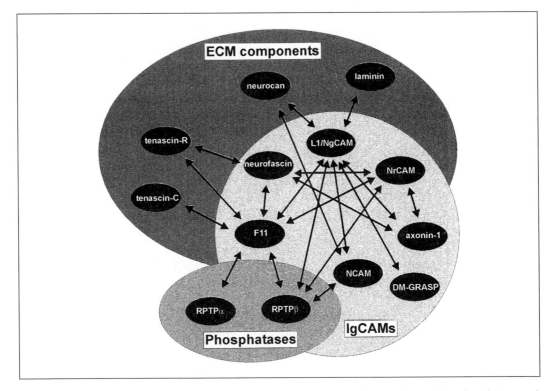

Fig. 4 Interaction patterns of IgCAMs, ECM components, and receptor protein tyrosine phosphatases. A collection of IgCAM interactions defined on the basis of binding studies. See text for interactions that are important for neural differentiation. Note that RPTPβ is also expressed as an ECM component termed phosphacan.

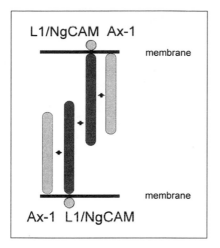

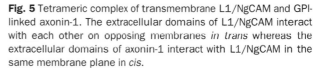

Fig. 5 Tetrameric complex of transmembrane L1/NgCAM and GPI-linked axonin-1. The extracellular domains of L1/NgCAM interact with each other on opposing membranes *in trans* whereas the extracellular domains of axonin-1 interact with L1/NgCAM in the same membrane plane in *cis*.

to allow neurite outgrowth of cultured dorsal root ganglion cells on an NgCAM substrate (37). Axonin-1 may interact with NgCAM within the same membrane plane in *cis* as shown by co-capping and crosslinking experiments (Fig. 5). Binding of axonin-1 to NgCAM in *cis* configuration can be maintained in the presence of a simultaneous homophilic NgCAM interaction in *trans*. These findings provide evidence that NgCAM binds to NgCAM homophilically in *trans* and that axonin-1 interacts with NgCAM heterophilically in *cis*. Both NgCAM and axonin-1 are expressed on opposing surfaces of fasciculating neurites. Thus, a tetrameric complex of two NgCAM and axonin-1 molecules could be established. In fact, a tetrameric complex of NgCAM and axonin-1 has been precipitated from fasciculating neural cultures and the extent of complex formation is dependent on the degree of fasciculation (38). Axons fasciculate with other axons to extend to their target area and establish nerve bundles by neuron–neuron interactions. The results described above suggest that tetrameric complexes of NgCAM and axonin-1 may participate in this process.

4.2 Neuron–glia interactions

The interactions of extending neurons with glial cells have been suggested to be important for neurite outgrowth. RPTPβ was found in radial glial and astrocytes that are implicated in axon guidance (39). Neural IgCAMs NCAM, NgCAM, NrCAM, and F11/contactin have been shown to bind to the receptor protein tyrosine phosphatase β (RPTPβ) *in vitro* (40–42). The extracellular portion of RPTPβ contains a domain related to carbonic anhydrase, a FNIII-like repeat, and a spacer region. The intracellular region contains two phosphatase domains. Three isoforms including two membrane-associated forms of different lengths and one soluble form were described. Both the soluble form and the long transmembrane form of RPTPβ may be expressed as a chondroitin sulfate proteoglycan. The soluble form is also known as phosphacan, a component of the extracellular matrix. Therefore, RPTPβ may interact

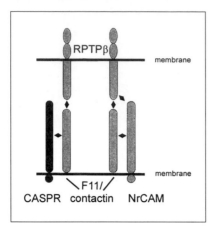

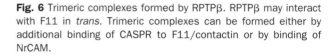

Fig. 6 Trimeric complexes formed by RPTPβ. RPTPβ may interact with F11 in *trans*. Trimeric complexes can be formed either by additional binding of CASPR to F11/contactin or by binding of NrCAM.

with F11/contactin either as a membrane-bound phosphatase or as a component of the extracellular matrix.

RPTPβ interacts with F11/contactin via the carbonic anhydrase domain (Fig. 6) and serves as a substrate to induce neurite outgrowth by interactions with receptor F11/contactin (41). As F11/contactin is linked to the membrane by a GPI anchor, it needs interactions with further transmembrane components to transmit signals into the interior of the cell. A complex of F11/contactin and the transmembrane cell surface molecule NrCAM can be formed within the same membrane plane in *cis* (Fig. 6) (42). This *cis* interaction may occur simultaneously with binding of F11/contactin to RPTPβ suggesting the formation of a tripartite complex among RPTPβ, F11/contactin, and NrCAM. In support of the view that the participation of NrCAM in this complex is functionally significant, anti-NrCAM antibodies have been shown to interfere with neurite outgrowth dependent on F11/contactin–RPTPβ interactions. NrCAM may link the F11/contactin–RPTPβ complex to the cytoskeleton as it was suggested to interact with ankyrin (34).

Another tripartite complex can be formed among RPTPβ, F11/contactin, and CASPR (contactin-associated protein) (43). CASPR is related to neurexin IV that was first identified in *Drosophila* (44). It is composed of a discoidin-like domain, several laminin G-like regions, EGF-like repeats, a proline-, glycine-, tyrosine-rich domain, and a transmembrane domain. The cytoplasmic domain of CASPR contains proline-rich sequences that may be recognized by SH3 domains of intracellular proteins important for signal transduction. Although the functional relevance of this triple complex is still unknown, the structural properties of CASPR make it an interesting candidate for signalling in axon guidance. Binding of NrCAM or CASPR to F11/contactin complexed to RPTPβ may be mutually exclusive or may happen simultaneously. However, the relationship of F11/contactin–RPTPβ–NrCAM and F11/contactin–RPTPβ–CASPR complexes also remains to be elucidated. The involvement of RPTPβ suggests bidirectional signalling. It may be possible that neurons signal back to the glial cells and induce phosphatase activity in glial cells via

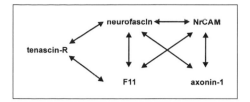

Fig. 7 Network of possible interactions as shown by *in vitro* binding studies. IgCAMs neurofascin, NrCAM, F11/contactin, and axonin-1, and extracellular matrix component tenascin-R may interact in many combinations.

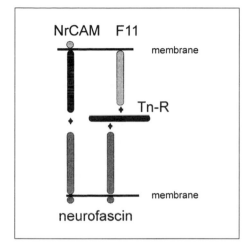

Fig. 8 Receptor switch in the presence of tenascin-R. NrCAM is used as a receptor for neurofascin to promote neurite outgrowth. In the presence of tenascin-R (TN-R) NrCAM is not used as receptor any more. Instead F11/contactin (F11) is involved in the recognition of a neurofascin/tenascin-R complex.

RPTPβ. Whether F11/contactin or the other RPTPβ ligands NCAM and NgCAM induce RPTPβ activity is still unknown.

4.3 Neuron–ECM interactions

Extending axons are not only confronted with the surface of surrounding cells but also with components of the extracellular matrix (ECM). Interaction with ECM components is another possibility to regulate IgCAM functions.

The L1-like IgCAMs, neurofascin and NrCAM, and the GPI-linked IgCAMs, F11/contactin and axonin-1, interact with each other in many different combinations (Fig. 7). Neurofascin, F11/contactin, and NrCAM are expressed by outgrowing neurites. Using chicken tectal cells, various substrate–receptor pairs have been described. For example, neurons use NrCAM as a cellular receptor on neurofascin or F11/contactin substrates (Fig. 8) (45, 46).

ECM-derived tenascin-R binds both neurofascin and F11/contactin and shows a partially overlapping expression pattern with F11/contactin and neurofascin (47). This suggests that tenascin-R may interfere with F11/contactin or neurofascin interactions in restricted areas. Tenascin-R, if complexed with either neurofascin or F11/contactin substrates, induces neurite outgrowth (47, 48). Actually, NrCAM is not the cellular receptor anymore if confronted with a neurofascin substrate that has bound tenascin-R (Fig. 8). Binding of tenascin-R may cover the respective NrCAM binding site on neurofascin. There is evidence from neurite outgrowth and competitive binding assays that cellular F11/contactin binds to tenascin-R within the neurofascin/tenascin-R substrate complex, thus contributing to cell adhesion but not neurite extension. The factors that are important for neurite extension are not yet defined, but axonin-1 may be involved at least indirectly.

The presence of tenascin-R in restricted areas of the developing nervous system may therefore influence the interactions of neuronal F11 or neurofascin with ligands in the environment.

4.4   Regulation of IgCAM interactions

Various IgCAM interactions were described as summarized in Fig. 4. IgCAMs are often co-expressed in many different axon tracts. This raises the question whether all these possible interactions occur simultaneously or whether they are regulated.

Different cell types may engage different IgCAM receptor–ligand pairs in order to promote neurite outgrowth. Neuronal F11/contactin is important for tectal neurons to grow on an NrCAM substrate (46). In contrast, dorsal root ganglion cells use axonin-1 but not F11/contactin as a cellular receptor for substrate-bound NrCAM (49). In this case, NrCAM–axonin-1 interactions may be important for neuron–glia interactions as NrCAM expression is found on peripheral glia cells of dorsal root ganglions (31).

Neurofascin interactions with the four ligands, NrCAM, F11/contactin, axonin-1, and tenascin-R can be modulated by alternative splicing (47). Binding of NrCAM and F11/contactin appears to be only minimally influenced by the presence or absence of alternatively spliced exons. In contrast, tenascin-R and axonin-1 binding was strongly regulated by the presence or absence of various alternatively spliced exons. Therefore, isoform switching may be a tool to regulate binding of neurofascin to tenascin-R and axonin-1. This mechanism may add further possibilities to regulate IgCAM functions by differential expression.

Alternative splicing may also provide a mechanism to control the biological activity of NCAM. An NCAM isoform that contains the VASE sequence was shown to reduce NCAM-dependent neurite outgrowth (50). Thus, the neurite inducing activity of NCAM is modulated by alternative splicing of the VASE exon. This finding is in accordance with the expression of the VASE exon in the more differentiated nervous system. Similarly, removal of PSA from neurons by enzymatic treatment reduces neurite outgrowth in response to NCAM indicating that PSA modification of NCAM is important for neurite outgrowth (51). Accordingly, PSA is expressed by embryonal neurons and is down-regulated at later stages of development.

5.   Intracellular interactions of IgCAMs with cytoskeletal components

The actin cytoskeleton is rearranged during neurite extension. Growth cones contain motile filopodia that quickly protrude or retract in response to signals from the environment. Actin filament assembly or disassembly is the basis of directed movement of filopodia. Interaction with a favoured environment leads to a stabilization of filopodial contacts and helps the growth cone to move in a particular direction. Linkage to the cytoskeleton may therefore be an important aspect in regulation of neural differentiation. Cytoskeletal association may also be important for the stabilization of differentiated neurons.

5.1   L1-like IgCAMs interact with ankyrin

L1-like molecules including L1, neurofascin, NrCAM, CHL1, and *Drosophila* neuroglian contain a conserved sequence FIGQY within the cytoplasmic domain that mediates binding to the cytoskeletal component ankyrin (Fig. 9) (52). Three forms of ankyrin, termed ank_E, ank_G, and ank_B, have been identified. Ank_G has been found to interact with NrCAM and neurofascin, whereas L1 is associated with ank_B (53). Ankyrin indirectly links IgCAMs to the actin cytoskeleton via interaction with spectrin. Ankyrin may direct IgCAMs to subcellular locations, as in the case of ankyrin$_G$, that may target neurofascin to axon initial segments of cerebellar Purkinje cells. This was tested in mutated mice that do not express a cerebellar isoform of ankyrin$_G$ (54). In contrast to wild-type mice, neurofascin is uniformly distributed

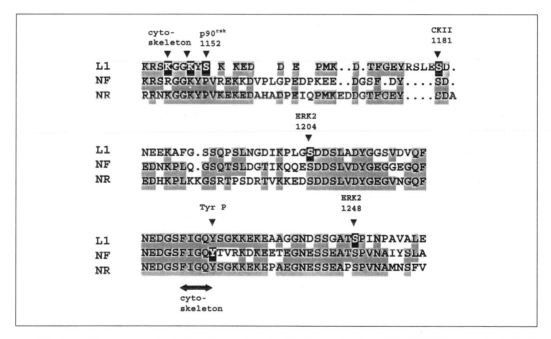

Fig. 9 Amino acid sequence of cytoplasmic domains of rat L1-like molecules L1/NgCAM (L1), neurofascin (NF), and NrCAM (NR). Conserved residues among L1-like molecules are shaded in grey. Residues that are important for cytoskeletal association (cytoskeleton) or phosphorylation are highlighted in black. Only residues that have been shown to be important in the respective IgCAM were considered. Numbers refer to positions within the L1/NgCAM polypeptide that are phosphorylated by kinases as depicted.

over the surface of Purkinje cells in mice devoid of ankyrin$_G$ in the cerebellum. On the other hand, evidence has also been provided that transmembrane IgCAMs may be involved in the redistribution of ankyrin. Extracellular homophilic axonin-1 interactions in *trans* may induce L1 to recruit intracellular ankyrin to L1 (55). Taken together, these results suggest a role for IgCAMs to couple cell adhesion to functions of the cytoskeleton.

5.2   Regulation of ankyrin binding

Interaction of neurofascin with ankyrin is regulated by tyrosine phosphorylation of the FIGQY sequence (Fig. 9) (56). Tyrosine phosphorylation prevents ankyrin binding to neurofascin and enhances lateral mobility early in development. Later in development, neurofascin is dephosphorylated, binds to ankyrin, and is presumably engaged in the maintenance of neuronal structures. This tyrosine residue is conserved within the L1-like subgroup of IgCAMs. A candidate phosphotyrosine kinase important for phosphorylation may be represented by the Eph receptor EphB2 (CEK5). The cytoplasmic domain of L1 has been shown to be a substrate for EphB2, but the precise location of phosphorylated tyrosine residue(s) has not yet been determined (57).

5.3   Additional links to the cytoskeleton

Mutational analysis revealed that L1 contains a second, membrane proximal sequence important for the association with the actin cytoskeleton (58). Mutation of two critical lysine residues impaired co-localization of neurofascin with cytoskeletal stress fibres (Fig. 9). This site allows association to the actin cytoskeleton independent of ankyrin binding by a so far unknown linker molecule.

6.   Signalling of IgCAMs

IgCAMs not only allow neural cells to adhere to a substratum but they also transduce extracellular signals to intracellular signalling pathways. For example, neurite extension is enhanced by L1/NgCAM when applied as a soluble ligand for neural cells adhering to poly-L-lysine. This indicates a role for L1/NgCAM as a signalling factor in the absence of L1/NgCAM-dependent adhesion (46, 59). IgCAM-dependent signalling may involve the function of different signalling molecules, including tyrosine phosphatases, tyrosine kinases, and serine/threonine kinases.

6.1   Non-receptor tyrosine kinases

Studies with genetically modified animal models have suggested the non-receptor tyrosine kinases of the Src family are important for IgCAM-dependent neurite outgrowth (60, 61). For example, L1-dependent neurite outgrowth was reduced in Src-deficient neurons and neurons lacking c-fyn showed reduced growth on NCAM.

This indicates that src family members may be important for neurite outgrowth on specific substrates. c-fyn has also been shown to be complexed with the GPI-linked IgCAMs axonin-1 and F11/contactin (38, 62). Complex formation of axonin-1 with NgCAM in the course of axon fasciculation leads to a reduction of axonin-1 associated c-fyn activity.

GPI-anchored IgCAMs are linked to the outer leaflet of the cytoplasmic membrane and non-receptor tyrosine kinases of the src family are intracellular proteins that are attached to the inner leaflet of the cytoplasmic membrane via myristoic acid. This raises the question how GPI-linked IgCAMs might interact with cytosolic signalling components. GPI-linked IgCAMs appear to require additional interaction partners to link IgCAM functions to the signalling cascades. Evidence has been provided that transmembrane RPTPα is an F11/contactin ligand in *cis* (63). Brain-derived RPTPα may activate non-receptor tyrosine c-fyn and could therefore provide a link between F11/contactin and associated c-fyn (64).

6.2 Receptor protein tyrosine phosphatases

Receptor protein tyrosine phosphatases (RPTPs) like RPTPδ, RPTPμ, and RPTPκ are IgCAMs that contain intracellular tyrosine phosphatase-related domains. They could combine adhesive properties of IgCAMs with intracellular signalling activity. They are expressed in the brain and confer adhesion by homophilic interactions (65). RPTPδ induces neurite induction of dissociated forebrain cells by presumably homophilic interactions (23). These proteins are discussed in detail in Chapter 7.

6.3 IgCAM signalling via the FGF receptor

Evidence has been provided for a link of IgCAMs to signalling pathways of the fibroblast growth factor receptor (FGFR), a receptor phosphotyrosine kinase (RPTK). L1- and NCAM-dependent signalling induces the phosphorylation of the FGFR independent of FGF ligand binding (66). Studies with different pharmacological inhibitors suggest a pathway including activation of phospholipase Cγ and generation of arachidonic acid that ultimately trigger an influx of extracellular calcium via N- and L-type calcium channels (66, 67). A CAM homology domain has been defined on FGFR, NCAM, and L1 that is suggested to be involved in IgCAM–FGFR interaction. L1- or NCAM-dependent neurite outgrowth can be blocked by peptides derived from this motif. However, the link between FGFR and IgCAMs is still debated. This is because a direct binding of IgCAMs to the FGFR has not yet been shown so far.

6.4 Serine/threonine kinases

Sequence analysis revealed that several putative serine/threonine phosphorylation sites are present in the cytoplasmic domain of L1/NgCAM and serine/threonine kinases were shown to be associated with L1/NgCAM (Fig. 9). A casein kinase II-related activity (CKII), a S6 kinase-related activity p90^{rsk}, ERK2, and Raf-1 have been shown to be associated with L1/NgCAM (38, 68). CKII-related activity phosphorylates L1 at Ser^{1181} and S6 kinase p90^{rsk}-related activity phosphorylates L1 at Ser^{1152} *in vitro*. Kinase p90^{rsk} is a distal member of the mitogen-activated protein kinase (MAPK) pathway involving sequential activation of Ras, Raf, MEK, ERK2, and p90^{rsk} (69). ERK2 that is upstream of p90^{rsk} can also phosphorylate L1 at Ser^{1204} and Ser^{1248} (70).

The NgCAM–axonin-1 tetramer formed by homophilic NgCAM interactions in *trans* and heterophilic NgCAM–axonin-1 interactions in *cis* is important for axonal fasciculation (Fig. 5). Concomitant with fasciculation, the amount of NgCAM phosphorylated by CKII-related activity increases within the cell. In contrast, phosphorylation of axonin-1 associated fyn is decreased in fasciculation (38). Serine phosphorylation is also important for neurite extension. Introduction of a peptide into primary neurons that contains Ser^{1152}, the p90^{rsk} phosphorylation site, has been shown to inhibit neurite outgrowth (68).

ERK activation has been shown to be linked to different neural functions including axonal growth, neuronal survival, and synaptic plasticity that overlap with known functions of IgCAMs (71). L1 activates ERK2 when crosslinked with antibodies after overexpression in non-neural NIH 3T3 cells. ERK co-localizes with L1 in endosomes. Inhibition of L1 internalization blocks ERK activation indicating that this interaction may be important for L1 trafficking (70).

Most interestingly, clustering of NCAM in B35 neuroblastoma cells also leads to ERK activation with kinetics similar to what occurs with L1 clustering, suggesting that different IgCAMs may induce common pathways (72). However, it has not yet been shown whether NCAM physically associates with ERK. ERK is important for the phosphorylation of the NCAM homologue in *Aplysia*, apCAM, and for apCAM internalization (73). ERKs also respond to other signalling pathways including receptor phosphotyrosine kinase activation, calcium influx, and cyclic adenosine monophosphate protein kinase A activation (71). Therefore, ERKs may integrate signals from different pathways in neuronal cells.

6.5   IgCAMs and cell death

IgCAMs may also be involved in apoptotic signalling. Apoptosis, also termed physio-logical cell death, has been shown to be important for neuronal cell death in response to deprivation of trophic factors (74). Application of soluble L1 or CHL1 enhances neural survival in the absence of serum, indicating that L1 may play a role as a survival factor (75). Survival of cerebellar neurons in the presence of L1 is accom-panied by an increase in intracellular bcl-2, an anti-apoptotic factor that prevents apoptotic cell death by inhibition of cytochrome *c* release from mitochondria. Cytoplasmic cytochrome *c* forms a complex with Apaf-1, dATP, and caspase-9 that becomes activated and may switch on the pro-apoptotic caspase cascade that finally leads to cell death. Bcl-2 is also involved in the regulation of a neuron's intrinsic ability to extend neurites (76). Therefore, it is unclear whether up-regulation of bcl-2 in response to soluble L1 acts on survival or neurite extension or both.

DCC (Deleted in Colorectal Cancer) previously described as a tumour suppressor gene has also been shown to be involved in cell death. Loss of DCC expression by deletion is associated with colonic carcinoma (24). In the absence of its ligand, netrin-1, DCC has been shown to induce apoptotic cell death in tumour cells (77). Apoptosis is dependent on caspase-3-dependent cleavage of the intracellular domain of DCC and netrin-1 ligand suppresses DCC-mediated cell death. This finding implies a crucial role for DCC in tumorigenesis. Metastatic cells are destined to die if they escape from their natural environment where netrin-1 expression is predominant. On the other hand, deletion of DCC may prevent DCC-dependent cell death and may lead to tumour progression.

7. IgCAM function *in vivo*

In vitro assays have considerably contributed to the understanding of molecular interactions involved in IgCAM function. However, they do not provide direct evidence for *in vivo* functions in a natural, more complex environment. *In vivo* studies have been performed using two approaches. First, injection of blocking antibodies into neural tissues has been used for perturbing IgCAM functions. These antibodies should mask specific IgCAMs involved in axon guidance, thereby preventing interactions with counter-receptors. Secondly, mice in which particular genes have been deleted (knock-out mice) were generated. Altogether, these studies underline an important role for IgCAMs in development.

7.1 Guidance of commissural axons

In the early development of the spinal cord, cell bodies of commissural axons reside in the dorsal horn of the spinal cord (Fig. 10). Axons extend from dorsal to ventral at the margin of the spinal cord. Arriving at the ventral spinal cord, commissural axons cross the floor plate and turn rostrally, joining the longitudinal tract. The floor plate is a transient organizing centre of the ventral neural tube that will disappear later in development. The floor plate is implicated in the specification of neurons and in the guidance of extending neurites (78).

Two mechanisms contribute to the guidance of commissural axons to the floor plate. A chemoattractant factor, netrin-1 is expressed and released from floor plate cells that direct commissural axons to the floor plate over long distances (79). Netrin-1 guides axons by binding to the IgCAM DCC on the surface of extending commissural axons (80). DCC has previously been described as a tumour suppressor and is also implicated in apoptosis as described above.

The function of DCC *in vivo* was clarified with the help of DCC-deficient mice. It was anticipated that these mice show a phenotype that correlates with a DCC function as a tumour suppressor and exhibit loss of controlled cell proliferation. Surprisingly, analysis of DCC-deficient mice did not support the idea that DCC functions as a tumour suppressor (81). Rather, pathfinding errors of commissural axons were found that were similar to those found in netrin-1 (−/−) mice implying

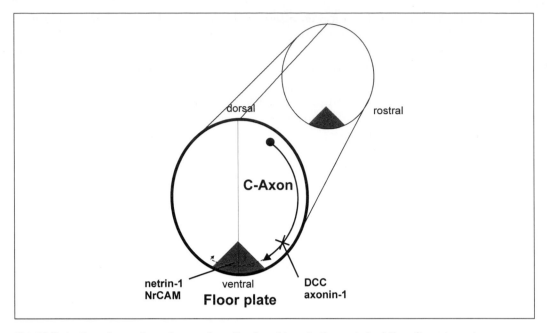

Fig. 10 Projection of commissural axons from the dorsal horn to the ventral midline. Commissural axons cross the floor plate and turn rostrally to proceed in the longitudinal tract. DCC receptor and axonin-1 are expressed by commissural axons whereas netrin-1 and NrCAM are expressed by the floor plate.

that DCC plays a pivotal role in the guidance of commissural axons to the floor plate (82).

In spite of the clear role of DCC and netrin-1 in axon guidance, the DCC–netrin-1 interactions appear not to be sufficient for commissural axons to enter the floor plate. This has been shown with the help of antibodies specific for the IgCAMs axonin-1 and NrCAM injected into the spinal cord of chick embryos *in ovo*. The IgCAMs axonin-1 and NrCAM that may interact with each other in *trans*, are candidates for axon guidance at the floor plate because NrCAM is expressed by the floor plate, whereas axonin-1 is found on commissural axons. Injection of antibodies to NrCAM or axonin-1 caused similar pathfinding errors. A subpopulation of commissural axons did not cross the floor plate but projected aberrantly (83). Moreover, as shown in an *in vitro* co-culture system, both NrCAM and axonin-1 are required for commissural axons to enter the floor plate (84).

7.2   NCAM-deficient mice

Mice deficient in NCAM expression were healthy and showed no apparent phenotype. However, some morphological changes were observed in the brain. The size of the olfactory bulb was reduced by 36% and the development of hippocampal mossy fibres was impaired (85, 86). NCAM-deficient mice showed reduced spatial learning ability when tested in the Morris water maze. Concomitant with loss of

NCAM, most of polysialic acid disappeared from the brain. This implies that modified NCAM is the major source for PSA. In fact, removal of PSA by Endo-N treatment causes similar effects to removal of NCAM. For example, loss of NCAM or PSA impairs long-term potentiation (LTP) in organotypic hippocampal cultures (87).

7.3 L1-deficient mice

Several groups have employed knock-out mice to study the function of L1 *in vivo*. Mutant mice lacking L1 display pathfinding errors in the cortico-spinal tract consistent with a role of L1 in neuronal guidance. Axons of the cortico-spinal tract project from the cortex to the motoneurons of the spinal cord. These axons ventrally approach the hindbrain/spinal cord transition and then turn dorsally, cross the midline, and thereby enter the dorsal column. Mice deficient in L1 contain cortico-spinal axons that fail to cross the midline correctly, indicating that lack of L1 influences axonal pathfinding (88). Accordingly, the size of the cortico-spinal tract at the level of the medulla oblongata is reduced (89). These mice display a delayed motor response, weakness of the hind limbs, and reduced sensitivity to touch and pain. Further abnormalities are enlarged ventricles and a hypoplastic corpus callosum. Most interestingly, a similar phenotype has been found in ank_B-deficient mice (90). L1 is coupled to the cortical cytoskeleton by interactions of its cytoplasmic domain with ank_B. This implies that L1 functions may require cytoskeletal association via ankyrin.

L1-deficient mice also develop alterations in the architecture of sensory tracts. Unmyelinated sensory neurons of the dorsal root ganglia fail to be contacted by Schwann cell processes. The failure of axon–Schwann cell interaction is due to axonal but not glial L1, indicating that the interaction between sensory axons and Schwann cells is mediated by heterophilic interactions with an unknown Schwann cell receptor (91). This is in accordance with a role for L1 for axon–Schwann cell interactions.

7.4 F11/contactin-deficient mice

F11/contactin (–/–) mice display a severe ataxic phenotype (92). Also, different features of cerebellar architecture are perturbed. For example, parallel fibres project parallel but not perpendicular to the plane of Purkinje cell dendritic branches in the outer molecular layer.

7.5 P_0-deficient mice

P_0 is a major protein of peripheral myelin and is expressed on Schwann cells in the wraps of the myelin sheath (for review see ref. 5). P_0 may be important for compaction through homophilic interactions. In P_0-deficient mice, peripheral axons are hypomyelinated (93). Schwann cells still contact axons but membrane compaction is severely reduced. Genetic mapping revealed the P_0 gene to be one locus that correlates with the outcome of Charcot–Marie–Tooth (CMT) type I and Dejerine–Sottas syndrome (DSS) (5). Mutations may therefore cause peripheral neuropathy by aberrant myelination.

8. L1 and pathogenesis

IgCAMs may be involved in human disorder. P_0, and DCC have been already mentioned in the sections above. Mutations of the L1 gene are also implicated in pathogenesis and will be discussed in the following sections.

8.1 Relationship of mutations in L1 (–/–) mice to human X-linked hydrocephalus

Defects found in L1-deficient mice like aberrant projection of cortico-spinal axons, enlarged ventricles, and corpus callosum hypoplasia are also associated with a human hereditary disease linked to mutations in the L1 gene (94). The disorder causes a syndrome including mental retardation, X-linked hydrocephalus, spastic paraplegia, corpus callosum hypoplasia, and adducted thumbs. Some abnormalities such as ventricular dilatation depend on the genetic background. The extent of hydrocephalus and the survival rate may therefore vary even within one affected family.

8.2 Classification of human L1 mutations

The L1 gene is located on chromosome X (Xq28) and more than 70 mutations have been mapped (for detailed information see http://dnalab-www.uia.ac.be/dnalab/l1). Mutations have been found in all parts of the L1 molecule including the extracellular and intracellular domains. The mutations were classified into distinct classes that correlate with the severity of the disease (95). Class I mutations are located in the extracellular domains of L1 and include nonsense and frame shift mutations that lead to a truncation of L1. These mutations lack membrane anchorage and are expected to be released from the cell surface resulting in a complete loss of function. Accordingly, these mutations are associated with a severe phenotype leading to hydrocephalus and death before the age of two.

Class II mutations are missense mutations that reside in the extracellular domain. The outcome of the disease may be either severe or mild, depending on the location of the mutation. For example, mutation of key residues important for proper folding of the whole molecule results in a more severe phenotype. In contrast, mutations of residues less critical for the conformational integrity result in a less severe phenotype.

As described above, L1 may interact homophilically with L1 on opposing membranes or heterophilically with F11/contactin and axonin-1. Surface mutations of human L1 show differential binding patterns to L1, F11/contactin, or axonin-1 ligands (96). Some mutations inhibit predominantly L1–L1 homophilic interactions or L1–F11/contactin and L1–axonin-1 heterophilic interactions. This indicates that L1 mutations may affect the interaction of L1 with different binding partners.

Mutations of the cytoplasmic domain of L1 including frame shifts, missense, or nonsense mutations are grouped in class III. Phenotypes are less severe, although such mutations may interfere with intracellular signalling of the L1 molecule. As an L1 molecule devoid of the cytoplasmic domain still retains adhesive properties, the class III phenotype may be generally mild.

Class IV mutations affect conserved sequence motifs that are implicated in the functioning of the splicing machinery and cause aberrant splicing patterns. As different L1-related peptides could result from such mutations, it is not easy to establish a genotype–phenotype correlation.

8.3   L1 and tumour cell migration

L1 is also implicated in the function of tumour cells. Melanoma cells may interact with substrate-bound L1 via integrin $\alpha v\beta 3$ (97). This interaction may contribute to adhesion, spreading, and haptotactic migration of melanoma cells that can also shed and deposit L1 in their environment. L1 may also interact with malignant glioma cells to regulate tumour migration and invasion (98).

9.   Emerging concepts

Different aspects of IgCAM function have been described, like complex formation, including different homo- and heterophilic interactions important for neural differentiation. A link to intracellular interactions with cytoskeletal components has been found, and first evidence for signalling pathways implicated in IgCAM functions has been provided. Nevertheless, the picture remains rather incomplete as many possible interactions were described by binding studies that could not yet be correlated with cellular functions (Fig. 4).

It could be speculated that there are multiprotein complexes at the cell surface involving IgCAMs that interact with other IgCAM complexes at opposing surfaces. Still, defining the composition of IgCAM complexes needs further investigations to clarify the relationship of different molecules that may participate in such a complex. To name just one, can we expect F11/contactin to participate with both CASPR and NrCAM in binding to RPTPβ or are CASPR and NrCAM interactions with F11/contactin mutually exclusive?

Further questions concern the signal transduction into the cell. Some evidence has been provided with respect to the activation of signalling molecules by clustering of IgCAMs at the cell surface. However, it remains unclear how neurons *in vivo* will react when these signalling pathways are activated. Some evidence was obtained from the study of apCAM function in *Aplysia*. In this case, evidence was provided that ERK functions may correlate with effects on long-term facilitation (73).

Still the most challenging aspect of IgCAM function is the understanding of how all these pieces of evidence obtained so far can be put together in order to give a coherent picture that explains how extracellular interactions may modulate neuronal functions in terms of cell migration, axon extension, pathfinding, and the stabilization of the nervous system. Additional signal pathways have also been described to modulate neuronal functions. Cell adhesion molecules from other protein families (for example cadherins and integrins, see Chapters 3 and 4) are also implicated in neural differentiation. Soluble molecules like neurotrophins, netrins, and semaphorins that act as chemoattractants or chemorepulsive factors in neural development may also trigger signalling systems. Integration of all these signals will finally regulate the cellular response to signals in its environment. Dissection of these mechanisms will be the major task for the future.

Acknowledgements

I thank Dr T. Brümmendorf and Dr C. Leibrock for helpful discussions and comments on the manuscript.

References

1. Brümmendorf, T. and Rathjen, F. G. (1995). Cell adhesion molecules 1: immunoglobulin superfamily. *Protein Profile*, **2**, 963.
2. Walsh, F. S. and Doherty, P. (1997). Neural cell adhesion molecules of the immunoglobulin superfamily: role in axon growth and guidance. *Annu. Rev. Cell Dev. Biol.*, **13**, 425.
3. Brümmendorf, T., Kenwrick, S., and Rathjen, F. G. (1998). Neural cell recognition molecule L1: from cell biology to human hereditary brain malformations. *Curr. Opin. Neurobiol.*, **8**, 87.
4. Hassel, B., Rathjen, F. G., and Volkmer, H. (1997). Organization of the neurofascin gene and analysis of developmentally regulated alternative splicing. *J. Biol. Chem.*, **272**, 28742.
5. Kamiguchi, H., Hlavin, M. L., Yamasaki, M., and Lemmon, V. (1998). Adhesion molecules and inherited diseases of the human nervous system. *Annu. Rev. Neurosci.*, **21**, 97.
6. Williams, A. F. and Barclay, A. N. (1988). The immunoglobulin superfamily—domains for cell surface recognition. *Annu. Rev. Immunol.*, **6**, 381.
7. Harpaz, Y. and Chothia, C. (1994). Many of the immunoglobulin superfamily domains in cell adhesion molecules and surface receptors belong to a new structural set which is close to that containing variable domains. *J. Mol. Biol.*, **238**, 528.
8. Holden, H. M., Ito, M., Hartshorne, D. J., and Rayment, I. (1992). X-ray structure determination of telokin, the C-terminal domain of myosin light chain kinase, at 2.8 Å resolution. *J. Mol. Biol.*, **227**, 840.
9. Osborn, L., Vasallo, C., Browning, B. G., Tizard, R., Haskard, D. O., Benjamin, C. D., *et al.* (1994). Arrangement of domains, and amino acid residues required for binding of vascular cell adhesion molecule-1 to its counter-receptor VLA-4 ($\alpha_4\beta_1$). *J. Cell Biol.*, **124**, 601.
10. Rader, C., Kunz, B., Lierheimer, R., Giger, R. J., Berger, P., Tittmann, P., *et al.* (1996). Implications for the domain arrangement of axonin-1 derived from the mapping of its NgCAM binding site. *EMBO J.*, **15**, 2056.
11. Su, X. D., Gastinel, L. N., Vaughn, D. E., Faye, I., Poon, P., and Bjorkman, P. J. (1998). Crystal structure of hemolin: a horseshoe shape with implications for homophilic adhesion. *Science*, **281**, 991.

12. Bateman, A., Jouet, M., MacFarlane, J., Du, J.-S., Kenwrick, S., and Chothia, C. (1996). Outline structure of the human L1 cell adhesion molecule and the sites where mutations cause neurological disorders. *EMBO J.*, **15**, 6050.

13. Ren, Q. and Bennett, V. (1998). Palmitoylation of neurofascin at a site in the membrane-spanning domain highly conserved among the L1 family of cell adhesion molecules. *J. Neurochem.*, **70**, 839.

14. Little, E. B., Edelman, G. M., and Cunningham, B. A. (1998). Palmitoylation of the cytoplasmic domain of the neural cell adhesion molecule N-CAM serves as an anchor to the cellular membranes. *Cell Adhes. Commun.*, **6**, 415.

15. Agrawal, H. C. and Agrawal, D. (1989). Effect of cycloheximide on palmitoylation of P_0 protein of the peripheral nervous system myelin. *Biochem. J.*, **263**, 173.

16. Jentoft, N. (1990). Why are proteins O-glycosylated? *Trends Biochem. Sci.*, **15**, 291.

17. Marg, A., Sirim, P., Spaltmann., F., Plagge, A., Kauselmann, G., Buck, F., *et al.* (1999). Neurotractin, a novel neurite outgrowth-promoting Ig-like protein that interacts with CEPU-1 and LAMP. *J. Cell Biol.*, **145**, 865.

18. Kim, D. S., Rhew, T. H., Moss, D. J., and Kim, J. Y. (1999). cDNA cloning of the CEPUS, a secreted type of neural glycoproteins belonging to the immunoglobulin-like opiod binding cell adhesion molecule (OBCAM) subfamily. *Mol. Cells*, **9**, 270.

19. Wang, J. and Bixby, J. L. (1999). Receptor tyrosine phosphatase-δ is a homophilic, neurite-promoting cell adhesion molecule for CNS neurons. *Mol. Cell. Neurosci.*, **14**, 370.

20. Gebbink, M. F., van Etten, I., Hateboer, G., Suijkerbuijk, R., Beijersbergen, R. L., Geurts van Kessel, A., *et al.* (1991). Cloning, expression and chromosomal localization of a new putative receptor-like protein tyrosine phosphatase. *FEBS Lett.*, **290**, 123.

21. Jiang, Y. P., Wang, H., D'Eustachio, P., Musacchio, J. M., Schlessinger, J., and Sap, J. (1993). Cloning and characterization of R-PTP-kappa, a new member of the receptor protein tyrosine phosphatase family with a proteolytically cleaved cellular adhesion molecule-like extracellular region. *Mol. Cell. Biol.*, **13**, 2942.

22. Beckman, G. and Bork, P. (1993). An adhesive domain detected in functionally divers receptors. *Trends Biochem. Sci.*, **18**, 40.

23. Wang, J. and Bixby, J. L. (1999). Receptor tyrosine phophatase-δ is a homophilic, neurite-promoting cell adhesion molecule for CNS neurons. *Mol. Cell. Neurosci.*, **14**, 370.

24. Fearon, E. R., Cho, K. R., Nigro, J. M., Kern, S. E., Simons, J. W., Ruppert, J. M., *et al.* (1990). Identification of a chromosome 18q gene that is altered in colorectal cancers. *Science*, **247**, 49.

25. Vielmetter, J., Kayyem, J. F., Roman, J., and Dreyer, W. J. (1994). Neogenin, an avian cell surface protein expressed during terminal neuronal differentiation, is closely related to the human tumor suppressor molecule deleted in colorectal cancer. *J. Cell Biol.*, **127**, 2009.

26. Cashman, N. R., Covault, J., Wollman, R. L., and Sanes, J. R. (1987). Neural cell adhesion molecule in normal, denervated, and myopathic human muscle. *Ann. Neurol.*, **21**, 481.

27. Reid, R. A. and Hemperly, J. J. (1992). Variants of human L1 cell adhesion molecule arise through alternate splicing of RNA. *J. Mol. Neurosci.*, **3**, 127.

28. Miura, M., Kobayashi, M., Asou, H., and Uyemura, K. (1991). Molecular cloning of cDNA encoding the rat neural cell adhesion molecule L1. Two L1 isoforms in the cytoplasmic region are produced by differential splicing. *FEBS Lett.*, **289**, 91.

29. Jouet, M., Rosenthal, A., and Kenwrick, S. (1995). Exon 2 of the gene for neural cell adhesion molecule L1 is alternatively spliced in B cells. *Brain Res. Mol. Brain Res.*, **30**, 378.

30. Muller, D., Wang, C., Skibo, G., Toni, N., Cremer, H., Calaora, V., *et al.* (1996). PSA-NCAM is required for activity-induced synaptic plasticity. *Neuron*, **17**, 413.

31. Suter, D. M., Pollerberg, G. E., Buchstaller, A., Giger, R. J., Dreyer, W. J., and Sonderegger, P. (1995). Binding between the neural cell adhesion molecule axonin-1 and NrCAM/Bravo is involved in neuron-glia interaction. *J. Cell Biol.*, **131**, 1067.

32. Krushel, L. A., Prieto, A. L., Cunningham, B. A., and Edelman, G. M. (1993). Expression pattern of the cell adhesion molecule Nr-CAM during histogenesis of the chick nervous system. *Neuroscience*, **53**, 797.

33. Collinson, J. M., Marshall, D., Gillespie, C. S., and Brophy, P. J. (1998). Transient expression of neurofascin by oligodendrocytes at the onset of myelinogenesis: implications for mechanisms of axon–glial interactions. *Glia*, **23**, 11.

34. Davis, J. Q., Lambert, S., and Bennett, V. (1996). Molecular composition of the node of Ranvier: identification of ankyrin-binding cell adhesion molecules neurofascin (mucin+/third FNIII domain-) and NrCAM at nodal axon segments. *J. Cell Biol.*, **13**, 1355.

35. Lemmon, V., Farr, K. L., and Lagenaur, C. (1989). L1-mediated axon outgrowth occurs via homophilic binding mechanisms. *Neuron*, **2**, 1597.

36. Williams, E. J., Doherty, P., Turner, G., Reid, R. A., Hemperly, J. J., and Walsh, F. S. (1992). Calcium influx into neurons can solely account for cell contact dependent neurite outgrowth stimulated by transfected L1. *J. Cell Biol.*, **119**, 883.

37. Buchstaller, A., Kunz, S., Berger, P., Kunz, B., Ziegler, U., Rader, C., *et al.* (1996). Cell adhesion molecules NgCAM and axonin-1 form heterodimers in the neuronal membrane and cooperate in neurite outgrowth promotion. *J. Cell Biol.*, **135**, 1593.

38. Kunz, S., Ziegler, U., Kunz, B., and Sonderegger, P. (1996). Intracellular signalling is changed after clustering of the neural cell adhesion molecules axonin-1 and NgCAM during neurite fasciculation. *J. Cell Biol.*, **135**, 253.

39. Peles, E., Schlessinger, J., and Grumet, M. (1998). Multi-ligand interactions with receptor-like protein tyrosine phosphatase β: implications for intercellular signalling. *Trends Biochem. Sci.*, **23**, 121.

40. Milev, P., Friedlander, D. R., Sakurai, T., Karthikeyan, L., Flad, M., Margolis, R. K., *et al.* (1994). Interactions of the chondroitin sulfate proteoglycan phosphacan, the extracellular domain of a receptor-type protein tyrosine phosphatase, with neurons, glia, and neural cell adhesion molecules. *J. Cell Biol.*, **127**, 1703.

41. Peles, E., Nativ, M., Campbell, P. L., Sakurai, T., Martinez, R., Lev, S., *et al.* (1995). The carbonic anhydrase domain of receptor tyrosine phosphatase beta is a functional ligand for the axonal cell recognition molecule contactin. *Cell*, **82**, 251.

42. Sakurai, T., Lustig, M., Nativ, M., Hemperly, J. J., Schlessinger, J., Peles, E., *et al.* (1997). Induction of neurite outgrowth through contactin and Nr-CAM by extracellular regions of glial receptor tyrosine phosphatase beta. *J. Cell Biol.*, **136**, 907.

43. Peles, E., Nativ, M., Lustig, M., Grumet, M., Schilling, J., Martinez, R., *et al.* (1997). Identification of a novel contactin-associated transmembrane receptor with multiple domains implicated in protein-protein interactions. *EMBO J.*, **16**, 978.

44. Peles, E., Joho, K., Plowman, G. D., and Schlessinger, J. (1997). Close similarity between *Drosophila* neurexin IV and mammalian Caspr protein suggests a conserved mechanism for cellular interactions. *Cell*, **88**, 745.

45. Morales, G., Hubert, M., Brümmendorf, T., Treubert, U., Tarnok, A., Schwarz, U., *et al.* (1993). Induction of axonal growth by heterophilic interactions between the cell surface proteins F11 and NrCAM/Bravo. *Neuron*, **11**, 1113.

46. Volkmer, H., Leuschner, R., Zacharias, U., and Rathjen, F. G. (1996). Neurofascin induces neurite outgrowth by heterophilic interactions with axonal NrCAM while NrCAM requires F11 on the axonal surface to extend neurites. *J. Cell Biol.*, **135**, 1059.

47. Volkmer, H., Zacharias, U., Nörenberg, U., and Rathjen, F. G. (1998). Dissection of complex molecular interactions of neurofascin with axonin-1, F11 and tenascin-R which promote attachment and neurite formation of tectal cells. *J. Cell Biol.*, **142**, 1083.

48. Zacharias, U., Norenberg, U., and Rathjen, F. G. (1999). Functional interactions of the immunoglobulin superfamily member F11 are differentially regulated by the extracellular matrix proteins tenascin-R and tenascin-C. *J. Biol. Chem.*, **274**, 24357.

49. Lustig, M., Sakurai, T., and Grumet, M. (1999). Nr-CAM promotes neurite outgrowth from peripheral ganglia by a mechanism involving axonin-1 as a neuronal receptor. *Dev. Biol.*, **209**, 340.

50. Doherty, P., Moolenaar, C., Ashton, S. V., Michalides, R. J., and Walsh, F. S. (1992). The VASE exon downregulates the neurite growth-promoting activity of NCAM 140. *Nature*, **356**, 791.

51. Doherty, P., Cohen, J., and Walsh, F. S. (1990). Neurite outgrowth in response to transfected NCAM changes during development and is modulated by polysialic acid. *Neuron*, **5**, 209.

52. Davis, J. Q., McLaughlin, T., and Bennett, V. (1993). Ankyrin-binding proteins related to nervous system cell adhesion molecules: candidates to provide transmembrane and inter-cellular connections in adult brain. *J. Cell Biol.*, **121**, 121.

53. Scotland, P., Zhou, D., Beneviste, H., and Bennett, V. (1998). Nervous system defects of ankyrinB (−/−) mice suggest functional overlap between the cell adhesion molecule L1 and 440 kD ankyrinB in premyelinated axons. *J. Cell Biol.*, **143**, 1305.

54. Zhou, D., Lambert, S., Malen, P. L., Carpenter, S., Boland, L. M., and Bennett, V. (1998). AnkyrinG is required for clustering of voltage-gated Na channels at axon initial segments and for normal action potential firing. *J. Cell Biol.*, **143**, 1295.

55. Malhotra, J. D., Tsiotra, P., Karagogeos, D., and Hortsch, M. (1998). Cis-activation of L1-mediated ankyrin recruitment by TAG-1 homophilic cell adhesion. *J. Biol. Chem.*, **273**, 33354.

56. Garver, T. D., Ren, Q., Tuvia, S., and Bennett, V. (1997). Tyrosine phosphorylation at a site highly conserved in the L1 family of cell adhesion molecules abolishes ankyrin binding and increases lateral mobility of neurofascin. *J. Cell Biol.*, **137**, 703.

57. Zisch, A. H., Stallcup, W. B., Chong, L. D., Dahlin-Huppe, K., Voshol, J., Schachner, M., *et al.* (1997). Tyrosine phosphorylation of the L1 family adhesion molecules: implications of the Eph kinase Cek5. *J. Neurosci. Res.*, **47**, 655.

58. Dahlin-Huppe, K., Berglund, E. O., Ranscht, B., and Stallcup, W. B. (1997). Mutational analysis of the L1 neuronal cell adhesion molecule identifies membrane-proximal amino acids of the cytoplasmic domain that are required for cytoskeletal anchorage. *Mol. Cell. Neurosci.*, **9**, 144.

59. Doherty, P., Williams, E., and Walsh, F. S. (1995). A soluble chimeric form of the L1 glyco-protein stimulates neurite outgrowth. *Neuron*, **14**, 57.

60. Ignelzi, M. A. Jr, Miller, D. R., Soriano, P., and Maness, P. F. (1994). Impaired neurite out-growth of src-minus cerebellar neurons on the cell adhesion molecule L1. *Neuron*, **12**, 873.

61. Beggs, H. E., Soriano, P., and Maness, P. F. (1994). NCAM-dependent neurite outgrowth is inhibited in neurons from Fyn-minus mice. *J. Cell Biol.*, **127**, 825.

62. Zisch, A. H., d'Alessandri, L., Amrein, K., Ranscht, B., Winterhalter, K. H., and Vaughan, L. (1995). The glypiated neuronal cell adhesion molecule contactin/F11 complexes with src-family protein tyrosine kinase fyn. *Mol. Cell. Neurosci.*, **6**, 263.

63. Zeng, L., d'Allessandri, L., Kalousek, M. B., Vaughan, L., and Pallen, C. J. (1999). Protein tyrosine phosphatase α (PTPα) and contactin form a novel neuronal receptor complex linked to the intracellular tyrosine kinase fyn. *J. Cell Biol.*, **147**, 707.

64. Bhandari, V., Lim, K. L., and Pallen, C. J. (1998). Physical and functional interactions between receptor-like protein-tyrosine phosphatase alpha and p59fyn. *J. Biol. Chem.*, **273**, 8691.

65. Zondag, G. C. M., Koningstein, G. M., Jiang, Y.-P., Sap, J., Moolenaar, W. H., and Gebbink, M. F. B. G. (1995). Homophilic interactions mediated by receptor tyrosine phosphatases μ and κ. *J. Biol. Chem.*, **270**, 14247.

66. Saffell, J. L., Williams, E. J., Mason, I. J., Walsh, F. S., and Doherty, P. (1997). Expression of a dominant negative FGF receptor inhibits axonal growth and FGF receptor phosphorylation stimulated by CAMs. *Neuron*, **18**, 231.

67. Williams, E. J., Furness, J., Walsh, F. G., and Doherty, P. (1994). Activation of the FGF receptor underlies neurite outgrowth stimulated by L1, N-CAM and N-Cadherin. *Neuron*, **13**, 583.

68. Wong, E. V., Schaefer, A. W., Landreth, G., and Lemmon, V. (1996). Casein kinase II phosphorylates the neural cell adhesion molecule L1. *J. Neurochem.*, **66**, 779.

69. Denhardt, D. T. (1996). Signal-transducing protein phosphorylation cascades mediated by Ras/Rho proteins in the mammalian cell: the potential for multiplex signalling. *Biochem. J.*, **318**, 729.

70. Schaefer, A. W., Kamiguchi, H., Wong, E. V., Beach, C. M., Landreth, G., and Lemmon, V. (1999). Activation of the MAPK signal cascade by the neural cell adhesion molecule L1 requires L1 internalization. *J. Biol. Chem.*, **274**, 37965.

71. Grewal, S. S., York, R. D., and Stork, P. J. S. (1999). Extracellular-signal-regulated kinase signalling in neurons. *Curr. Opin. Neurobiol.*, **9**, 544.

72. Schmid, R. F., Graff, R. D., Schaller, M. D., Chen, S., Schachner, M., Hemperly, J. J., *et al.* (1999). NCAM stimulates the Ras-MAPK pathway and CREB phosphorylation in neuronal cells. *J. Neurobiol.*, **38**, 542.

73. Bailey, C. H., Kaang, B. K., Chen, M., Martin, K. C., Lim, C. S., Casadio, A., *et al.* (1997). Mutation in the phosphorylation sites of MAP kinase blocks learning-related internalization of apCAM in *Aplysia* sensory neurons. *Neuron*, **18**, 913.

74. Pettmann, B. and Henderson, C. E. (1998). Neuronal cell death. *Neuron*, **20**, 633.

75. Chen, S., Mantei, N., Dong, L., and Schachner, M. (1999). Prevention of neuronal death by neural cell adhesion molecules L1 and CHL1. *J. Neurobiol.*, **38**, 428.

76. Holm, K. and Isacson, O. (1999). Factors intrinsic to the neuron can induce and maintain its ability to promote axonal outgrowth: a role for bcl2? *Trends Neurosci.*, **22**, 269.

77. Mehlen, P., Rabizadeh, S., Snipas, S. J., Assa-Munt, N., Salvesen, G. S., and Bredesen, D. E. (1998). The DCC gene product induces apoptosis by a mechanism requiring receptor proteolysis. *Science*, **395**, 801.

78. Dodd, J., Jessell, T. M., and Placzek, M. (1998). The when and where of floor plate induction. *Science*, **282**, 1654.

79. Kennedy, T. E., Serafini, T., de la Torre, J. R., and Tessier-Lavigne, M. (1994). Netrins are chemotropic diffusible factors for commissural axons in the embryonic spinal cord. *Cell*, **78**, 425.

80. Keino-Masu, K., Masu, M., Hinck, L., Leonardo, E. D., Chan, S. S., Culotti, J. G., *et al.* (1996). Deleted in Colorectal Cancer (DCC) encodes a netrin receptor. *Cell*, **87**, 175.

81. Fazeli, A., Dickinson, S. L., Hermiston, M. L., Tighe, R. V., Steen, R. G., Small, C. G., *et al.* (1997). Phenotype of mice lacking functional Deleted in Colorectal Cancer (DCC) gene. *Nature*, **386**, 796.

82. Serafini, T., Colamarino, S. A., Leonardo, E. D., Wang, H., Beddington, R., Skarnes, W. C., *et al.* (1996). Netrin-1 is required for commissural axon guidance in the developing vertebrate nervous system. *Cell*, **87**, 1001.

83. Stoeckli, E. T. and Landmesser, L. T. (1995). Axonin-1, Nr-CAM, and Ng-CAM play different roles in the *in vivo* guidance of chick commissural neurons. *Neuron*, **14**, 1165.

84. Stoeckli, E. T., Sonderegger, P., Pollerberg, G. E., and Landmesser, L. T. (1997). Interference with axonin-1 and NrCAM unmasks a floor-plate activity inhibitory for commissural axons. *Neuron*, **18**, 209.

85. Cremer, H., Lange, R., Christoph, A., Plomann, M., Vopper, G., Roes, J., et al. (1994). Inactivation of the N-CAM gene in mice results in size reduction of the olfactory bulb and in deficits in spatial learning. *Nature*, **367**, 455.

86. Cremer, H., Chazal, G., Goridis, C., and Represa, A. (1997). NCAM is essential for axonal growth and fasciculation in the hippocampus. *Mol. Cell. Neurosci.*, **8**, 323.

87. Muller, D., Wang, C., Skibo, G., Toni, N., Cremer, H., Calaora, V., et al. (1996). PSA-NCAM is required for activity-induced synaptic plasticity. *Neuron*, **17**, 413.

88. Cohen, N. R., Taylor, J. S. H., Scott, I. B., Guillery, R. W., Soriano, P., and Furley, A. J. W. (1997). Errors in the corticospinal axon guidance in mice lacking the neural cell adhesion molecule L1. *Curr. Biol.*, **8**, 26.

89. Dahme, M., Bartsch, U., Martini, R., Anliker, B., Schachner, M., and Mantei, N. (1997). Disruption of the mouse L1 gene leads to malformations of the nervous system. *Nature Genet.*, **17**, 346.

90. Scotland, P., Zhou, D., Benveniste, H., and Bennett, V. (1998). Nervous system defects of ankyrin B (−/−) mice suggest functional overlap between the cell adhesion molecule L1 and 440 kD ankyrinB in premyelinated axons. *J. Cell Biol.*, **143**, 1305.

91. Haney, C. A., Sahenk, Z., Li, C., Lemmon, V. P., Roder, J., and Trapp, B. D. (1999). Heterophilic binding of L1 on unmyelinated sensory axons mediates Schwann cell adhesion and is required for axonal survival. *J. Cell Biol.*, **146**, 1173.

92. Berglund, E. O., Murai, K. K., Fredette, B., Sekerkova, G., Marturano, B., Weber, L., et al. (1999). Ataxia and abnormal cerebellar microorganization in mice with ablated contactin gene. *Neuron*, **24**, 739.

93. Giese, K., Martini, R., Lemke, G., Soriano, P., and Schachner, M. (1992). Mouse P_0 gene disruption leads to hypomyelination, abnormal expression of recognition molecules, and degeneration of myelin and axons. *Cell*, **71**, 565.

94. Kamiguchi, H., Hlavin, M. L., and Lemmon, V. (1998). Role of L1 in neural development: what the knockouts tell us. *Mol. Cell. Neurosci.*, **12**, 48.

95. Fransen, E., van Camp, G., d'Hooge, R., Vits, L., and Willems, P. J. (1998). Genotype-phenotype correlation in L1 associated diseases. *J. Med. Genet.*, **35**, 399.

96. De Angelis, E., MacFarlane, J., Du, J. S., Yeo, G., Hicks, R., Rathjen, F. G., et al. (1999). Pathological missense mutations of neural cell adhesion molecule L1 affect homophilic and heterophilic binding activities. *EMBO J.*, **18**, 4744.

97. Montgomery, A. M. P., Becker, J. C., Siu, C. H., Lemmon, V. P., Cherech, D. A., Pancook, J. D., et al. (1996). Human neural cell adhesion molecule L1 and rat homologue NILE are ligands for integrin $\alpha_v\beta_3$. *J. Cell Biol.*, **132**, 475.

98. Izumoto, S., Ohnishi, T., Arita, N., Hiraga, S., Taki, T., and Hyakawa, T. (1996). Gene expression of neural cell adhesion molecule L1 in malignant gliomas and biological significance of L1 in glioma invasion. *Cancer Res.*, **56**, 1440.

2 | Selectins

RODGER P. MCEVER

1. Introduction

Most eukaryotic cells adhere stably to other cells or to extracellular matrix in order to maintain the architectures of specific organs. Changes in cell adhesion occur relatively slowly over minutes to hours. By contrast, circulating blood cells are usually non-adhesive, yet they can attach to other blood cells or to the vascular wall in a matter of seconds. This dynamic form of cell adhesion occurs in a shear flow field that applies significant force to adhesive bonds.

Circulating leukocytes patrol the body to guard against injurious external agents. Some lymphocytes move between the circulation and secondary lymphoid organs, where they develop immunological memory after encountering processed antigens. In response to inflammatory mediators, other leukocytes migrate into extravascular sites to engulf and destroy invading pathogens and to repair tissue damage. The emigration of leukocytes usually occurs in post-capillary venules, where shear stresses are lower than in other blood vessels. The critical first step is the tethering of a flowing leukocyte to the vessel wall. This tether is transient; the cell either detaches back into the fluid stream or begins to roll along the vessel wall. Rolling is a form of reversible adhesion in which the leukocyte must continuously form new adhesive bonds at the leading edge of the cell to replace bonds that dissociate at the trailing edge. The rolling cell may eventually detach, or it may become activated by inflammatory chemokines or lipid mediators, after which it arrests and spreads. In response to chemotactic gradients, it then migrates between endothelial cells into the underlying tissues. This multistep process of tethering, rolling, firm adhesion, and transendothelial migration requires the co-ordinated expression of different adhesion and signalling molecules (1, 2).

This chapter provides an overview of the selectins, a group of three specialized molecules that mediate the tethering and rolling adhesion of leukocytes through interactions with cell surface glycoconjugates. Earlier reviews provide additional primary references (3, 4). L-selectin, expressed on most leukocytes, binds to ligands on endothelial cells and on other leukocytes. P-selectin, expressed on activated platelets and endothelial cells, binds to ligands on leukocytes and, in some cases, on endothelial cells. E-selectin, expressed on activated endothelial cells, binds to ligands on leukocytes. Thus, selectins initiate interactions of leukocytes with other blood

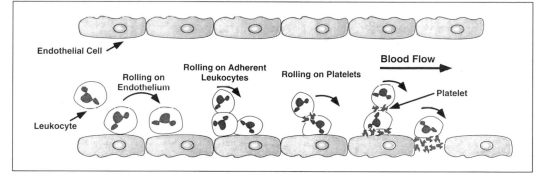

Fig. 1 Multicellular adhesive interactions mediated by selectins under hydrodynamic flow. Within the lumen of a blood vessel, leukocytes tether to and roll on E-selectin, P-selectin, or L-selectin ligands that are expressed on activated endothelial cells. Leukocytes roll on adherent leukocytes through bidirectional interactions of L-selectin with carbohydrate ligands; some of these leukocytes then translate to and roll on the endothelial cells. Leukocytes roll on P-selectin expressed on activated adherent platelets. These platelets may adhere to subendothelial tissues at sites of vascular injury (far right). Activated platelets also use P-selectin to roll on ligands expressed on high endothelial venules in lymph nodes or on activated endothelial cells in other tissues; by simultaneously binding to leukocytes, they may bring leukocytes in contact with the endothelial cell surface. Unactivated platelets may use a less well-characterized ligand to roll on P- or E-selectin that is expressed on activated endothelial cells.

cells and with the endothelial cells lining blood vessels (Fig. 1). The selectins have attracted great interest because of their biological importance for leukocyte trafficking and because they are the best characterized class of animal lectins that mediate cell adhesion.

2. Structure of selectins

Each selectin has an N-terminal domain characteristic of calcium-dependent (C-type) lectins, followed by an epidermal growth factor (EGF)-like motif, a series of short consensus repeats, a transmembrane domain, and a cytoplasmic tail (Fig. 2). The selectins share significant amino acid similarity in the lectin and EGF domains, suggesting that both domains contribute to ligand recognition. The lectin domain is thought to bind directly to carbohydrates in a Ca^{2+}-dependent manner. The three-dimensional structure of the lectin and EGF domains of E-selectin has been solved by X-ray crystallography (5). Molecular modelling suggests that P- and L-selectin have similar but not identical structures. The lectin domain of E-selectin has a single Ca^{2+} binding site located on the face opposite to where the EGF domain is attached. Substitution of residues on this surface impairs binding of E- or P-selectin to leukocytes. A model, based on co-crystallization of an oligomannose ligand with the C-type lectin mannose binding protein, suggests that sialyl Lewis x (sLe^x), a tetrasaccharide bound by selectins, could dock to the residues on E- or P-selectin that were identified by mutational analysis. The fucose is predicted to bind to a site analogous to where mannose binds to mannose binding protein, and the carboxylate

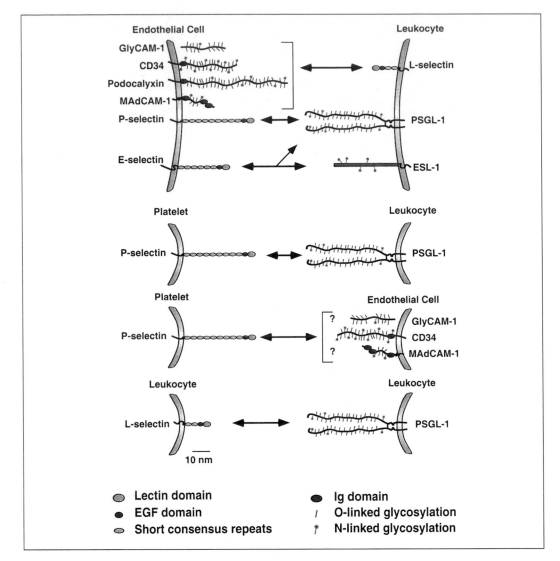

Fig. 2 Selectins and their best-characterized glycoprotein ligands. The estimated lengths of the selectins and of PSGL-1 are based on hydrodynamic data and electron microscopy. The lengths of GlyCAM-1, CD34, podocalyxin, and MAdCAM-1 are modelled from the dimensions of another sialomucin, CD43. ESL-1, E-selectin glycoprotein ligand-1; GlyCAM-1, glycosylation cell adhesion molecule-1; MAdCAM-1, mucosal addressin cell adhesion molecule-1; PSGL-1, P-selectin glycoprotein ligand-1.

moiety of the sialic acid is predicted to bind to adjacent, positively charged residues. Substitution of certain residues on this surface of mannose binding protein alters its carbohydrate binding specificity such that it binds sLe^x rather than mannose (6, 7). A co-crystal structure of sLe^x complexed to one of these chimeric lectins confirms docking of fucose to the predicted region overlapping the Ca^{2+} coordination site, but does not reveal binding of sialic acid to the protein (8). The chimeric lectins have provided valuable insights into the roles of particular residues in carbohydrate recognition. However, the structural basis by which authentic selectins bind to specific carbohydrate ligands requires further study.

Deletion of the EGF domain appears to impair the folding of the lectin domain of selectins. An L-selectin chimera containing the EGF domain of P-selectin binds with altered kinetics to carbohydrate ligands (9). The structural basis for this altered binding is unknown. Other selectin chimeras with substituted EGF domains have no obvious differences in ligand binding, and there is no direct evidence that the EGF domain binds directly to cell surface ligands. It has been argued that the short consensus repeats of E- and L-selectin enhance ligand binding affinity (10, 11), but the studies did not exclude whether some selectin constructs formed oligomers, which would enhance binding avidity. A soluble monomeric form of P-selectin containing only the lectin and EGF domains binds to a specific ligand on leukocytes with the same kinetics and affinity as soluble P-selectin that contains all the consensus repeats (12). Thus, the available data suggest that the lectin and EGF domains are sufficient for ligand binding. The EGF domain may contribute indirectly to binding by subtly altering the orientation of the lectin domain.

3. Selectin ligands

Like other mammalian lectins, the selectins bind selectively, but with low affinity, to particular oligosaccharides (13, 14). All three selectins bind to glycans that contain lactosamine units with $\alpha2,3$-linked sialic acid and $\alpha1,3$- or $\alpha1,4$-linked fucose. The prototype of these structures is sLe^x, a terminal component of oligosaccharides attached to glycoproteins and glycolipids on most leukocytes and some endothelial cells (Fig. 3A). Selectin ligands on leukocytes and some endothelial cells require $\alpha2,3$-linked sialic acid and $\alpha1,3$-linked fucose for function. However, it is now clear that sLe^x does not contain all the information necessary to constitute a preferred selectin ligand. The K_d for binding of E-selectin to sLe^x is only 0.1–1 mM, and the affinities of P- and L-selectin for sLe^x appear to be even lower. E-selectin binds better to some oligosaccharides that contain linear, sialylated dimeric Le^x structures. P- and L-selectin, but not E-selectin, also bind to sulfatides, subsets of heparan sulfate, and other sulfated glycans that lack sialic acid and fucose (13, 14). Finally, the selectins bind with higher affinity or avidity to only a subset of membrane glycoproteins on leukocytes or endothelial cells (Fig. 2). A key question is why certain glycosylated structures bind better to selectins than others do. An equally important question is whether a ligand that binds better to a selectin in solution also preferentially mediates leukocyte adhesion to a selectin under shear flow.

E-selectin binds avidly to E-selectin ligand-1 (ESL-1), a membrane glycoprotein that, on leukocytes, expresses sialylated and fucosylated N-glycans that confer selectin recognition (15). In all other cases, the best described glycoprotein ligands are mucins, that is, glycoproteins that have multiple Ser / Thr-linked oligosaccharides (O-glycans). In the mucins studied to date, selectin recognition requires the O-glycans rather than the relatively few N-glycans. The sialomucins bind particularly well to

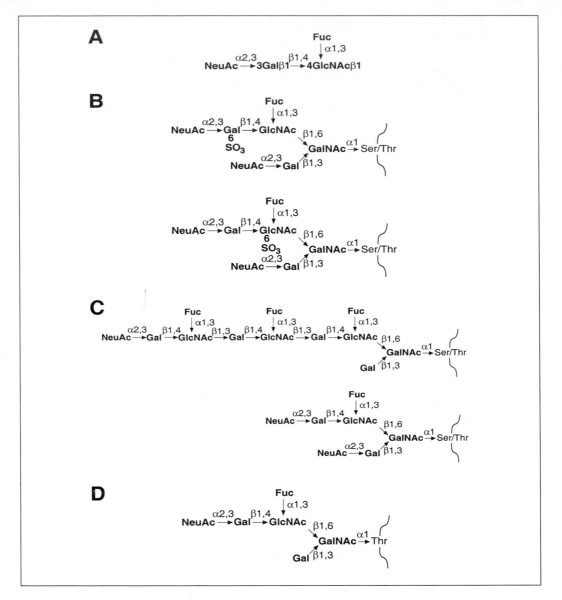

Fig. 3 Selected sialylated and fucosylated glycans that bind to selectins. (A) Sialyl Lewis x (sLe^x), a tetrasaccharide that is a terminal component of some *N*- and *O*-glycans. (B) Structures of identified fucosylated, core-2 *O*-glycans from murine GlyCAM-1. Larger core-2 structures that are multiply fucosylated or sulfated may also exist. Note that sulfate can be attached at the C-6 position of either Gal or GlcNAc. (C) Structures of the two fucosylated core-2 *O*-glycans on PSGL-1 from human HL-60 cells. Neither contains sulfate. One of these *O*-glycans is probably attached to Thr-57, located near the N-terminus of mature PSGL-1 (see Fig. 4). (D) A fucosylated core-2 *O*-glycan attached to a threonine in a synthetic glycosulfopeptide that binds to P-selectin (see Fig. 4). The amino acid sequence of the glycosulfopeptide corresponds to the N-terminal 22 residues of mature PSGL-1.

P- and L-selectin, but they may also bind to E-selectin. They must be appropriately sialylated and fucosylated to bind any of the selectins, and they must also be sulfated to bind to P- and L-selectin. The post-translational modifications of glycosylated cell adhesion molecule-1 (GlyCAM-1) and P-selectin glycoprotein ligand-1 (PSGL-1) provide two contrasting methods to create binding sites for selectins.

GlyCAM-1 is one of several L-selectin binding mucins that are synthesized by endothelial cells of high endothelial venules in murine lymph nodes (16, 17). The other mucins include CD34, podocalyxin, and a less well-characterized 200 kDa glycoprotein (Fig. 2). GlyCAM-1 lacks a transmembrane domain and is secreted into the plasma. Rather than mediate cell adhesion, it may bind to and activate lymphocytes that circulate through lymph nodes. The structures of the O-glycans of GlyCAM-1 have been determined (18–20). Most have a branched core-2 structure, in which a GlcNAc is linked via a β1,6 branch to GalNAc. This branched structure is required for extension and further modification of the oligosaccharide. Some core-2 O-glycans have a capping group that includes sLe^x and a sulfate linked to the C-6 position of the Gal or GlcNAc (Fig. 3B). There may be larger versions of this structure that are multiply fucosylated or sulfated; whether such structures enhance binding to L-selectin is unknown. The described structures have relatively little polylactosamine, that is, repeating disaccharides of the type $[\rightarrow3Gal\beta1\rightarrow GlcNAc\beta\rightarrow]_n$ on the β1,6 branch. The isolated sialylated, fucosylated O-glycans, even with attached sulfate, still bind with low affinity to L-selectin. Therefore the organization of the O-glycans on GlyCAM-1 must enhance binding. The most likely possibility is that many clustered O-glycans increase the avidity of binding of GlyCAM-1 to L-selectin. However, specific groupings of O-glycans may juxtapose sulfates, hydroxyl groups, or other groups to create composite recognition sites with higher affinity for L-selectin. Multimerization of GlyCAM-1 may also allow it to bind to L-selectin on cell surfaces with higher avidity. Similar principles may be used to construct effective ligands for L-selectin on other mucins expressed in murine endothelial cells.

PSGL-1 is a disulfide-bonded homodimeric mucin expressed on human and murine leukocytes (21). When appropriately modified, it binds to all three selectins. Most of the biochemical analysis has been performed on the human molecule. The extracellular domain has many serines, threonines, and prolines, most of which are present in a series of 15 or 16 decameric consensus repeats (22). O-glycans are attached to many of the serines and threonines (23, 24). In the cells examined to date, the three N-glycans on PSGL-1 are not required for selectin binding. PSGL-1 requires α2,3 sialylation and α1,3 fucosylation of branched core-2 O-glycans to bind to selectins. The structures of the O-glycans on PSGL-1 from human promyelocytic HL-60 cells have been determined (25). Like those from GlyCAM-1, the O-glycans are mostly branched core-2 structures. A major difference is that the O-glycans of PSGL-1 are not detectably sulfated. The structures of the fucosylated O-glycans also differ from those of GlyCAM-1 (Fig. 3C). The major fucosylated species has an extended, trifucosylated polylactosamine on the β1,6 branch. Only a minority of the O-glycans is fucosylated. This suggests that PSGL-1 does not simply present many fucosylated O-glycans that increase binding avidity to selectins. It further suggests that the relatively few O-glycans containing fucose and/or polylactosamine are attached to specific regions of the protein backbone. Consistent with this notion, a monoclonal antibody to an N-terminal protein epitope on PSGL-1 blocks binding to P- and L-selectin, although not to E-selectin (26).

The N-terminal region of PSGL-1 has a group of three tyrosines in an anionic sequence that favours tyrosine sulfation (Fig. 4). Sulfation of one or more of these tyrosines is absolutely required for PSGL-1 to bind to P- and L-selectin, but not to E-selectin (27–29). Mutational analysis suggests that a critical fucosylated O-glycan must be attached to a specific nearby threonine for binding to P- and L-selectin (30). A small N-terminal proteolytic fragment of PSGL-1 also binds well to P-selectin (31). Indeed, a synthetic glycosulfopeptide containing the N-terminal 22 residues of PSGL-1 is sufficient to bind to P-selectin (32). Binding of the glycosulfopeptide requires sulfation on the tyrosine residues and construction of a core-2 O-glycan capped by sLe^x on the specific threonine (Fig. 3D and Fig. 4); either modification alone does not confer detectable binding. Remarkably, an isomeric glycosulfopeptide that contains sLe^x on an extended, non-branched O-glycan on the same threonine does not bind P-selectin. This indicates that binding requires presentation of tyrosine sulfate and an O-glycan in a specific orientation. In summary, PSGL-1, like GlyCAM-1, requires both sulfation and specific O-glycosylation to bind P- and/or L-selectin. However, sulfate is attached to O-glycans on GlyCAM-1 (and probably on other mucin ligands for L-selectin in murine endothelial cells) but to tyrosines on PSGL-1. The C-type lectin domain of each selectin may contain a specific region that binds sialylated and fucosylated glycans, and the lectin domains of P- and L-selectin may contain an additional region that binds anionic structures such as sulfated saccharides or sulfated tyrosines. Elucidation of the molecular contacts between a selectin and a specific ligand will require solving the structures of co-crystallized complexes of these molecules.

As mentioned previously, PSGL-1 requires neither tyrosine sulfation nor its N-terminal region to bind E-selectin, although it does require at least one sialylated and fucosylated O-glycan. Depending on their state of differentiation, human or murine T cells express forms of PSGL-1 that bind only P-selectin or bind both P- and E-selectin (33, 34). The ability of PSGL-1 to bind E-selectin is correlated with its ability to bind a monoclonal antibody that recognizes a subset of oligosaccharides that includes sLe^x. This observation may mean that PSGL-1 must present fucosylated glycans in a specific orientation to bind E-selectin. It will be interesting to determine whether fucose must be attached to many different O-glycans, or whether polyfucosylation of a limited number of extended core-2 structures is sufficient to enhance binding.

In most cases, cell adhesion molecules probably bind to their ligands with low affinity so that multivalent cell attachment does not become irreversible. Consistent with this generalization, monomeric L-selectin binds to GlyCAM-1 with low affinity ($K_d \approx 100 \ \mu M$) (35). In marked contrast, soluble monomeric P-selectin binds to PSGL-1 with a K_d of ≈ 300 nM, an affinity some 300-fold greater than that of L-selectin for GlyCAM-1, and some three orders of magnitude higher than the affinities of most

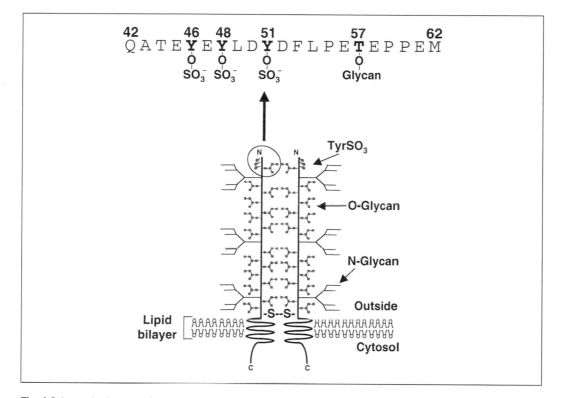

Fig. 4 Schematic diagram of the homodimeric structure of human PSGL-1. Each subunit has a single extracellular cysteine at residue 320 near the transmembrane domain that forms a disulfide bond. After cleavage of a signal peptide and a propeptide, the N-terminus of the mature subunit begins at residue 42 and ends at residue 412 in the cytoplasmic domain. The extracellular domain of each subunit has three potential *N*-glycans and up to 70 *O*-glycans. P- and L-selectin bind to the extreme N-terminal region of PSGL-1. Indeed, P-selectin binds to a synthetic glycosulfopeptide derived from this region (shown at top) with the same affinity as it binds to native PSGL-1. The structure of the *O*-glycan in the glycosulfopeptide is shown in Fig. 3D.

monomeric lectins for small oligosaccharides (12). These two examples suggest that selectins may interact with mucins through multiple low-affinity but high-avidity contacts, through single higher-affinity contacts, or by some combination of the two mechanisms. As discussed in the next section, binding in either case requires very fast association and dissociation rates to permit rolling cell adhesion.

Although several glycoproteins bind well to selectins in solution, only PSGL-1 has been clearly shown to mediate rolling cell adhesion when expressed on cell surfaces (21). Antibodies to the N-terminal glycosulfopeptide region of PSGL-1 block the ability of flowing leukocytes to tether to and roll on surfaces containing either P- or L-selectin, both *in vitro* and *in vivo* (26). Remarkably, PSGL-1 contains only a small fraction of the total sLe^x on the surface of leukocytes (24). Yet it represents all of the high-affinity binding sites for P-selectin on leukocytes, and it is essential for leukocytes to tether to and roll on P-selectin under shear flow (26). As mentioned earlier, GlyCAM-1 is a secreted protein, so it probably does not mediate cell adhesion directly. The role of other known glycoprotein ligands for selectins in cell adhesion is much less certain. As discussed below, the specific modifications and cell surface presentations of these molecules may determine whether they participate in cell adhesion under flow. It is also possible that leukocyte tethering or rolling through L- or E-selectin requires binding to more than one type of ligand.

4. Mechanisms for selectin-mediated cell adhesion under flow conditions

During inflammation, selectin–ligand interactions initiate the rolling adhesion of leukocytes on the vessel wall in shear flow. To tether to a surface, a flowing leukocyte must form adhesive bonds very rapidly. For the cell to roll, these bonds must dissociate quickly at the trailing edge of the cell as new bonds form at the leading edge of the cell (Fig. 5). Thus, the kinetics of bond association and dissociation are important determinants of rolling adhesion (36). Under flow conditions, forces are applied to adhesive bonds, which alter their rates of dissociation. At least two parameters are required to characterize bond dissociation under force: the intrinsic dissociation rate in the absence of force and the sensitivity of the intrinsic dissociation rate to force. The latter parameter is inversely proportional to the mechanical strength of the bond, that is, its ability to resist accelerated dissociation by force. These parameters are most relevant for the bonds at the trailing edge of the cell, because the dissociation of these bonds leads either to detachment of the cell into the fluid stream or stepwise rotation of the cell if it continues to roll through formation of new bonds. Premature dissociation results in detachment of the cell before new bonds can form.

The biochemical and mechanical features of selectin–ligand interactions have evolved to meet the specialized requirements for rolling cell adhesion under shear stress. Leukocytes perfused over very low densities of immobilized P-selectin, E-selectin, or a mucin ligand for L-selectin form transient tethers that do not convert to rolling adhesions (37, 38). The number of tethers is linearly related to the selectin or selectin ligand density, and the tether lifetimes obey first-order kinetics. This suggests that the transient tethers represent quantal units, perhaps single selectin–ligand bonds. The average durations of the tethers are very short (≤ 1 sec), particularly those mediated by L-selectin. But physiological shear forces do not markedly decrease tether durations, suggesting that the bonds have tensile strength that resists dissociation by applied force.

Leukocytes form rolling adhesions at higher densities of selectin and/or selectin ligand, but they must be perfused above a minimum shear stress if they are to roll (39, 40). If shear is too low, the cell does not roll, but instead detaches from the surface. The shear threshold requirement is particularly marked for rolling through L-selectin. Faster rotation of cells at higher shear may increase selectin–ligand collisions at the leading edge of the cell; this may help to form new bonds to replace those that dissociate at the trailing edge. However, other still poorly characterized

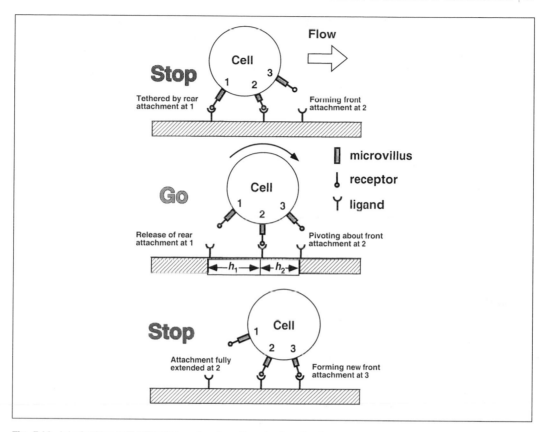

Fig. 5 Model of rolling cell adhesion under shear flow. Leukocytes have many microvilli near the surface that are capable of forming new attachments. For clarity, only the microvilli that actually form attachments at positions 1, 2, and 3 are shown. Rolling step distances ($h1$ and $h2$) may vary, depending on where attachments form. The relative sizes of the cell, the microvilli, and the individual molecules are not drawn to scale. Microvilli are much longer than even highly extended molecules such as P-selectin and PSGL-1. Microvilli are probably also flexible, compressible, and extensible. There could be multiple bonds per microvillus. There could also be a cluster of microvilli at either the leading or trailing edge of the cell.

factors must also contribute to the shear threshold phenomenon. Mild periodate treatment of CD34, a mucin ligand for L-selectin, selectively alters the side chains of its sialic acids, enhances the mechanical stability of its interaction with L-selectin, and reduces the shear threshold required for L-selectin-dependent cell rolling (41). It is not known whether the enhanced mechanical stability of the bond is responsible for the reduction in the shear threshold for rolling. Cells expressing L- or P-selectin roll less efficiently on recombinant forms of PSGL-1 in which any two of the three N-terminal tyrosines are altered (42). Remarkably, some of the tyrosine mutations increase the unstressed dissociation rate but do not alter the mechanical strength of the L-selectin bond, whereas they decrease the mechanical strength but do not alter the unstressed dissociation rate of the P-selectin bond.

Leukocyte rolling usually involves a series of irregular jerky motions, punctuated by skips in which the cell momentarily detaches before reattaching downstream. This phenomenon suggests that rolling involves a small number of bonds, perhaps alternating between one and two in some cases, with the duration of each jerky motion dependent on the lifetime of the bond, or bond cluster, at the trailing edge of the cell (Fig. 5). Although there is an average lifetime for a population of bonds at a given applied force, the lifetime of each individual bond will vary randomly according to a probability distribution. A bond cluster probably allows force to be evenly distributed among each bond, delaying release at the trailing edge of the cell. Sequential dissociation of each bond in the cluster increases the force applied to the remaining bonds, but the trailing edge of the cell is released only after the final bond dissociates. Thus, clustering of bonds can significantly slow rolling velocity.

Flowing cells also form rolling adhesions through interactions of α4 integrins with their immunoglobulin-like counter-receptors, or of anti-carbohydrate antibodies with their carbohydrate ligands (43–45). Where examined, these molecular pairs have rapid association and dissociation rates and at least some degree of mechanical strength. But rolling through these molecules is relatively unstable, and is confined to a relatively narrow range of receptor or ligand densities and shear stresses. At lower shears or higher densities, cells arrest rather than roll. As shear is increased, cells roll progressively faster and then detach. By contrast, the velocities of cells rolling through selectins tend to plateau as shear is increased, and yet the cells detach at low shear forces unless selectin and ligand densities are markedly increased. In other words, selectins allow cells to roll over a wide range of shear stresses and molecular densities, but they do not promote firm cell adhesion. These data suggest that selectin–ligand interactions require more than rapid binding kinetics and high tensile strength to mediate stable rolling adhesion. There is indirect evidence that increasing shear stress elevates the average number of selectin–ligand bonds on a rolling cell, despite the fact that increasing shear also accelerates the dissociation of these bonds (46). No mechanism for this putative shear-augmented increase in bond formation has been established. This phenomenon could serve to stabilize rolling adhesion as shear stress increases, and to destabilize adhesion as shear stress decreases below a threshold level.

In addition to these intrinsic features of selectin–ligand interactions, the cell surface organization of selectins or their ligands contributes to the efficiency of leukocyte adhesion under flow. For example, P-selectin is a long protein, with its nine consensus repeats extending the C-type lectin domain approximately 40 nm above the cell surface (47). Flowing neutrophils tether to and roll on wild-type P-selectin expressed on a monolayer of transfected Chinese hamster ovary (CHO) cells (48). However, very few rolling neutrophils accumulate on CHO cells expressing shortened forms of P-selectin that have less than five of the nine consensus repeats. Furthermore, those cells that adhere to shortened P-selectin roll faster and detach sooner in response to increasing shear stress. These data suggest that P-selectin projects its lectin domain sufficiently far above the plasma membrane to mediate optimal attachment of leukocytes under flow. This may increase its effective radius of contact with PSGL-1 on a flowing cell. By extending above most of the glycocalyx, it may also interact with PSGL-1 under conditions that reduce electrostatic repulsion between cells. The lengths of other selectins or selectin ligands may also be important for function. E-selectin is a long protein (49), and PSGL-1 is an extended mucin that projects the N-terminal binding domain for P- and L-selectin approximately 50 nm above the cell (50). It has been hypothesized that mucins lengthen further in response to force, but there is no direct evidence to support this conjecture.

Leukocytes have a highly irregular surface that contains many microvilli. Both L-selectin and PSGL-1 are concentrated on microvillous tips, the sites of earliest contact between a flowing leukocyte and another cell (26, 51). When expressed in transfected murine pre-B cells, wild-type L-selectin is concentrated on microvilli, where it supports tethering and rolling of the cells on immobilized L-selectin ligands (52). A chimera with the extracellular domain of L-selectin fused to the transmembrane and cytoplasmic domains of another membrane glycoprotein is distributed on the cell body rather than on the microvilli (53). Relatively few cells expressing the chimera tether to L-selectin ligands under flow; however, the cells that do tether roll with velocities and adhesive strengths that are similar to those of cells expressing wild-type L-selectin. Thus, the microvillar localization of L-selectin is particularly important for mediating the initial tethering event, but not the subsequent rolling event. Microvilli stretch in response to force, which may allow selectin–ligand bonds to persist longer before breaking (54). It is not known whether stretching is important for selectins or selectin ligands that are not clustered on microvillous tips.

As mentioned earlier, clustering of selectins or their ligands may also regulate leukocyte adhesion under flow. A form of L-selectin lacking most of the cytoplasmic tail is still targeted to microvilli of transfected pre-B cells (55). However, cells expressing this tailless L-selectin roll poorly on L-selectin ligands (52). Leukocyte activation transiently increases the adhesive function of L-selectin, perhaps by enhancing dimerization through interactions with the cytoskeleton (56). The cytoplasmic tail of L-selectin binds the actin crosslinking protein, α-actinin (55), and treatment of leukocytes with cytochalasins, which induce the disassembly of actin filaments, abrogates the ability of L-selectin to mediate cell adhesion under shear flow (52). PSGL-1 may also interact with the cytoskeleton; it is redistributed from microvilli to the uropods of activated leukocytes, but pre-treatment with cyto-chalasin D prevents this redistribution (57). Cytoskeletal interactions may enhance the adhesive function of selectins or selectin ligands in two ways. First, they may cluster the molecules, delaying the dissociation of selectin–ligand bonds. Secondly, they may prevent forced extraction of the molecules from the lipid bilayer during cell–cell detachment (37).

Unlike L-selectin, the cytoplasmic domains of P- and E-selectin have not been demonstrated to bind to the cytoskeletons of unstimulated cells. As discussed below, the cytoplasmic domain of P-selectin contains signals that mediate its rapid endo-cytosis in clathrin-coated pits (58). To test whether interactions of the cytoplasmic domain of P-selectin with coated pits enhance cell adhesion, transfected CHO cells were prepared that express wild-type P-selectin or P-selectin constructs with alterations in the cytoplasmic domain that increase or decrease its internalization rate (59). Under flow, neutrophils tether equivalently to all constructs expressed at matched densities. However, neutrophils roll on the internalization-competent constructs with greater adhesive strength, at slower velocities, and with more uniform motion. Confocal immunofluorescence microscopy demonstrates co-localization of α-adaptin, a component of clathrin-coated pits, with wild-type P-selectin but not with internalization-defective P-selectin lacking the cytoplasmic domain. Treatment of transfected CHO cells or endothelial cells with hypertonic medium reversibly impairs both the adhesive function and the internalization of P-selectin.

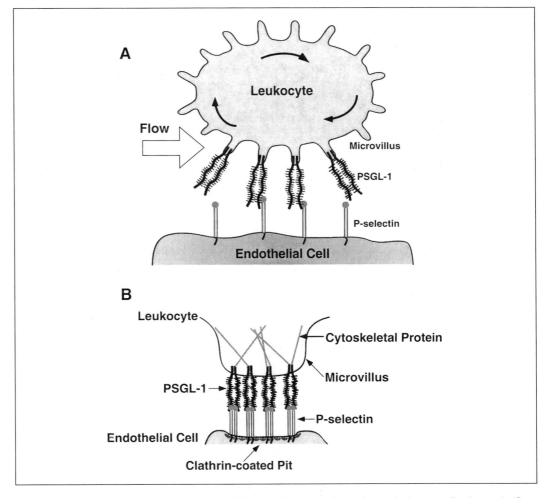

Fig. 6 Model for cell surface presentations of PSGL-1 and P-selectin that enhance leukocyte adhesion under flow. (A) The microvillous localization of PSGL-1 and the extended lengths of PSGL-1 and P-selectin enhance the initial tethering of flowing leukocytes to endothelial cells. (B) Dimerization of P-selectin and PSGL-1 through self-association, clustering of PSGL-1 through cytoskeletal interactions, and clustering of P-selectin in clathrin-coated pits facilitate leukocyte rolling. See text for discussion.

Therefore, interactions of P-selectin with clathrin-coated pits provide an alternative to cytoskeletal interactions to enhance adhesive function. The association of P-selectin with clathrin-coated pits may delay dissociation of P-selectin–PSGL-1 bond clusters and/or prevent forced extraction of P-selectin from the membrane.

Selectins or their ligands may also self-associate in the plane of the membrane. A disulfide bond formed at a single extracellular cysteine mediates covalent dimerization of PSGL-1. It has been suggested that only disulfide-bonded dimers of PSGL-1 bind P-selectin (60). However, crosslinking studies indicate that PSGL-1 forms non-covalent dimers even when the single extracellular cysteine is mutated; these non-covalent dimers clearly bind P-selectin in several assays. Furthermore, a small N-terminal fragment of PSGL-1 binds to P-selectin (31). It remains possible that non-covalent or covalent dimerization of PSGL-1 enhances binding avidity, which may be particularly important for cell adhesion under flow conditions. P-selectin isolated from platelet membranes forms dimers and oligomers in non-ionic detergent above the critical micellar concentration. Self-association may be mediated by the transmembrane domain, because recombinant P-selectin lacking this domain is monomeric (47).

Figure 6 illustrates a model of how the cell surface presentations of P-selectin and PSGL-1 may regulate leukocyte adhesion under flow. The microvillous localization of PSGL-1 allows early contacts with P-selectin. This promotes formation of the initial P-selectin–PSGL-1 bond, which tethers the free-flowing leukocyte to the vascular surface. The extended lengths of PSGL-1 and P-selectin may also promote tethering, and may facilitate rolling by minimizing the stretch on bonds at the trailing edge of the cell. Clustering of PSGL-1, perhaps through cytoskeletal interactions, and of P-selectin in clathrin-coated pits increases the local concentrations of both molecules in microdomains of the plasma membrane. This may increase bond clustering, which will decrease the average force on each bond and delay the average time for the last bond to dissociate. Dimerization or oligomerization of PSGL-1 and P-selectin might also delay bond dissociation. Interactions of PSGL-1 with the cytoskeleton and of P-selectin with clathrin coated lattices may prevent their extraction from the lipid bilayer by the forces applied to the P-selectin–PSGL-1 bond. Further studies are required to test all the predictions of this model, and to extend the model to other selectin–ligand pairs.

5. Regulation of expression of selectins and selectin ligands

The expression of selectins and their ligands is normally tightly regulated to ensure that leukocytes tether to and roll on the vascular wall only at appropriate locations. In lymphoid tissues, lymphocytes roll on the endothelium through interactions of L-selectin with constitutively expressed ligands on high endothelial venules; this provides the first step in the continuous trafficking of lymphocytes into these tissues. Upon activation by inflammatory mediators in other tissues, endothelial cells (primarily in post-capillary venules) transiently express E-selectin, P-selectin, and/

or ligands for L-selectin that allow flowing leukocytes to form rolling adhesions. Myeloid cells constitutively express selectin ligands. Subsets of lymphocytes and natural killer cells bind to selectins, but only after they differentiate in response to specific antigenic challenge. Leukocytes roll on P-selectin expressed on adherent activated platelets, and activated platelets use P-selectin to roll on L-selectin ligands expressed on endothelial cells. Leukocytes also roll on L-selectin ligands on adherent leukocytes. Unstimulated platelets use other ligands to roll on P- or E-selectin on activated endothelial cells. Thus, the regulation of expression of selectins and selectin ligands controls the initial adhesive interactions of leukocytes, platelets, and endothelial cells at the vascular surface.

P-selectin is constitutively synthesized by megakaryocytes and endothelial cells, where it is sorted to and concentrated in regulated storage granules: the α-granules of platelets and the Weibel–Palade bodies of endothelial cells (61). Mediators such as thrombin, histamine, complement components, or oxygen-derived radicals cause

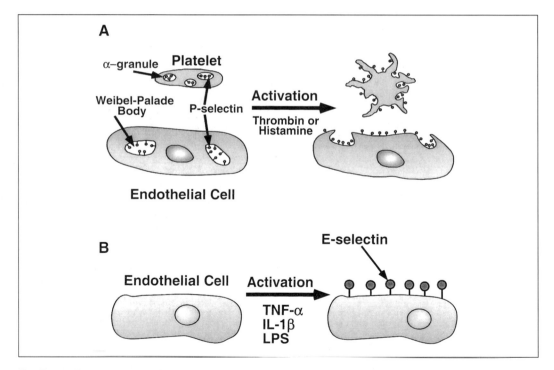

Fig. 7 Inducible expression of P-selectin and E-selectin in platelets and/or endothelial cells. (A) P-selectin is constitutively synthesized by megakaryocytes and endothelial cells, where it is sorted into the membranes of α-granules in platelets and Weibel–Palade bodies in endothelial cells. Within seconds to minutes after cellular activation by mediators such as thrombin or histamine, P-selectin is redistributed to the plasma membrane. (B) Most endothelial cells do not constitutively synthesize E-selectin. Inflammatory mediators such as TNF-α, IL-1β, or LPS induce the transient synthesis of E-selectin, which is then transported to the cell surface. Because of the time required for protein synthesis and vesicular transport, newly synthesized E-selectin requires at least 30 minutes to reach the plasma membrane.

rapid redistribution of P-selectin to the cell surface through fusion of granule membranes with the plasma membrane (Fig. 7A). This mobilizable pool of pre-formed molecules allows P-selectin to appear on the cell surface within seconds to minutes after an inflammatory challenge. In contrast, endothelial cells normally synthesize E-selectin only in response to mediators such as tumour necrosis factor α (TNF-α), interleukin-1 β (IL-1β), or bacterial lipopolysaccharide (LPS) (62). Newly synthesized E-selectin travels directly to the cell surface rather to secretory granules (Fig. 7B). Because of the time required for new mRNA and protein synthesis, E-selectin requires a few hours to accumulate on the cell surface after the initial stimulation of the endothelial cells. Thus, depending on the agonist, activated endothelial cells express different selectins with distinct kinetics. Interleukin-4 or oncostatin M, which signal through different mechanisms than TNF-α or IL-1β, increase P-selectin mRNA and protein in human endothelial cells in a delayed and sustained fashion (63). This may provide a mechanism to mobilize additional P-selectin to the cell surface in chronic inflammatory conditions.

In human endothelial cells, TNF-α or IL-1β does not increase P-selectin mRNA, and IL-4 inhibits the TNF-α-induced expression of E-selectin mRNA (64). In murine endothelial cells, however, TNF-α or IL-1β activates transcription of both P- and E-selectin mRNA with similar kinetics (65). These data suggest that the mechanisms for transcriptional regulation of the P- and E-selectin genes have diverged in mice and humans. TNF-α activates the human E-selectin gene by mobilizing nuclear factor κB (NF-κB), activating transcription factor-2 (ATF-2), c-jun, and other transcription factors that bind to conserved DNA elements to form a highly organized enhance-osome (62). Similar DNA elements are present in the murine E-selectin gene and the murine P-selectin gene, but not in the human P-selectin gene (66, 67).

Removal of selectins from the cell surface helps to limit the inflammatory response. One method for removal is proteolysis of the extracellular domains of L- and P-selectin. Calmodulin binds to the cytoplasmic domain of L-selectin (68). Activation of leukocytes induces dissociation of calmodulin from L-selectin, prob-ably because increased intracellular calcium alters the conformation of calmodulin. Dissociation of calmodulin somehow alters the structure of L-selectin, exposing a site in the extracellular domain near the membrane that is recognized by a surface metalloproteinase. This leads to rapid cleavage of the extracellular domain near the membrane, releasing a soluble form of L-selectin. Leukocytes rolling on L-selectin ligands also shed L-selectin, which may accelerate rolling velocity (69, 70). *In vitro*, P-selectin is not proteolytically shed, but *in vivo*, it is cleaved from the surfaces of both activated platelets and endothelial cells (71, 72). The kinetics of cleavage after activa-tion have not been examined in detail, but appear to be slower than those of L-selectin. In some situations, plasma levels of L-selectin, but not of P-selectin, may be sufficient to competitively inhibit leukocyte rolling (47, 73).

P- and E-selectin are removed from the activated endothelial cell surface by endocytosis (64). P-selectin is rapidly internalized through clathrin-coated pits (58, 59). E-selectin is internalized somewhat more slowly, perhaps also in clathrin-coated pits. Both proteins recycle from early endosomes to the cell surface, where they are reinternalized, a process that may be repeated many times. Unlike many receptors that follow this pathway, however, P- and E-selectin spend a relatively brief time on the cell surface because they are much more frequently diverted from early endosomes to late endosomes. From this compartment they may return to the *trans*-Golgi network, where P-selectin may be reincorporated into new Weibel–Palade bodies (74). Alternatively, they may move from late endosomes to lysosomes, where they are degraded (75). The cytoplasmic domain of P-selectin contains signals that direct sorting into secretory granules (76), endocytosis in clathrin-coated pits (58), and movement from early to late endosomes (75). The transmembrane domain of P-selectin may modulate sorting, perhaps by promoting dimerization of the molecule (77). Although less well studied, the cytoplasmic domain of E-selectin also mediates endocytosis and probably endosomal sorting, but not sorting into secretory granules (78).

The steady state distribution of P- and E-selectin in endothelial cells reflects the balance between their rates of synthesis, their rates of sorting into various subcellular compartments, and their rates of cleavage from the cell surface (Fig. 8). For example, cultured human endothelial cells treated with IL-4 express some P-selectin on the cell surface, probably because the increased synthesis saturates the sorting pathway from the *trans*-Golgi network to Weibel–Palade bodies and diverts some newly synthesized P-selectin to the plasma membrane (63). Subsequent stimulation with histamine further increases surface P-selectin during fusion of Weibel–Palade bodies with the cell surface. Surface levels rapidly decline as P-selectin is internalized and sorted away from early endosomes, but will remain somewhat elevated until the cytokine stimulus for increased protein synthesis is removed. Mice expressing a form of P-selectin that lacks the cytoplasmic domain fail to sort P-selectin into Weibel– Palade bodies (72). As a result, endothelial cells from the mice express more P-selectin on the cell surface. The mice have more soluble P-selectin in plasma, probably because the higher cell surface levels of P-selectin increase the amount subjected to proteolytic cleavage. Strikingly, platelets from these mice still sort P-selectin into α-granules, indicating that the cytoplasmic domain of P-selectin is not required for sorting into secretory granules in megakaryocytes. Although platelets have some capacity to internalize membrane proteins, the internalization rates appear to be very slow. Proteolysis is probably the major mechanism to remove P-selectin from the surface of activated platelets.

Since post-translational modifications are essential to construct glycoconjugate ligands for selectins, the cell must express the relevant glycosyltransferases or sulfotransferases in the Golgi complex and the *trans*-Golgi network. In some cases, the cell must also express a specific protein, such as PSGL-1, on which the modifications are made. The tyrosyl protein sulfotransferases that modify PSGL-1 are expressed constitutively on many cells (79). By contrast, the expression of sulfotransferases that transfer sulfate to the C-6 position of Gal and GlcNAc residues on carbohydrates is regulated in some tissues. One GlcNAc-6-sulfotransferase is expressed constitutively in high endothelial venules of lymph nodes but not in most other endothelial cells (80, 81). Inflammatory mediators cause extralymphoid endothelial cells to express sulfated carbohydrate epitopes found on L-selectin ligands, perhaps

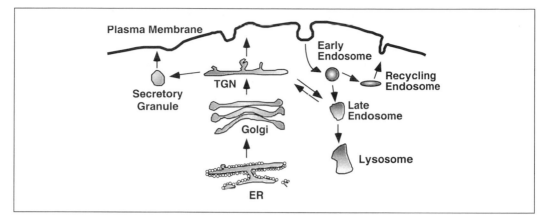

Fig. 8 Subcellular trafficking of P-selectin and E-selectin. Following translation, both proteins are glycosylated as they pass from the endoplasmic reticulum (ER) through the Golgi complex to the *trans*-Golgi network (TGN). E-selectin then proceeds directly to the cell surface. P-selectin is sorted into secretory granules, where it can be rapidly redistributed to the cell surface in response to secretagogues. P-selectin and perhaps E-selectin can be proteolytically cleaved at the plasma membrane. Both proteins are also internalized, probably in clathrin-coated pits, and then recycle to the plasma membrane. In addition, P-selectin and probably E-selectin are sorted more efficiently than most membrane proteins to late endosomes, from where they move to lysosomes for degradation. A fraction of P- and E-selectin in late endosomes may return to the TGN, where P-selectin may again be sorted into secretory granules. The steady state distribution of the proteins reflects the balance between the rates of synthesis, sorting to various compartments, and cleavage from the cell surface.

because they induce expression of the GlcNac-6-sulfotransferase (82). Consistent with this hypothesis, transfection of plasmids encoding a GlcNAc-6-sulfotransferase and an appropriate α1,3 fucosyltransferase promotes expression of L-selectin ligands on some cells (83). Myeloid leukocytes constitutively express two α1,3 fucosyl-transferases, Fuc-TVII and Fuc-TIV, that add the necessary fucose residues to PSGL-1 such that it can bind P-selectin (84). Most T lymphocytes express PSGL-1 but not Fuc-TVII. Differentiation to the T helper-1 subset of memory cells induces expression of Fuc-TVII, which fucosylates PSGL-1 to allow binding to P-selectin (85). Activation of extralymphoid endothelial cells induces expression of Fuc-TVII, which is also required for construction of L-selectin ligands (86). The synthesis of other glycosyl-transferases may be regulated in some cells, providing an additional means to control the expression of selectin ligands.

6. Signalling through selectins or selectin ligands

Selectins and their ligands, like other adhesion receptors, may transduce intracellular signals as well as mediate cell adhesion. Antibody-mediated crosslinking of L-selectin or PSGL-1, or leukocyte adhesion to L-selectin ligands or to P-selectin, respectively, initiates signalling cascades that include calcium mobilization and tyrosine phosphorylation of mitogen-activated protein kinases and other proteins (87–89). The intensity of signalling in leukocytes probably depends on the number of molecules of L-selectin or PSGL-1 that are engaged and the duration of engagement. In some cases, signalling through L-selectin or PSGL-1 is sufficient to induce effector responses such as activation of β1 or β2 integrins that stabilize adhesion (90–92). In most situations, effector responses require integration of signals through L-selectin or PSGL-1 with those generated by binding of chemokines or lipid autacoids to specific receptors on leukocytes. For example, monocytes mobilize the transcription factor NF-κB and synthesize the cytokines TNF-α and monocyte chemotactic protein-1 when the cells adhere to immobilized P-selectin and platelet-activating factor, but not to either molecule alone (93). Monocytes secrete a different profile of cytokines when they are exposed to P-selectin and the platelet-derived chemokine, RANTES, but not to either protein alone (94). This co-operative signalling ensures that leukocytes are fully activated only at specific vascular sites. Co-operative signalling may be particularly important during rolling adhesion, which limits encounter times of selectins with their ligands and chemokines with their receptors. Signalling through E-selectin ligands has received less attention, in part because the physiologically relevant glycoprotein ligands for E-selectin on leukocytes have not yet been identified.

Antibody-mediated crosslinking of E- or P-selectin, or leukocyte adhesion to E- or P-selectin on endothelial cells, induces transient increases in cytosolic calcium that are associated with rearrangements of cytoskeletal proteins, cellular contraction, and association of the cytoplasmic domain of E-selectin with α-actinin and other cytoskeletal proteins (95, 96). These selectin-mediated signalling events may alter endothelial cell junctions, enabling leukocytes to migrate between endothelial cells in response to chemotactic gradients.

7. Functions of selectins *in vivo*

Leukocyte trafficking into lymphoid or extralymphoid tissues begins with reversible, rolling adhesion on the endothelium, usually in post-capillary venules where shear stress is lowest and where P- and E-selectin and L-selectin ligands are primarily expressed. If the rolling cell encounters a chemokine or other mediator at a sufficiently high concentration, it becomes activated. It then arrests, spreads, and ultimately emigrates between endothelial cells into the underlying tissues, presumably in response to chemotactic gradients. Many studies using animal models have established that selectins initiate the rolling adhesion of leukocytes on the vessel wall, whereas leukocyte integrins help slow rolling velocities, stabilize adhesion, and then promote migration of the cells (Fig. 9). The available evidence suggests that combinatorial diversity in the expression of selectins, selectin ligands, integrins, integrin ligands, chemokines, and chemokine receptors dictates which subclasses of leukocytes emigrate into extravascular tissues at a given location (1, 2). A major experimental tool in these studies is intravital microscopy, which allows direct visualization of the microcirculation in mesentery, cremaster muscle, peripheral lymph node, or other thin tissues that can be displayed on a microscope stage. The functions of individual molecules have been studied with specific blocking

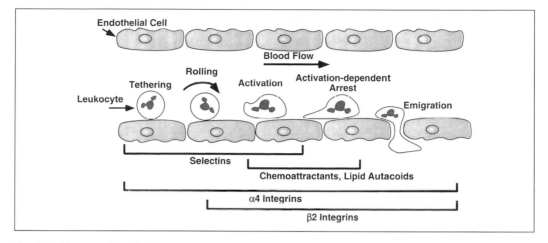

Fig. 9 Multistep model of leukocyte recruitment. Free-flowing leukocytes tether to and then roll on post-capillary venules at inflammatory sites or on high endothelial venules of lymphoid tissues. The rolling leukocyte may be activated by specific chemokines or lipid autacoids expressed on the endothelial cell surface. The activated leukocyte then arrests, spreads, and finally emigrates between endothelial cells into the underlying tissues. Selectins have a critical function in mediating the initial tethering and rolling of leukocytes, whereas β2 leukocyte integrins mediate arrest, spreading, and emigration. Under some conditions, α4 and perhaps α9 integrins expressed on subsets of leukocytes mediate both rolling and arrest, although they are probably less efficient than selectins in mediating tethering. Signalling of leukocytes through L-selectin or ligands for P- and E-selectin may be co-operative with signalling through chemokine receptors to slow rolling velocities and stabilize adhesion through integrins. Signalling of endothelial cells through P- and E-selectin may alter cell junctions, enhancing the ability of leukocytes to emigrate into the subendothelial tissues. Not illustrated are selectin- and integrin-mediated adhesive interactions of leukocytes with other leukocytes or with activated platelets, which may amplify leukocyte recruitment to the endothelial cell surface.

monoclonal antibodies or other inhibitors, or with mice in which the genes encoding specific adhesion molecules or chemokines have been disrupted by homologous recombination (97). Here the studies using gene disruption approaches will be emphasized.

Mice made genetically deficient in each of the three selectins appear healthy, but they have obvious defects in leukocyte trafficking in response to specific challenges (98). Lymphocytes from L-selectin-deficient mice home less efficiently to peripheral lymph nodes (99). Mice lacking L- or P-selectin demonstrate impaired rolling of leukocytes in venules of exteriorized mesentery and other extralymphoid tissues (99, 100). The defect in rolling is observed earlier after tissue exteriorization in P-selectin-deficient mice, consistent with the rapid mobilization of P-selectin to the endothelial cell surface in response to histamine released from perivascular mast cells (101). Comparison of the onset, duration, and quality of leukocyte rolling among wild-type, L-selectin-deficient, and P-selectin-deficient mice suggests that L- and P-selectin function co-operatively during acute inflammatory responses. This may reflect the overlapping expression of both P-selectin and an L-selectin ligand on the activated endothelial cell surface. P-selectin-dependent platelet–leukocyte contacts may amplify the number of flowing leukocytes that tether to the vessel wall (102). L-selectin-dependent leukocyte–leukocyte contacts may also augment leukocyte delivery to the endothelium, although some evidence suggests that these interactions are less important *in vivo* (103).

Mice deficient in either L- or P-selectin have impaired leukocyte recruitment into tissues in models of acute and chronic inflammation. Such defects are less obvious in E-selectin-deficient mice, but can be elicited by blocking P-selectin function by infusion of a monoclonal antibody (104). Consistent with this observation, mice lacking both P- and E-selectin have frequent severe infections and shortened survival (105, 106). Leukocytes roll much faster in post-capillary venules of E-selectin-deficient mice (107). This suggests that P- and L-selectin may be particularly important for mediating the initial tethering of flowing leukocytes, whereas E-selectin may function primarily to slow rolling velocities. Slower rolling increases the encounter time with chemokines expressed at the site of infection or injury, enhancing the probability for activation (108).

Mice deficient in PSGL-1 have defects in leukocyte rolling and leukocyte recruitment into inflamed tissues that resemble those in P-selectin-deficient mice (109). These results confirm the importance of PSGL-1 as the predominant ligand for P-selectin on leukocytes. Myeloid cells lacking PSGL-1 roll on E-selectin *in vivo*, indicating that PSGL-1 on these cells is not an essential ligand for E-selectin (109). However, Th1 cells may use PSGL-1 and other unidentified ligands to roll on E-selectin (110).

Leukocytes from mice deficient in Fuc-TVII fail to roll on post-capillary venules in either lymphoid or extralymphoid tissues, and have markedly decreased ability to emigrate into extravascular tissues in response to inflammatory stimuli (111). This establishes the importance of this enzyme for constructing fucosylated glycoconjugates for selectins. The absence of Fuc-TVII decreases the expression of fucosylated structures on leukocytes, which interact primarily with P- and E-selectin, and it also decreases expression of fucosylated ligands for L-selectin on lymph node high endothelial venules and on endothelial cells at sites of inflammation. Fuc-TIV, the other $\alpha1,3$ fucosyltransferase expressed in leukocytes, may also contribute to the construction of selectin ligands. *In vitro* studies with simple oligosaccharide acceptors suggest that Fuc-TVII and Fuc-TIV add fucose residues to GlcNAc residues at different positions (112), but it is not known whether these preferential additions occur on glycoproteins in the Golgi complex of intact cells.

The enzyme core-2 $\beta1,6$ *N*-acetylglucosaminyltransferase (C2GnT) constructs core-2 *O*-glycans by adding a GlcNAc residue in $\beta1,6$ linkage to the GalNAc that is attached to a Ser or Thr residue (Fig. 3B–D). Mice deficient in the C2GnT expressed in leukocytes have significant defects in leukocyte trafficking in response to inflammatory challenge, although the defects are not as severe as those caused by Fuc-TVII deficiency (113). Leukocytes from these mice roll poorly on P-selectin in venules of inflamed tissues. These results suggest that P-selectin binds primarily to core-2 *O*-glycans on leukocytes, and are consistent with the requirement for core-2 branching as a scaffold for extension and further modifications, including sialylation and fucosylation, of *O*-glycans. Surprisingly, lymphocyte homing to lymph nodes is normal. This is probably because lymph node high endothelial venules express a second C2GnT that constructs the appropriate branched *O*-glycans on mucins in these cells (114, 115).

The importance of selectins in humans is underscored by the discovery of a congenital disorder of fucose metabolism, termed leukocyte adhesion deficiency 2 (LAD-2) (116–118). Because patients with LAD-2 lack all fucosylated glycoconjugates, they do not express functional selectin ligands on leukocytes or, presumably, on endothelial cells. Leukocytes from these patients do not tether to and roll on P- or E-selectin surfaces. Clinically, the patients have more infectious diseases, supporting the concept that the selectins have important functions in initiating leukocyte recruitment in humans.

In vitro, P- and L-selectin bind to human haematopoietic progenitor cells, at least in part through PSGL-1 (119, 120). *In vitro* studies also suggest that E-selectin contributes to angiogenesis, consistent with its observed expression on proliferating endothelial cells of new blood vessels (121, 122). Neither selectin-deficient mice nor LAD-2 patients have obvious defects in haematopoiesis, angiogenesis, or wound healing. However, haematopoietic stem cells home less efficiently to the bone marrows of lethally irradiated mice that are deficient in P- and E-selectin (123). Selectins may therefore contribute to haematopoiesis and perhaps to angiogenesis, but other molecules may compensate for their absence unless these functions are subject to major challenge.

8. Contributions of selectins to disease

Excessive accumulation of leukocytes contributes to the pathogenesis of inflammatory disorders such as ischaemia–reperfusion injury, organ allograft rejection, asthma, and rheumatoid arthritis. Tissue injury results from leukocyte release of oxygen-derived radicals, proteases, and other mediators. Dysregulated expression of selectins has been implicated in several forms of leukocyte-mediated tissue injury.

Activated complement and oxygen radicals, which are elaborated during the early stages of sepsis or ischaemia–reperfusion syndromes, mobilize P-selectin to the surface of endothelial cells *in vitro* (124). Oxygen radicals prolong the expression of P-selectin on the endothelial cell surface, perhaps by inhibiting endocytosis. Endothelial dysfunction decreases formation of nitric oxide, an oxygen radical scavenger that may normally dampen the expression of P-selectin. Hypoxia also translocates P-selectin to the surface of endothelial cells. Consistent with these observations, ischaemia–reperfusion mobilizes P-selectin to the surface of endothelial cells *in vivo* (125). Monoclonal antibodies to P- and L-selectin significantly reduce neutrophil accumulation and tissue damage in models of ischaemia–reperfusion, sepsis, and lung injury (126). Soluble forms of PSGL-1, which bind to P- and L-selectin, attenuate acute rejection of kidney allografts in mice (127). Thus, just as P- and L-selectin function co-operatively in physiological leukocyte recruitment, they may co-operatively enhance acute leukocyte-mediated tissue injury in some diseases. Selectin expression has been observed on the surfaces of venular endothelial cells from patients with some chronic or allergic inflammatory diseases, suggesting that they may contribute to the pathogenesis of these disorders (128, 129).

Several lines of evidence suggest that P- and E-selectin contribute to the pathogenesis of atherosclerosis and to thrombotic disease arising from atherosclerosis or other processes. Oxidized low-density lipoprotein activates both platelets and endothelial cells, promoting P-selectin-dependent platelet–leukocyte aggregates and leukocyte adhesion to the arterial endothelium *in vitro* and *in vivo* (130). Cigarette smoke, which causes release of oxygen radicals, produces similar effects *in vivo* (131). Some viral infections prolong the expression of P-selectin on the surface of cultured endothelial cells. These insults may allow monocytes and other leukocytes to emigrate beneath the endothelium during the early stages of atherosclerosis. In the later stages of atherosclerosis in humans, P-selectin is observed on the apical surface of the endothelium (132), perhaps in response to the local synthesis of IL-4 or oncostatin M by subendothelial macrophages or T cells (63). Endothelially expressed P- and E-selectin may promote recruitment of additional monocytes, particularly in areas of arterial bifurcation where shear stresses are lower. Genetic deficiencies in P- and E-selectin reduce the size and number of atherosclerotic lesions in mice lacking the receptor for low-density lipoprotein or apolipoprotein E (133, 134). Rupture of advanced atherosclerotic plaques promotes platelet aggregation and thrombus formation. Leukocytes accumulating on adherent platelets may express tissue factor, further augmenting thrombin and fibrin generation (135). Monoclonal antibodies to P-selectin accelerate pharmacological thrombolysis in a primate model of arterial thrombosis (136), and reduce thrombus formation and ischaemic injury in a model of endothelial injury in coronary arteries (137). Monoclonal antibodies to P-selectin also reduce infiltration of inflammatory cells in a rat model of venous thrombosis (138).

Some cancer cells express selectins or selectin ligands normally used for leukocyte or platelet adhesion. Genetic deficiency of P-selectin in mice significantly reduces the size and number of pulmonary metastases resulting from subcutaneous injection of human carcinoma cells (139). The anti-metastatic effect appears to be due to a reduction in the formation of circulating aggregates of tumour cells bound to P-selectin on activated platelets.

These data suggest that inhibitors of selectin function or expression might be effective therapeutics in some inflammatory or thrombotic disorders, and might also inhibit metastatic spread of some cancers. A potential risk of such agents is interference with the physiological recruitment of leukocytes required to combat infections. However, antibodies to P- or L-selectin do not significantly increase infections in some experimental models, suggesting that other adhesion molecules may suffice in the absence of a major infectious challenge. Clinical trials of anti-selectin therapies are ultimately required to validate their safety and therapeutic efficacy.

9. Conclusions

The importance of selectins in initiating cell–cell interactions in the vascular system has been confirmed by many approaches. Nevertheless, many interesting questions remain to be answered.

The selectins are closely related C-type lectins, but they do not bind identically to glycoconjugates. Determining the basis for the fine specificities of carbohydrate recognition will require solving the three-dimensional structures of P- and L-selectin and comparing them with those of E-selectin and other C-type lectins. The structures of different selectin ligands must also be determined, and ultimately, the sites of interactions between selectins and specific ligands must be determined, probably by a combination of crystallographic and nuclear magnetic resonance approaches. The structures will allow formulation of hypotheses about selectin–ligand interactions that can be tested by mutating specific amino acids or by altering post-translational modifications. Both biophysical and biochemical approaches are required to understand how selectin–ligand interactions mediate cell adhesion under shear flow. In this regard it will be important to determine which specific ligands, other than PSGL-1, contribute most directly to cell adhesion.

The contributions of selectins and selectin ligands to signalling deserve further study. Emerging data suggest that the transition from rolling adhesion to firm adhesion is a continuous process that requires co-operation between selectins, integrins, and conventional signalling molecules such as chemokines.

These *in vitro* studies need to be complemented by additional studies *in vivo*. The power of homologous recombination to delete genes in mice should be used to study the relative contributions of specific selectins, glycoprotein ligands for selectins, or glycosyltransferases or sulfotransferases that modify selectin ligands. Homologous recombination can also be used to modify the expression patterns of these genes or to make specific changes in a gene product. These studies will allow a much more detailed analysis of the contributions of selectins in models of inflammation, haemostasis, haematopoiesis, wound healing, atherogenesis, and tumour metastasis.

References

1. Springer, T. A. (1994). Traffic signals for lymphocyte recirculation and leukocyte emigration: the multistep paradigm. *Cell*, **76**, 301.
2. Butcher, E. C. and Picker, L. J. (1996). Lymphocyte homing and homeostasis. *Science*, **272**, 60.
3. McEver, R. P., Moore, K. L., and Cummings, R. D. (1995). Leukocyte trafficking mediated by selectin-carbohydrate interactions. *J. Biol. Chem.*, **270**, 11025.
4. Kansas, G. S. (1996). Selectins and their ligands: current concepts and controversies. *Blood*, **88**, 3259.
5. Graves, B. J., Crowther, R. L., Chandran, C., Rumberger, J. M., Li, S., Huang, K.-S., *et al.* (1994). Insight into E-selectin/ligand interaction from the crystal structure and mutagenesis of the lec/EGF domains. *Nature*, **367**, 532.
6. Blanck, O., Iobst, S. T., Gabel, C., and Drickamer, K. (1996). Introduction of selectin-like binding specificity into a homologous mannose-binding protein. *J. Biol. Chem.*, **271**, 7289.

7. Torgersen, D., Mullin, N. P., and Drickamer, K. (1998). Mechanism of ligand binding to E- and P-selectin analyzed using selectin/mannose-binding protein chimeras. *J. Biol. Chem.*, **273**, 6254.

8. Ng, K. K. S. and Weis, W. I. (1997). Structure of a selectin-like mutant of mannose-binding protein complexed with sialylated and sulfated Lewis[x] oligosaccharides. *Biochemistry*, **36**, 979.

9. Kansas, G. S., Saunders, K. B., Ley, K., Zakrzewicz, A., Gibson, R. M., Furie, B. C., *et al.* (1994). A role for the epidermal growth factor-like domain of P-selectin in ligand recognition and cell adhesion. *J. Cell Biol.*, **124**, 609.

10. Watson, S. R., Imai, Y., Fennie, C., Geoffrey, J., Singer, M., Rosen, S. D., *et al.* (1991). The complement binding-like domains of the murine homing receptor facilitate lectin activity. *J. Cell Biol.*, **115**, 235.

11. Li, S. H., Burns, D. K., Rumberger, J. M., Presky, D. H., Wilkinson, V. L., Anostario, M. Jr, *et al.* (1994). Consensus repeat domains of E-selectin enhance ligand binding. *J. Biol. Chem.*, **269**, 4431.

12. Mehta, P., Cummings, R. D., and McEver, R. P. (1998). Affinity and kinetic analysis of P-selectin binding to P-selectin glycoprotein ligand-1. *J. Biol. Chem.*, **273**, 32506.

13. Varki, A. (1994). Selectin ligands. *Proc. Natl. Acad. Sci. USA*, **91**, 7390.

14. Varki, A. (1997). Selectin ligands: will the real ones please stand up? *J. Clin. Invest.*, **99**, 158.

15. Steegmaier, M., Levinovitz, A., Isenmann, S., Borges, E., Lenter, M., Kocher, H. P., *et al.* (1995). The E-selectin-ligand ESL-1 is a variant of an FGF-receptor. *Nature*, **373**, 615.

16. Rosen, S. D., Hwang, S. T., Giblin, P. A., and Singer, M. S. (1997). High-endothelial-venule ligands for L-selectin: identification and functions. *Biochem. Soc. Trans.*, **25**, 428.

17. Sassetti, C., Tangemann, K., Singer, M. S., Kershaw, D. B., and Rosen, S. D. (1998). Identification of podocalyxin-like protein as a high endothelial venule ligand for L-selectin: Parallels to CD34. *J. Exp. Med.*, **187**, 1965.

18. Hemmerich, S., Bertozzi, C. R., Leffler, H., and Rosen, S. D. (1994). Identification of the sulfated monosaccharides of GlyCAM-1, an endothelial-derived ligand for L-selectin. *Biochemistry*, **33**, 4820.

19. Hemmerich, S., Leffler, H., and Rosen, S. D. (1995). Structure of the *O*-glycans in GlyCAM-1, an endothelial-derived ligand for L-selectin. *J. Biol. Chem.*, **270**, 12035.

20. Hemmerich, S. and Rosen, S. D. (1994). 6′-sulfated sialyl Lewis x is a major capping group of GlyCAM-1. *Biochemistry*, **33**, 4830.

21. McEver, R. P. and Cummings, R. D. (1997). Role of PSGL-1 binding to selectins in leukocyte recruitment. *J. Clin. Invest.*, **100**, 485.

22. Sako, D., Chang, X.-J., Barone, K. M., Vachino, G., White, H. M., Shaw, G., *et al.* (1993). Expression cloning of a functional glycoprotein ligand for P-selectin. *Cell*, **75**, 1179.

23. Moore, K. L., Stults, N. L., Diaz, S., Smith, D. L., Cummings, R. D., Varki, A., *et al.* (1992). Identification of a specific glycoprotein ligand for P-selectin (CD62) on myeloid cells. *J. Cell Biol.*, **118**, 445.

24. Norgard, K. E., Moore, K. L., Diaz, S., Stults, N. L., Ushiyama, S., McEver, R. P., *et al.* (1993). Characterization of a specific ligand for P-selectin on myeloid cells. A minor glycoprotein with sialylated *O*-linked oligosaccharides. *J. Biol. Chem.*, **268**, 12764.

25. Wilkins, P. P., McEver, R. P., and Cummings, R. D. (1996). Structures of the *O*-glycans on P-selectin glycoprotein ligand-1 from HL-60 cells. *J. Biol. Chem.*, **271**, 18732.

26. Moore, K. L., Patel, K. D., Bruehl, R. E., Fugang, L., Johnson, D. A., Lichenstein, H. S., *et al.* (1995). P-selectin glycoprotein ligand-1 mediates rolling of human neutrophils on P-selectin. *J. Cell Biol.*, **128**, 661.

27. Wilkins, P. P., Moore, K. L., McEver, R. P., and Cummings, R. D. (1995). Tyrosine sulfation of P-selectin glycoprotein ligand-1 is required for high affinity binding to P-selectin. *J. Biol. Chem.*, **270**, 22677.
28. Pouyani, T. and Seed, B. (1995). PSGL-1 recognition of P-selectin is controlled by a tyrosine sulfation consensus at the PSGL-1 amino terminus. *Cell*, **83**, 333.
29. Sako, D., Comess, K. M., Barone, K. M., Camphausen, R. T., Cumming, D. A., and Shaw, G. D. (1995). A sulfated peptide segment at the amino terminus of PSGL-1 is critical for P-selectin binding. *Cell*, **83**, 323.
30. Liu, W.-J., Ramachandran, V., Kang, J., Kishimoto, T. K., Cummings, R. D., and McEver, R. P. (1998). Identification of N-terminal residues on P-selectin glycoprotein ligand-1 required for binding to P-selectin. *J. Biol. Chem.*, **273**, 7078.
31. Epperson, T. K., Patel, K. D., McEver, R. P., and Cummings, R. D. (2000). Noncovalent dimerization of P-selectin glycoprotein ligand-1 and minimal determinants for binding to P-selectin. *J. Biol. Chem.*, **275**, 7839.
32. Leppanen, A., Mehta, P., Ouyang, Y.-B., Ju, T., Helin, J., Moore, K. L., *et al.* (1999). A novel glycosulfopeptide binds to P-selectin and inhibits leukocyte adhesion to P-selectin. *J. Biol. Chem.*, **274**, 24838.
33. Borges, E., Pendl, G., Eytner, R., Steegmaier, M., Zöllner, O., and Vestweber, D. (1997). The binding of T cell-expressed P-selectin glycoprotein ligand-1 to E- and P-selectin is differentially regulated. *J. Biol. Chem.*, **272**, 28786.
34. Fuhlbrigge, R. C., Kieffer, J. D., Armerding, D., and Kupper, T. S. (1997). Cutaneous lymphocyte antigen is a specialized form of PSGL-1 expressed on skin-homing T cells. *Nature*, **389**, 978.
35. Nicholson, M. W., Barclay, A. N., Singer, M. S., Rosen, S. D., and Van der Merwe, P. A. (1998). Affinity and kinetic analysis of L-selectin (CD62L) binding to glycosylation-dependent cell-adhesion molecule-1. *J. Biol. Chem.*, **273**, 763.
36. Lawrence, M. B. and Springer, T. A. (1991). Leukocytes roll on a selectin at physiologic flow rates: Distinction from and prerequisite for adhesion through integrins. *Cell*, **65**, 859.
37. Alon, R., Hammer, D. A., and Springer, T. A. (1995). Lifetime of the P-selectin: carbohydrate bond and its response to tensile force in hydrodynamic flow. *Nature*, **374**, 539.
38. Alon, R., Chen, S. Q., Puri, K. D., Finger, E. B., and Springer, T. A. (1997). The kinetics of L-selectin tethers and the mechanics of selectin-mediated rolling. *J. Cell Biol.*, **138**, 1169.
39. Finger, E. B., Puri, K. D., Alon, R., Lawrence, M. B., Von Andrian, U. H., and Springer, T. A. (1996). Adhesion through L-selectin requires a threshold hydrodynamic shear. *Nature*, **379**, 266.
40. Lawrence, M. B., Kansas, G. S., Kunkel, E. J., and Ley, K. (1997). Threshold levels of fluid shear promote leukocyte adhesion through selectins (CD62L,P,E). *J. Cell Biol.*, **136**, 717.
41. Puri, K. D., Chen, S., and Springer, T. A. (1998). Modifying the mechanical property and shear threshold of L-selectin adhesion independently of equilibrium properties. *Nature*, **392**, 930.
42. Ramachandran, V., Nollert, M. U., Qiu, H., Liu, W., Cummings, R. D., Zhu, C., *et al.* (1999). Tyrosine replacement in P-selectin glycoprotein ligand-1 affects distinct kinetic and mechanical properties of bonds with P- and L-selectin. *Proc. Natl. Acad. Sci. USA*, **96**, 13771.
43. Berlin, C., Bargatze, R. F., Campbell, J. J., Von Andrian, U. H., Szabo, M. C., Hasslen, S. R., *et al.* (1995). α4 integrins mediate lymphocyte attachment and rolling under physiologic flow. *Cell*, **80**, 413.

44. Alon, R., Kassner, P. D., Carr, M. W., Finger, E. B., Hemler, M. E., and Springer, T. A. (1995). The integrin VLA-4 supports tethering and rolling in flow on VCAM-1. *J. Cell Biol.*, **128**, 1243.

45. Chen, S. Q., Alon, R., Fuhlbrigge, R. C., and Springer, T. A. (1997). Rolling and transient tethering of leukocytes on antibodies reveal specializations of selectins. *Proc. Natl. Acad. Sci. USA*, **94**, 3172.

46. Chen, S. Q. and Springer, T. A. (1999). An automatic braking system that stabilizes leukocyte rolling by an increase in selectin bond number with shear. *J. Cell Biol.*, **144**, 185.

47. Ushiyama, S., Laue, T. M., Moore, K. L., Erickson, H. P., and McEver, R. P. (1993). Structural and functional characterization of monomeric soluble P-selectin and comparison with membrane P-selectin. *J. Biol. Chem.*, **268**, 15229.

48. Patel, K. D., Nollert, M. U., and McEver, R. P. (1995). P-selectin must extend a sufficient length from the plasma membrane to mediate rolling of neutrophils. *J. Cell Biol.*, **131**, 1893.

49. Moore, K. L., Eaton, S. F., Lyons, D. E., Lichenstein, H. S., Cummings, R. D., and McEver, R. P. (1994). The P-selectin glycoprotein ligand from human neutrophils displays sialylated, fucosylated, O-linked poly-N-acetyllactosamine. *J. Biol. Chem.*, **269**, 23318.

50. Li, F., Erickson, H. P., James, J. A., Moore, K. L., Cummings, R. D., and McEver, R. P. (1996). Visualization of P-selectin glycoprotein ligand-1 as a highly extended molecule and mapping of protein epitopes for monoclonal antibodies. *J. Biol. Chem.*, **271**, 6342.

51. Picker, L. J., Warnock, R. A., Burns, A. R., Doerschuk, C. M., Berg, E. L., and Butcher, E. C. (1991). The neutrophil selectin LECAM-1 presents carbohydrate ligands to the vascular selectins ELAM-1 and GMP-140. *Cell*, **66**, 921.

52. Kansas, G. S., Ley, K., Munro, J. M., and Tedder, T. F. (1993). Regulation of leukocyte rolling and adhesion to high endothelial venules through the cytoplasmic domain of L-selectin. *J. Exp. Med.*, **177**, 833.

53. Von Andrian, U. H., Hasslen, S. R., Nelson, R. D., Erlandsen, S. L., and Butcher, E. C. (1995). A central role for microvillous receptor presentation in leukocyte adhesion under flow. *Cell*, **82**, 989.

54. Shao, J. Y., Ting-Beall, H. P., and Hochmuth, R. M. (1998). Static and dynamic lengths of neutrophil microvilli. *Proc. Natl. Acad. Sci. USA*, **95**, 6797.

55. Pavalko, F. M., Walker, D. M., Graham, L., Goheen, M., Doerschuk, C. M., and Kansas, G. S. (1995). The cytoplasmic domain of L-selectin interacts with cytoskeletal proteins via α-actinin: Receptor positioning in microvilli does not require interaction with α-actinin. *J. Cell Biol.*, **129**, 1155.

56. Li, X., Steeber, D. A., Tang, M. L. K., Farrar, M. A., Perlmutter, R. M., and Tedder, T. F. (1998). Regulation of L-selectin-mediated rolling through receptor dimerization. *J. Exp. Med.*, **188**, 1385.

57. Lorant, D. E., McEver, R. P., McIntyre, T. M., Moore, K. L., Prescott, S. M., and Zimmerman, G. A. (1995). Activation of polymorphonuclear leukocytes reduces their adhesion to P-selectin and causes redistribution of ligands for P-selectin on their surfaces. *J. Clin. Invest.*, **96**, 171.

58. Setiadi, H., Disdier, M., Green, S. A., Canfield, W. M., and McEver, R. P. (1995). Residues throughout the cytoplasmic domain affect the internalization efficiency of P-selectin. *J. Biol. Chem.*, **270**, 26818.

59. Setiadi, H., Sedgewick, G., Erlandsen, S. L., and McEver, R. P. (1998). Interactions of the cytoplasmic domain of P-selectin with clathrin-coated pits enhance leukocyte adhesion under flow. *J. Cell Biol.*, **142**, 859.

60. Snapp, K. R., Craig, R., Herron, M., Nelson, R. D., Stoolman, L. M., and Kansas, G. S. (1998). Dimerization of P-selectin glycoprotein ligand-1 (PSGL-1) required for optimal recognition of P-selectin. *J. Cell Biol.*, **142**, 263.

61. McEver, R. P. (2000). P-selectin. In *Platelets, thrombosis and the vessel wall* (ed. M. C. Berndt), p. 209. Harwood Academic Publishers, London.

62. Collins, T., Read, M. A., Neish, A. S., Whitley, M. Z., Thanos, D., and Maniatis, T. (1995). Transcriptional regulation of endothelial cell adhesion molecules: NF-κB and cytokine-inducible enhancers. *FASEB J.*, **9**, 899.

63. Yao, L., Pan, J., Setiadi, H., Patel, K. D., and McEver, R. P. (1996). Interleukin 4 or oncostatin M induces a prolonged increase in P-selectin mRNA and protein in human endothelial cells. *J. Exp. Med.*, **184**, 81.

64. McEver, R. P. (1997). Regulation of expression of E-selectin and P-selectin. In *The selectins: initiators of leukocyte endothelial adhesion* (ed. D. Vestweber), p. 31. Harwood Academic Publishers, Amsterdam.

65. Yao, L., Setiadi, H., Xia, L., Laszik, Z., Taylor, F. B., and McEver, R. P. (1999). Divergent inducible expression of P-selectin and E-selectin in mice and primates. *Blood*, **94**, 3820.

66. Pan, J., Xia, L., and McEver, R. P. (1998). Comparison of promoters for the murine and human P-selectin genes suggests species-specific and conserved mechanisms for transcriptional regulation in endothelial cells. *J. Biol. Chem.*, **273**, 10058.

67. Pan, J., Xia, L., Yao, L., and McEver, R. P. (1998). Tumor necrosis factor-α- or lipopolysaccharide-induced expression of the murine P-selectin gene in endothelial cells involves novel κB sites and a variant ATF/CRE element. *J. Biol. Chem.*, **273**, 10068.

68. Kahn, J., Walcheck, B., Migaki, G. I., Jutila, M. A., and Kishimoto, T. K. (1998). Calmodulin regulates L-selectin adhesion molecule expression and function through a protease-dependent mechanism. *Cell*, **92**, 809.

69. Walcheck, B., Kahn, J., Fisher, J. M., Wang, B. B., Fisk, R. S., Payan, D. G., *et al.* (1996). Neutrophil rolling altered by inhibition of L-selectin shedding *in vitro*. *Nature*, **380**, 720.

70. Hafezi-Moghadam, A. and Ley, K. (1999). Relevance of L-selectin shedding for leukocyte rolling *in vivo*. *J. Exp. Med.*, **189**, 939.

71. Michelson, A. D., Barnard, M. R., Hechtman, H. B., MacGregor, H., Connolly, R. J., Loscalzo, J., *et al.* (1996). *In vivo* tracking of platelets: circulating degranulated platelets rapidly lose surface P-selectin but continue to circulate and function. *Proc. Natl. Acad. Sci. USA*, **93**, 11877.

72. Hartwell, D. M., Mayadas, T. N., Berger, G., Frenette, P. S., Rayburn, H., Hynes, R. O., *et al.* (1998). Role of P-selectin cytoplasmic domain in granular targeting *in vivo* and in early inflammatory responses. *J. Cell Biol.*, **143**, 1129.

73. Schleiffenbaum, B., Spertini, O., and Tedder, T. F. (1992). Soluble L-selectin is present in human plasma at high levels and retains functional activity. *J. Cell Biol.*, **119**, 229.

74. Subramaniam, M., Koedam, J. A., and Wagner, D. D. (1993). Divergent fates of P- and E-selectins after their expression on the plasma membrane. *Mol. Biol. Cell*, **4**, 791.

75. Green, S. A., Setiadi, H., McEver, R. P., and Kelly, R. B. (1994). The cytoplasmic domain of P-selectin contains a sorting determinant that mediates rapid degradation in lysosomes. *J. Cell Biol.*, **124**, 435.

76. Disdier, M., Morrissey, J. H., Fugate, R. D., Bainton, D. F., and McEver, R. P. (1992). Cytoplasmic domain of P-selectin (CD62) contains the signal for sorting into the regulated secretory pathway. *Mol. Biol. Cell*, **3**, 309.

77. Fleming, J. C., Berger, G., Guichard, J., Cramer, E. M., and Wagner, D. D. (1998). The transmembrane domain enhances granular targeting of P-selectin. *Eur. J. Cell Biol.*, **75**, 331.

78. Chuang, P. I., Young, B. A., Thiagarajan, R. R., Cornejo, C., Winn, R. K., and Harlan, J. M. (1997). Cytoplasmic domain of E-selectin contains a non-tyrosine endocytosis signal. *J. Biol. Chem.*, **272**, 24813.

79. Ouyang, Y. B., Lane, W. S., and Moore, K. L. (1998). Tyrosylprotein sulfotransferase: Purification and molecular cloning of an enzyme that catalyzes tyrosine *O*-sulfation, a common posttranslational modification of eukaryotic proteins. *Proc. Natl. Acad. Sci. USA*, **95**, 2896.

80. Bistrup, A., Bhakta, S., Lee, J. K., Belov, Y. Y., Gunn, M. D., Zuo, F. R., *et al.* (1999). Sulfotransferases of two specificities function in the reconstitution of high endothelial cell ligands for L-selectin. *J. Cell Biol.* **145**, 899.

81. Hiraoka, N., Petryniak, B., Nakayama, J., Tsuboi, S., Suzuki, M., Yeh, J. C., *et al.* (1999). A novel, high endothelial venule-specific sulfotransferase expresses 6-sulfo sialyl Lewis[x], an L-selectin ligand displayed by CD34. *Immunity*, **11**, 79.

82. Onrust, S. V., Hartl, P. M., Rosen, S. D., and Hanahan, D. (1996). Modulation of L-selectin ligand expression during an immune response accompanying tumorigenesis in transgenic mice. *J. Clin. Invest.*, **97**, 54.

83. Kimura, N., Mitsuoka, C., Kanamori, A., Hiraiwa, N., Uchimura, K., Muramatsu, T., *et al.* (1999). Reconstitution of functional L-selectin ligands on a cultured human endothelial cell line by cotransfection of α1→3 fucosyltransferase VII and newly cloned GlcNAcβ:6-sulfotransferase cDNA. *Proc. Natl. Acad. Sci. USA*, **96**, 4530.

84. Natsuka, S., Gersten, K. M., Zenitas, K., Kannagi, R., and Lowe, J. B. (1994). Molecular cloning of a cDNA encoding a novel human leukocyte α1,3-fucosyltransferase capable of synthesizing the sialyl Lewis x determinant. *J. Biol. Chem.*, **269**, 16789.

85. Wagers, A. J., Waters, C. M., Stoolman, L. M., and Kansas, G. S. (1998). Interleukin 12 and interleukin 4 control cell adhesion to endothelial selectins through opposite effects on α1,3-fucosyltransferase VII gene expression. *J. Exp. Med.*, **188**, 2225.

86. Tu, L., Delahunty, M. D., Ding, H., Luscinskas, F. W., and Tedder, T. F. (1999). The cutaneous lymphocyte antigen is an essential component of the L-selectin ligand induced on human vascular endothelial cells. *J. Exp. Med.*, **189**, 241.

87. Waddell, T. K., Fialkow, L., Chan, C. K., Kishimoto, T. K., and Downey, G. P. (1995). Signaling functions of L-selectin. Enhancement of tyrosine phosphorylation and activation of MAP kinase. *J. Biol. Chem.*, **270**, 15403.

88. Brenner, B., Gulbins, E., Schlottman, K., Koppenhoefer, U., Busch, G. L., Walzog, B., *et al.* (1996). L-selectin activates the Ras pathway via the tyrosine kinase p56[lck]. *Proc. Natl. Acad. Sci. USA*, **93**, 15376.

89. Hidari, K. I. P. J., Weyrich, A. S., Zimmerman, G. A., and McEver, R. P. (1997). Engagement of P-selectin glycoprotein ligand-1 enhances tyrosine phosphorylation and activates mitogen-activated protein kinases in human neutrophils. *J. Biol. Chem.*, **272**, 28750.

90. Simon, S. I., Burns, A. R., Taylor, A. D., Gopalan, P. K., Lynam, E. B., Sklar, L. A., *et al.* (1995). L-selectin (CD62L) cross-linking signals neutrophil adhesive functions via the Mac-1 (CD11b/CD18) β$_2$-integrin. *J. Immunol.*, **155**, 1502.

91. Giblin, P. A., Hwang, S. T., Katsumoto, T. R., and Rosen, S. D. (1997). Ligation of L-selectin on T lymphocytes activates β$_1$ integrins and promotes adhesion to fibronectin. *J. Immunol.*, **159**, 3498.

92. Evangelista, V., Manarini, S., Sideri, R., Rotondo, S., Martelli, N., Piccoli, A., *et al.* (1999). Platelet/polymorphonuclear leukocyte interaction: P-selectin triggers protein-tyrosine phosphorylation-dependent CD11b/CD18 adhesion: Role of PSGL-1 as a signaling molecule. *Blood*, **93**, 876.

93. Weyrich, A. S., McIntyre, T. M., McEver, R. P., Prescott, S. M., and Zimmerman, G. A. (1995). Monocyte tethering by P-selectin regulates monocyte chemotactic protein-1 and tumor necrosis factor-α secretion. *J. Clin. Invest.*, **95**, 2297.

94. Weyrich, A. S., Elstad, M. R., McEver, R. P., McIntyre, T. M., Moore, K. L., Morrissey, J. H., *et al.* (1996). Activated platelets signal chemokine synthesis by human monocytes. *J. Clin. Invest.*, **97**, 1525.

95. Yoshida, M., Westlin, W. F., Wang, N., Ingber, D. E., Rosenzweig, A., Resnick, N., *et al.* (1996). Leukocyte adhesion to vascular endothelium induces E-selectin linkage to the actin cytoskeleton. *J. Cell Biol.*, **133**, 445.

96. Lorenzon, P., Vecile, E., Nardon, E., Ferrero, E., Harlan, J. M., Tedesco, F., *et al.* (1998). Endothelial cell E- and P-selectin and vascular cell adhesion molecule-1 function as signaling receptors. *J. Cell Biol.*, **142**, 1381.

97. Ley, K. (1996). Molecular mechanisms of leukocyte recruitment in the inflammatory process. *Cardiovasc. Res.*, **32**, 733.

98. Frenette, P. S. and Wagner, D. D. (1997). Insights into selectin function from knockout mice. *Thromb. Haemost.*, **78**, 60.

99. Arbones, M. L., Ord, D. C., Ley, K., Ratech, H., Maynard-Curry, C., Otten, G., *et al.* (1994). Lymphocyte homing and leukocyte rolling and migration are impaired in L-selectin-deficient mice. *Immunity*, **1**, 247.

100. Mayadas, T. N., Johnson, R. C., Rayburn, H., Hynes, R. O., and Wagner, D. D. (1993). Leukocyte rolling and extravasation are severely compromised in P selectin-deficient mice. *Cell*, **74**, 541.

101. Ley, K., Bullard, D. C., Arbonés, M. L., Bosse, R., Vestweber, D., Tedder, T. F., *et al.* (1995). Sequential contribution of L- and P-selectin to leukocyte rolling *in vivo*. *J. Exp. Med.*, **181**, 669.

102. Diacovo, T. G., Puri, K. D., Warnock, R. A., Springer, T. A., and Von Andrian, U. H. (1996). Platelet-mediated lymphocyte delivery to high endothelial venules. *Science*, **273**, 252.

103. Kunkel, E. J., Chomas, J. E., and Ley, K. (1998). Role of primary and secondary capture for leukocyte accumulation *in vivo*. *Circ. Res.*, **82**, 30.

104. Labow, M. A., Norton, C. R., Rumberger, J. M., Lombard-Gillooly, K. M., Shuster, D. J., Hubbard, J., *et al.* (1994). Characterization of E-selectin-deficient mice: demonstration of overlapping function of the endothelial selectins. *Immunity*, **1**, 709.

105. Frenette, P. S., Mayadas, T. N., Rayburn, H., Hynes, R. O., and Wagner, D. D. (1996). Susceptibility to infection and altered hematopoiesis and mice deficient in both P- and E-selectin. *Cell*, **84**, 563.

106. Bullard, D. C., Kunkel, E. J., Kubo, H., Hicks, M. J., Lorenzo, I., Doyle, N. A., *et al.* (1996). Infectious susceptibility and severe deficiency of leukocyte rolling and recruitment in E-selectin and P-selectin double mutant mice. *J. Exp. Med.*, **183**, 2329.

107. Kunkel, E. J. and Ley, K. (1996). Distinct phenotype of E-selectin-deficient mice—E-selectin is required for slow leukocyte rolling *in vivo*. *Circ. Res.*, **79**, 1196.

108. Jung, U., Norman, K. E., Scharffetter-Kochanek, K., Beaudet, A. L., and Ley, K. (1998). Transit time of leukocytes rolling through venules controls cytokine-induced inflammatory cell recruitment *in vivo*. *J. Clin. Invest.*, **102**, 1526.

109. Yang, J., Hirata, T., Croce, K., Merrill-Skoloff, G., Tchernychev, B., Williams, E., *et al.* (1999). Targeted gene disruption demonstrates that P-selectin glycoprotein ligand 1 (PSGL-1) is required for P-selectin-mediated but not E-selectin-mediated neutrophil rolling and migration. *J. Exp. Med.*, **190**, 1769.

110. Hirata, T., Merrill-Skoloff, G., Aab, M., Yang, J., Furie, B. C., and Furie, B. (2000). An important role for PSGL-1 in Th1 lymphocyte migration: *in vivo* studies with PSGL-1-deficient mice. *J. Exp. Med.*, **192**, 1669.

111. Maly, P., Thall, A. D., Petryniak, B., Rogers, G. E., Smith, P. L., Marks, R. M., *et al.* (1996). The α(1,3)Fucosyltransferase Fuc-TVII controls leukocyte trafficking through an essential role in L-, E-, and P-selectin ligand biosynthesis. *Cell*, **86**, 643.

112. Niemelä, R., Natunen, J., Majuri, M. L., Maaheimo, H., Helin, J., Lowe, J. B., *et al.* (1998). Complementary acceptor and site specificities of Fuc-TIV and Fuc-TVII allow effective biosynthesis of sialyl-TriLex and related polylactosamines present on glycoprotein counterreceptors of selectins. *J. Biol. Chem.*, **273**, 4021.

113. Ellies, L. G., Tsuboi, S., Petryniak, B., Lowe, J. B., Fukuda, M., and Marth, J. D. (1998). Core 2 oligosaccharide biosynthesis distinguishes between selectin ligands essential for leukocyte homing and inflammation. *Immunity*, **9**, 881.

114. Yeh, J. C., Ong, E., and Fukuda, M. (1999). Molecular cloning and expression of a novel β1,6-*N*-acetylglucosaminyltransferase that forms core 2, core 4, and I branches. *J. Biol. Chem.*, **274**, 3215.

115. Schwientek, T., Nomoto, M., Levery, S. B., Merkx, G., van Kessel, A. G., Bennett, E. P., *et al.* (1999). Control of *O*-glycan branch formation. Molecular cloning of human cDNA encoding a novel β1,6-*N*-acetylglucosaminyltransferase forming core 2 and core 4. *J. Biol. Chem.*, **274**, 4504.

116. Von Andrian, U. H., Berger, E. M., Ramezani, L., Chambers, J. D., Ochs, H. D., Harlan, J. M., *et al.* (1993). *In vivo* behavior of neutrophils from two patients with distinct inherited leukocyte adhesion deficiency syndromes. *J. Clin. Invest.*, **91**, 2893.

117. Etzioni, A., Frydman, M., Pollack, S., Avidor, I., Phillips, M. L., Paulson, J. C., *et al.* (1992). Brief report: recurrent severe infections caused by a novel leukocyte adhesion deficiency. *N. Engl. J. Med.*, **327**, 1789.

118. Philips, M. L., Schwartz, B. R., Etzioni, A., Bayer, R., Ochs, H. D., Paulson, J. C., *et al.* (1995). Neutrophil adhesion in leukocyte adhesion deficiency syndrome type 2. *J. Clin. Invest.*, **96**, 2898.

119. Zannettino, A. C. W., Berndt, M. C., Butcher, C., Butcher, E. C., Vadas, M. A., and Simmons, P. J. (1995). Primitive human hematopoietic progenitors adhere to P-selectin (CD62P). *Blood*, **85**, 3466.

120. Laszik, Z., Jansen, P. J., Cummings, R. D., Tedder, T. F., McEver, R. P., and Moore, K. L. (1996). P-selectin glycoprotein ligand-1 is broadly expressed in cells of myeloid, lymphoid, and dendritic lineage and in some nonhematopoietic cells. *Blood*, **88**, 3010.

121. Nguyen, M., Strubel, N. A., and Bischoff, J. (1993). A role for sialyl Lewis-X/A glycoconjugates in capillary morphogenesis. *Nature*, **365**, 267.

122. Kräling, B. M., Razon, M. J., Boon, L. M., Zurakowski, D., Seachord, C., Darveau, R. P., *et al.* (1996). E-selectin is present in proliferating endothelial cells in human hemangiomas. *Am. J. Pathol.*, **148**, 1181.

123. Frenette, P. S., Subbarao, S., Mazo, I. B., Von Andrian, U. H., and Wagner, D. D. (1998). Endothelial selectins and vascular cell adhesion molecule-1 promote hematopoietic progenitor homing to bone marrow. *Proc. Natl. Acad. Sci. USA*, **95**, 14423.

124. Lefer, A. M. and Lefer, D. J. (1996). The role of nitric oxide and cell adhesion molecules on the microcirculation in ischaemia-reperfusion. *Cardiovasc. Res.*, **32**, 743.

125. Grisham, M. B., Granger, D. N., and Lefer, D. J. (1998). Modulation of leukocyte-endothelial interactions by reactive metabolites of oxygen and nitrogen: Relevance to ischemic heart disease. *Free Radic. Biol. Med.*, **25**, 404.

126. Sharar, S. R., Winn, R. K., and Harlan, J. M. (1995). The adhesion cascade and anti-adhesion therapy: An overview. *Springer Semin. Immunopathol.*, **16**, 359.

127. Takada, M., Nadeau, K. C., Shaw, G. D., Marquette, K. A., and Tilney, N. L. (1997). The cytokine-adhesion molecule cascade in ischemia/reperfusion injury of the rat kidney. Inhibition by a soluble P-selectin ligand. *J. Clin. Invest.*, **99**, 2682.

128. Grober, J. S., Bowen, B. L., Ebling, H., Athey, B., Thompson, C. B., Fox, D. A., *et al.* (1993). Monocyte-endothelial adhesion in chronic rheumatoid arthritis: *in situ* detection of selectin and integrin-dependent interactions. *J. Clin. Invest.*, **91**, 2609.

129. Symon, F. A., Walsh, G. M., Watson, S. R., and Wardlaw, A. J. (1994). Eosinophil adhesion to nasal polyp endothelium is P-selectin-dependent. *J. Exp. Med.*, **180**, 371.

130. Lehr, H.-A., Olofsson, A. M., Carew, T. E., Vajkoczy, P., Von Andrian, U. H., Hübner, C., *et al.* (1994). P-selectin mediates the interaction of circulating leukocytes with platelets and microvascular endothelium in response to oxidized lipoprotein *in vivo*. *Lab. Invest.*, **71**, 380.

131. Lehr, H.-A., Frei, B., and Arfors, K.-E. (1994). Vitamin C prevents cigarette smoke-induced leukocyte aggregation and adhesion to endothelium *in vivo*. *Proc. Natl. Acad. Sci. USA*, **91**, 7688.

132. Johnson-Tidey, R. R., McGregor, J. L., Taylor, P. R., and Poston, R. N. (1994). Increase in the adhesion molecule P-selectin in endothelium overlying atherosclerotic plaques. Coexpression with intercellular adhesion molecule-1. *Am. J. Pathol.*, **144**, 952.

133. Dong, Z. M., Chapman, S. M., Brown, A. A., Frenette, P. S., Hynes, R. O., and Wagner, D. D. (1998). Combined role of P- and E-selectins in atherosclerosis. *J. Clin. Invest.*, **102**, 145.

134. Collins, R. G., Velji, R., Guevara, N. V., Hicks, M. J., Chan, L., and Beaudet, A. L. (2000). P-selectin or intercellular adhesion molecule (ICAM)-1 deficiency substantially protects against atherosclerosis in apolipoprotein E-deficient mice. *J. Exp. Med.*, **191**, 189.

135. Palabrica, T., Lobb, R., Furie, B. C., Aronovitz, M., Benjamin, C., Hsu, Y.-M., *et al.* (1992). Leukocyte accumulation promoting fibrin deposition is mediated *in vivo* by P-selectin on adherent platelets. *Nature*, **359**, 848.

136. Toombs, C. F., DeGraaf, C. L., Martin, J. P., Geng, J. G., Anderson, D. C., and Shebuski, R. J. (1995). Pretreatment with a blocking monoclonal antibody to P-selectin accelerates pharmacological thrombolysis in a primate model of arterial thrombosis. *J. Pharmacol. Exp. Ther.*, **275**, 941.

137. Ikeda, H., Ueyama, T., Murahara, T., Yasukawa, H., Haramaki, N., Eguchi, H., *et al.* (1999). Adhesive interaction between P-selectin and sialyl Lewis[x] plays an important role in recurrent coronary arterial thrombosis in dogs. *Arterioscler. Thromb. Vasc. Biol.*, **19**, 1083.

138. Wakefield, T. W., Strieter, R. M., Downing, L. J., Kadell, A. M., Wilke, C. A., Burdick, M. D., *et al.* (1996). P-selectin and TNF inhibition reduce venous thrombosis inflammation. *J. Surg. Res.*, **64**, 26.

139. Kim, Y. J., Borsig, L., Varki, N. M., and Varki, A. (1998). P-selectin deficiency attenuates tumor growth and metastasis. *Proc. Natl. Acad. Sci. USA*, **95**, 9325.

3 | Cadherins

GLENN L. RADICE and MASATOSHI TAKEICHI

1. Introduction

The ability of dissociated cells to reassemble into recognizable anatomical structures was observed in the marine sponge nearly a hundred years ago (1). While not appreciated at the time, it was later demonstrated in the amphibian system that embryonic cells prefer to interact with like cells thus facilitating cell sorting and rearrangements leading to the reconstruction of specific tissue structures (2). The 'differential adhesion hypothesis' was later proposed to explain the ability of cells to sort out and segregate into embryonic tissues (3). Closer examination of the effects of calcium on proteolysis and subsequent cellular aggregation demonstrated that cell–cell adhesion could be functionally divided into two distinct mechanisms: the Ca^{2+}-dependent and Ca^{2+}-independent systems (4). The identification of cell adhesion molecules on the surface of cells showing different specificities as well as affinities provided molecular evidence consistent with many of the early hypotheses.

Since the initial identification of the calcium-dependent cell adhesion molecule, known as uvomorulin/LCAM/E-cadherin, in the late 1970s approximately 30 novel cadherins have since been identified in vertebrates. The specific roles of the different cadherins in the generation of the complex organization of the multicellular organism remains to be elucidated in most cases. There is a strong correlation between the complexity of the organism and the number of cadherin subtypes necessary to generate the animal form. As expected no cadherin sequences were identified in the genome of the unicellular organism, the budding yeast, *Saccharomyces cerevisiae*. The completion of the genomic sequence of the first multicellular organism, the nematode worm *Caenorhabditis elegans*, reveals one classical cadherin and 11 protocadherins consistent with its relative simplicity (5; M. Costa and J. Pettit, unpublished data). With the expected completion of the human genome project in the next few years, it will be interesting to see how many cadherins are necessary for the generation of the human form. As genomic sequences become available from additional organisms a comparison of cadherin utilization and organismal complexity will shed light on the evolution of morphogenetic processes. Understanding the function(s) of the different cadherins will provide important insight into the aetiology of human diseases including cancer.

A growing number of mutations in cadherin and its associated proteins, catenins, have been generated and characterized in several model organisms including *Drosophila melangaster*, *C. elegans*, Zebrafish, and mice. These studies have shed light on the roles of specific cadherins in various morphogenetic processes. Interestingly, a comparison of cadherin mutant phenotypes from different organisms indicate the most dramatic cell adhesion defects are associated with tissues undergoing active cellular rearrangements and/or increased mechanical stress. The fact that most cells, especially in vertebrates, express multiple cadherin subtypes on their surface suggests a complex co-operative mechanism of cell adhesion/signalling. The unique spatial patterns of cadherin expression in the brain suggests a possible role in the selectivity and connectivity of neurons in the developing CNS. While numerous studies have focused on the role of cadherins as cell adhesion molecules, increasing evidence suggests cadherins can provide specific signals affecting numerous cellular processes including the following:

- morphogenesis
- apoptosis
- cell proliferation
- migration
- differentiation
- invasiveness

2. Molecular architecture of a cadherin

The cadherin superfamily can be classified into four categories: classical cadherins, desmosomal cadherins, protocadherins, and cadherin-related proteins. The desmosomal cadherins include the desmocollins and desmogleins are discussed in Chapter 10. The one common feature of cadherin family members is the presence of multiple cadherin-specific repeats (CR) in the extracellular domain (Fig. 1). The cadherin motif is approximately 110 amino acids long with several highly conserved amino acid sequences. The number of cadherin repeats varies from four to more than 30 among the different cadherin family members. The cadherin repeat also has distinctive characteristics that allow the family members to be subdivided into smaller groups, as discussed below.

2.1 Classical cadherin family

The classical cadherins such as E-, P-, and N-cadherin mediate strong Ca^{2+}-dependent, homotypic cell–cell adhesion when localized to the zonula adherens or adherens junction. The overall structures of all classical cadherins are essentially equivalent. The signal sequence at the N-terminus is flanked by a prosequence that contains a protease processing signal sequence, K/RRXKR, at the C-terminal end. The proteolytic processing of the precursor appears to be essential for activation of

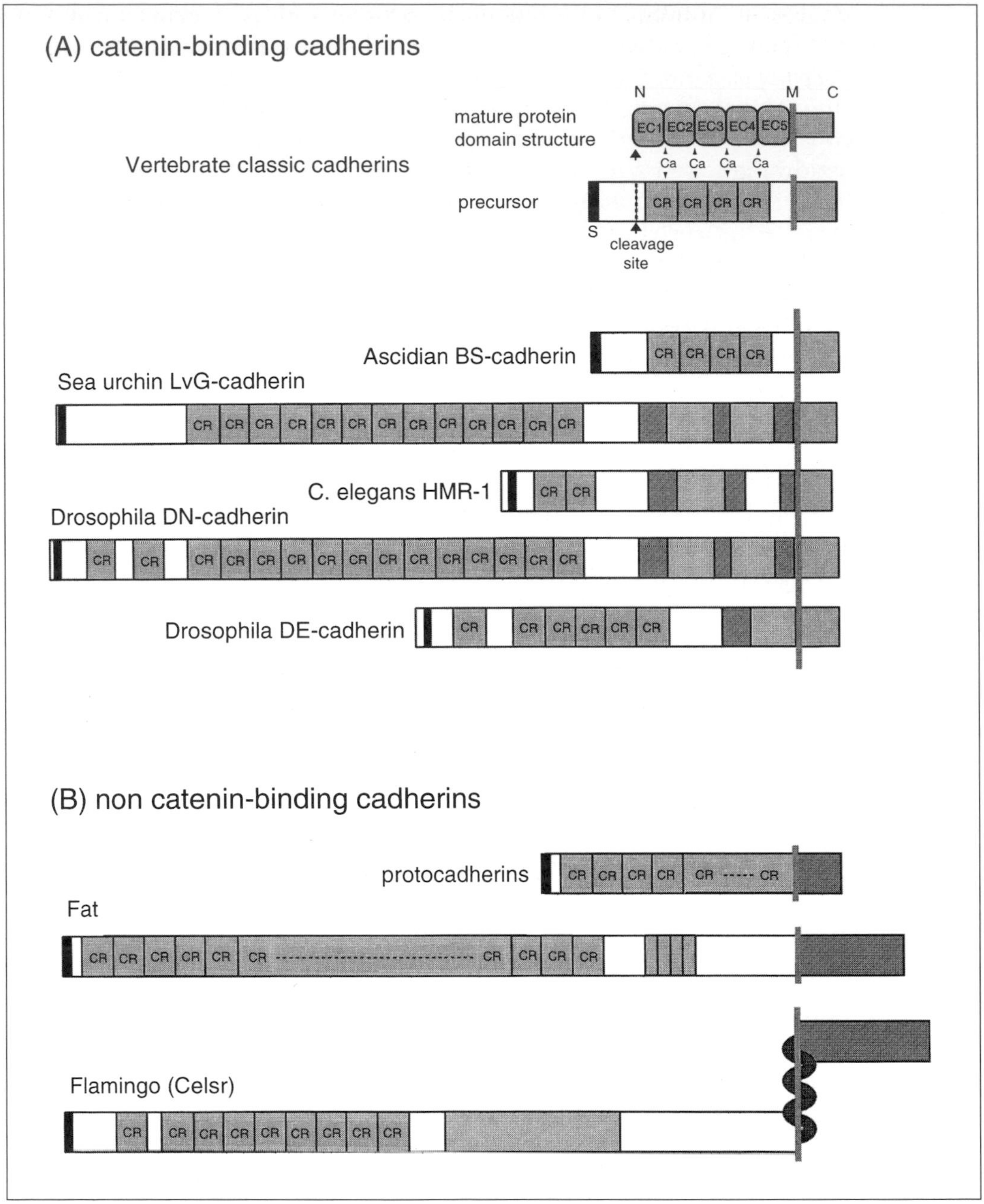

Fig. 1 The primary structure of cadherin superfamily members from various organisms. The sequence similarity of the cytoplasmic domains including the β-catenin binding site distinguishes the catenin binding (A) and non-catenin binding (B) cadherin subfamilies. EC, extracellular domain; N, amino terminus; M, membrane; C, carboxy terminus; Ca, calcium; CR, cadherin repeat; S, signal peptide.

the classical cadherin. Following the prosequence are five extracellular domains, four of which contain the characteristic cadherin repeat. The fifth domain lacks the DRE and DXNDNXPXF sequences common to the cadherin repeat, and in addition, contains four distinct cysteine residues. Classical cadherins have a single transmembrane domain and a highly conserved cytoplasmic domain consisting of approximately 150 amino acids. Several distinct functional domains have been identified within the classical cadherins. Consensus calcium binding sites are present within each of the four cadherin repeats in the extracellular domain. Calcium binding is thought to be required for dimerization and rigidification of classical cadherins and provides protection from proteolytic degradation (6). The EC1 amino terminal subdomain contains a cell adhesion recognition sequence containing a conserved core, His–Ala–Val (HAV), plus additional flanking sequences, and also the domain with which the HAV interacts. Amino acids sequences surrounding HAV are involved in determining the specificity of cadherin self-association as determined by mutational analysis (7). The ability of cadherin subtypes to mediate cell sorting was demonstrated by transfecting L cells with different cadherin cDNAs and performing aggregation assays (8, 9). Cells expressing E-cadherin sorted out and formed distinct aggregates from P- and N-cadherin positive cells and vice versa (Fig. 2). Structural analysis of the amino terminal domain suggests cadherins may form a zipper-like structure across the junction discussed in Chapter 8. The alignment of amino acid sequences of various cadherins shows that classical cadherins can be classified into two types. Many characteristic amino acids, as well as short amino acid deletions and additions, are present in each particular group. Furthermore, the overall amino acid identity among classical cadherins of the same group is higher than that between the two groups. The Type I classical cadherins include E-, N-, P-, and R-cadherin. The more recently identified classical cadherins belong to the Type II group including cadherin-5 (also called VE-cadherin) through cadherin-12 (10). The cytoplasmic domains of Type I and Type II classical cadherins interact with catenins providing linkage to the cytoskeleton.

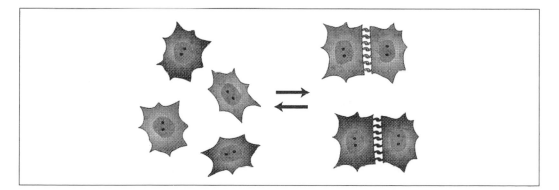

Fig. 2 Homophilic binding of cadherin subtypes mediates cell sorting. Ectopic expression of different cadherin subtypes, including E-, N-, and P-cadherin, in L cells leads to cell sorting and increased cell–cell adhesion.

T-cadherin, for truncated cadherin, belongs to a small group of unique cadherins that lack the cytoplasmic domain and part of the transmembrane domain. T-cadherin is anchored to the membrane through a glycosyl phosphatidylinositol (GPI) moiety (11), and it mediates calcium-dependent, homophilic binding *in vitro* (12). The adhesion mechanism of truncated cadherins is poorly understood.

The cadherin cytoplasmic domain interacts with a group of proteins termed catenins, which link the cadherin to the actin cytoskeleton (Fig. 3) (13, 14). Deletion of the cytoplasmic domain of E-cadherin results in a dramatic reduction of cadherin-mediated cell adhesion (15). Further examination of the cytoplasmic domain of classical cadherin indicates at least two distinct protein binding sites; a carboxy terminal region for β- or γ-catenin (plakoglobin) and a juxtamembrane region for p120ctn binding. Either β-catenin or plakoglobin, which are members of the armadillo family of proteins (16), binds directly to the cadherin. The β-catenin protein consists of a 130 amino acid amino terminal domain followed by 12 imperfect repeats of 42 amino acids (referred to as ARM repeats) and a carboxy terminal domain of 100 amino acids. The ARM repeats are necessary for interaction with the cytoplasmic domain of cadherin as well as other proteins to be discussed later, including adenomatous polyposis coli (APC) and lymphoid enhancer factor (LEF)/T cell factor (TCF) family members. The amino terminus of β-catenin contains the α-catenin binding site as well as multiple phosphorylation sites that are correlated with its inactivation. The carboxy terminus contains the transcriptional activation domain.

α-Catenin, which shares homology with the cytoskeleton-associated protein vinculin, binds the cadherin indirectly through β-catenin or plakoglobin. α-Catenin interacts with the actin-based cytoskeleton directly or indirectly, involving other cytoplasmic proteins such as α-actinin and vinculin (17–19). Thus, α-catenin serves to link the cadherin/catenin complex to the actin cytoskeleton. The interaction with the actin cytoskeleton may be regulated by different α-catenin subtypes. Whereas αE-catenin is widely expressed, αN-catenin is expressed predominantly in neural tissue (20). It is presently unknown how the α-catenin subtypes influence the cadherin/catenin adhesion complex in different cell types. The spatial pattern of cadherins in the CNS, particularly their localization to the synapse suggests that modification of the cadherin–cytoskeletal linkage may be important for neural connectivity. Neural-specific accessory proteins such as catenins may play an important role in the synaptic junction.

The ability of cadherins to rapidly respond to different cellular signals is likely mediated through the cytoplasmic tail. Thus, altering the cadherin interaction with the cytoskeleton, or its lateral dimerization within the plane of the plasma membrane leads to increased or decreased cell adhesion. Phosphorylation of β-catenin can lead to increased instability of the cadherin/catenin complex and decreased adhesion. In contrast, the function of p120ctn and related proteins in the cadherin adhesion complex is less well understood. The p120ctn phosphoprotein was originally identified as a target of p60v-Src protein kinase. It contains ARM repeats, and binds to the cadherin complex (21). There is growing evidence that p120 may act as a inhibitor of cadherin function by binding to the membrane proximal region of the cadherin

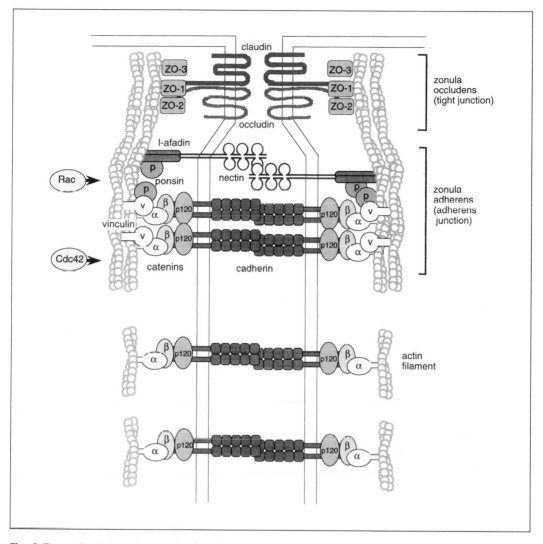

Fig. 3 The cadherin/catenin complex in epithelial cells. Tight junctions are located near the apical surface followed by the zonula adherens or adherens junctions. The cadherin is anchored to the actin cytoskeleton via a series of protein–protein interactions which may be modified depending on the cellular context. ZO, zonula occludin.

cytoplasmic domain (22, 23). However, the function of p120ctn may depend on the cellular context and its phosphorylation state, since a positive role for p120ctn in cell adhesion also has been reported (24).

A colon carcinoma cell line, Colo 205, with an intact E-cadherin/catenin adhesion system was studied for its inability to form compact aggregates (23). Treatment of these cells with the kinase inhibitor, staurosporine, induced cell aggregation within a few hours, accompanied by a shift in the electrophoretic mobility of p120ctn.

Additional experiments indicated that the change in p120ctn size was due to dephosphorylation. To determine whether p120ctn was directly involved in the increased adhesiveness of the Colo 205 cells, a series of truncated p120ctn constructs were transfected into the cells. A p120ctn amino terminal deletion resulted in restored adhesion presumably by competing with the endogenous p120ctn for cadherin binding. To confirm that the cellular aggregation was inhibited by p120ctn interaction with the cytoplasmic domain, a full-length or juxtamembrane deleted N-cadherin was introduced into Colo 205 cells. The wild-type N-cadherin had no effect on cellular aggregation, while the mutant N-cadherin with the deleted p120ctn binding site restored cell adhesion demonstrating that p120ctn inhibits cadherin-mediated adhesion in Colo 205 cells by binding to the cadherin juxtamembrane region.

There also is evidence for a positive role for p120ctn in lateral dimerization and strengthen cell adhesion (24). CHO cells were transfected with wild-type and various cytoplasmic deletions of *Xenopus* C-cadherin to determine the role of the juxta-membrane region in cadherin clustering and adhesion. CHO cells expressing wild-type C-cadherin cluster in a ligand-specific manner to substrata covered with the extracellular domain of C-cadherin (25). Specific deletion of the juxtamembrane region abolished clustering, whereas the juxtamembrane region by itself was sufficient to induce protein clustering when it was attached to either a cadherin or a heterologous membrane protein. Surprisingly, the cadherin mutant containing the membrane proximal region alone (i.e. capable of clustering) had adhesive activity when tested in laminar flow detachment and aggregation assays.

A novel ARM repeat protein, δ-catenin, was discovered independently by two groups, in mouse (26) and human (27). Human δ-catenin was identified in a yeast two-hybrid screen by its ability to bind to the loop region of presenilin-1 (28), which is encoded by the gene most commonly mutated in familial Alzheimer's disease (29, 30). δ-Catenin has greatest similarity with the ARM repeats of desmosomal protein, p0071 (69% identity), compared to p120ctn (48% identity). Interestingly, δ-catenin appears to be almost exclusively expressed in the nervous system similar to αN-catenin, suggesting that differential regulation of the cadherin complex in neurons may occur via alternative catenin usage. Ectopic expression of δ-catenin in MDCK cells led to changes in cell morphology from a polygonal to irregular shape, including a fibroblastic appearance (27). In addition, the cells became more responsive to hepatocyte growth factor (HGF)-induced cell scattering. Together, the high expression of δ-catenin in the ventricular zone of the developing brain, a region of active neuronal differentiation and migration, its effect on cell scattering, and its putative relationship with non-receptor tyrosine kinases, suggest δ-catenin may act as a regulator of neuronal migration (27).

Cell adhesion is a co-operative process requiring co-ordinated interactions between different cell adhesion systems and the actin cytoskeleton. The recently identified NAP adhesion system composed of nectin, afadin, and ponsin co-localizes with the cadherin/catenin complex in polarized epithelial cells (31–33). Disruption of the NAP complex by targeted ablation of the afadin gene in mice results in disorganized ectodermal cells and early embryonic lethality indicating its importance in maintaining the structural integrity of cell–cell junctions (34).

2.2 β-Catenin: a mediator of Wnt signalling

In addition to playing structural roles in cell–cell adhesion, β-catenin and plakoglobin, along with armadillo, can function as signalling molecules, although these roles appear separable as demonstrated by experiments performed in *Drosophila* (35). β-Catenin is an essential component of the Wnt/Wingless signalling pathway that is involved in a variety of developmental processes, including segment polarity in *Drosophila* (36), axis specification in *Xenopus* (37), mesoderm induction in *C. elegans* (38), and mammary tumour development in mice (39). β-Catenin is a downstream effector of the wnt signalling pathway that begins at the cell surface with the binding of the cysteine-rich wnt protein to its cognate frizzled receptor (40). Normally, most of the β-catenin in the cell is localized to the adherens junctions with cadherins. However, in response to a wnt signal, β-catenin is stabilized (i.e. not degraded) and its cytoplasmic and nuclear levels increase. This allows β-catenin to interact with a family of transcription factors referred to as LEF or TCF, and activate target genes. Stabilization of β-catenin is a key step in the transduction of the wnt signal to the nucleus. Several proteins are involved in this important regulation of steady state β-catenin levels; however, the exact mechanism is not fully understood. One player is the serine/threonine kinase, zeste-white 3 in *Drosophila* or glycogen synthase kinase 3 (GSK3) in vertebrates, which phosphorylates β-catenin, leading to its degradation by the ubiquitin–proteasome pathway. Deletion or mutation of specific amino acids in the amino terminus of β-catenin leads to increased stability and constitutive activity (41). The tumour suppressor protein, APC, binds both GSK3 and β-catenin, as does the protein axin/conductin thus facilitating the targeting of β-catenin for destruction. Interestingly, elevated levels of β-catenin have been observed in several colon carcinoma (42) and melanoma cell lines (43), in some cases, mutations have been observed in β-catenin. The mutations are thought to interfere with phosphorylation of β-catenin and its subsequent degradation thus causing an increase in free β-catenin in the cell. This signalling pathway is discussed in more detail in Chapter 8.

2.3 Invertebrate cadherins

Cadherins have now been identified in many invertebrate organisms, including sea squirts, worms, and fruit flies (Fig. 1). Many of these cadherins have more than four cadherin repeats in their extracellular domain that are more similar to protocadherin than classical vertebrate cadherin repeats. They also contain additional sequences in the extracellular domain, including EGF-like repeats, laminin A domains, novel cysteine-rich domains, and in *Drosophila* a fly classic cadherin box (Fcc) (44). The function of these additional subdomains in the extracellular portion of the protein is presently unclear. The *Drosophila* cadherins, DE- and DN-cadherin, can interact with catenins and promote homotypic cell adhesion *in vitro* (44, 45).

2.4   Protocadherin family

This cadherin subfamily was discovered in mammals using a polymerase chain reaction (PCR) strategy with degenerate oligonucleotide primers based on a conserved region of the extracellular domain (46). Protocadherins lack the prosequences found in classical cadherins, and have more than four cadherin repeats that are most similar to EC2 or EC4 in classical cadherins. The highly divergent cytoplasmic tail among protocadherins suggests the family is quite heterogeneous with numerous subfamilies. The lack of a classical cadherin cytoplasmic domain, including the β-catenin binding site, indicates a different adhesion mechanism (i.e. non-catenin binding). Cell adhesion is mediated by protocadherins but appears weaker than classical cadherins based on *in vitro* aggregation assays (46).

The protocadherin subfamily is larger and more diverse than the classical cadherin group, and includes several hybrid cadherins discovered in invertebrates that are apparently absent in vertebrates. The hybrid molecules contain protocadherin-type cadherin repeats and a cytoplasmic domain homologous to classical cadherins (i.e. catenin binding). This intriguing phylogenetic relationship of the protocadherin and its relatives suggests protocadherin may represent the primordial cadherin molecule (47).

2.5   Signalling in the CNS via protocadherin superfamily members

The identification of a novel hybrid cadherin in vertebrates provides an attractive candidate for cadherin-mediated cell–cell signalling during CNS development (48). This cadherin consists of protocadherin extracellular domains anchored to the membrane via a seven-pass transmembrane G protein-coupled receptor that has homology to the peptide hormone binding group of receptors, family B (49). The presence of homologues of this unique cadherin in both *C. elegans* (50) and *Drosophila* (51, 52) promises to yield important insight into a possible cadherin-mediated signalling pathway. In addition to CNS abnormalities, *Drosophila* mutant for the hybrid cadherin, flamingo, exhibit defects in planar cell polarity in the wing which is dependent on the cell surface receptor, frizzled (51).

In a search for molecules coupled to the Src family member of protein tyrosine kinases, Fyn, the yeast two-hybrid system identified a novel family of cadherin-related neuronal receptors, CNR (53). The Cnr gene family consists of approximately 20 members, eight of which have been sequenced. The CNR molecule consists of six cadherin repeats in the extracellular domain, a transmembrane domain, and a highly conserved cytoplasmic domain unique to the CNR family (54). The intracellular domain has five PXXP proline-rich motifs, a minimum consensus sequence for SH3 binding sites, which mediate interaction with Fyn and would be expected to support subsequent signal transduction. The cytoplasmic and the transmembrane regions of the CNR family members contain five conserved cysteine residues appearing at regular intervals, plus a lysine-rich sequence found at the carboxy terminal region. Genomic analysis suggests the diversity of the Cnr gene family arises from differential usage of 5' exons with the 3' exons held in common resembling the genomic organization of the immunoglobulin and T cell receptor gene clusters (55, 56). Large exons encoding the extracellular and transmembrane domains of individual Cnrs (variable region) are tandemly arranged followed by three exons encoding the cytoplasmic domain (constant region). A similar genomic organization was observed for two additional protocadherin gene families located nearby on human chromosome 5. The CNRs are expressed in a neuron-specific manner with localization to the synaptic junction, suggesting these chimeric cell adhesion/signalling molecules may play an important role in synaptic connectivity and plasticity. Exciting new biochemical data indicate that CNRs bind Reelin (57), an extracellular protein involved in the cortical layering and positioning of neurons in the developing brain.

3. Developmental regulation of cadherin subtypes

Changes in cadherin expression often are associated with morphogenetic processes. The predominant classical cadherin of all embryonic and adult epithelia is E-cadherin, which is present in the oocyte and subsequently activated zygotically shortly after fertilization. At gastrulation, E-cadherin is down-regulated in the primitive streak as cells undergo an epithelial–mesenchymal transition leading to expression of N-cadherin in the migrating mesoderm. The E-/N-cadherin switch at gastrulation has been conserved from flies to mammals. During neurulation, a similar change in expression occurs in the developing neuroepithelium when E-cadherin is lost during the invagination of the neural tube and replaced by N-cadherin (Fig. 4). A dynamic pattern of cadherin expression also is observed in the dorsal neural tube and its progenitors, the neural crest cells. In the chick, N-cadherin is down-regulated in the dorsal neural tube and cadherin-6B is turned on. This expression pattern is maintained during neural tube closure, and subsequently lost after neural crest cells have emigrated from the neural tube (58). A subset of neural crest cells migrating lateral to the neural tube begin to express cadherin-7. These cells migrate to restricted regions, including the dorsal root, ventral root, and spinal nerve. It has been postulated that cadherins may function in maintenance of specific populations of neural crest cells during their journey toward particular regions of the developing embryo. Injection of adenoviral vectors containing different cadherin constructs into developing chick embryos demonstrated that overexpression of N-cadherin or cadherin-7 blocked the segregation of neural crest cells from the dorsal neural tube, resulting in accumulation of the crest cells in the lumen of the neural tube (59). These experiments suggest that the regulation of cadherin expression is critical for the emigration of neural crest cells from the neural tube.

3.1 A role for cadherins in selective adhesion in the brain

The correct intercellular connections between neurons are essential for normal brain function. The growth cone of the neuron must recognize the correct target and form a functional synapse with its designated partner. Many families of cell surface

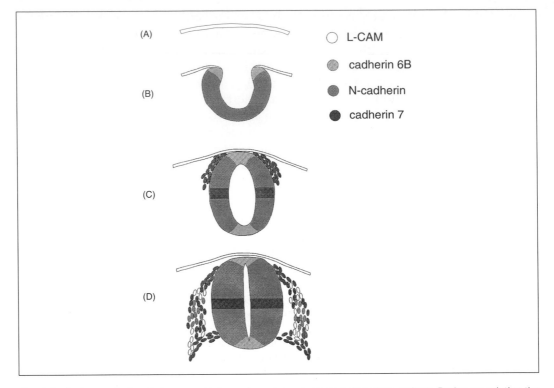

Fig. 4 Cadherin expression during neural tube and crest development in the chick embryo. During neurulation the LCAM/E-cadherin expressing ectoderm (A) invaginates to form the neural tube (B). LCAM expression in the neuro-epithelium is replaced by N-cadherin and cadherin-6B in the dorsal and ventral region. Following neural tube closure (C), neural crest cells begin to migrate from the dorsal region, cadherin-6B expression ceases, and cadherin-7 is expressed in the migrating neural crest cells and medial neural tube. The differential expression of cadherins (D) may facilitate the clustering and migration of specific populations of neural crest cells to particular locations.

molecules have been implicated in this selective adhesion process between neurons, including cadherins. The localization of cadherins and catenins at the synaptic junction of a variety of neurons indicates that cadherin-mediated adhesion between the pre- and postsynaptic membranes may be critical for interneuronal connections (60, 61). It was demonstrated that function-blocking antibodies against N- or E-cadherin could significantly reduce long-term potentiation (LTP) in hippocampal slices (62). The unique spatial patterns of different cadherin subtypes in the developing brain suggest cadherins may play an important role in establishing specific neuronal circuits. The distinct expression pattern of cadherin-6, a Type II cadherin, in the postnatal forebrain is illustrated below (Fig. 5) (63, 64). The domains of expression often coincide with neuronal connections. For example, the principal auditory thalamic nucleus, MG, expresses cadherin-6, and the same cadherin is expressed in layer IV of the auditory cortex, to which MG axons project.

Many members of the protocadherin subfamily are highly expressed in the brain (46). For example, the recently identified OL-protocadherin exhibits specific expres-

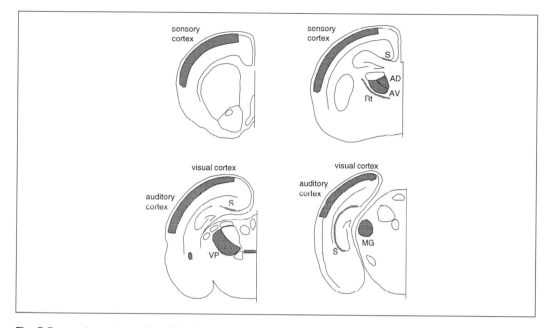

Fig. 5 Expression pattern of cad6 in the postnatal forebrain at four different frontal levels. Upper left to lower right, rostral to caudal. Areas or nuclei expressing cad6 are shaded. The differential expression pattern of a number of different cadherin subtypes including cad6 suggest cadherin-mediated adhesion may be involved in the selective association between particular neurons. For example, the principal auditory thalamic nucleus, MG, expresses cad6, and the same cadherin is expressed in layer IV of the auditory cortex, to which MG axons project (lower right). AD, anterodorsal thalamic nucleus; AV, anteroventral thalamic nucleus; MG, medial geniculate nucleus; Rt, reticular thalamic nucleus; S, subiculum; VP, ventral posterior thalamic nucleus. (Adapted from ref. 55, with permission.)

sion in the olfactory system (65). Similar to non-neural cells, neurons may express multiple cadherin subtypes, providing complexity and specificity to the recognition process. The less adhesive protocadherins may function more in the selective recognition process, whereas classical cadherins with their catenin binding may establish and maintain synaptic junctions. Understanding the role of cadherins in the precise wiring of the mammalian brain will require mutational analysis in mice, accompanied by careful examination of the neuronal phenotype.

4. Mutational analysis of the cadherin/catenin adhesion complex in invertebrates

4.1 Humpback and Hammerhead worms: a lesson in body language

The nematode, *Caenorhabditis elegans* (*C. elegans*), provides a model system to study how cells move and change shape to generate body form. Together, the relatively small number of cells (959), characteristic lineage map, and translucent nature of the worm facilitate the analysis of cell movements *in vivo*. At hatching, the embryo consists of an outer monolayer of epithelial cells called hypodermal cells that are linked together by adherens junctions. During morphogenesis, hypodermal cells migrate to enclose the embryo in an epithelium and, subsequently change shape co-ordinately to elongate the embryo. These morphological processes change the ellipsoidal mass of embryonic cells into a long slender worm. A genetic screen uncovered mutant embryos defective in body elongation and having abnormally bulged dorsal surfaces, hence the name humpback (hmp) (66). An additional phenotype was observed in which the hypodermis failed to fully enclose the embryo, causing internal structures to be extruded from an opening at the anterior and resulting in a hammerhead (hmr) appearance. While these mutations affect the shape of the developing embryo, there is no apparent cell adhesion defect, and the embryonic cells can differentiate into various tissues, including muscle and neuron.

The hmp-1, hmp-2, and hmr-1 alleles encode α-catenin, β-catenin, and a classical cadherin, respectively (66). Normally actin filament bundles are located circumferentially in the hypodermal cells. In hmp-1 mutants, the actin bundles become detached from the adherens junctions in the dorsal hypodermis. However, the actin cytoskeleton appears normal in the lateral and ventral hypodermal cells where the cadherin/catenin complex also is present. Hence the hunchback phenotype may be due to a specific weakness in the dorsal hypodermis. The larger dorsal hypodermal cells may undergo more mechanical stretching during elongation, possibly explaining the specificity of the cytoskeletal defect. The cadherin/catenin complex is not essential for general cell adhesion in the *C. elegans* embryo, but rather important during specific cell shape change involving action of the actin cytoskeleton in the hypodermis during the elongation process of the worm.

4.2   Cellular specificity of cadherin function in *C. elegans*: hyp10

A cadherin superfamily member (cdh-3) belonging to the *fat* subfamily is involved in morphogenesis of a single cell in the tail of the nematode (67). The original protocadherin family member, *fat*, was identified as a recessive lethal mutation in *Drosophila* which cause hyperplastic, tumour-like overgrowth of larval imaginal discs (68). The extracellular portion of the cdh-3 protein consists of 19 cadherin domains most closely related to the *fat* type, two EGF-like repeats, and a laminin A G-domain repeat. The unique cytoplasmic domain is very short (93 aa) and has little similarity to either *fat* or classical cadherin intracellular domains. The expression pattern of cdh-3 suggests an important role in cell adhesion and recognition in several morphological processes; however, mutation of this gene has a very specific phenotype. During nematode embryonic development a small cluster of cells gives rise to the tip of the tail (hyp8–11). The elongation of one of the cells, the binucleated hyp10 cell, is necessary for tail formation in the early larvae. cdh-3 mutants show morphological defects, including kinked and bifarcated tails, that are most prominent during the early larvae when the tip is formed by only hyp10. It is postulated that cdh-3 facilitates cellular interactions between hyp10 and its neighbours, hyp9 and hyp11, that are critical for elongation of hyp10 along the anterior–posterior axis. Despite widespread expression of cdh-3 in the developing embryo, the specificity of the phenotype suggests other cell adhesion molecules can compensate for loss of cdh-3 in most tissues.

4.3   Genetic analysis of cadherin function in *Drosophila*

Following the identification of armadillo as a homologue of the cadherin-associated protein, γ-catenin or plakoglobin, it became clear that *Drosophila* likely utilized a cadherin-based cell adhesion system similar to vertebrates (69). However, due to divergence in the cadherin family between vertebrates and invertebrates, it took several years after the cloning of armadillo before DE-cadherin was identified in *Drosophila* (44). *Drosophila* genetics provides a powerful system to analyse the function of cadherins *in vivo*. The use of multiple alleles, P-element-mediated transgenesis, inducible promoters (i.e. heat shock), and mosaic analysis make *Drosophila* an ideal system to study the role of cadherins during invertebrate development. The *shotgun* (*shg*) mutation revealed the gene encoding DE-cadherin, opening the door to a series of elegant experiments demonstrating the requirement for DE-cadherin in several morphological processes (70, 71). Morphogenetic movements of epithelia include various processes such as invaginations, delaminations of individual cells, and convergence-extension. Interestingly, the *shg* phenotype is not as severe as one might predict for a cell adhesion molecule expressed in all epithelia of the developing embryo. The most strongly affected tissues in *shg* mutants are ones undergoing active morphogenetic movements. When the gene is inactivated zygotically (i.e. in the embryo), a maternal pool of DE-cadherin is sufficient to maintain most epithelial structures. However, tissues undergoing dynamic epithelial reorganization such as the delamination of neuroblast cells in the ventral epidermis, lengthening of the Malpighian tubule, and generation of the tracheal branch network are strongly affected later in development as the maternal supply is exhausted. The absence of DE-cadherin during oogenesis results in disruption of oocyte production, leading to sterility. A weak *shg* allele encoding a partially functional protein results in the production of a few eggs. However, the resulting embryos exhibit severe defects in most epithelia.

The ability of cells to sort out into distinct aggregates based on differences in cell–cell adhesion has been demonstrated clearly *in vitro*, but whether this process occurs during normal development was less clear. Recent analysis of the role of cadherin-mediated cell adhesion in the positioning of the *Drosophila* oocyte demonstrates that differential adhesion also plays an important role *in vivo* (72, 73). During *Drosophila* oogenesis the germ cell divides four times generating a cyst of 16 cells, consisting of 15 nurse cells and one oocyte (Fig. 6A). As the cyst moves down the germarium, the oocyte becomes localized posterior to the other 15 germ cells, generating an anterior–posterior asymmetry that polarizes the axis for the rest of development. Somatic follicle cells migrate and surround the cyst and form a short stalk that separates it from the adjacent cysts. The mislocalization of the oocyte in germline clones mutant for ARM or *shg* shows that DE-cadherin-mediated adhesion is necessary for either localization of the oocyte in the cyst or maintenance of its position as the egg chamber grows. DE-cadherin is express to varying degrees in all germ cells and follicle cells, with highest expression observed in the oocyte and anterior and posterior follicle cells (Fig. 6C). It is possible to remove DE-cadherin in a tissue-specific and temporal manner using the yeast FLP/FRT recombinase system (74) in conjunction with an inducible heat shock promoter (Fig. 6B). Analysis of germline clones mutant in *shg* revealed mislocalization of the oocyte early in matura-

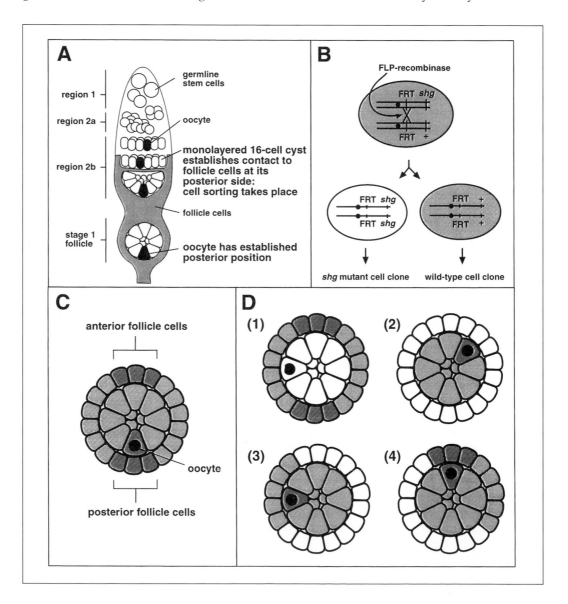

tion, indicating DE-cadherin is important for the initial positioning of the oocyte at the posterior. Similar experiments performed later after the oocyte had localized to the posterior region indicate that DE-cadherin is required for the maintenance of the oocyte in the posterior follicle. The lack of *shg* in either the germline cells, or the surrounding follicle cells, resulted in mislocalization of the oocyte, suggesting a requirement for DE-cadherin in both cell lineages (Fig. 6D). If the surrounding follicular epithelium is mosaic for DE-cadherin, the DE-cadherin positive oocyte interacts preferentially with follicle cells expressing the highest levels of DE-cadherin, demonstrating that the localization of the oocyte is mediated by the increased adhesion between the oocyte and follicle cells (Fig. 6D). These elegant experiments demonstrate for the first time the importance of differential cell adhesion in a morphogenetic process.

During morphogenesis cells rearrange themselves to obtain the correct position amongst their neighbours. This can require cells to move a considerable distance. During the movement of cells transient contacts are made between surrounding cells and the extracellular environment. Recent evidence indicates that DE-cadherin is required for cellular motility during *Drosophila* oogenesis (75, 76). The egg chamber consists of germline cells surrounded by somatic follicle cells. During oogenesis two subpopulations of follicle cells, the border cells and the centripetal cells undergo specific migration movements on the surface of the germline cells. The border cells, located anteriorly, migrate between the nurse cells through the centre of the follicle towards the oocyte. The centrally located centripetal cells migrate inward along the interface between the oocyte and nurse cells, and cover the anterior end of the oocyte. Using the FLP/FRT system, examination of chimeric egg chambers containing *shg* mutant (null allele) follicle cell clones and wild-type germline cells showed that DE-cadherin is required for the migration of border cells between the nurse cells. The DE-cadherin negative border cell clusters could not penetrate the nurse cells and

Fig. 6 (A) Schematic overview of a *Drosophila* germarium. Follicles are assembled in region 2b in which a cyst of 16 germline cells becomes surrounded by follicle cells. One of the 16 germline cells is the oocyte that assumes the posterior most position in the follicle during follicle formation. (B) Induction of *shotgun* (*shg*)/DE-cadherin mutant cell clones with the yeast FLP/FRT recombination system. Stage-specific expression of a heat shock promoter controlled FLP-recombinase gene is induced by heat shock treatment at early stages of ovary development. FLP-recombinase elicits homologous recombination between FRT sites in *shg* heterozygous cells. Subsequent cell division gives rise to a homozygous wild-type (+) cell and a homozygous *shg* mutant cell. (C) Early follicle illustrating DE-cadherin expression levels. Oocyte (o), anterior, and posterior follicle cells express high levels of DE-cadherin (dark grey) whereas the remaining germline cells and lateral follicle cells express lower levels of DE-cadherin. The oocyte is in contact with posterior follicle cells in normal follicles. (D) Mosaic follicles containing *shg* mutant (white) and wild-type cells (grey). Oocyte mislocalization is observed in follicles that lack DE-cadherin expression in either germline (1) or all follicle cells (2), or both (not shown). These findings show that DE-cadherin-mediated contact between germline and follicle cells is required for posterior oocyte localization. Furthermore, in follicles with a mosaic follicular epithelium a sorting process between oocyte and other germline cells takes place, which causes the oocyte to establish contact to the remaining DE-cadherin expressing follicle cells (3). In follicles with a mosaic follicular epithelium, in which some of the DE-cadherin positive follicle cells express high levels of DE-cadherin and others express low levels of DE-cadherin, in four out of five cases the oocyte will attach to the follicle cells that express higher levels of DE-cadherin (4). These findings suggest that differences in DE-cadherin concentration promote a cell sorting process among germline cells that establishes contact between the oocyte and those follicle cells that express the highest levels of DE-cadherin.

remained in the anterior tip of the follicle. In a weaker *shg* allele, border cells express low levels of DE-cadherin coincident with reduced motility. In addition, the cells often migrate along different pathways. Interestingly, several border cell clusters containing DE-cadherin positive and negative cells were able to migrate between the nurse cells towards the oocyte. The DE-cadherin expressing cells were always found at the leading edge of the cluster, whereas the DE-cadherin negative cells were trailing behind. This result suggests that the DE-cadherin positive cells are actively migrating while the other cells in the cluster are being pulled along. It also implies that the high expression of DE-cadherin throughout the border cell cluster is not required to maintain adhesion between the border cells themselves. Furthermore, using an intermediate *shg* allele, it was shown that reduced DE-cadherin levels slowed down the migration of border cells between the nurse cells. In the converse experiment, wild-type border cell clusters did not penetrate the nurse cells if they did not express DE-cadherin (i.e. germline *shg* clones). Centripetal cell migration also was dependent on DE-cadherin expression in both the germline and soma. Hence, DE-cadherin-dependent migration depends on heterotypic cell adhesion between the border cell cluster and the nurse cells. The migration of clusters of vertebrate neural crest cells also may depend on cadherin-mediated interactions for motility and guidance along their correct pathways. The role of cadherins in cell motility may have important implications for how tumour cells utilize their cadherin repertoire during invasion and metastasis. A basic understanding of cadherin function in different cellular contexts is critical to determine how cadherin regulation may influence cell behaviour.

In *shg* mutants, a dramatic reduction of catenins was observed in the epithelia, consistent with DE-cadherin providing the cadherin-mediated adhesion in that tissue. However, axons remained enriched for catenins, indicative of the presence of an additional cadherin in the CNS. The cloning and identification of mutations in the neural cadherin, DN-cadherin, has implicated this cadherin in axonal interactions critical for generation of the correct neuronal circuitry in the fly (45). DN-cadherin is expressed in nascent *Drosophila* mesoderm, similar to vertebrates. However, although DN-cadherin was expressed in mesoderm-derived myoblasts and myotubes, it was not found in cardiac cells of the fruit fly. In contrast, N-cadherin is the predominant cadherin in embryonic and adult heart muscle in vertebrates. Instead, it appears that DE-cadherin is utilized in the construction of a simple heart tube in *Drosophila*. DN-cadherin is widely expressed in the nervous system, specifically in the neurons, with glial cells apparently negative. A null mutation of DN-cadherin affects the overall configuration of the axon tracts, presumably due to decreased fasciculation. Abnormal trajectory of specific axons was detected in DN-cadherin-deficient embryos, whereas the remaining nervous system appeared normal.

5. Genetic analysis of cadherin/catenin complex in mice

Homologous recombination in murine embryonic stem (ES) cells provides a powerful tool to examine the function of cadherins and their associated proteins *in vivo*. The

Table 1 Cadherin/catenin mutations in mice

Knock-outs	Stage[a]	Phenotype	Reference
E-cadherin	E4	Pre-implantation lethality, defects in trophectoderm	78, 79
N-cadherin	E10	Myocytes dissociate, severe cardiovascular defect	82
P-cadherin	V, F	Precocious differentiation of the mammary gland	94
VE-cadherin	E10	Vascular defects, most severe in yolk sac and allantois	87, 88
α-Catenin	E4	Pre-implantation lethality, defects in trophectoderm	81
β-Catenin	E8	Epiblast dissociates, no mesoderm formation	89
γ-Catenin (plakoglobin)	E12–17	Cardiac rupture with desmosomal defects, epidermal blistering	91–93

[a] E, embryonic lethal (day of lethality); V, viable; F, fertile.

first generation of mutations, discussed below, are germline null or constitutive mutations (Table 1). The next generation of mutations which are currently being produced will involve tissue-specific ablation of the gene of interest, thus overcoming embryonic lethality in many cases.

5.1 Pre-implantation embryonic lethal mutations: E-cadherin, αE-catenin

E-cadherin-mediated adhesion provides structural integrity via adherens junctions for all epithelia in the body. Furthermore, down-regulation of E-cadherin expression frequently is observed during the progression of tumours of epithelial origin (carcinomas). Early studies indicated that E-cadherin function blocking antibodies disrupt the compaction process of the mouse pre-implantation embryo, resulting in loss of both cell polarity and blastomere adhesion (77). To examine the function of E-cadherin in embryonic development and cancer, mutant mice were generated using gene targeting technology (78, 79). In comparison to the *in vitro* experiments, embryos carrying a null mutation in the E-cadherin gene compacted and developed to the blastocyst stage, apparently due to maternal E-cadherin. Although maternally-derived E-cadherin is sufficient to facilitate the compaction process, E-cadherin-deficient embryos failed to form a trophectodermal epithelium and a blastoceol cavity thus interfering with hatching from the zona pellucida and subsequent implantation in the uterus. Loss of E-cadherin does not affect cell survival or cellular differentiation of trophectoderm cells. As discussed later in the chapter, down-regulation of E-cadehrin is observed in a variety of malignant carcinomas; however, no increase in tumour incidence has been observed in E-cadherin heterozygous mice. The analysis of E-cadherin function during later developmental stages, or in adults, will require the generation of a conditional mutation using more advanced genetic technology such as the bacteriophage P1 Cre/loxP recombinase system (80).

A mutant allele of αE-catenin was discovered in a gene trap mutation screen performed in ES cells (81). The gene trap vector containing the lacZ reporter gene generated a fusion protein missing the carboxy terminal third of αE-catenin resulting in a non-functional protein. Similar to the E-cadherin mutant phenotype, αE-catenin mutants cannot form trophectoderm, and embryonic development ceases at the blastocyst stage. The αE-catenin-deficient embryos also appear to be temporarily rescued by maternal αE-catenin. However, there appears to be an absolute requirement for αE-catenin in cell adhesion of the pre-implantation embryo.

5.2 Post-implantation embryonic lethal mutations: N-cadherin, VE-cadherin, β-catenin, plakoglobin

N-cadherin is first expressed in the nascent mesoderm migrating from the primitive streak in the gastrulating mouse embryo (82). A day later (neurula stage), there is widespread expression of N-cadherin in the neural tube, notochord, somites, and precardiac mesoderm. Interestingly, despite the widespread expression at this stage, embryos lacking N-cadherin undergo gastrulation, neurulation, and somitogenesis, however, the neural tube and somites often are malformed (82). The expression of other cell adhesion molecules, including cadherin-11, may partially substitute for lack of N-cadherin in the paraxial mesoderm (83). However, N-cadherin appears to play an important role in the coalesce of smaller presomitic clusters of cells into a well-organized epithelial somite structure (84). As discussed later, the Zebrafish paraxial protocadherin (Papc) also is involved in mesodermal patterning (85). In addition, cadherin-6 present in the neural folds may be sufficient for neural tube closure in N-cadherin-deficient embryos (58).

While N-cadherin is expressed in various cell lineages of the early embryo, the most severe cell adhesion defect is observed in the developing myocardium (82). The N-cadherin-deficient myocardial cells surrounding the intact endocardium (N-cadherin negative) lose their adhesion for one another, resulting in a misshapened heart tube. In contrast to other tissues, no other adhesion molecule can substitute for N-cadherin in the myocardium. N-cadherin-mediated adhesion is required for a functional myocardium in the primitive heart tube. The severity of the adhesion defect in the heart may be partly due to the mechanical stress placed on the myocytes during contraction of the primitive heart tube. However, defective cell adhesion is also apparent when N-cadherin null myocytes are cultured *in vitro*. In the normal animal, N-cadherin continues to be expressed in the developing myocardium, and its localization to the intercalated disc structure suggests a critical role in the maintenance of the working heart. The function of N-cadherin later in heart development, as well as other tissues including the brain, awaits the generation of a conditional mutation.

The genetic ablation of VE-cadherin in ES cells, as well as the mouse germline, demonstrated the importance of VE-cadherin in vascular branching and morphogenesis (86–88). The severe vascular defect in null animals resulted in embryonic lethality around midgestation (similar stage as N-cadherin mutation); however, no apparent cell adhesion defect was observed in the endothelium, as determined by the normal localization of the cell adhesion molecule, PECAM-1, and electron microscopy analysis. The extraembryonic allantois and yolk sac vasculature was especially affected in the VE-cadherin null animals, in contrast to the intraembryonic vessels, which exhibited more restricted vascular defects. VE-cadherin-mediated adhesion was not required for vascular endothelial cell differentiation, as evidenced by expression of multiple endothelial-specific markers. The apparently normal adhesion in the vascular endothelium suggests that other cell adhesion molecules can compensate for the loss of VE-cadherin, but not for the patterning of endothelial cells into yolk sac vasculature.

β-Catenin is a multifunctional protein with important roles in both the cadherin/catenin adhesion complex and the Wnt signalling pathway. β-Catenin or plakoglobin can bind to the cytoplasmic domain of cadherin mediating interaction with α-catenin and subsequent connection to the actin cytoskeleton. β-Catenin and E-cadherin co-localize to areas of cell–cell contact between the blastomeres of the early pre-implantation embryo. Interestingly, β-catenin mutant embryos can implant in the uterus; however, shortly thereafter, embryonic ectodermal cells lose their adhesiveness, leading to developmental arrest at the egg cylinder stage (89). In contrast to the cell adhesion defect observed in embryonic ectoderm or epiblast, the extraembryonic cell layers remain intact. The ability of the β-catenin mutant embryos to develop further than the E-cadherin and α-catenin-deficient embryos is likely due to a combination of maternal β-catenin, similar to early rescue of the E-cad and α-catenin mutant embryos, and the partial redundant function of plakoglobin. The unique requirement for β-catenin in the epiblast is intriguing, suggesting that in addition to its role in the cadherin adhesion complex, β-catenin may provide a specific signal to the primitive ectoderm necessary for gastrulation not provided by plakoglobin. Alternatively, the inability of plakoglobin to rescue the primitive ectoderm may reflect a limited amount of plakoglobin for use in both adherens junctions and desmosomes. The ability of plakoglobin to, at least temporarily, replace β-catenin in the cadherin/catenin complex likely explains why the mutant embryos develop to the egg cylinder stage. The lack of mesoderm differentiation in the β-catenin mutants is consistent with experiments performed in *Xenopus* where depletion of β-catenin inhibited dorsal mesoderm formation (90).

Plakoglobin can interact with the cytoplasmic domain of classical, as well as desmosomal, cadherins. Most mice lacking plakoglobin die from severe haemorrhaging in the heart around E14; however, a subset of animals live to E17 and exhibit the additional phenotype of epidermal blistering (91–93). Interestingly, β-catenin can interact with desmosomal cadherins in the plakoglobin mutant embryos; however, the morphology of the desmosomal plaque is abnormal. This phenotype is discussed in more detail in Chapter 10.

5.3 Viable mutation: P-cadherin

P-cadherin (placental) is the first reported classical cadherin mutation to result in a viable phenotype (94). Despite the high level expression of P-cadherin in placenta and testis, P-cadherin null females and males are both fertile. Many P-cadherin positive tissues also express E-cadherin, possibly explaining the lack of a cell adhesion defect in these animals. The intriguing expression pattern of P-cadherin in the basal undifferentiated cell layer of several tissues including the epidermis, prostate, and mammary gland, and its subsequent down-regulation suggest a possible role in maintaining cells in an undifferentiated state. In contrast to many tissues, P- and E-cadherin have a distinctive expression pattern in the murine mammary gland as described below.

The mammary gland develops postnatally under the proper hormonal stimuli during puberty and adolescence. The morphogenesis of the mammary ductal tree occurs when the end buds of the duct invade the surrounding fatty stroma until they reach the edge of the fat pad. P-cadherin is normally expressed in the highly proliferative and invasive cells of the terminal end bud, called cap cells, and absent in the lumenal cells. E-cadherin expression is restricted to the lumenal or secretory epithelium. P-cadherin expression continues in the cap cell progenitors, including the myoepithelial cells surrounding the duct and small clusters of cells that migrate into the lumenal compartment. The nature of these migrating cells is unknown, but because cap cells are the least differentiated of identifiable mammary cell types, it has been postulated that they may represent a source of mammary stem cells that are dispersed throughout the mammary tree (95). During early pregnancy lateral buds differentiate from the ducts and during the second half of pregnancy these alveoli develop into fully differentiated secretory lobules capable of synthesizing and secreting milk proteins. During lactation, the myoepithelium surrounding the ducts and alveoli contracts expelling the milk from the gland. Function-blocking antibodies were used *in situ* to examine the role of E- and P-cadherin in maintaining the tissue integrity of the end bud (96). Anti-E-cadherin induced disruption of the body epithelium resulting in epithelial cells freely floating in the lumen. Interestingly, decreased DNA synthesis was observed in the disaggregated lumenal cells. Proliferation returned to normal after the cells reaggregated 72 hours later. These data suggest that mammary epithelial cells require normal cell–cell associations to permit passage through the cell cycle. In contrast, anti-P-cadherin had no effect on the lumenal layer, but partially disrupted the basally located cap cell layer. However, the lack of a cell adhesion defect in the mammary glands of the P-cadherin null mice suggest other cell adhesion molecules can substitute for P-cadherin. The ability of the P-cadherin-deficient females to nurse their pups indicates that loss of P-cadherin from the myoepithelium did not compromise its contractile function. Nevertheless, an interesting phenotype was observed in the mammary glands of these animals. The virgin P-cadherin null females display precocious differentiation of the mammary epithelium resembling early pregnancy. The mutant mice develop focal hyperplasia and dysplasia of the mammary epithelium with age, however, no mammary tumours were observed. The coincident down-regulation of P-cadherin in several cell types undergoing differentiation and the phenotype of the P-cadherin mutant mice suggest P-cadherin may provide a signal necessary for maintenance of cells in the undifferentiated state. The mechanism by which loss of P-cadherin causes the mammary gland phenotype is unclear.

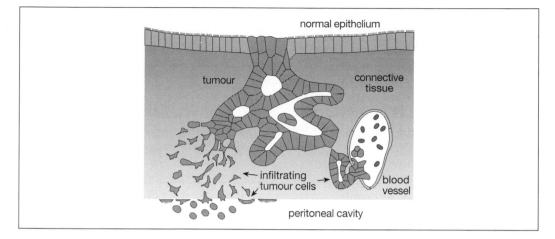

Fig. 7 Diagram depicting the multiple stages of cancer. The ability of cells to escape from the primary tumour mass and invade surrounding tissues requires changes in cell adhesive properties.

6. Role of cadherins in cancer

Cancer cells differ from normal cells largely in their ability to proliferate and migrate (i.e. metastasize) in an uncontrolled manner (Fig. 7). Examination of the cadherin profile of various tumour types suggests a more complex role for cadherins in tumour progression than previously appreciated. The dissemination of cells from a solid tumour requires reduced cell–cell adhesion, which can occur by a variety of mechanisms. The reduction of cell–cell adhesion primarily due to perturbation of the E-cadherin/catenin complex has been well documented in various carcinomas (97, 98). Such studies have demonstrated that mutations or deletions that affect the function of E-cadherin, or one of its associated catenins, contribute to the tumorigenic phenotype. Interestingly, more recent data indicate that tumour cells can utilize their cadherin repertoire to alter cellular behaviour, leading to more aggressive tumours. N- and P-cadherin have been implicated in this latter category.

Somatic mutations in E-cadherin have been identified in sporadic histologically diffuse gastric carcinoma (99–101), lobular breast cancers (102–104), and carcinomas of the endometrium and ovary (105). Inactivation of both E-cadherin alleles by mutation and LOH is common in sporadic tumours. Epigenetic inactivation of the E-cadherin gene by CpG hypermethylation also has been observed in several tumour types (106–108). Reduced E-cadherin expression has been correlated with extensive lymph node metastasis and distant metastasis in breast cancer (109–111).

For the first time, germline mutations in E-cadherin were found to predispose individuals to diffuse gastric carcinoma in several New Zealand families (112). Over the past 30 years, over 25 family members have died of this disease without evidence of elevation of cancer in other organs. In contrast to sporadic gastric cancer in the population, which generally affects individuals over 60 years of age, the majority of death from the early onset familial form occurred from 14 to 40 years of age. Several mutations were found in the E-cadherin gene including a transversion mutation (G–T) that affects the splice donor site. This results in the use of an adjacent cryptic splice site in the intron, thus introducing an addition seven base pairs and leading to premature termination of the protein in the extracellular domain. The identical somatic mutation previously had been found in a histologically diffuse gastric carcinoma (113).

A causal role for E-cadherin in the transition from adenoma to carcinoma was demonstrated using a well-defined mouse model of pancreatic β-cell carcinogenesis (114). In this animal model (Rip1Tag2), the SV40 T antigen is under the transcriptional control of the insulin promoter, directing expression of the oncogene to the β-cells of pancreatic islets of Langerhans (115). E-cadherin is normally expressed on the surface of islet β-cells and in the acini cells of the exocrine pancreas, and its expression was unaltered in adenoma cells. In contrast, transcriptional down-regulation of the E-cadherin gene was observed in the carcinomas. The expression of two other cadherins (N- and R-cadherin) were not affected by the transition from adenoma to carcinoma. To determine whether E-cadherin could block tumour progression *in vivo*, mice expressing E-cadherin from the same promoter (Rip1E-cad) were bred to the Rip1Tag2 mice to generate double transgenic mice (Rip1Tag2/ Rip1E-cad). Histological comparison of tumours in Rip1Tag2 alone and Rip1Tag2/ Rip1E-cad littermates indicated that tumour numbers did not change. However, the incidence of carcinoma development was significantly reduced in the double transgenic mice. To determine whether loss of E-cadherin was sufficient to promote tumour progression, the same group utilized transgenic mice expressing a dominant-negative E-cadherin molecule under the control of the insulin promoter, Rip1dnE-cad. The dominant-negative E-cadherin lacks the extracellular domain, but contains the cytoplasmic and transmembrane domain. As above, double transgenic mice were generated (Rip1Tag2/Rip1dnE-cad), and tumour progression was examined. While tumour incidence and tumour volume were not significantly changed in these animals, carcinoma formation was significantly increased. In addition, double transgenic mice developed pancreatic lymph node metastases, an invasive phenotype which was never observed in single transgenic (Rip1Tag2) mice. Therefore, loss of cadherin-mediated cell adhesion is sufficient to induce tumour progression, as well as invasion and metastasis. These elegant *in vivo* experiments demonstrate that loss of E-cadherin is a critical step in the progression from adenoma to carcinoma consistent with observations made in human tumours.

Recently a novel cadherin, H-cadherin, was found to be expressed in human mammary epithelial cells (116). H-cadherin lacks a cytoplasmic domain and is most similar to T-cadherin. Similar to E-cadherin, its expression was found to be significantly reduced in human breast carcinoma cell lines and primary breast tumours. Furthermore, transfection of H-cadherin into the breast cancer cell lines led to a decreased cell growth rate and loss of anchorage-independent growth in soft agar.

The function of different cadherin subtypes depends on the cellular context. For example, N-cadherin is expressed in less adhesive migratory fibroblastic cells. It also mediates strong cell–cell adhesion in the myocardium. These different adhesive characteristics are likely controlled by modification of protein interactions with the cytoplasmic domain, thus affecting linkage with the actin cytoskeleton and adhesion strength. The cellular morphology and behaviour of different squamous cell carcinoma cell lines has been investigated with respect to cadherin expression (117). The cadherin profile of normal squamous cells of the skin consists of E- and P-cadherin (118). Several cell lines, as well as squamous cell carcinomas *in situ*, were found to up-regulate N-cadherin. The expression of N-cadherin correlated with a more scattered (i.e. invasive) phenotype both *in vitro* and *in vivo*. Interestingly, E-cadherin was down-regulated in the cell lines exhibiting the scattered phenotype consistent with reduced cell–cell adhesion. To determine whether N-cadherin was responsible for this phenotype, N-cadherin under the control of an inducible promoter was transfected into a squamous cell line with normal morphology. Induction of N-cadherin expression caused the cells to scatter and acquire a phenotype similar to the invasive squamous cell lines. The level of E-cadherin decreased concomitantly with the increase in N-cadherin and coincident with a change in cell behaviour. Conversely, transfection of invasive squamous cell lines with an antisense N-cadherin construct resulted in a dramatic change in morphology, consisting of a more cohesive, less scattered appearance. The reversion to a more normal morphology was accompanied by a reduction of N-cadherin protein along with an increase in E-cadherin consistent with the ectopic studies mentioned above. Hence, inappropriate expression of a non-epithelial cadherin (i.e. N-cadherin) in a squamous epithelial cell line results in a cell with a more scattered and less adhesive phenotype, which is typical of invasive tumour cells. Additional *in vitro* studies support a role for N-cadherin in motility and invasiveness of breast cancer cells independent of E cadherin expression (119). It will be very interesting to determine which specific domain(s) in the N-cadherin molecule is required for the cell scattering phenotype in this cell type. Increasing evidence suggests that cadherins may interact with growth factor receptors such as fibroblast growth factor (FGF) receptor, perhaps facilitating dimerization and subsequent activation of a signalling pathway. Cross-talk between specific members of the cadherin and growth factor receptor families may partly explain why cadherins can influence cell behaviour in different ways depending on the cellular context.

As mentioned above, P-cadherin is not expressed in lumenal epithelial cells, but restricted to the less differentiated cap cells of the end bud and the myoepithelial cells surrounding the ducts. Aberrant expression of P-cadherin has been observed in a subset of breast carcinomas apparently in the absence of myoepithelial differentiation (120, 121). The significance of the up-regulation of P-cadherin in this subset of carcinomas is presently unknown. However, recently, a more comprehensive study discussed below, indicates that inappropriate expression of P-cadherin may play an important role in breast cancer progression.

The expression of P-, E-, and N-cadherin, as well as α- and β-catenin was examined in 183 cases of invasive breast carcinoma by immunohistochemistry using specific antibodies (122). Approximately half of the breast cancer specimens (95) were P-cadherin positive whereas the remaining specimens were negative. Interestingly, five years after surgery, 90% of the patients with P-cadherin negative tumours were alive, in contrast to only 59% of patients with P-cadherin positive tumours. A significant difference in survival was already apparent two years after surgery. P-cadherin expression was independent of tumour size and lymph node metastases, but correlated inversely with oestrogen/progesterone receptor status. In ductal carcinomas, positive P-cadherin expression correlated with a higher histological grade, whereas E-cadherin expression was low, but negative tumours were un-common. In lobular carcinomas, E-cadherin expression was frequently negative or low, and P-cadherin was always negative. The expression of E-, N-cadherin, α- and β-catenin did not correlate with patient survival. The finding that P-cadherin is an independent indicator of poor survival in breast cancer is intriguing since P-cadherin expression is normally restricted to the cap cells and their progenitors, the myo-epithelial cells. Since these are not myoepithelial-derived tumours, it would suggest a histogenetic origin in cap cells or some rare P-cadherin positive cell population that hasn't been identified, for example, mammary stem cells. Alternatively, the tumours may acquire cellular characteristics similar to cap cells, resulting in up-regulation of P-cadherin expression.

The specific expression of P-cadherin in the undifferentiated basal cell layer of several glandular tissues including salivary, mammary, and prostate suggests a potential role for this adhesion receptor in the critical process of secretory cell differentiation. While P-cadherin expression is restricted to the basal layer, E-cadherin is found in all epithelial layers of the prostatic gland. The basal layer of the prostate is thought to give rise to the lumenal secretory epithelial cells, hence functioning as a stem cell population. Prostate tumour specimens are generally negative for P-cadherin (123, 124), consistent with the hypothesis that prostate cancers originate from an epithelial cell rather than a basal cell. However, a subset of tumours were positive for P-cadherin and negative for prostate-specific antigen (PSA), a common serum marker used in the diagnosis and treatment of prostate cancer. The authors suggest that P-cadherin may serve as an useful marker for diagnosis of prostate tumours with low PSA levels. A proteolytic fragment of E-cadherin can be detected in the serum of cancer patients. It will be interesting to determine whether P-cadherin serum levels also will be useful in diagnosis of specific cancers. The differential expression of cadherins may prove useful in tracing the histogenesis of tumour cells in various cancers (125–127).

7. Genetic regulation of cadherin function

The dynamic expression pattern of cadherins during morphogenetic movements of cells is likely controlled by complex extracellular and intracellular stimuli leading to the regulation of cadherin transcription. Initial molecular analysis of the promoter regions of classical cadherins indicates a complex regulation, including the presence of enhancer elements located in intronic sequences. Meanwhile, genetic analysis of lower organisms has uncovered mutations in transcription factors that directly or indirectly regulate cadherin function.

DE-cadherin expression was found to be regulated by the zinc finger gene, escargot (esg), in the tip cells of a subset of tracheal branches (128). The promoter activity of DE-cadherin was examined using a lacZ reporter, thus allowing the temporal and tissue-specific expression to be monitored in detail. In the absence of esg, the reporter construct was not activated in the specific tracheal cells affected in the esg mutant. Expression of DE-cadherin from an inducible promoter can rescue the fusion defect in the tracheal tip cells in the esg mutant indicating that DE-cadherin is downstream of esg activity.

Improved genetic mapping in the Zebrafish with its transparent embryos and short gestation period make it an excellent model system to study vertebrate morphogenesis. There have been several cadherins cloned in Zebrafish, including N-cadherin (129), VN-cadherin (130), R-cadherin (131), and protocadherins (85). Recent evidence from Zebrafish (85) and *Xenopus* (132), implicate paraxial protocadherin (PAPC) in convergence of mesodermal cells during gastrulation. Genetic analysis indicates that activation of papc gene is controlled by spadetail (spt), a transcription factor of the T-box homeodomain family (133). In spt mutants, cells of the lateral mesoderm fail to converge properly toward the dorsal midline, leading to decreased numbers of somitic cells in the trunk and to an accumulation of the corresponding cells in the tail region (134). The spt mutation affects predominantly the paraxial (somitic) mesoderm, whereas mutations in other genes, such as no tail (ntl, a T-box transcription factor homologous to mouse Brachyury) and floating head (flh, a homeobox gene homologous to the organizer-specific *Xenopus* gene Xnot), affect the axial (notochordal) mesoderm (135–138). The strikingly similar expression patterns of spt and papc in the early gastrula and the reduction of papc in spt mutant embryos

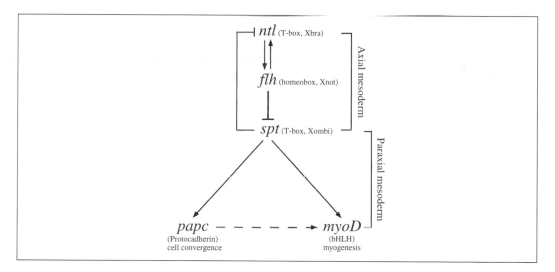

Fig. 8 Model of genetic interactions involved in axial and paraxial mesoderm specification in Zebrafish. *Solid arrows* indicate genetic interactions based on loss-of-function studies in mutant fish. Gain-of-function studies performed in *Xenopus* further support some of these gene interactions. The results from the dominant-negative papc experiments are indicated by the dashed line. (From ref. 75, with permission.)

demonstrate that papc is downstream of the T-box transcription factor, spt. Interestingly, injection of a dominant-negative papc mRNA into 2–4 cell embryos caused laterally expanded somites or loss of somite structures and myoD expression similar to that observed in spt mutants (85). Together with results from *Xenopus*, papc appears to play an important role in cell convergence movements during gastrulation. Many genes implicated in mesodermal patterning encode transcription or secretory factors, hence, it is very exciting to identify a cell adhesion molecule in this genetic pathway (Fig. 8).

8. Cadherin-mediated adhesion and tissue differentiation

The ability of cadherins to mediate cell–cell adhesion is well established. However, in contrast to integrins, intracellular signals resulting from the homophilic binding of cadherins is less well understood. Evidence suggests that cadherins themselves may elicit specific signals that regulate tissue differentiation. A cadherin-mediated signalling pathway is an attractive model for regulating the differentiation of groups of like cells (i.e. community effect). The ability of cadherins to influence cell fate was elegantly demonstrated using ES cells lacking E-cadherin and asking what happens when the cells utilize a different cadherin during differentiation (139). ES cells lacking both E-cadherin alleles display a disaggregated phenotype. Interestingly, the loss of E-cadherin resulted in the expression of the transcription factor, T-brachyury, demonstrating that cadherins can influence gene expression. The differentiation ability of ES cells can be examined *in vitro*, as well as by injection into syngenic animals to generate benign teratomas. Wild-type ES cells can differentiate into many cell types with a remarkable degree of tissue organization, including derivatives of all three germ layers, including for example, epithelium, muscle, neuroepithelium, bone, teeth, and hair follicles. In contrast, teratomas formed by E-cadherin null ES cells were very disorganized. However, several early and late differentiation markers were observed, indicating the cells were still competent to differentiate. The mutant ES cells were rescued by expressing E-cadherin under the control of a constitutive promoter. Interestingly, teratomas formed from the E-cadherin transfected mutant ES cells were almost exclusively comprised of epithelia. To determine whether a different cadherin subtype could influence the differentiation process, teratomas established from E-cadherin minus ES cells constitutively expressing N-cadherin were examined. Now the teratomas were comprised mostly of neuro-epithelium and cartilage. Alternatively, the expression of R-cadherin led to the formation of striated muscle and epithelia (140). Since all three cadherin subtypes can rescue the cell adhesion defect in the E-cadherin-minus ES cells, the different teratoma phenotypes indicate that cadherins provide additional information besides mediating cell adhesion.

Using an *in vitro* model system, N-cadherin can induce skeletal muscle differentiation coincident with cellular aggregation (141). This effect was not induced by a adhesion-defective N-cadherin molecule. However, wild-type E-cadherin could suffice suggesting that a common cadherin-dependent adhesion mechanism was required. The binding of the N-cadherin receptor itself may be sufficient to stimulate this signalling pathway as demonstrated by aggregating cells with N-cadherin-coated beads (142). Although the mechanism of cadherin-mediated cell signalling is not understood, it is intriguing that the cadherin/catenin complex is providing more than just structural information.

9. Cadherins and cell survival

Programmed cell death, or apoptosis, is fundamental to development and disease processes (143). Anchorage of cells to the extracellular matrix through integrins (144, 145), as well as cell–cell interactions mediated by cadherins (146), are thought to play an important role in cell survival. Recent evidence suggests that induction of apoptosis in endothelial cells coincides with cleavage of β-catenin and plakoglobin, but not α-catenin, and shedding of VE-cadherin (147). The truncated β-catenin no longer interacts with α-catenin hence it may facilitate retraction and cytoskeletal disruption in cells undergoing apoptosis. Evidence suggests that caspases and metalloproteinases may contribute to dissolution of adherens junctions during endo-thelial apoptosis by cleaving catenins and cadherins, respectively. The mechanism by which VE-cadherin promotes cell survival was examined in embryonic endothelial cells derived from VE-cadherin mutant embryos (87). Endothelial cells lacking a functional VE-cadherin receptor exhibited increased apoptosis. Survival was not rescued by the endothelial survival factor, vascular endothelial growth factor type A (VEGF-A), even though mutant cells express the VEGF-A receptor, vascular endothelial growth factor receptor type 2 (VEGFR-2), on their cell surface. In contrast, basic fibroblast growth factor (bFGF) was capable of rescuing the VE-cadherin mutant cells from death. VE-cadherin-mediated survival appears specific to VEGF-A since other growth factors had no effect on survival even though their receptors were present (87). Furthermore, inhibitors of phosphoinositide 3 (PI3)-kinase, a downstream effector of VEGF signalling, enhance apoptosis of wild-type cells but not of the VE-cadherin mutant endothelial cells. Co-immunoprecipitation experiments revealed an association between VE-cadherin, β-catenin, PI3-kinase, and VEGFR-2 in wild-type but not mutant cells. The experiments performed in this study provide evidence for a novel role for VE-cadherin/β-catenin in clustering proteins involved in VEGF-A-mediated survival signalling. Cadherin-mediated adhesion is also important for survival of other cells, including N-cadherin in granulosa cells (148, 149) and E-cadherin in prostate and mammary cells (150).

10. Conclusions

Recent genetic analysis of the cadherin/catenin adhesion complex in several model organisms has provided novel insight into the *in vivo* function of specific cadherin subtypes. Some findings are consistent with previous *in vitro* analysis using function-blocking antibodies, whereas in other cases antibody perturbation experiments appear inconsistent with the genetic ablation studies. Interestingly, a comparison of cadherin mutant phenotypes from different organisms indicate the most dramatic cell adhesion defects are associated with tissues undergoing active cellular re-arrangements and/or increased mechanical stress. In comparison, understanding cadherin function in static tissues expressing multiple cadherin subtypes may require analysis of organisms lacking two or more cadherin receptors in order to observe a phenotype. There is much we don't know about the seemingly ever expanding protocadherin gene family. For example, with their highly divergent cytoplasmic tails (i.e. non-catenin binding) how do these molecules facilitate cell–cell adhesion? Do they interact with the actin cytoskeleton via some novel mechanism? The complex expression pattern of protocadherin/CNR family members in the brain involves a novel mechanism of gene regulation that may prove critical for neuronal recognition and connectivity in the CNS. The use of genetic models is providing a framework to place cell adhesion in a developmental programme for the first time. Mutations that alter the complex temporal and spatial expression patterns of cadherins will yield clues into the transcriptional regulation of this gene family. The signals a cell receives from its neighbour cell and the extracellular environment influence its behaviour, including its survival. The ability of cadherins to alter the function of other cell surface proteins can influence, for example, cell–ECM inter-actions (via integrins), gap junction communication (via connexins), proliferation and motility (via growth factor receptors). Due to the fact that cadherin-mediated adhesion alters the function of other proteins on the cell surface, it is difficult to examine potential downstream signalling events. Thus, a major question in the field is whether cadherin interactions themselves initiate a signalling cascade resulting in gene transcription. The gain-of-function experiments using E-cadherin-minus ES cells as a model system present the most compelling evidence to date. However, the loss-of-function experiments in intact animals have provided only limited support of a role for cadherins in tissue differentiation. Therefore, further experiments are necessary to clarify this important issue. The dissemination of tumour cells from the original tumour mass involves the modification of many cellular processes besides reduction of cell adhesive properties. The down-regulation of the E-cadherin/catenin adhesion complex in tumour progression is well established. However, alternative means of perturbing adhesion appear to exist since not all tumours exhibit this phenomena. As we learn more about the regulation of the cadherin associated proteins, catenins/p120, novel mechanisms may emerge. Although it may sound contradictory, tumour cells also can up-regulate or misexpress cadherins, leading to changes in cell behaviour (i.e. migration). The behaviour of a tumour cell changes dramatically during cancer progression. When cells detach from the primary tumour, invade the surrounding tissue, and metastasize to a distant site (i.e. reattachment), it is not surprising that the cadherin repertoire might change concomitantly. The challenge for the future is to better understand how these changes relate to the pathology of a given tumour and might impact potential therapies. Whereas E-cadherin is currently used as a prognostic indicator, additional information might be gleaned by examination of a more extensive cadherin profile. Despite many exciting findings, there are still many unanswered questions in the cadherin field to keep investigators busy for some time.

Acknowledgements

We thank Ulrich Tepass for preparing Fig. 6 and Karen Knudsen for helpful discussion and comments on the manuscript. G. L. R. is supported by grants from NIH (HL57554) and American Heart Association.

References

1. Wilson, H. V. (1907). On some phenomena of coalescence and regeneration in sponges. *J. Exp. Zool.*, **5**, 245.
2. Townes, P. L. and Holtfreter, J. (1955). Directed movements and selective adhesion of embryonic amphibian cells. *J. Exp. Zool.*, **128**, 53.
3. Steinberg, M. S. (1963). Reconstruction of tissues by dissociated cells. *Science*, **141**, 401.
4. Takeichi, M. (1977). Functional correlation between cell adhesive properties and some cell surface proteins. *J. Cell Biol.*, **75**, 464.
5. Consortium, T. C. e. S. (1998). Genome sequence of the nematode *C. elegans*: a platform for investigating biology. *Science*, **282**, 2012.
6. Nagar, B., Overduin, M., Ikura, M., and Rini, J. M. (1996). Structural basis of calcium-induced E-cadherin rigidification and dimerization. *Nature*, **380**, 360.
7. Nose, A., Tsuji, K., and Takeichi, M. (1990). Localization of specificity determining sites in cadherin cell adhesion molecules. *Cell*, **61**, 147.
8. Nose, A., Nagafuchi, A., and Takeichi, M. (1988). Expressed recombinant cadherins mediate cell sorting in model systems. *Cell*, **54**, 993.
9. Miyatani, S., Shimamura, K., Hatta, M., Nagafuchi, A., Nose, A., and Matsunaga, M. (1989). Neural cadherin: role in selective cell–cell adhesion. *Science*, **245**, 631.
10. Suzuki, S., Sano, K., and Tanihara, H. (1991). Diversity of the cadherin family: evidence for eight new cadherins in nervous tissue. *Cell Regul.*, **2**, 261.
11. Ranscht, B. and Dours-Zimmerman, M. T. (1991). T-cadherin, a novel cadherin cell adhesion molecule in the nervous system lacks the conserved cytoplasmic domain. *Neuron*, **7**, 391.
12. Vestal, D. J. and Ranscht, B. (1992). Glycosyl phosphatidylinositol-anchored T-cadherin mediates calcium-dependent, homophilic cell adhesion. *J. Cell Biol.*, **119**, 451.
13. Ozawa, M., Baribault, H., and Kemler, R. (1989). The cytoplasmic domain of the cell adhesion molecule uvomorulin associates with three independent proteins structurally related in different species. *EMBO J.*, **8**, 1711.
14. Nagafuchi, A. and Takeichi, M. (1989). Transmembrane control of cadherin-mediated cell adhesion: a 94 kDa protein functionally associated with a specific region of the cytoplasmic domain of E-cadherin. *Cell Regul.*, **1**, 37.
15. Nagafuchi, A. and Takeichi, M. (1988). Cell binding function of E-cadherin is regulated by the cytoplasmic domain. *EMBO J.*, **7**, 3679.
16. Peifer, M. (1995). Cell adhesion and signal transduction: the Armadillo connection. *Trends Cell Biol.*, **5**, 224.

17. Knudsen, K. A., Soler, A. P., Johnson, K. R., and Wheelock, M. J. (1995). Interaction of α-actinin with the cadherin/catenin cell–cell adhesion complex via α-catenin. *J. Cell Biol.*, **130**, 67.

18. Rimm, D. L., Koslov, E. R., Kebriaei, P., Cianci, C. D., and Morrow, J. S. (1995). α(E)-catenin is an actin-binding and -bundling protein mediating the attachment of F-actin to the membrane adhesion complex. *Proc. Natl. Acad. Sci. USA*, **92**, 8813.

19. Watabe-Uchida, M., Uchida, N., Imamura, Y., Nagafuchi, A., Fujimoto, K., Uemura, T., *et al.* (1998). Alpha-catenin-vinculin interaction functions to organize the apical junctional complex in epithelial cells. *J. Cell Biol.*, **142**, 847.

20. Hirano, S., Kimoto, N., Shimoyama, Y., Hirohashi, S., and Takeichi, M. (1992). Identification of a neural alpha-catenin as a key regulator of cadherin function and multicellular organization. *Cell*, **70**, 293.

21. Reynolds, A. B., Daniel, J., McCrea, P. D., Wheelock, M. J., Wu, J., and Zhang, Z. (1994). Identification of a new catenin: the tyrosine kinase substrate p120cas associates with E-cadherin complexes. *Mol. Cell. Biol.*, **14**, 8333.

22. Ozawa, M. and Kemler, R. (1998). The membrane-proximal region of the E-cadherin cytoplasmic domain prevents dimerization and negatively regulates adhesion activity. *J. Cell Biol.*, **142**, 1605.

23. Aono, S., Nakagawa, S., Reynolds, A. B., and Takeichi, M. (1999). p120ctn acts as an inhibitory regulator of cadherin function in colon carcinoma cells. *J. Cell Biol.*, **145**, 551.

24. Yap, A. S., Niessen, C. M., and Gumbiner, B. M. (1998). The juxtamembrane region of the cadherin cytoplasmic tail supports lateral clustering, adhesive strengthening, and interaction with p120ctn. *J. Cell Biol.*, **141**, 779.

25. Yap, A. S., Brieher, W. M., Pruschy, M., and Gumbiner, B. M. (1997). Lateral clustering of the adhesive ectodomain: a fundamental determinant of cadherin function. *Curr. Biol.*, **7**, 308.

26. Paffenholz, R. and Franke, W. W. (1997). Identification and localization of a neurally expressed member of the plakoglobin/armadillo multigene family. *Differentiation*, **61**, 293.

27. Lu, Q., Paredes, M., Medina, M., Zhou, J., Cavallo, R., Peifer, M., *et al.* (1999). Gamma-catenin, and adhesive junction-associated protein which promotes cell scattering. *J. Cell Biol.*, **144**, 519.

28. Zhou, J., Liyanage, U., Medina, M., Ho, C., Simmons, A. D., Lovett, M., *et al.* (1997). Presenilin 1 interacts in brain with a novel member of the Armadillo family. *Neuroreport*, **8**, 1489.

29. Clark, R. F., Hutton, M., Fuldner, R. A., Froelich, S., Karran, E., Talbot, C., *et al.* (1995). The structure of the presenilin 1 (S182) gene and identification of six novel mutations in early onset AD families. *Nature Genet.*, **11**, 219.

30. Sherrington, R., Rogaev, E. I., Liang, Y., Rogaeva, E. A., Levesque, G., Ikeda, M., *et al.* (1995). Cloning of a novel gene bearing missense mutations in early onset familial Alzheimer disease. *Nature*, **375**, 754.

31. Takahashi, K., Nakanishi, H., Miyahara, M., Mandai, K., Satoh, K., Satoh, A., *et al.* (1999). Nectin/PRR: an immunoglobulin-like cell adhesion molecule recruited to cadherin-based adherens junctions through interaction with afadin, a PDZ domain-containing protein. *J. Cell Biol.*, **145**, 539.

32. Mandai, K., Nakanishi, H., Satoh, A., Takahashi, K., Satoh, K., Nishioka, H., *et al.* (1999). Ponsin/SH3P12: an l-afadin- and vinculin-binding protein localized at cell–cell and cell–matrix adherens junctions. *J. Cell Biol.*, **144**, 1001.

33. Mandai, K., Nakanishi, H., Satoh, A., Obaishi, H., Wada, M., Nishioka, H., *et al.* (1997). Afadin: a novel actin filament-binding protein with one PDZ domain localized at cadherin-based cell-to-cell adherens junction. *J. Cell Biol.*, **139**, 517.

34. Ikeda, W., Nakanishi, H., Miyoshi, J., Mandai, K., Ishizaki, H., Tanaka, M., *et al.* (1999). Afadin: a key molecule essential for structural organization of cell–cell junctions of polarized epithelia during embryogenesis. *J. Cell Biol.*, **146**, 1117.

35. Orsulic, S. and Peifer, M. (1996). An *in vivo* structure–funtion study of armadillo, the beta-catenin homologue, revelas both separate and overlapping regions of the protein required for cell adhesion and for wingless signaling. *J. Cell Biol.*, **134**, 1283.

36. Riggleman, B., Schedl, P., and Wieschaus, E. (1990). Spatial expression of the *Drosophila* segment polarity gene armadillo is post-transcriptionally regulated by wingless. *Cell*, **63**, 549.

37. McCrea, P. D., Brieher, W. M., and Gumbiner, B. M. (1993). Induction of a secondary body axis in *Xenopus* by antibodies to β-catenin. *J. Cell Biol.*, **123**, 477.

38. Thorpe, C. J., Schlesinger, A., Carter, J. C., and Bowerman, B. (1997). Wnt signaling polarizes an early *C. elegans* blastomere to distinguish endoderm from mesoderm. *Cell*, **90**, 695.

39. Tsukamoto, A. S., Grosschedl, R., Guzman, R. C., Parslow, T., and Varmus, H. E. (1988). Expression of the int-1 gene in transgenic mice is associated with mammary gland hyperplasia and adenocarcinomas in male and female mice. *Cell*, **55**, 619.

40. Willert, K. and Nusse, R. (1998). β-catenin: a key mediator of Wnt signaling. *Curr. Opin. Gen. Dev.*, **8**, 95.

41. Munemitsu, S., Albert, I., Rubinfeld, B., and Polakis, P. (1996). Deletion of an amino-terminal sequence stabilizes β-catenin *in vivo* and promotes hyperphosphorylation of the adenomatous polyposis coli tumor suppressor protein. *Mol. Cell. Biol.*, **16**, 4088.

42. Korinek, V., Barker, N., Morin, P. J., van Wichen, D., de Weger, R., Kinzler, K. W., *et al.* (1997). Constitutive transcriptional activation by a β-catenin-Tcf complex in APC−/− colon carcinoma. *Science*, **275**, 1784.

43. Rubinfeld, B., Robbins, P., El-Gamil, M., Albert, I., Porfiri, E., and Polakis, P. (1997). Stabilization of β-catenin by genetic defects in melanoma cell lines. *Science*, **275**, 1790.

44. Oda, H., Uemura, T., Harada, Y., Iwai, Y., and Takeichi, M. (1994). A *Drosophila* Homolog of cadherin associated with Armadillo and essential for embryonic cell–cell adhesion. *Dev. Biol.*, **165**, 716.

45. Iwai, Y., Usui, T., Hirano, S., Steward, R., Takeichi, M., and Uemura, T. (1997). Axon patterning requires DN-cadherin, a novel neuronal adhesion receptor, in the *Drosophila* embryonic CNS. *Neuron*, **19**, 77.

46. Sano, K., Tanihara, H., Heinmark, R. L., Obata, S., Davidson, M., St. John, T., *et al.* (1993). Protocadherins: a large family of cadherin-related molecules in central nervous system. *EMBO J.*, **12**, 2249.

47. Suzuki, S. T. (1996). Structural and functional diversity of cadherin superfamily: are new members of cadherin superfamily involved in signal transduction pathway? *J. Cell. Biochem.*, **61**, 531.

48. Hadjantonakis, A., Sheward, W. J., Harmar, A. J., de Galan, L., Hoovers, J. M. N., and Little, P. F. R. (1997). Celsr1, a neural-specific gene encoding an unusual seven-pass transmembrane receptor, maps to mouse chromosome 15 and human chromosome 22qter. *Genomics*, **45**, 97.

49. Kolakowski, L. F. (1994). GCRDb-A G protein-coupled receptor database. *Receptors Channels*, **2**, 1.

50. Hadjantonakis, A., Formstone, C. J., and Little, P. F. R. (1998). MCelsr1 is an evolutionarily conserved seven-pass transmembrane receptor and is expressed during mouse embryonic development. *Mech. Dev.*, **78**, 91.

51. Usui, T., Shima, Y., Shimada, Y., Hirano, S., Burgess, R. W., Schwartz, T. L., *et al.* (1999). Flamingo, a seven-pass transmembrane cadherin, regulates planar cell polarity under the control of frizzled. *Cell*, **98**, 585.

52. Lu, B., Usui, T., Uemura, T., Jan, L., and Jan, Y.-N. (1999). Flamingo controls the planar polarity of sensory bristles and asymmetric division of sensory organ precursors in *Drosophila. Curr. Biol.*, **9**, 1247.

53. Kai, N., Mishina, M., and Yagi, T. (1997). Molecular cloning of Fyn-associated molecules in the mouse central nervous system. *J. Neurosci. Res.*, **48**, 407.

54. Kohmura, N., Senzaki, K., Hamada, S., Kai, N., Yasuda, R., Watanabe, M., *et al.* (1998). Diversity revealed by a novel family of cadherins expressed in neurons at a synaptic complex. *Neuron*, **20**, 1137.

55. Sugino, H., Hamada, S., Yasuda, R., Tuji, A., Matsuda, Y., Fujita, M., *et al.* (2000). Genomic organization of the family of CNR cadherin genes in mice and humans. *Genomics*, **63**, 75.

56. Wu, Q. and Maniatis, T. (1999). A striking organization of a large family of human neural cadherin-like cell adhesion genes. *Cell*, **97**, 779.

57. Senzaki, K., Ogawa, M., and Yagi, T. (1999). Proteins of the CNR family are multiple receptors for Reelin. *Cell*, **99**, 635.

58. Nakagawa, S. and Takeichi, M. (1995). Neural crest cell–cell adhesion controlled by sequential and subpopulation-specific expression of novel cadherins. *Development*, **121**, 1321.

59. Nakagawa, S. and Takeichi, M. (1998). Neural crest emigration from the neural tube depends on regulated cadherin expression. *Development*, **125**, 2963.

60. Fannon, A. M. and Colman, D. R. (1996). A model for central synaptic junctional complex formation based on the differential adhesive specificities of the cadherins. *Neuron*, **17**, 423.

61. Uchida, N., Honjo, Y., Johnson, K. R., Wheelock, M. J., and Takeichi, M. (1996). The catenin/cadherin adhesion system is localized in synaptic junctions bordering transmitter release zones. *J. Cell Biol.*, **135**, 767.

62. Tang, L., Hung, C. P., and Schuman, E. M. (1998). A role for the cadherin adhesion molecules in hippocampal long-term potentiation. *Neuron*, **20**, 1165.

63. Inoue, T., Tanaka, T., Suzuki, S. C., and Takeichi, M. (1998). Cadherin-6 in the developing mouse brain: expression along restricted connection systems and synaptic localization suggest a potential role in neuronal circuitry. *Dev. Dyn.*, **211**, 338.

64. Suzuki, S. C., Inoue, T., Kimura, Y., Tanaka, T., and Takeichi, M. (1997). Neuronal circuits are subdivided by differential expression of type-II classic cadherins in postnatal mouse brains. *Mol. Cell. Neurosci.*, **9**, 433.

65. Hirano, S., Yan, Q., and Suzuki, S. T. (1999). Expression of a novel protocadherin, OL-protocadherin, in a subset of functional systems of the developing mouse brain. *J. Neurosci.*, **19**, 995.

66. Costa, M., Raich, W., Agbunag, C., Leung, B., and Hardin, J. (1998). A putative catenin-cadherin system mediates morphogenesis of the *Caenorhabditis elegans* embryo. *J. Cell Biol.*, **141**, 297.

67. Pettitt, J., Wood, W. B., and Plasterk, R. H. A. (1996). cdh-3, a gene encoding a member of the cadherin superfamily, functions in epithelial cell morphogenesis in *Caenorhabditis elegans. Development*, **122**, 4149.

68. Mahoney, P., Weber, U., Onofrechcuk, P., Biessmann, H., Bryant, P., and Goodman, C. (1991). The fat tumor suppressor gene in *Drosophila* encodes a novel member of the cadherin gene superfamily. *Cell*, **67**, 853.

69. Peifer, M. and Wieschaus, E. (1990). The segment polarity gene armadillo encodes a functionally modular protein that is the *Drosophila* homolog of human plakoglobin. *Cell*, **63**, 1167.

70. Tepass, U., Gruszynski-DeFeo, E., Haag, T. A., Omatyar, L., Torok, T., and Hartenstein, V. (1996). shotgun encodes *Drosophila* E-cadherin and is preferentially required during cell rearrangement in the neuroectoderm and other morphogenetically active epithelia. *Genes Dev.*, **10**, 672.

71. Uemura, T., Oda, H., Kraut, R., Hayashi, S., Kataoka, Y., and Takeichi, M. (1996). Zygotic *Drosophilia* E-cadherin expression is required for processes of dynamic epithelial cell rearrangement in the *Drosophila* embryo. *Genes Dev.*, **10**, 659.

72. Godt, D. and Tepass, U. (1998). *Drosophila* oocyte localization is mediated by differential cadherin-based adhesion. *Nature*, **396**, 387.

73. Gonzalez-Reyes, A. and St. Johnston, D. (1998). The *Drosophila* AP axis is polarized by the cadherin-mediated positioning of the oocyte. *Development*, **125**, 3635.

74. Chou, T. B. and Perrimon, N. (1992). Use of a yeast site-specific recombinase to produce female germline chimeras in *Drosophila*. *Genetics*, **131**, 643.

75. Oda, H., Uemura, T., and Takeichi, M. (1997). Phenotypic analysis of null mutants of DE-cadherin and Armadillo in *Drosophila* ovaries reveals distinct aspects of their functions in cell adhesion and cytoskeletal organization. *Genes Cells*, **2**, 29.

76. Niewiadomska, P., Godt, D., and Tepass, U. (1999). DE-cadherin is required for inter-cellular motility during *Drosophila* oogenesis. *J. Cell Biol.*, **144**, 533.

77. Kemler, R., Babinet, C., Eisen, H., and Jacob, F. (1977). Surface antigen in early differen-tiation. *Proc. Natl. Acad. Sci. USA*, **74**, 4449.

78. Larue, L., Ohsugi, M., Hirchenhain, J., and Kemler, R. (1994). E-cadherin null mutant embryos fail to form a trophectoderm epithelium. *Proc. Natl. Acad. Sci. USA*, **91**, 8263.

79. Riethmacher, D., Brinkmann, V., and Birchmeier, C. (1995). A targeted mutation in the mouse E-cadherin gene results in defective preimplantation development. *Proc. Natl. Acad. Sci. USA*, **92**, 855.

80. Sauer, B. and Henderson, N. (1988). Site-specific DNA recombination in mammalian cells by the Cre recombinase of bacteriophage P1. *Proc. Natl. Acad. Sci. USA*, **85**, 5166.

81. Torres, M., Stoykova, A., Huber, O., Chowdhury, K., Bonaldo, P., Mansouri, A., *et al.* (1997). An alpha-E-catenin gene trap mutation defines its function in preimplantation development. *Proc. Natl. Acad. Sci. USA*, **94**, 901.

82. Radice, G. L., Rayburn, H., Matsunami, H., Knudsen, K. A., Takeichi, M., and Hynes, R. O. (1997). Developmental defects in mouse embryos lacking N-cadherin. *Dev. Biol.*, **181**, 64.

83. Kimura, Y., Matsunami, H., Inoue, T., Shimamura, K., Uchida, N., Ueno, T., *et al.* (1995). Cadherin-11 expressed in association with mesenchymal morphogenesis in the head, somite, and limb bud of early mouse embryos. *Dev. Biol.*, **169**, 347.

84. Linask, K. K., Ludwig, C., Han, M., Liu, X., Radice, G. L., and Knudsen, K. A. (1998). N-cadherin/catenin-mediated morphoregulation of somite formation. *Dev. Biol.*, **202**, 85.

85. Yamamoto, A., Amacher, S., Kim, S., Geissert, D., Kimmel, C. B., and De Roberts, E. M. (1998). Zebrafish paraxial protocadherin is a downstream target of spadetail involved in morphogenesis of gastrula mesoderm. *Development*, **125**, 3389.

86. Vittet, D., Buchou, T., Schewitzer, A., Dejana, E., and Huber, P. (1997). Targeted null-mutation in the vascular endothelial-cadherin gene impairs the organization of vascular-like structures in embryoid bodies. *Proc. Natl. Acad. Sci. USA*, **94**, 6273.

87. Carmeliet, P., Lampugnani, M.-G., Moons, L., Breviaro, F., Compernolle, V., Bono, F., *et al.* (1999). Targeted deficiency or cytoplasmic truncation of the VE-cadherin gene in mice impairs VEGF-mediated endothelial survival and angiogenesis. *Cell*, **98**, 147.

88. Gory-Faure, S., Prandini, M., Pointu, H., Roullot, V., Pignot-Paintrant, I., Vernet, M., *et al.* (1999). Role of vascular endothelial-cadherin in vascular morphogenesis. *Development*, **126**, 2093.

89. Haegel, H., Larue, L., Ohsugi, M., Fedorov, L., Herrenknecht, K., and Kemler, R. (1995). Lack of beta-catenin affects mouse development at gastrulation. *Development*, **121**, 3529.

90. Heasman, J., Ginsberg, D., Geiger, B., Goldstone, K., Pratt, T., Yoshida-Noro, C., *et al.* (1994). A functional test for maternally inherited cadherin in *Xenopus* shows its importance in cell adhesion at the blastula stage. *Development*, **120**, 49.

91. Ruiz, P., Brinkmann, V., Ledermann, B., Behrend, M., Grund, C., Thalhammer, C., *et al.* (1996). Targeted mutation of plakoglobin in mice reveals essential functions of desmosomes in the embryonic heart. *J. Cell Biol.*, **135**, 215.

92. Bierkamp, C., McLaughlin, K. J., Schwarz, H., Huber, O., and Kemler, R. (1996). Embryonic heart and skin defects in mice lacking plakoglobin. *Dev. Biol.*, **180**, 780.

93. Bierkamp, C., Schwarz, H., Huber, O., and Kemler, R. (1999). Desmosomal localization of beta-catenin in the skin of plakoglobin null-mutant mice. *Development*, **126**, 371.

94. Radice, G. L., Ferreira-Cornwell, M. C., Robinson, S. D., Rayburn, H., Chodosh, L. A., Takeichi, M., *et al.* (1997). Precocious mammary gland development in P-cadherin-deficient mice. *J. Cell Biol.*, **139**, 1025.

95. Faulkin, L. J. Jr. and DeOme, K. B. (1960). Regulation of growth and spacing of gland elements in the mammary fat pad of the C3H mouse. *J. Natl. Cancer Inst.*, **24**, 953.

96. Daniel, C. W., Strickland, P., and Friedmann, Y. (1995). Expression and functional role of E- and P-cadherins in mouse mammary ductal morphogenesis and growth. *Dev. Biol.*, **169**, 511.

97. Berx, G., Nollet, F., and van Roy, F. (1998). Dysregulation of the E-cadherin/catenin complex by irreversible mutations in human carcinomas. *Cell Adhes. Commun.*, **6**, 171.

98. Birchmeier, W. and Behrens, J. (1994). Cadherin expression in carcinomas: role in the formation of cell junctions and the prevention of invasiveness. *Biochim. Biophys. Acta*, **1198**, 11.

99. Muta, H., Noguchi, M., Kanai, Y., Ochiai, A., Nawata, H., and Hirohashi, S. (1996). E-cadherin mutations in signet ring cell carcinoma of the stomach. *Jap. J. Cancer Res.*, **87**, 843.

100. Tamura, G., Sakata, K., Nishizuka, S., Maesawa, C., Suzuki, Y., Iwaya, T., *et al.* (1996). Inactivation of the E-cadherin gene in primary gastric carcinomas and gastric carcinoma cell lines. *Jap. J. Cancer Res.*, **87**, 1153.

101. Becker, K. F., Atkinson, M. J., Reich, U., Becker, I., Nekarda, H., Siewert, J. R., *et al.* (1994). E-cadherin gene mutations provide clues to diffuse type gastric carcinomas. *Cancer Res.*, **54**, 3845.

102. Berx, G., Cleton-Jansen, A. M., Nollet, F., de Leeuw, W. J., van de Vijver, M., Cornelisse, C., *et al.* (1995). E-cadherin is a tumour/invasion suppressor gene mutated in human lobular breast cancers. *EMBO J.*, **14**, 6107.

103. Berx, G., Cleton-Jansen, A. M., Strumane, K., de Leeuw, W. J., Nollet, F., van Roy, F., *et al.* (1996). E-cadherin is inactivated in a majority of invasive human lobular breast cancers by truncation mutations throughout its extracellular domain. *Oncogene*, **13**, 1919.

104. Moll, R., Mitze, M., Frixen, U. H., and Birchmeier, W. (1993). Differential loss of E-cadherin expression in infiltrating ductal and lobular breast carcinomas. *Am. J. Pathol.*, **143**, 1731.

105. Risinger, J. I., Berchuck, A., Kohler, M. F., and Boyd, J. (1994). Mutations of the E-cadherin gene in human gynecologic cancers. *Nature Genet.*, **7**, 98.

106. Yoshiura, K., Kanai, Y., Ochiai, A., Shimoyama, Y., Sugimura, T., and Hirohashi, S. (1995). Silencing of the E-cadherin invasion-suppressor gene by CpG methylation in human carcinomas. *Proc. Natl. Acad. Sci. USA*, **92**, 7416.

107. Graff, J. R., Herman, J. G., Lapidus, R. G., Chopra, H., Xu, R., Jarrard, D. F., *et al.* (1995). E-cadherin expression is silenced by DNA methylation in human breast and prostate carcinomas. *Cancer Res.*, **55**, 5195.

108. Graff, J. R., Greenberg, V. E., Herman, J. G., Westra, W. H., Boghaert, E. R., Ain, K. B., *et al.* (1998). Distinct patterns of E-cadherin CpG island methylation in papillary, follicular, Hurthle's cell, and poorly differentiated human thyroid carcinoma. *Cancer Res.*, **58**, 2063.

109. Gamallo, C., Palacios, J., Suarez, A., Pizarro, A., Navarro, P., Quintanilla, M., *et al.* (1993). Correlation of E-cadherin expression with differentiation grade and histological type in breast carcinoma. *Am. J. Pathol.*, **142**, 987.

110. Lindblom, A., Rotstein, S., Skoog, L., Nordenskjold, M., and Larsson, C. (1993). Deletions on chromosome 16 in primary familial breast carcinomas are associated with development of distant metastases. *Cancer Res.*, **53**, 3707.

111. Oka, H., Shiozaki, H., Kobayashi, K., Inoue, M., Takara, H., Kobayashi, T., *et al.* (1993). Expression of E-cadherin cell adhesion molecules in human breast cancer tissues and its relationship to metastasis. *Cancer Res.*, **53**, 1697.

112. Guilford, P., Hopkins, J., Harraway, J., McLeod, M., McLeod, N., Harawira, P., *et al.* (1998). E-cadherin germline mutations in familial gastric cancer. *Nature*, **392**, 402.

113. Oda, T., Kanai, Y., Oyama, T., Yoshiura, K., Shimoyama, Y., Birchmeier, W., *et al.* (1994). E-cadherin gene mutations in human gastric carcinoma cell lines. *Proc. Natl. Acad. Sci. USA*, **91**, 1858.

114. Perl, A., Wilgenbus, P., Dahl, U., Semb, H., and Christofori, G. (1998). A causal role for E-cadherin in the transition from adenoma to carcinoma. *Nature*, **392**, 190.

115. Hanahan, D. (1985). Heritable formation of pancreatic beta-cell tumours in transgenic mice expressing recombinant insulin/simian 40 oncogenes. *Nature*, **315**, 115.

116. Lee, S. M. (1996). H-cadherin, a novel cadherin with growth inhibitory functions and diminished expression in human breast cancer. *Nature Med.*, **2**, 776.

117. Islam, S., Carey, T. E., Wolf, G. T., Wheelock, M. J., and Johnson, K. R. (1996). Expression of N-cadherin by human squamous carcinoma cells induces a scattered fibroblastic phenotype with disrupted cell–cell adhesion. *J. Cell Biol.*, **135**, 1643.

118. Wheelock, M. J. and Jensen, P. J. (1992). Regulation of keratinocyte intercellular junction organization and epidermal morphogenesis by E-cadherin. *J. Cell Biol.*, **117**, 415.

119. Nieman, M. T., Prudoff, R. S., Johnson, K. R., and Wheelock, M. J. (1999). N-cadherin promotes motility in human breast cancer cells regardless of their E-cadherin expression. *J. Cell Biol.*, **147**, 631.

120. Palacios, J., Benito, N., Pizarro, A., Suarez, A., Espada, J., Cano, A., *et al.* (1995). Anomalous expression of P-cadherin in breast carcinoma: correlation with E-cadherin expression and pathological features. *Am. J. Pathol.*, **146**, 605.

121. Rasbridge, S. A., Gillett, C. E., Sampson, S. A., Walsh, F. S., and Millis, R. R. S. (1993). Epithelial (E-) and placental (P-) cadherin cell adhesion molecule expression in breast carcinoma. *J. Pathol.*, **169**, 245.

122. Peralta Soler, A., Knudsen, K. A., Salazar, H., Han, A. C., and Keshgegian, A. A. (1999). P-cadherin expression in breast cancer indicates poor survival. *Cancer*, **86**, 1263.

123. Jarrard, D. F., Paul, R., van Bokhoven, A., Nguyen, S. H., Bova, G. S., Wheelock, M. J., *et al.* (1997). P-cadherin is a basal cell-specific epithelial marker that is not expressed in prostate cancer. *Clin. Cancer Res.*, **3**, 2121.

124. Peralta Soler, A., Harner, G. D., Knudsen, K. A., McBrearty, F. X., Grujic, E., Salazar, H., *et al.* (1997). Expression of a P-cadherin identifies prostate-specific-antigen-negative cells in epithelial tissues of male sexual accessory organs and in prostatic carcinomas: implications for prostate cancer biology. *Am. J. Pathol.*, **151**, 471.

125. Peralta Soler, A., Knudsen, K. A., Jaurand, M.-C., Johnson, K. R., Wheelock, M. J., Klein-Szanto, A. J. P., *et al.* (1995). The differential expression of N-cadherin and E-cadherin distinguishes pleural mesotheliomas from lung adenocarcinomas. *Hum. Pathol.*, **26**, 1363.

126. Peralta Soler, A., Knudsen, K. A., Tecson-Miguel, A., McBrearty, F. X., Han, A. C., and Salazar, H. (1997). Expression of E-cadherin and N-cadherin in surface epithelial-stromal tumors of the ovary distinguishes mucinous from serous and endometrioid tumors. *Hum. Pathol.*, **28**, 734.

127. Han, A. C., Peralta-Soler, A., Knudsen, K. A., Wheelock, M. J., Johnson, K. R., and Salazar, H. (1997). Differential expression of N-cadherin in pleural mesotheliomas and E-cadherin in lung adenocarcinomas in formalin-fixed, paraffin-embedded tissues. *Hum. Pathol.*, **28**, 641.

128. Tanaka-Matakatsu, M., Uemura, T., Oda, H., Takeichi, M., and Hayashi, S. (1996). Cadherin-mediated cell adhesion and cell motility in *Drosophila* trachea regulated by the transcription factor Escargot. *Development*, **122**, 3697.

129. Bitzur, S., Kam, Z., and Beiger, B. (1994). Structure and distribution of N-cadherin in developing zebrafish embryos: morphogenetic effects of ectopic over-expression. *Dev. Dyn.*, **201**, 121.

130. Franklin, J. and Sargent, T. (1996). Ventral neural cadherin, a novel cadherin expressed in a subset of neural tissues in the zebrafish embryo. *Dev. Dyn.*, **206**, 121.

131. Liu, Q., Sanborn, K. L., Cobb, N., Raymond, P. A., and Marrs, J. A. (1999). R-cadherin expression in the developing and adult zebrafish visual system. *J. Comp. Neurol.*, **410**, 303.

132. Kim, S., Yamamoto, A., Bouwmeester, T., Agius, E., and De Robertis, E. M. (1998). The role of paraxial protocadherin in selective adhesion and cell movements of the mesoderm during *Xenopus* gastrulation. *Development*, **125**, 4681.

133. Griffin, K. J. P., Amacher, S. L., Kimmel, C. B., and Kimelman, D. (1998). Interactions between T-box genes regulate zebrafish trunk and tail formation. *Development*, **125**, 3379.

134. Kimmel, C. B., Kane, D. A., Walker, C., Warga, R. M., and Rothman, M. B. (1989). A mutation that changes cell movement and cell fate in the zebrafish embryo. *Nature*, **337**, 358.

135. Schulte-Merker, S., van Eeden, F. J., Halpern, M. E., Kimmel, C. B., and Nusslein-Volhard, C. (1994). no tail (ntl) is the zebrafish homologue of the mouse T (Brachyury) gene. *Development*, **120**, 1009.

136. Talbot, W. S., Trevarrow, B., Halpern, M. E., Melby, A. E., Farr, G., Postlethwait, J. H., *et al.* (1995). A homeobox gene essential for zebrafish notochord development. *Nature*, **378**, 150.

137. Halpern, M. E., Ho, R. K., Walker, C., and Kimmel, C. B. (1993). Induction of muscle pioneers and floor plate is distinguished by the zebrafish no tail mutation. *Cell*, **75**, 99.

138. Halpern, M. E., Thisse, C., Ho, R. K., Thisse, B., Riggleman, B., Trevarrow, B., *et al.* (1995). Cell-autonomous shift from axial to paraxial mesodermal development in zebrafish floating head mutants. *Development*, **121**, 4257.

139. Larue, L., Antos, C., Butz, S., Huber, O., Delmas, V., Dominis, M., *et al.* (1996). A role for cadherins in tissue formation. *Development*, **122**, 3185.

140. Rosenberg, P., Esni, F., Sjodin, A., Lionel, L., Carlsson, L., Gullberg, D., *et al.* (1997). A potential role of R-cadherin in striated muscle formation. *Dev. Biol.*, **187**, 55.

141. Redfield, A., Nieman, M. T., and Knudsen, K. A. (1997). Cadherins promote skeletal muscle differentiation in three-dimensional cultures. *J. Cell Biol.*, **138**, 1323.

142. Goichberg, P. and Geiger, B. (1998). Direct involvement of N-cadherin-mediated signaling in muscle differentiation. *Mol. Biol. Cell*, **9**, 3119.

143. Thompson, C. B. (1995). Apoptosis in pathogenesis and treatment of disease. *Science*, **267**, 1456.

144. Chen, C. S., Mrksich, M., Huang, S., Whitesides, G. M., and Ingber, D. E. (1997). Geometric control of cell life and death. *Science*, **276**, 1425.

145. Frisch, S. M. and Francis, H. (1994). Disruption of epithelial cell–matrix interactions induces apoptosis. *J. Cell Biol.*, **124**, 619.

146. Hermiston, M. L. and Gordon, J. I. (1995). *In vivo* analysis of cadherin function in the mouse intestinal epithelium: essential roles in adhesion, maintenance of differentiation, and regulation of programmed cell death. *J. Cell Biol.*, **129**, 489.

147. Herren, B., Levkau, B., Raines, E. W., and Ross, R. (1998). Cleavage of beta-catenin and plakoglobin and shedding of VE-cadherin during endothelial apoptosis: evidence for a role for caspases and metalloproteinases. *Mol. Biol. Cell*, **9**, 1589.

148. Makrigiannakis, A., Coukos, G., Christofidou-Solomidou, M., Four, B., Radice, G., Blaschuk, O., *et al.* (1999). N-cadherin-mediated human granulosa cell adhesion prevents apoptosis: a role in follicular atresia and luteolysis? *Am. J. Pathol.*, **154**, 1391.

149. Peluso, J. J., Pappalardo, A., and Trolice, M. P. (1996). N-cadherin-mediated cell contact inhibits granulosa cell apoptosis in a progesterone-independent manner. *Endocrinology*, **137**, 1196.

150. Day, M. L., Zhao, X., Vallorosi, C. J., Putzi, M., Powell, C. T., Lin, C., *et al.* (1999). E-cadherin mediates aggregation-dependent survival of prostate and mammary epithelial cells through the retinoblastoma cell cycle control pathway. *J. Biol. Chem.*, **274**, 9656.

4 | The integrin family of cell adhesion molecules

BETTE J. DZAMBA, MARGARET A. BOLTON, and DOUGLAS W. DESIMONE

1. Introduction

The integrins are transmembrane glycoproteins that mediate the adhesion of cells to the extracellular matrix (ECM) and to one another. They comprise the largest and most important group of receptors for ECM proteins and include several counter-receptors for select members of other families of cell adhesion molecules. Integrins provide a critical functional linkage between the extracellular environment and the cytoskeleton. This 'integration' is of fundamental importance to the mechanics of cell adhesion and migration, and to the propagation of various cell signals. Together, the integrins represent an ancient family of adhesion molecules common to all metazoa and, as such, participate in a wide range of normal physiological processes, developmental events, and disease states.

In the 14 years that have elapsed since the current system of nomenclature for this receptor family was first proposed (1), there has been an explosion of information concerning integrin structure and function. As of this writing, over 18,000 citations (PubMed search, April 2001) containing the word integrin in the abstract and/or title have appeared in print. Approximately 90 integrin-related papers per week were published during the first quarter of 2001 alone! While an exhaustive overview of this literature is all but impossible, the goal of this chapter is to highlight historical milestones that led to the discovery of the integrins, to provide a primer on integrin structure and functions, and to consider significant recent advances in our understanding of the biology of these receptors.

1.1 The identification of integrins as cell–matrix adhesion receptors

The discovery of integrins was driven in large part by a series of early observations suggesting that adhesion to ECM is mediated at the cell surface by specific receptors. Some of the most compelling evidence came from investigators studying the ECM glycoprotein fibronectin (2). By 1980, it became clear that fibronectin was one of a group of ECM proteins present in serum that could promote the adhesion of cells to tissue culture dishes. Fibronectin was shown to be absent from the surfaces of many cells following transformation (3, 4). Surface association of fibronectin was correlated with increased cell adhesion and spreading in non-transformed cells, whereas transformed cells attached poorly and appeared rounded. When purified fibronectin was added back to transformed cells they became more adherent to the substrate, flattened, and assembled an extensive cytoskeleton of actin microfilaments typical of non-transformed cells (5, 6). This suggested that fibronectin binding to the cell surface was a critical step in the regulation of cell-adhesive behaviour. Actin microfilaments were subsequently shown to co-localize with fibronectin fibrils at the cell surface (7, 8), thus providing additional support for the hypothesis that transmembrane molecules are responsible for linking the ECM to the cytoskeleton.

The search for candidate ECM receptors was advanced by two distinct research strategies. The first involved screens for monoclonal antibodies (mAbs) that perturbed cell–substrate adhesion. The CSAT (9) and JG22 (10) mAbs were discovered in independent screens for reagents that interfered with avian myogenic cell adhesion *in vitro*. These antibodies recognized a 140 kDa complex of proteins from muscle and a variety of other cell types that co-localized with actin filaments at sites of cell attachment to the substrate (11, 12). A similar immunological approach was used to identify a putative 140 kDa fibronectin receptor from mammalian fibroblasts (13).

The second general strategy utilized affinity chromatography to isolate ECM receptors from mammalian cells. A key to the success of this approach was to come from studies undertaken to define the major cell binding site on fibronectin. Using a series of short overlapping synthetic peptides, Pierschbacher and Ruoslahti (14) determined that the tetrapeptide sequence Arg–Gly–Asp–Ser (RGDS), located in a predicted hydrophilic loop within the centrally located cell-attachment region of fibronectin, could mimic the adhesive activity of the intact molecule. Subsequent studies by these and other investigators showed that soluble RGD-containing peptides, like the CSAT and JG22 mAbs, would inhibit the attachment of cells to intact fibronectin and a number of other RGD-containing ligands.[1] Control peptides containing a single amino acid substitution (RGDS → RGES) lacked these activities. These studies represented a significant conceptual advance, that a simple linear sequence of amino acids could be used to isolate and characterize cell surface receptors for RGD-containing adhesive glycoproteins of the ECM. Pytela and colleagues (15) passed detergent extracts of human osteosarcoma cells over affinity columns containing an immobilized cell binding fragment of fibronectin. A 140 kDa glycoprotein complex was retained on the column after washing and specifically eluted with RGD peptides. This strategy was used to obtain similar complexes of membrane

---

[1] RGD-like sequences are present in a structurally diverse range of biologically active peptides, ECM molecules, and other proteins involved in adhesion, indicating that this peptide motif is an evolutionarily well-conserved cell recognition site. While the presence of an RGD site is typically thought of as being involved in promoting adhesion, there are some interesting exceptions. In the case of the snake venom disintegrins, the apparent physiological function of these peptides is to antagonize normal integrin function (see Chapter 6 for additional information).

proteins from other cell types (16, 17) and to obtain receptors for other RGD-containing ECM molecules such as vitronectin (18), fibrinogen (19), and collagen (20).

The results of both the antibody- and affinity-based approaches were intriguing in their similarities but raised many additional questions. Each method resulted in the identification of a complex of membrane proteins in the 140 kDa range by SDS–PAGE under reducing conditions. Under non-reducing conditions the complex was resolved into additional bands on these gels. In the case of the putative mammalian receptor for fibronectin, two major bands were revealed at 120 kDa and 140 kDa (15) whereas the avian complex obtained using the CSAT and JG22 mAbs resolved into at least three proteins over a similar size range (21, 22). This suggested that the avian and mammalian receptors might represent different proteins or were assembled into functional complexes with differing stoichiometries. It soon became apparent, however, that all integrins functioned as heterodimers and that the multiple bands purified from avian cell extracts with CSAT and JG22 represented several distinct receptors, which shared a common subunit recognized by the antibodies (23). These studies mark an interesting chapter in the history of integrins that is covered in detail in a number of excellent reviews that appeared during this period (1, 24, 25). Sections 1.2 and Section 2 provide a summary of the key data that help define the integrin family and the basic cell biology of these receptors.

1.2   Emergence of the integrin family

Early research on integrins represented the convergence of several major areas of investigation by the mid 1980s in addition to the cell–ECM adhesion field. These included studies of platelet activation and aggregation, leukocyte activation and cell–cell interactions, and *Drosophila* development. In each of these cases, available information suggested that some of the proteins involved in these processes were structurally and functionally related to the cell–ECM receptors. For example, the glycoprotein complex IIb/IIIa found on the platelet cell surface was shown to be a heterodimer of two transmembrane proteins, gpIIb and gpIIIa (26), which behaved similarly to the ECM receptors by SDS–PAGE. The IIb/IIIa complex was also shown to bind to a number of plasma proteins involved in platelet adhesion: fibrinogen, fibronectin, vitronectin, thrombospondin, and von Willebrand factor. Each of these proteins contain RGD sequences and RGD peptides were shown to interfere with platelet aggregation and adhesion. The very late antigens (VLA) are another group of surface glycoproteins initially identified on T lymphocytes where they would appear *very late* after activation of these cells. Five VLA heterodimers were identified each of which share a common 130 kDa subunit (27). Another well studied group of receptors found on leukocytes include the lymphocyte function-associated antigen LFA-1, the p150/95 complex, and the Mac-1 complement (C3bi) receptor on macrophages (28). These receptor complexes also share a common subunit, and play important roles in inflammation and various lymphocyte functions as discussed in Section 6. Finally, there were strong indications that a group of antigens expressed in a temporal and position-specific manner during imaginal disc development in

Table 1 Alternative names for integrins[a]

	Gene name	CD antigen name	Other names
α1	ITGA1	CD49a	VLA1
α2	ITGA2	CD49b	VLA2, Spla, ECMR II
α3	ITGA3	CD49c	VLA3, ECMR I
α4	ITGA4	CD49d	VLA4
α5	ITGA5	CD49e	VLA5, FNRα, ECMRVI
α6	ITGA6	CD49f	
α7	ITGA7		
α8	ITGA8		
α9	ITGA9		αRLC
α10	ITGA10		
α11	ITGA11		
αv	ITGAV	CD51	VNRα
αIIb	ITGA2B	CD41	gpIIb
αL	ITGAL	CD11a	LFA1α
αM	ITGAM	CD11b	Mac1α, CR3α, Mo1
αX	ITGAX	CD11c	p150, 95α
αD	ITGAD	CD11d	
αE	ITGAE	CD103	HMLA-1
β1	ITGB1	CD29	VLAβ, FNRβ
β2	ITGB2	CD18	p150, 95β
β3	ITGB3	CD61	gpIIb, VNRβ
β4	ITGB4	CD104	
β5	ITGB5		
β6	ITGB6		
β7	ITGB7		
β8	ITGB8		

[a] Individual α and β subunits are listed but note that many of the 'other' names define heterodimers. For example, VLA5 refers to α5β1 but the β1 subunit is common to all of the VLAs, therefore, a specific VLA designation is defined by the α subunit. In some cases the name is followed by an α or β (e.g. LFA1α, FNRβ) to denote an individual subunit.

Drosophila (PS antigens, Section 5.1) were also related to the vertebrate ECM receptors (29).

With the availability of initial cDNA sequence information (30–33) it became clear that the avian, mammalian, and *Drosophila* proteins were members of a family of structurally related receptors. The term 'integrin' was introduced to denote that the receptors are made up of integral membrane glycoproteins involved in the transmembrane association of the ECM and cytoskeleton (30). The Gordon Research Conference on 'Fibronectins', held in Santa Barbara, California, in 1987, represented a turning point in the field that brought together each of the research areas discussed above. Around this time, Hynes (1) proposed a system of nomenclature that categorized the individual subunits of each integrin heterodimer as either α or β based on their sequence similarities, as discussed below. This involved the 'renaming' of a number of well-studied proteins. Table 1 includes a list of these earlier and alternative names, most of which appear in the literature prior to 1987.

2. Integrin structure and ligand binding functions

All integrins are heterodimers composed of non-covalently associated α and β subunits (Fig. 1). Both α and β subunits are transmembrane glycoproteins and cell surface expression of each subunit is dependent on heterodimer formation. Specific αβ subunit combinations (Fig. 2) define ligand binding activities (Table 2) and various signalling functions. The bulk of the integrin heterodimer is exposed to the extracellular environment, while interactions with the actin cytoskeleton and signalling machinery generally occur via the relatively short cytoplasmic domains of each subunit. Because most cells express multiple integrins, the adhesion-dependent behaviours of a given cell typically reflect a combination of receptor–ligand associations.

2.1 Structural features of integrin α and β subunits

Aside from overall similar topology as type-1 transmembrane proteins, integrin α and β subunits are distinct proteins with little sequence similarity. At present, 18 α subunits and eight β subunits have been identified in mammals. With the exception of β4 (see Section 2.3), all vertebrate integrin β subunits are between 760–790 amino acids in length, while α subunits range in size from 1000–1200 amino acids. Analyses

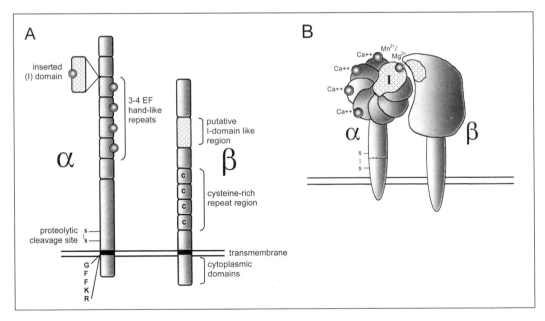

Fig. 1 Generalized integrin structure. (A) Schematic of integrin α and β subunits with approximate locations of domain structures. Note not all α subunits are proteolytically processed. (B) Cartoon of integrin heterodimer showing stylized features of an included I-domain containing α subunit. Individual β-sheets of a putative β propeller are shown, including the I-domain, which is predicted to lie at the top of the propeller with the midas motif (with bound Mg^{2+} or Mn^{2+}) along a solvent exposed face. Putative EF hand domains indicated by Ca^{2+}. Presence of I-domain-like region in β subunit represented by stippled area. Approximate dimensions of globular head domains are based on electron microscopic imaging of rotary shadowed heterodimers (see text).

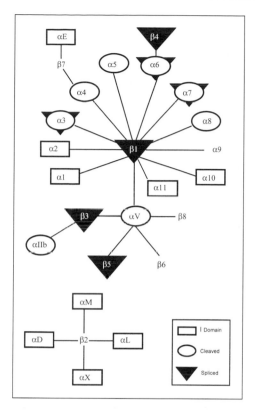

Fig. 2 Integrin subunit associations and processing. The α and β subunit associations of the known vertebrate (human) integrins are indicated with general structural features noted: (ovals) α subunit extracellular cleavage, (rectangles) I-domain containing α subunits and, (filled triangles) alternatively spliced isoforms. The β1, αv, and β2 subunits define the three largest subfamilies of integrin heterodimers.

of the deduced amino acid sequences of both α and β subunits have revealed a number of interesting features.

The extracellular domains of all β subunits contain 56 cysteines (with minor variations in β4 and β6–8), the majority of which are located within a stretch of approximately 260 amino acids separated approximately 50 residues from the transmembrane domain. Within this cysteine-rich stretch is a characteristic five-fold repeat of eight cysteines (30), the first four of which are highly similar (with identical spacing between cysteines). The C-terminal halves of these four repeats are similar to the EGF-like repeats found in laminin, which are predicted to form rod-like structural domains (34).

Integrin α subunits contain seven N-terminal repeats of about 60 amino acids each, three to four of which include divalent cation binding EF hand motifs with the consensus sequence DxDxDGxxD (x = any amino acid). These metal binding domains are important for integrin ligand binding function and conformational changes that affect receptor affinity for ligand. As indicated in Fig. 2, α subunits can be grouped on the basis of shared structural features. One group includes those subunits with a post-translational proteolytic cleavage site within the extracellular domain. Typically, this cleavage site is located proximal to the plasma membrane and is bridged by a disulfide bond. The α4 subunit, however, is cleaved some

Table 2 Reported integrin ligands

Integrin	Cell–matrix ligands									Cell–cell ligands
	FN	LN	Col	TN	VN	OPN	FBG	TSP	Other	
α1β1		X	X							
α2β1		X	X						Echovirus 1	
α3β1	X	X	X					X	FBN	
α4β1	X									VCAM-1
α4β7	X									VCAM-1, MadCAM-1
α5β1	X									ADAM 15
α6β1		X								ADAM 2
α6β4		X								
α7β1	X	X								
α8β1	X	?		X	X	X				
α9β1				X		X				VCAM-1
α10β1			X							
α11β1			X							
αvβ1	X				X					
αvβ3	X			X	X	X	X	X	FBN	ADAM 15, ADAM 23
αvβ5	X				X					
αvβ6	X			X					TGFβ, LAP	
αvβ8	X	X	X							
αIIbβ3	X						X	X	VWF	
αDβ2										ICAM-3, VCAM-1
αLβ2										ICAM-1,2,3,5
αMβ2							X		NIF	ICAM-1, C3bi
αXβ2							X			C3bi
αEβ7										E-Cadherin, MadCAM-1

Abbreviations: ADAM, a disintegrin and a metalloprotease; CAM, cell adhesion molecule (Ig superfamily); FN, fibronectin; LN, laminin; Col, collagen; TN, tenascin; VN, vitronectin; NIF, neutrophil inhibitory factor (hookworm product); OPN, osteopontin; FBG, fibrinogen; TSP, thrombosondin; FBN, fibrillin; VWF, von Willebrand factor; LAP, latency-associated peptide.

distance from the membrane and is not joined by a disulfide linkage. Half of all vertebrate α subunits (i.e. α1, α2, α10, α11, αE, αL, αM, αD, and αX) lack an extracellular cleavage site but contain instead an 'inserted' I-domain (Fig. 1) near the N-terminus, which is also referred to as an A-domain based on similarity to the collagen binding 'A3' motif found in the plasma protein, von Willebrand factor (VWF). X-ray crystallography has been used to obtain structures for the α2, αM, and αL subunit I-domains (35–37). Each I-domain forms a Rossman fold comprised of a β-sheet surrounded by seven α-helices. An unusual divalent cation binding site, termed a MIDAS (metal ion-dependent adhesion site) motif, is located in each I-domain and this feature is critical for the ligand binding function of subunits that contain it. Bound cation forms a total of five coordinations and three of the residues involved are included within the sequence DxSxS, which is conserved in the MIDAS motif (36). Interestingly, integrin β subunits also contain the DxSxS sequence within a modified I-domain-like region located in a conserved stretch of 240 amino acids near the N-terminus. Mutations in this sequence abolish ligand binding (38). One subunit, α9, contains neither a post-translational cleavage site nor an I-domain.

High resolution structural information for intact integrin αβ heterodimers is unavailable, however, a generalized model of integrin structure can be pieced together based on a variety of analyses (Fig. 1B). Electron microscopy of purified, rotary shadowed integrins reveals two 'comma'-shaped subunits joined via their globular heads ($\sim 80 \times 120$ Å when joined) (34, 39). Two tails (~ 180–200 Å long) are also distinguishable under appropriate detergent conditions, consistent with the presence of hydrophobic α and β transmembrane domains. The observed lengths of these individual α and β subunit tails are in good agreement with secondary structure predictions obtained from the individual subunit sequences. While detailed structural information is available only for I-domains, alignments of α subunit sequences predict the folding of a β propeller with the blades corresponding to each of the seven 60 amino acid repeats within the N-terminus (40). The I-domain is situated between the second and third N-terminal repeats and is predicted to sit on top of the β propeller with the MIDAS motif located along the solvent-exposed face of the domain (Fig. 1B).

The integrins represent an ancient family of proteins and, not surprisingly, both α and β subunits have been identified in several invertebrate species as discussed in Section 5.1 and reviewed in Burke (41). Of particular note, however, is the apparent lack of I-domains in all invertebrate α subunits sequenced to date, suggesting that integrin I-domains arose sometime following the appearance of the vertebrates. While invertebrate integrin subunits bear considerable sequence similarities to vertebrate integrins, there is a comparatively small number of α and β subunits in invertebrate species, as evidenced by analyses of the recently 'completed' *Caenorhabditis elegans* and *Drosophila melanogaster* genome sequences. Only two α subunits (α ina-1 and α pat-2) and a single β subunit (β pat-3) are encoded in the *C. elegans* genome, with five α subunits (αPS1 through αPS5) and two β subunits (βPS and βv) identified thus far for *Drosophila* (see Section 5.1). An α and β subunit have also been cloned from sponges (42, 43) and a β subunit from a coral species (42), however, little is known about the functions integrins may play in these organisms.

2.2 Extracellular ligand binding and receptor specificity

As a group, integrins recognize a wide assortment of extracellular ligands and do so with varied specificities. For example, α5β1 binds exclusively to the ECM protein fibronectin, whereas αvβ3 is reported to bind at least seven different ECM ligands. Table 2 lists a number of reported ECM ligands and counter-receptors for vertebrate integrins. What is at once apparent from this assembled information is the diversity of integrin binding interactions and, given that most cells express multiple integrins, the possibility for substantial functional 'redundancy' and/or compensatory adhesive mechanisms. Why some integrins recognize specific ligands while others display more 'promiscuous' specificities remains unclear. However, a combination of available structural information, mutational analyses, crosslinking studies, and the use of monoclonal antibodies has provided several important insights into the nature of the binding sites of both integrins and their ligands.

The most obvious determinant of integrin ligand binding specificity is the αβ composition of each heterodimer (Table 2 and Fig. 2). The β1 subfamily provides a convincing demonstration of this fact; each of these 12 receptors share a common β1 subunit but differ widely in terms of both the number of ligands and types of ligand binding sites recognized. Thus, the α subunit can be thought of as defining the binding specificity of a given receptor, particularly in the context of the β1 and β2 integrins.[2] A deeper understanding of ligand binding specificity requires analyses of both specific receptors and the ligands that are recognized by them. As discussed earlier, a number of integrins were identified on the basis of their ability to recognize RGD sequences in proteins like fibronectin and vitronectin. However, a receptor that recognizes RGD in fibronectin (e.g. α5β1) may not necessarily recognize the RGD in other ECM molecules. This highlights the probable importance of flanking sequences and the differing conformations of both the adhesive sites within ligands and the binding 'pockets' of individual integrins. Some receptors, however, are able to bind RGD sequences presented in a variety of contexts as in the case of αvβ3, which recognizes at least seven RGD-containing ligands (Table 2). The factors that contribute to integrin ligand binding, therefore, represent an important biological problem. We will briefly consider several of these factors and their consequences for integrin function.

The best understood integrin–ligand binding interactions involve those of I-domain-containing integrins, in large part because recombinantly expressed I-domains retain ligand binding activity (44–47). In many cases, ligand binding to purified (recombinant) I-domain is indistinguishable from that of the intact heterodimeric receptor from which the I-domain was derived. In addition, many mAbs identified as ligand 'blockers' of integrins recognize epitopes within the α subunit I-domain (44, 48). For these reasons, high resolution crystal structures of individual I-domains have provided a number of useful insights regarding mechanisms of integrin ligand binding, as discussed above. Both ECM molecules (e.g. collagen and laminin) and counter-receptors of the immunoglobulin superfamily (e.g. intercellular adhesion molecules; I-CAMs) are recognized by integrins that contain α subunit I-domains. I-domains are also involved in binding pathogens such as the yeast *Candida* and human echovirus 1, hookworm neutrophil inhibitory factor, and some toxins (48–51).

Less information is available concerning the ligand binding sites of non-I-domain-containing α subunits. While only nine integrin α subunits possess I-domains, it is important to recall that all β subunits contain a conserved sequence with significant

---

[2] Initial indications suggested that the integrins could be organized into subfamilies based on common β subunit sequences (1, 315). However, as additional receptors were identified, it became apparent that at least some α subunits could pair with more than one β subunit, as in the case of αv (Fig. 2). In this instance it is arguably more appropriate to think of αv as defining its own subfamily while bearing in mind that the 'subfamily' designation is a somewhat arbitrary distinction. In any event, it is clear that only a subset of all possible αβ combinations occurs in nature.

similarity to this region. It is likely that proper folding of the putative β subunit I-domain requires association with the α subunit (52) and, not surprisingly, functionally active recombinant β subunit I-domains have not been reported.[3] Nevertheless, the importance of this conserved I-domain-like sequence has been suggested by crosslinking studies (53, 54), phage display (55), and by mutations in this region that abolish ligand binding (38, 56, 57). Interestingly, similar mutations in the β2 subunit, which forms heterodimers with only I-domain-containing α subunits (Fig. 2), also abrogates ligand binding (58). These data indicate that the α subunit I-domain alone is not sufficient to mediate binding of ligand in the context of the intact heterodimeric receptor.

There is considerable evidence available to suggest that integrins maintain different functional 'activation' states. Activation in this instance may be defined as a change in the ligand binding affinity of an individual receptor. This process has been termed 'affinity modulation' (Fig. 3), and represents an important regulatory step in the function of receptors such as the αIIbβ3 receptor on platelets and the β2 integrins expressed by leukocytes (and perhaps many others, e.g. see Section 5). As discussed in Section 3.2, a change in the conformation of these receptors is brought about by an 'inside-out' mechanism that modulates extracellular ligand binding. Therefore, affinity modulation may be thought of as a rapid way to up-regulate (or down-regulate) the adhesive functions of a cell in response to various physiological insult or stimuli (e.g. vascular injury or inflammation).

Integrin affinity modulation may be demonstrated experimentally using mAbs that function as 'ligand-mimetics' or as reporters of other conformational changes that accompany receptor activation. The latter are termed anti-LIBS (ligand-induced binding site) or anti-CLIBS (cation-and-ligand-induced binding site) antibodies because they recognize integrin epitopes exposed following ligand binding or in response to the 'artificial' activation of the receptor by divalent cations such as Mn^{2+}. Metal ions play a critical role in the ligand binding functions of all integrin heterodimers as suggested by the presence of both EF hand and the MIDAS motifs. The requirement for divalent cations is readily demonstrated by the addition of chelators such as EDTA, which results in the loss of integrin ligand binding activity and reduced cell substrate adhesion. However, evidence suggests that divalent cations may also have an inhibitory effect on integrin ligand binding. For example, while Mn^{2+} leads to a demonstrable up-regulation in the affinity of most integrins (i.e. 'activated state'), many receptors are stabilized in a low affinity conformation (i.e. 'resting state') in the presence of Ca^{2+} (59, 60). Accordingly, 'activated' receptors will be recognized by anti-LIBS Abs while 'resting' receptors will not. While the physiological significance of the Mn^{2+} effect on receptor activation is doubtful, it has been suggested by many investigators that Mn^{2+} likely stabilizes the receptor in what approximates a normal ligand-bound conformation. Indeed, ligand binding to

---

[3] Synthetic peptides corresponding to sequences within the putative I-domain region of β3 are reported to block fibrinogen binding to αIIbβ3 (316), indicating that this region may retain some biological activity in isolation.

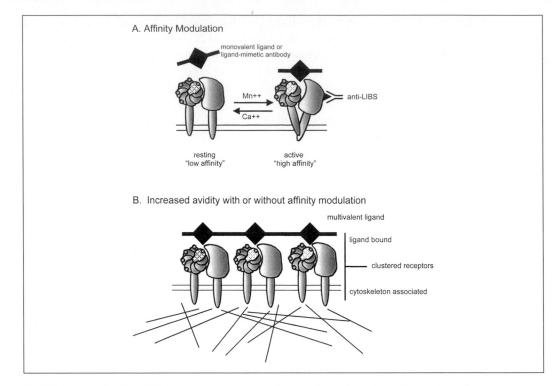

Fig. 3 Integrin activation. (A) Integrin receptors can undergo a change in conformation that results in an increase in receptor affinity for ligand. Non-ligand associated receptors may be stabilized in a 'resting' state by divalent cations like Ca^{2+} or rapidly activated in the presence of Mn^{2+}, which may keep the receptor 'stuck' in a high affinity conformation. Affinity may also be modulated by inside-out signalling mechanisms and by binding to some extracellular ligands, including ligand-mimetic mAbs. Readout of conformational change can often be made with anti-cation-and-ligand-induced binding site Abs (anti-CLIBS, anti-LIBS). (B) Integrin functional 'activation' may also be achieved by clustering integrins without requiring a change in receptor affinity for ligand. In this model, avidity is responsible for the apparent increase in adhesive activity.

integrin receptors alone is often enough to increase individual receptor affinity, as demonstrated by RGD peptide binding to αIIbβ3 (61).

It is important to consider additional ways in which an increase in integrin ligand binding activity may be brought about independently of integrin conformational changes (62). The most obvious of these is the increase in cell avidity that could occur upon clustering of the receptor (Fig. 3). In this model, both multivalent extracellular ligands and/or receptor association with the cytoskeleton would limit receptor diffusion in the membrane and thus increase the apparent strength of an adhesive interaction.

2.3 Transmembrane and cytoplasmic domain interactions

Integrin α and β subunit cytoplasmic tails are relatively short, yet there is a surprisingly large number of cytosolic and cytoskeletal proteins reported to interact with integrin receptors ($>$ 20 thought to interact directly with β subunits alone) (63). At least three general types of integrin cytoplasmic domain interactions can be defined. First, interactions between the individual α and β subunit tails are involved in regulating conformational changes associated with, for example, affinity modulation (Section 3). Secondly, integrins have no intrinsic catalytic activity but these receptors play key roles in the propagation of a variety of signals by associating with signalling molecules and/or various adaptor proteins such as FAK (64), ILK (65), cytohesin-1 (66), paxillin (64, 67), Shc, and Grb2 (68). Thirdly, integrins perform a critical mechanical function by serving as transmembrane links between the ECM and other adhesion molecules, and structural components of the cytoskeleton such as talin (69), α-actinin (70), and filamin (71). The reader is referred to Chapter 9 (Burridge) for further details regarding integrin interactions at sites of focal contact.

The general importance of integrin cytoplasmic tail sequences to receptor function is illustrated by a number of studies. For example, cytoplasmic tail-deleted $\beta 1$ subunit constructs are able to associate with intact α subunits, transit to the cell surface, and bind extracellular ligands, but these heterodimers localize poorly to focal adhesion sites (72, 73). Moreover, chimeric constructs composed of integrin β subunit tail sequences fused to the transmembrane and extracellular domains of 'irrelevant' reporter proteins will localize to focal adhesions, further implicating the importance of β subunit tail interactions to intact receptor functions (74). Over-expression of the latter constructs can also result in the 'dominant-negative' inhibition of endogenous integrin activities, such as matrix assembly, cell spreading, and migration (75), presumably by competing for a limited supply of functionally critical cytosolic and/or cytoskeletal components. Interestingly, expression of β subunit cytoplasmic tail (e.g. $\beta 1$, $\beta 3$, and $\beta 5$) chimeras is sufficient to stimulate some integrin-dependent signals such as tyrosine phosphorylation of FAK (76), the effect of which can be enhanced by clustering the chimera with antibodies (77). Different β subunit tails display significant sequence similarities, highlighting the importance of this domain in mediating the basic functions shared by many different integrins such as linkage to the actin cytoskeleton. The notable exception to the latter being the $\beta 4$ subunit, which has a large ($\sim$ 1000 amino acid) intracellular domain (78–80) that mediates interactions with the intermediate filament cytoskeleton at the hemidesmosome (81, 82). The cytoplasmic tail sequences of α subunits are considerably more divergent than the β tails and, with the exception of the membrane proximal GFFKR sequence (see Section 3), there is little overall similarity. However, specific α subunit sequences are well conserved among different species (83–87) suggesting conservation of specific α tail functions as well.

Compared to the β subunits, relatively few molecules are reported to interact with α subunit cytoplasmic tail sequences (e.g. GFFKR). Calreticulin (88), the guanine nucleotide exchange factor Mss4 (89), the calcium binding protein CIB (90), and the LIM-only protein DRAL/FHL2 (91) have each been shown to bind to α tails. As with several reported β subunit tail interactions, however, many of these associations were identified using *in vitro* binding assays or yeast two-hybrid screens so the physiological relevance of these purported associations is not entirely clear in some cases. For example, calreticulin normally resides within the lumen of the endoplasmic reticulum. How this calcium binding protein might gain access to α subunit tail target sequences is uncertain. Nevertheless, defects in integrin-mediated adhesion and signalling are observed in cells derived from calreticulin-deficient mice (92). Integrin α subunit transmembrane domains have also been shown to participate in receptor signalling by interacting with the adaptor protein calveolin (93). Calveolin mediates the binding of the tyrosine kinase Fyn, which signals through the Ras-ERK/MAPK pathway to promote progression through the cell cycle. A nine amino acid stretch of the α4 cytoplasmic tail was shown recently to support the specific binding of paxillin (67), which may function to regulate α4β1 adhesion to V-CAM and/or influence the activation of additional integrin receptor cell signalling pathways (94).

Integrin α subunit tails can influence the functions of intact integrin heterodimers in critical but often subtle ways. A striking example of this comes from α subunit cytoplasmic domain swapping experiments with α4 (94–96). Integrin α4β1 mediates cell migration in a variety of cellular contexts, and when joined with other α subunit ectodomains (e.g. α2, αIIb), the α4 cytoplasmic tail can result in increased cell migration. Other α subunit cytoplasmic tails are reported to influence matrix contraction (94), integrin activation kinetics (97), and responsiveness to R-Ras-induced cell migration (98). As discussed in Section 3.2, integrin α subunit tails are critical to the regulation of integrin activation by affinity modulation.

Integrin cytoplasmic tail functions may be complicated further by the expression of several possible alternatively spliced isoforms of both α and β subunits (99). Subunits with reported cytoplasmic tail splice variants are highlighted in Fig. 2. While specific isoforms do not appear to differ in terms of their subunit associations or contributions to extracellular ligand binding specificities, it is clear that subcellular receptor localization, and/or differences in receptor adhesive and signalling properties can result. For example, the β1A and β1D variants localize to focal contacts (100), while β1B (101) and β1C (100) do not. Some of these β1 splice variants can act in a dominant-negative fashion, further suggesting their involvement in the regulation of select integrin functions (102, 103). Differences in the adhesive and migratory properties of cells on laminin can also be demonstrated by transfecting cells with either the α6A (pro-migratory) or α6B subunit (104). In addition, α6A has been implicated in the regulation of myoblast proliferation and differentiation (105). Another interesting demonstration of isoform-specific subunit function includes the inhibition of prostate carcinoma cell proliferation by β1C (106). This is in contrast to the well established roles of β1A integrins in supporting anchorage-dependent growth and normal progression through the cell cycle (107).

2.4 Additional integrin-associated proteins

Although a detailed handling of this subject is beyond the scope of the current review, it is important to acknowledge the critical roles that other integrin-associated proteins may play in regulating the functional activities of integrin receptors. Two representative groups of molecules in this category include the tetraspan family (transmembrane-4, TM4SF) and members of the immunoglobulin (Ig) superfamily (108, 109). Integrin-associated protein (IAP or CD47) contains an extracellular Ig domain and five transmembrane segments (110). IAP is widely expressed and associates with both αvβ3 and αIIβb3. IAP-deficient mice have defects in neutrophil adhesion and migration (111), which may reflect their likely involvement in modulating αvβ3 ligand binding functions in these cells (112). The tetraspans (e.g. CD9, CD53, CD81, CD151) are known to associate with several integrins, particularly in migratory and invasive cells where they may function by recruiting signalling molecules such as PKC (113) and PI4-kinase (114) into close proximity with integrins. Thus, the tetraspan–integrin association may represent an unusual signalling complex important in phosphoinositide metabolism and PKC-dependent phosphorylation. See also Chapter 6 for additional discussion of CD9 involvement in ADAM 2 binding to α6β1 during fertilization (115).

3. Integrin signalling

The role of integrins in cell signalling has been the object of intense investigation during the past decade and the significant progress in this area has been covered in a number of reviews (107, 116–119). The two sections that follow, therefore, focus largely on the functional significance of integrin-mediated signalling rather than the molecular details of the signalling molecules involved (Fig. 4). The two most widely recognized categories of integrin signalling ('outside-in' and 'inside-out') are considered.

3.1 Outside-in signalling

The integrin-mediated adhesion of cells to the extracellular matrix results in the transduction of signals that influence cell cycle progression, cell survival, changes in gene expression, cell migration, and cytoskeletal organization. Interestingly, the pathways that lead to these events are not unique to integrin signalling but also may involve (or parallel) the activities of various growth factors (Fig. 4). One of the most significant concepts to emerge from studies of integrins during the past several years is that the integration of integrin and growth factor receptor signalling is critical; cross-talk between these multiple intersecting pathways is often required to elicit a wide range of cellular responses. A second general consideration is that integrins are not known to have intrinsic enzymatic activity so it is necessary for these receptors to associate with signal transduction machinery in order to effect various downstream functions. Among the physiological changes brought about by integrin ligand binding (and receptor clustering) are increases in intracellular pH and cAMP levels (120–124) changes in cytosolic Ca^{2+} levels (125, 126), increased tyrosine phosphorylation and MAP kinase activation (127), and changes in gene expression (see Section 4.3).

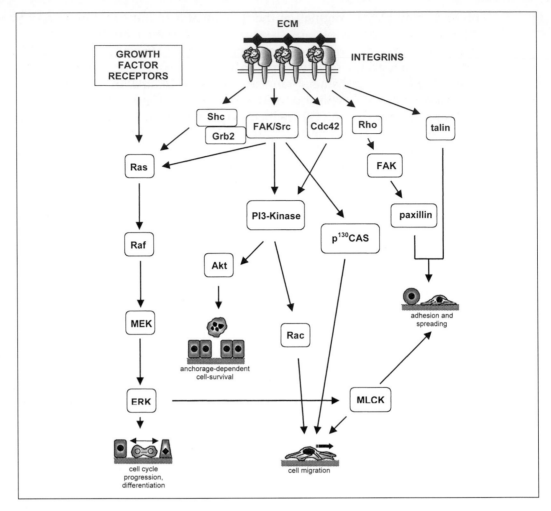

Fig. 4 Some integrin-dependent signalling pathways and their functional consequences. A subset of the known signalling pathways involving integrins is shown. Note that signalling through the Ras/ERK pathway involves both integrin and growth factor receptor inputs. Refer to text for additional details.

Rapid phosphorylation of cytosolic proteins on tyrosine is observed following cell attachment to FN or integrin receptor clustering by antibodies. One well-studied target of this activity is pp[125]FAK (FAK), a 125 kDa focal adhesion-associated non-receptor tyrosine kinase (128–131). FAK is localized to sites of focal adhesion in cultured cells, and is phosphorylated in response to integrin-mediated adhesion to fibronectin or in response to antibody crosslinking of β1 integrins (129, 132). FAK is reported to bind the integrin β1 cytoplasmic domain (64) and form associations with a variety of other molecules including Src family members (133, 134), Grb2/SOS (135), and paxillin (136). The molecular details of how integrins regulate the activity of FAK is not well understood but the clustering of integrins in focal adhesions appears critical and may reflect a requirement for oligomerization and transphos-phorylation of the kinase in a mechanism similar to some growth factor receptor-dependent phosphorylation events. This initial phosphorylation would lead to the recruitment of Src family kinases and further phosphorylation of FAK. This in turn leads to the recruitment of other signalling and cytoskeletal molecules to the focal adhesion complex. For example, FAK phosphorylation and subsequent binding of Grb2 provides a link to PI3-kinase, the Ras pathway, and MAP kinase (135). In addition, phosphorylation of the adaptor protein pp130^{Cas} by FAK associated Src leads to activation of MAP kinase (137) and also to cell migration (138, 139). As mentioned earlier, calveolin binding to β subunit transmembrane domain sequences can stimulate RAS/MAPK through activation of the tyrosine kinase Fyn and binding to the adaptor protein Shc (93). Thus, the FAK and Shc pathways are essential components of the integrin-dependent 'outside-in' signalling cascade important in the relay of information from the external environment to the cytoplasm.

Recent evidence supports a bidirectional link between integrins and the Rho family of small GTPases, which are involved in regulating actin cytoskeleton dynamics and adhesive behaviour (see also Chapter 9, Fig. 7). Ligation of integrins may lead to the activation of Rho (140, 141), Cdc42, and Rac (141, 142). In turn, activated Rac, Rho, and Cdc42 can lead to the assembly of adhesion complexes containing integrins and associated proteins such as paxillin and FAK. Rho can also stimulate the phosphorylation of FAK. It is presently unclear how ligation of integrins results in Rho, Rac, and Cdc42 activation but pp130^{CAS}, a downstream target of FAK, has been implicated (139, 143). Adhesion-dependent activation of Rho family GTPases provides another example of the probable synergy between integrin/extracellular matrix interactions and other signalling pathways. While integrin binding can stimulate Rho, this stimulation is weak compared to the activation of Rho by soluble agonists such as lysophosphatidic acid (LPA).

Although there is ample evidence that growth factors and integrins act co-ordinately to direct cellular behaviours, the molecular basis for this synergistic relationship remains unclear. One mechanism may be the spatial regulation of signalling activities through the assembly of focal adhesions. Both integrin ligation and growth factor signalling can activate FAK while disruption of the cytoskeleton inhibits focal adhesion formation, growth factor receptor activation, and subsequent events such as cell growth. Integrin ligation promotes the accumulation of both integrins and EGF, PDGF, and bFGF receptors in focal adhesion complexes along with various downstream signalling molecules (132, 144–146). This proximal associa-tion may facilitate intermolecular interactions between signalling components. In the case of FAK or MAP kinase, both integrin–ECM and growth factor–receptor inter-actions are stimulatory. It may be that threshold levels of both signals are required for subsequent activity. Another point of convergence in the activities of integrin and growth factor signalling was thought to be phosphorylation of Src family kinases but recent evidence from triple knock-out mice null for Src, Yes, and Fyn suggests that these kinases are required for integrin, but not PDGF signalling (147). A third type of signalling control mechanism could be mediated by adhesion-dependent cell shape changes. A continuum of tension from the extracellular matrix through integrins to the cytoskeleton and ultimately to the nuclear matrix and chromatin might help mediate transcriptional control (see also Section 4.3). In reality, each of these mechanisms is likely to participate in 'outside-in' signalling and the downstream regulation of various cellular behaviours.

Many of the same signalling components that are activated by integrin occupation are stimulated by growth factors, and in many cases, adhesion to the ECM is required for growth factor-induced mitogenesis (148). Platelet-derived growth factor (PDGF), for example, has mitogenic effects on adherent fibroblasts, but fails to stimulate proliferation of cells in suspension despite the efficient activation of PLCγ in both cell populations. This is explained by the observation that adhesion to FN stimulates production of PIP2, the substrate of PLCγ (149). Adhesion to FN therefore serves to 'prime' the cell for PDGF stimulation by generating substrate for a signalling component of that pathway. In addition, PDGF treatment of adherent fibroblasts results in the tyrosine phosphorylation of FAK and paxillin, which is also phosphorylated on tyrosine after integrin ligation (150). Furthermore, Grb2 binding links FAK to the Ras/MAPK pathway (135) and the αvβ3 integrin has been reported to associate with a protein that mediates insulin signalling following ligation of the insulin receptor (151). Thus not only do integrin and some growth factor-mediated signal transduction pathways display similar readouts, but they can also be thought of as functionally interdependent systems.

Integrin-dependent outside-in signalling has also been shown to mediate protection from apoptosis, which is known to be triggered in many cell types that lose their adherence to the ECM (152). This process has also been termed 'anoikis' from the Greek, for 'homelessness' (153). The ability to survive under appropriate substrate adherent conditions is an important mechanism that ensures that cells 'know their place' and do not end up surviving in inappropriate tissues and locations. The most familiar example of the latter being the proliferation of malignant cells that exhibit anchorage-independent growth. Integrins play a key role in transducing signals that are required for normal anchorage-dependent growth. For example, α5β1 adhesion to FN contributes protection from apoptosis by stimulating the expression of the cell survival factor Bcl-2 (154). It has also been suggested that the binding of FAK to PI3-kinase in response to integrin ligation may result in activation of B/Akt, a kinase involved in the inactivation of the cell death programme (155).

3.2 Inside-out signalling

Modulation of integrin extracellular ligand binding by inside-out signals has been documented in many systems, but has been especially well-studied in platelets. Platelets must flow freely in the bloodstream yet be able to respond rapidly to tissue injury by adhering to vessel walls and aggregating to form a haemostatic plug. While the initial binding of platelets to the vessel wall involves a complex repertoire of adhesive interactions, platelet aggregation is dependent on integrin αIIbβ3 interactions with the plasma protein fibrinogen. The dependence of platelet aggregation and haemostasis on αIIbβ3 is dramatically illustrated by patients with Glanzmann's thrombasthenia who suffer from bleeding disorders as a result of insufficient levels of functional αIIbβ3. The possibility that normal platelet activation involves changes in the functional state of αIIbβ3 is supported by the observation that resting platelets express the receptor at their surfaces yet are unable to bind fibrinogen until they are activated. As discussed in Section 2.2, ligand-mimetic and anti-LIBS mAbs have been used to demonstrate that αIIbβ3, like other integrin receptors, undergoes a conformational change when 'activated', which correlates with changes in receptor affinity (156). The identification of mAbs that can recognize and thereby 'report' changes in receptor conformation opened the door for numerous studies designed to address mechanisms of integrin activation by inside-out signals in several systems.

Constitutive fibrinogen or ligand-mimetic mAb binding to αIIbβ3 can be induced by deleting the cytoplasmic tail of αIIb, or by mutating the conserved α subunit membrane proximal sequence GFFKR (156, 157) or the β subunit membrane proximal sequence LLv-iHDR (158). Activation of integrins by disruption of these sequences is energy- and cell type-independent, suggesting the involvement of intrinsic changes in the subunit associations of the heterodimer rather than interactions with other cellular effectors. One model to explain these data is that a salt bridge between the membrane proximal sequences forms a structural constraint that keeps the integrin in a default low affinity state; when the bridge is interrupted a conformational change is propagated to the extracellular ligand binding domains. This intersubunit association has been likened to a 'hinge' and mutations that disrupt the putative association have been termed 'hinge mutants' (148).

The importance of α subunit cytoplasmic tails in regulating integrin binding activity is further underscored by experiments employing chimeric integrins in which the cytoplasmic tail of αIIb was replaced by the cytoplasmic tails of various other α subunits (156, 157). The cytoplasmic tails of α5, α6A, α6B, and α2 conferred a high affinity binding state on αIIbβ3, as assessed by the binding of fibrinogen or a ligand-mimetic mAb. In contrast to hinge mutants, modulation of the chimeric receptors required energy and was cell type specific. This suggests that additional cellular components may interact with the cytoplasmic tails to effect subunit associations and propagate changes in the extracellular conformation of the receptor. Expression of isolated cytoplasmic domains can inhibit ligand binding of active αIIbβ3 chimeras by a mechanism termed 'dominant suppression' suggesting that putative cellular components required to maintain the high affinity state are available in limiting amounts (159). A model of how conformational changes might enhance ligand binding has been proposed by Loftus and Liddington (160) based on structural information available for the αIIb and β3 subunits. In resting platelets the I-domain-like region near the N-terminus of β3 would block the ligand binding β propeller region of αIIb. Following platelet activation, the ligand binding site of αIIb would be exposed and the I-domain of β3 would assume an active ligand binding conformation. The intracellular signals responsible for the modification of αIIbβ3 ligand binding activity are still incompletely understood. Agonists include thrombin, thromboxane A_2, norepinephrine, adenosine diphosphate, and collagen.

While these molecules act through different receptors it appears that one point of convergence in their signalling pathways is that they induce phosphoinositide hydrolysis and the production of IP_3 and diacylglycerol.

Given the importance of the αIIbβ3 cytoplasmic tails in affinity modulation, investigators have focused on potential cytoplasmic tail binding proteins that might regulate receptor functions. One such protein, β3-endonexin, interacts specifically with the β3 cytoplasmic tail and this association increases the affinity state of αIIbβ3 (161). IAP (CD47, Section 2.4) can also associate with β3 but appears to act as a co-stimulator of platelet activation rather than acting directly on αIIbβ3 (162). Molecules other than those that interact directly with integrin cytoplasmic tails also play roles in the modulation of receptor binding activity. H-Ras and R-Ras are highly homologous GTP binding proteins that seem to have opposing effects on integrin activation. Activated R-Ras is reported to activate integrins (163), whereas H-Ras suppresses integrin activation (164). These opposing effects may be mediated by R-Ras antagonization of H-Ras suppression rather than the direct activation of integrins by R-Ras (165).

Platelet activation and the modulation of αIIbβ3 binding activity provide a good model for the regulation of integrin binding activity in general. The regulation must be rapid to allow for haemostasis and tightly controlled to prevent thrombosis. The increased adhesive activity is effected by a number of different agonists and is likely to involve contributions from both changes in affinity and avidity of the receptor. However, platelet activation is only one of many examples of the importance of inside-out integrin signalling in biological processes that involve the dynamic regulation of adhesive interactions. Other examples include leukocyte activation and extravasation, cell migration, keratinocyte differentiation, *Drosophila* development, and *Xenopus laevis* gastrulation (discussed in following sections).

While it appears that affinity modulation is important to the regulation of integrin binding activity, it is also likely that changes in avidity brought about by receptor clustering help to determine adhesive behaviour (e.g. Fig. 3). When αIIbβ3 constructs containing one or two copies of the FKBP binding protein fused to the cytoplasmic tail of αIIb were induced to cluster following the addition of the membrane permeable reagent AP1510, ligand binding increased (166). However, the increase in ligand binding was markedly less than that which could be achieved through affinity modulation alone. In the case of platelets, therefore, it would appear affinity modulation is mostly responsible for activation-dependent changes in ligand binding, whereas changes in avidity contribute but are likely more important for subsequent signalling events that depend on integrin clustering. Interestingly, NPXY is a conserved sequence found in most β subunit cytoplasmic tails that appears to be involved in regulating adhesive activity (167) by recruiting integrins to focal adhesions (168). This further suggests a role in avidity-modulation as opposed to affinity-modulation.

4. Cell biological consequences of integrin functions

The elucidation of specific integrin contributions to various cellular functions is complicated by the size and complexity of the integrin family, and the often overlapping patterns of expression of distinct receptors with 'redundant' ligand binding specificities. A cataloguing of the spatial and temporal expression patterns of integrin subunits in embryos (169), adult tissues, and cells has been an important initial step in approaching questions of integrin involvement in specific processes. Direct analyses of integrin functions, however, have generally proceeded in two general directions. The first is to introduce antibodies or other reagents (e.g. synthetic peptides, and 'dominant-negative' constructs) into a biological system in order to alter endogenous integrin functions. The potential limitations of this approach include incomplete inhibition of function and/or a lack of specificity leading to a loss-of-function of more than a single target heterodimer. Gain-of-function experiments (e.g. overexpression of cDNAs encoding full-length and chimeric subunits), while often informative, are open to questions of physiological significance. The second involves the use of powerful genetic analyses afforded by systems such as the worm, fly, and mouse, which make it possible to investigate function in the absence of a specific integrin subunit (Section 5). Even in this instance, however, the precise functions of a specific receptor combination may be masked by additional integrin-dependent or -independent mechanisms that are able to compensate for the loss-of-function mutation. Nevertheless, both approaches have provided significant insights into the biological roles of integrins. The following sections provide an overview of select integrin functions pieced together from work done in a variety of systems. The reader is also referred to Chapter 9 (Burridge) for additional information related to the general functions of integrins in regulating cell adhesion and spreading.

4.1 Integrin regulation of cell migration

Cell migration is a complex process that involves the co-ordinate regulation of the cytoskeleton and cell–substrate contact at the leading and trailing edges of a cell. Simply put, cells must adhere to an extracellular substrate, generate directed protrusions to initiate movement, and subsequently release adhesions to allow net translocation. Thus, we can think of the migrating cell as performing a delicate balancing act (Fig. 5). If substrate adhesive strength is too great, then the ability to release attachments is compromised and leading (i.e. extension of protrusions) and trailing (i.e. release and retraction) edge behaviours are inhibited. Correspondingly, if adhesion is too low, the cell is unable to spread and form stable protrusive contacts with the substrate. This somewhat paradoxical relationship between adhesion and motility is an area of active investigation that, in recent years, has focused largely on the roles of integrins (170). In this context, integrins function not only as key mechanical links between the ECM and cytoskeleton but also as a site for the regulated assembly/disassembly of adhesive contacts crucial for motility.

The dual nature of integrins as molecular linkers between the extracellular matrix and the cytoskeleton, and as participants in various cell signalling pathways, suggests that these transmembrane receptors are ideally suited for controlling cell migration. The net strength of a given cell's adherence to substrate can be influenced by several integrin-related variables including the types of integrins expressed, their density at

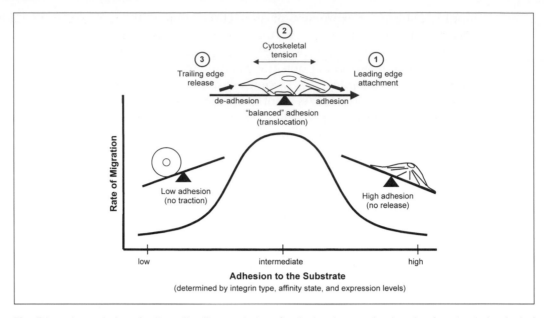

Fig. 5 Integrin regulation of cell motility. The regulation of cell migration may be thought of as the 'balancing' of three identifiable cellular events, each of which involves integrins. (1) A cell must form adhesions with the extracellular matrix at its leading edge. (2) Contractile force must be generated by the cytoskeleton. (3) Adhesion must be released at the rear of the cell to allow net movement to occur. The relationship between cell adhesive strength and migration speed can be envisioned as a bell-shaped curve (171). Cell speed is greatest at intermediate adhesive strength. When adhesive strength is low cells cannot generate the traction needed to locomote. In contrast, at high adhesive strengths a cell becomes anchored to the substrate and is unable to translocate.

the cell surface, and the 'affinity state' of individual receptors (171). Adhesion is also influenced by the structure and composition of the extracellular matrix, which in turn is influenced by integrin-mediated extracellular matrix assembly.

The actin cytoskeleton exerts force on focal adhesions that are formed at the leading edge of the cell where integrin-mediated signals may, in turn, influence cytoskeletal contractility. Integrin ligation can result in myosin light chain phosphorylation by myosin light chain kinase (MLCK) (Fig. 4) and increased contractility both by activation of Rho (172) and/or MAP kinase (138). Cytoskeletal tension (173) and the ligation of integrins on rigid substrates (174) increases the strength of the integrin cytoskeletal linkage. This linkage enables the cell to generate forces necessary for movement. Interestingly, focal adhesions are not stationary structures in immotile cells but, instead, travel centripetally (175). In migratory cells, however, the focal adhesions at the leading edge remain fixed as the cell moves over them, while adhesions at the rear of the cell are motile. Smilenov *et al.* (175) have suggested an intriguing mechanism whereby integrin-containing focal adhesions act like a 'clutch' to regulate cell motility; cytoskeletal tension allows the cell to move forward only when the clutch is engaged.

Focal adhesion turnover is critical to cell migration. Fibroblasts obtained from FAK-deficient mice show increased focal adhesions and a corresponding decrease in motility (176, 177). As discussed, cell movement forward requires that adhesions at the rear of the cell be released to ensure that the cell does not remain tethered in place. The release of adhesions may, in part, be regulated by calcium signalling. Transient increases in intracellular calcium have been reported to be induced by stretch-activated calcium channels (178). This calcium influx may help to mediate detachment by activating the calcium-dependent protease calpain and allowing it to cleave integrin cytoskeletal linkages (179). Increased intracellular calcium might also influence migration by activating other signal transduction molecules such as the phosphatase calcineurin (180).

4.2 Integrin involvement in extracellular matrix assembly

Integrins are important not only for the adhesion of cells but also the macromolecular assembly of ECM (181). The best understood example of this is integrin regulation of fibronectin fibrillogenesis. Fibronectin fibril assembly is initiated when integrins on the cell surface bind soluble fibronectin. Antibodies or peptides that interfere with the binding of fibronectin to integrin $\alpha5\beta1$ prevent the assembly of fibronectin into the matrix (182–184). Cells that lack $\alpha5\beta1$ are unable to assemble fibronectin fibrils but transfection of $\alpha5\beta1$ into these cells is sufficient to allow fibronectin to be incorporated into the matrix (185).

While these studies suggest that $\alpha5\beta1$ is the primary integrin responsible for fibronectin matrix assembly, cells derived from mice lacking $\alpha5$ (186) or $\beta1$ (187) are still able to form fibronectin fibrils. This demonstrates that other integrin hetero-dimers may be able to mediate assembly. Indeed, $\alpha v\beta3$ (187, 188), $\alpha IIb\beta3$ (189), and $\alpha4\beta1$ (181) have each been shown to support fibronectin matrix assembly in the absence of functional $\alpha5\beta1$. None of these receptors, however, is able to function in this capacity unless it is first exogenously activated using either antibodies (i.e. anti-LIBS) or Mn^{2+}, or by mutations in the α subunit cytoplasmic tail that lead to constitutive receptor activation. The activation state of $\alpha5\beta1$ is also likely to be important for the regulation of fibronectin assembly *in vivo*. Sechler *et al.* (190) found that the 'synergy site' located in the ninth type III repeat of fibronectin was necessary for the RGD-dependent incorporation of fibronectin into fibrils. This requirement could be overcome by exogenous activation of integrins using Mn^{2+}, possibly by mimicking the synergy-bound conformation of $\alpha5\beta1$.

Integrins appear to promote fibronectin fibrillogenesis by enhancing fibronectin homotypic binding following the initial binding step. Several sites of fibronectin–fibronectin interaction have been identified including the first five type I repeats, the first type III repeat, the RGD-containing type III 10 repeat, and the C-terminal heparin binding domain composed of the type III 12–14 repeats (191–195). Many of these interaction sites are cryptic in the intact fibronectin molecule where they remain 'hidden' between modules or within the folds of the modules themselves (196). Thus, one way in which integrins may mediate fibrillogenesis is by transmitting cytoskeletally generated tension to bound fibronectin molecules in order to expose these cryptic sites and allow multimerization to occur. A link between the actin cytoskeleton and fibronectin assembly has long been recognized (197). Cells cannot form a fibronectin matrix unless they are able to generate tension on the matrix (198). Recent measurements indicate that cells can stretch fibronectin fibrils up to four times their relaxed length (199). Activation of cytoskeletal contractility via Rho has also been shown to increase fibronectin assembly (200, 201) and to expose cryptic sites within fibronectin (201). This cytoskeletal tension may be transmitted to the fibronectin molecule via the tensin-mediated translocation of occupied α5β1 integrins from focal contacts to matrix contacts (202).

Basement membranes are specialized sheet-like extracellular matrices that underlie epithelial cells. The molecular components of the basement membrane are derived from both epithelial and mesenchymal cells (203). In *C. elegans*, the cells that synthesize type IV collagen are not the same as those that assemble it into basement membranes (204). This suggests that the cells responsible for assembling the basement membrane must have a mechanism for concentrating the components at their cell surfaces. Mouse blastocysts from β1 null embryos or teratomas derived from β1 null embryonic stem cells do not assemble normal basement membranes (205, 206). The conditional knock-out of β1 integrins from keratinocytes in mice leads to a loss of basement membrane at the dermal–epidermal junction (207, 208). The precise roles of integrins in the assembly of basement membranes remain to be established, however, they appear to co-operate with dystroglycan, a non-integrin cell surface receptor that is known to be required for formation of the basement membrane (209).

4.3   Integrin regulation of gene expression, differentiation, and cell proliferation

Integrins have been shown to influence tissue-specific gene expression and differentiation in a number of cellular contexts. There are two different, but not mutually exclusive, models to explain how integrins may affect gene expression. The first model asserts that physical forces can activate transcription (210). In this model integrin ligation and clustering by the ECM causes reorganization of the cytoskeleton, resulting in a change in cell shape and architecture to ultimately produce mechanical forces that regulate such responses as cell growth and gene expression. How mechanical forces generated by the cytoskeleton activate gene expression is not completely understood, but is thought to involve tension-induced changes in the nuclear matrix that regulate physical association of transcription factors and other regulatory proteins with the DNA to be transcribed. The second model posits that integrin–ECM interactions ultimately regulate transcription through biochemical signalling which can occur independently of changes in cell shape (116). As discussed, these interactions, often in synergy with growth factor signals, can activate several different signal transduction pathways that include MAP kinase (211, 212), FAK (213), and GTPases (214) leading to changes in gene expression (see also Section 3). Investigations of ECM-dependent integrin signalling in a variety of cell and tissue types have provided evidence that the 'physical and biochemical' signals discussed above are intimately linked. Physical changes in cell architecture may be thought of as forming a framework along which biochemical signals travel, thus constituting a putative signalling hierarchy (215). The specific contributions of integrin signalling to gene regulation and differentiation have been studied in a number of systems. These include the differentiation of mammary epithelial cells into functional milk-producing glands, the association of human salivary gland cells into ducts and glands that secrete digestive enzymes, the expression of metalloproteinases by rabbit synovial fibroblasts, and the expression of involucrin and subsequent terminal differentiation of human keratinocytes.

Mammary gland epithelial cells cultured on reconstituted basement membrane (Matrigel), or on purified laminin in the presence of lactogenic hormones such as prolactin, will organize into spherical alveoli, differentiate, and express β-casein (216–218). These and other data suggest that both ECM- and soluble factor-induced signals co-operate to effect gene transcription in mammary epithelial cells. Indeed, an enhancer element has been located in the promoter of the β-casein gene that regulates prolactin- and ECM-dependent transcription (219). The details of how ECM- and prolactin-dependent signalling leads to transcription are still being unravelled, but it appears that basement membrane-dependent signals inactivate phosphatases that allow prolactin-dependent phosphorylation of the prolactin receptor, thus initiating a cascade of phosphorylation events ultimately leading to transcriptional activation (220).

Both morphological and biochemical 'readouts' accompany the ECM-dependent activation of β-casein expression (217). The morphological signal induces rounding of the cells, which is necessary for a second biochemical signal needed to activate gene transcription. Blockage of either the morphological or biochemical signals inhibits β-casein expression. Inhibition studies using purified laminin E3 fragments and anti-E3 antibodies showed that cell adhesion to E3 mediates the morphological signal necessary for β-casein transcription (218, 221). The biochemical signal requires integrin clustering and tyrosine phosphorylation. The tyrosine kinase inhibitor genistein is able to block this signal but the cell shape change still occurs, suggesting that tyrosine phosphorylation is not required for the observed change in morphogenesis. Anti-α6 or anti-β1 antibodies also block β-casein expression in rounded cells, implicating α6β4 and a β1-containing integrin in the transduction of laminin-induced signals (221).

Human salivary gland (HSG) epithelial cells grow as undifferentiated cells *in vitro* (222), but when cultured in Matrigel they differentiate, organize into ducts and acini, and express cystatin and α-amylase (223, 224). HSG cell differentiation was inhibited under these conditions by blocking integrin–ECM interactions with antibodies directed against either collagen IV, laminin, integrin α6, or β1. Neutralizing antibodies directed against TGF-β3 are also reported to decrease acinar formation, suggesting that both matrix components and growth factors work in concert to influence HSG differentiation (224). Significant alterations in HSG gene expression also accompany α5β1-mediated adhesion to FN or α2β1- and α3β1-mediated adhesion to collagen I (225). More than 30 genes were shown to be up-regulated upon adhesion to FN or collagen I, a number of which were either FN- or collagen-specific. Some of these same genes were also up-regulated following anti-β1 antibody binding to HSG cells, further suggesting that the regulation of gene expression is integrin-dependent in this system (225).

Fibroblast remodelling of ECM is accompanied by changes in matrix-dependent gene expression, which represent key steps in the regulation of morphogenesis, migration, tissue organization, and invasion and metastasis. Expression of matrix metalloproteinases by rabbit synovial fibroblasts involves FN–integrin interactions (226). Plating of synovial fibroblasts on the 120 kDa fragment of FN, RGD peptides, mixed substrates of FN and tenascin, or antibodies to α5β1 caused the cells to spread and initiate collagenase, gelatinase B, and stromelysin gene expression (226, 227). However, plating of cells on full-length FN inhibited metalloproteinase gene expression suggesting that other sites in FN can negatively regulate expression. Indeed, further analysis showed that adhesion of α4β1 to the CS-1 peptide of the V region suppressed the expression of metalloproteinases induced by the RGD site. This suggests that the balance of signals from α4β1 and α5β1 ultimately influences FN-dependent gene expression (228). A model of α5β1-dependent regulation of collagenase gene expression in spread fibroblasts has been proposed (229). In this scenario, α5β1 binding to the 120 kDa fragment of FN or to immobilized anti-α5β1 antibodies induces transcription of the *c-fos* and *c-jun* transcription factors and directs the localization of these proteins to the nucleus. Once in the nucleus, these two proteins heterodimerize to form the AP1 transcription factor complex, which then binds to the AP1 site in the collagenase gene promoter to drive transcription. Ligation of α5β1 integrins on the surface of synovial fibroblasts in suspension has also been shown to induce collagenase expression (226). These cells remained rounded and displayed a disorganized actin cytoskeleton, in contrast to the spread cells plated on FN fragments or α5β1 antibody. It has since been determined that spread and rounded fibroblasts use two distinct signalling pathways to activate collagenase expression (230). Activation of a Rac-dependent pathway is involved in the α5β1-dependent expression of collagenase in rounded but not spread cells, again highlighting the importance of cell architecture to gene response.

Keratinocyte proliferation and differentiation are also influenced by integrin-dependent interactions with the ECM (231). Keratinocytes in contact with the basement membrane are proliferative but do not express markers of terminal differentiation such as involucrin. Differentiating keratinocytes lose contact with the basement membrane and begin to express differentiation markers as they move suprabasally. Proliferative keratinocytes of the basal layer in contact with the basement membrane can be subdivided into two populations; the stem cells that continue to divide throughout life, and transit amplifying cells, which divide only three to five times before giving rise to daughters that exit the cell cycle, move suprabasally, and terminally differentiate. These populations can be distinguished by their surface expression of β1 integrins with the stem cell population expressing two- to three-fold more β1 integrin than the transit amplifying cells (232). Integrin expression appears to be necessary for keratinocyte proliferation as the conditional knock-out of β1 from mouse skin leads to a reduction in keratinocyte proliferation (207). Expression of a dominant-negative β1 construct in cultured keratinocytes also reduces proliferative potential. This reduction is blocked by constitutive MAP kinase activation (233), suggesting that integrin signalling through MAP kinase may normally be responsible for maintaining the proliferative state. The ectopic expression of β1 in suprabasal keratinocytes of transgenic mice leads to hyperproliferation of the keratinocytes providing further evidence for integrin involvement in the regulation of proliferation. The importance of integrin–ECM interactions in cell survival and anchorage-dependent cell growth has been well-documented in a number of systems. The reader is referred to Section 3 and several excellent reviews for a more detailed handling of this subject (107, 234, 235).

The ligation of β1 integrins has been implicated in maintaining keratinocytes in the undifferentiated state, and several observations suggest that loss of integrin–ECM interaction promotes the terminal differentiation of these cells. Upon commitment to terminal differentiation, α5β1 loses affinity for its ligand fibronectin (236). Differentiating keratinocytes subsequently down-regulate transcription, glycosylation, and cell surface transport of the α5 and β1 subunits (237, 238). Moreover, involucrin expression is induced in keratinocytes grown in suspension, and can be blocked in the presence of soluble fibronectin or anti-β1 antibodies (239). While β1 integrins are clearly involved in these processes, recent studies using β1 conditional knock-outs in mouse skin suggest that down-regulation of β1 expression *per se* is not required to trigger keratinocyte terminal differentiation (207, 208).

5. Integrins in development

The period of metazoan development that begins shortly after fertilization is distinguished by intervals of active cell proliferation, by the precisely timed execution of cell fate decisions, and by the co-ordination of complex morphogenetic movements. This developmental sequence culminates in the establishment of the body plan of the embryo and the formation of major organ systems. Cell adhesive interactions have traditionally been associated with the 'cellular mechanics' of morphogenesis. It is now clear, however, that many adhesion molecules also participate in the specification and regulation of cell growth and differentiation (see also Chapters 3 and 8; Takeichi and Gumbiner, respectively). It is important to consider that early embryonic development is a highly dynamic process that involves extensive cell and tissue rearrangements and often short-term cellular associations. The transient nature of many embryonic cell interactions dictates that adhesive functional activities will, on occasion, vary in both space and time. This is in contrast to the more long-term and stable types of adhesion (e.g. junctional) observed later in development and in many adult tissues (Chapter 8). Both stable and transient cell adhesive associations, however, employ many of the same molecules (e.g. integrins and cadherins).

5.1 Invertebrate integrins: functional insights from sea urchins, worms, and flies

Three integrin β subunits (βG, βL, and βC) (41, 240–242) and three α subunits (41) have been cloned from sea urchins but only one, αSU2, has been characterized functionally (243). Regulation of αSU2 expression correlates with changes in the adhesive behaviour of cells. For example, during gastrulation, primary mesenchyme cells undergo an epithelial–mesenchymal transition and lose their affinity for laminin. This change in adhesive behaviour is correlated with a reduction in expression of αSU2. Thus, regulation of the expression of a given integrin subunit in these embryos may help define adhesive specificity during morphogenesis. The distinct expression patterns of βG and βL during early development suggest that each mediates a different subset of functions. For example βL is expressed on the tips of mesenchyme cells whereas βG is expressed on the endoderm during gastrulation. Antibodies directed against βL interrupt gastrulation presumably by blocking the organization of actin filaments in the bottle cells that normally drive mesenchyme ingression.

As discussed earlier, one beta subunit (β pat-3) and two alpha subunits (α ina-2 and α pat-2) have been identified in the nematode *C. elegans*. Embryos lacking the β pat-3 subunit display the 'paralysed, arrested development at two folds' phenotype (244). The β pat-3 protein is expressed in many locations during embryonic development and in the adult, but is especially prominent in muscle cells, consistent with the phenotype of the mutant embryos. The α ina-1 mutants have defects in axon fasiculation as well as in the morphogenesis of the head, pharynx, vulva, gonad, uterus, and male tail (245). The results are surprising in two respects. Antibody perturbation experiments and integrin localization studies in other organisms predict that integrins are required early in development for gastrulation yet the defects in the α ina-1 and β pat-3 embryos occur following gastrulation. Similarly, antibody inhibition studies (246, 247) had implicated integrins in axon outgrowth, yet it is not growth but the bundling of axons, a process previously unexpected to involve integrins, that is affected in the α ina-1 mutants (245).

Further support for the concept that integrins may confer dynamic tissue-specific adhesive identities at various stages of development comes from studies with *Drosophila*. The *Drosophila* integrins were described originally as 'position-specific antigens' (PS) because of their distribution patterns in imaginal discs. Five α and two β subunits have been identified from the *Drosophila* genome database (248). The βPS subunit is predicted to pair with all five of the PS α subunits, although only three heterodimers consisting of βPS and either αPS1, αPS2, or αPS3 have been purified biochemically (249–251). The second β subunit, βv, is expressed specifically in the midgut endoderm but the identity of the associated α subunit is unknown (252). The loci *multiple edematous wings, inflated, scab/volado,* and *myospheroid* encode αPS1, αPS2, αPS3, and βPS respectively. Initial analyses of embryos mutant at these loci suggested that one of the main functions of the PS integrins is to mechanically mediate attachment between tissues. For example, null alleles of *Myospheroid* result in embryonic lethality. As muscular contractions begin, these embryos rupture at the site of dorsal closure. The muscles lose their attachments to the cuticle and retract resulting in spheroidal embryos (253). In the developing wing, αPS1-containing integrins are found on the dorsal surface of the wing imaginal disc while αPS2 integrins are located on the ventral side. Localized patches of wing epithelia lacking these integrins (i.e. genetic mosaics of mutant cells) result in decreased adhesion between the dorsal and ventral surfaces and the formation of wing blisters (254–256). Interestingly, mutations in the *volado* locus impair olfactory short-term memory, implicating αPS3 in learning (257). It is unclear what function integrin αPS3βPS serves in the generation of memory, although it may act mechanically to generate or maintain synaptic connections or transmit signals important for this purpose.

The control of integrin-dependent adhesive activity in *Drosophila* is likely to involve more than just the 'simple' transcriptional regulation of subunit expression in specific tissues. As reported for some vertebrate developmental systems (236, 258), PS integrin adhesive activity appears to be temporally and spatially regulated post-translationally. For example, expression of a cytoplasmic tail-deleted form of αPS2 results in the appearance of additional, aberrantly located muscle attachments (259). This gain-of-function phenotype suggests that the deleted PS2 integrin can function in a manner analogous to similar integrin hinge-mutations in vertebrates (156). Thus, changes in integrin adhesive function via an inside-out signalling mechanism likely constitute a key regulatory step in the development of normal muscle attachments in *Drosophila*. More recently, integrin signalling has also been implicated in *Drosophila* differentiation and the regulation of gene expression. Adhesion-defective chimeric constructs containing the cytoplasmic tail of the βPS integrin subunit and an irrelevant extracellular and transmembrane domain is sufficient to regulate the expression of at least two integrin responsive genes, presumably by passing an integrin-dependent signal (260). Multimerization of the expressed chimera is required for propagation of this signal.

5.2 Integrin functions in aves and amphibia

Some of the earliest experiments undertaken to analyse the roles of cell–ECM interactions in development made use of chick and frog embryos because of their easy accessibility to experimental manipulation. Typical approaches have included the introduction of function-blocking antibodies, expression of introduced cDNAs and transcripts, and antisense strategies to perturb endogenous protein expression. The results of these studies, while frequently offering novel insights, sometimes differ from that observed following genetic ablation of a specific integrin receptor or its ligand (e.g. in mice, see Section 5.3). In some cases, this may reflect true functional differences among distantly related species and/or differences in the timing or location of an intended perturbation. On the other hand, different compensatory mechanisms may come into play, depending upon the specificity and/or efficacy of the method used to perturb expression.

As discussed in Section 2, studies of chick myoblast adhesion resulted in the preparation of mAbs that blocked cell–ECM adhesion and led, ultimately, to the identification of chick integrins. These and other antibodies together with RGD peptides and antisense strategies have been used to probe the functions of integrins in a variety of chick developmental processes that include somitogenesis (261–263), myogenesis (264), neural crest cell migration (265–268), and angiogenesis–vasculogenesis (269, 270).

Studies of chick myogenesis in particular have provided important insights into how integrin and growth factor mediated signals co-operate to regulate proliferation and differentiation. Both α6β1 (271) and α5β1 (272) are expressed in myoblasts where they mediate different biological responses. Myoblast differentiation is accompanied by the down-regulation of α5β1 functional activity (272), whereas overexpression of this receptor maintains these cells in a proliferative state (105). In contrast, increased expression of α6β1 can induce differentiation while antisense

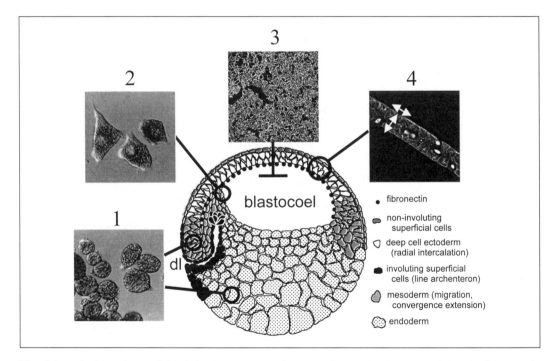

Fig. 6 Integrin-dependent cellular behaviours are spatio-temporally regulated at gastrulation in *Xenopus*. In *Xenopus*, there are at least four integrin α5β1-dependent behaviours that contribute to gastrulation and can be defined in both space and time. (1) Pre-involuted mesoderm, and all other cells in the embryo prior to and during gastrulation, can attach to the RGD site of FN using the α5β1 receptor. (2) As mesoderm involution begins at gastrulation, a putative inside-out signalling mechanism results in α5β1 recognition of the synergy site of FN and an accompanying switch in behaviour of these cells from 'attached' to spread and migratory. (3) The endogenous substrate for mesodermal cell migration is FN, which is assembled preferentially at the surfaces of the cells that line the roof of the blastocoel. α5β1 is required for the assembly of FN fibrils by these cells. (4) α5β1 interactions with FN fibrils are required not only for adhesion and migration of involuting mesoderm but also for the radial intercalation of blastocoel roof cells which helps drive epiboly (*white arrows*). Integrin signalling in this case, likely involves the maintenance of deep cell polarity possibly by influencing the non-canonical wnt signalling pathway through dishevelled (292). (dl) dorsal lip, early gastrula stage embryo depicted.

inhibition of endogenous α6β1 shifts cells to the proliferative state (105). In each case, growth factor responses are critically influenced by the integrin-dependent adhesion of these cells to substrate and the relative ratios of the integrins expressed by them.

The cell and tissue movements associated with gastrulation in several amphibia (esp. *Xenopus*) (273) have been well-documented and, in recent years, both the mechanics of these movements and the signals that control them have become subjects of intense interest. Cadherins and integrins are key players in this morphogenetic event, and cross-talk between these two adhesive systems may be critical to the appropriate regulation of cell motility at gastrulation. Gastrulation movements initiate on the dorsal side of the *Xenopus* embryo with the appearance of bottle cells and the slit-like invagination they produce, termed the dorsal lip of the blastopore (Fig. 6). Superficial cells involute through the blastopore and, coincident with this, there is a dramatic rearrangement of deep cells in this region, which drives the movements of axial and paraxial mesodermal tissues through the process of convergence extension. Cells in this region intercalate mediolaterally and extend at the midline along the anterior–posterior axis. The involuting superficial layer of cells spreads over the inner surface of the blastocoel to form the lining of the primitive gut (archenteron). At the anterior edge of the dorsal involuting tissue is the head mesoderm, which migrates along the roof of the blastocoel. A third movement important in amphibian gastrulation is epiboly, which is the process by which the non-involuting superficial cells spread over the outside of the embryo to form skin and CNS. The cellular basis for this movement is the interdigitation of multilayered deep cells in the radial direction, which results in the spreading of the non-intercalating superficial cell layer (Fig. 6). The extent to which integrin–ECM interactions participate in each of these processes has become increasingly clear in recent years, as discussed below.

Coincident with the onset of gastrulation in amphibia is the appearance of FN fibrils that are preferentially deposited along the roof of the blastocoel, but which are synthesized from maternal transcripts present in all cells of the embryo (274, 275). Both the timing of FN expression and the spatial localization of FN fibrils suggested a possible role for this ECM protein in mesoderm migration at gastrulation (276). This was supported subsequently by a number of studies, in which injection of RGD peptides or blocking antibodies into the blastocoels of these embryos disrupted gastrulation movements (258, 277–279). In addition, *Xenopus* mesodermal cells will spread and migrate on FN substrates *in vitro* (280), and attachment in these assays can be inhibited with RGD peptides or antibodies to FN (280, 281). Not surprisingly, antibodies directed against β1 integrins yield similar results (282). One intriguing observation was that pre-involuting mesoderm cells at the dorsal lip, like all cells in the embryo prior to gastrulation, are able to attach to FN in an RGD-dependent manner (258, 279). However, as these cells involute through the blastopore they gain the further ability to spread and migrate on FN. Moreover, animal cap ectodermal cells, which are not normally fated to form mesoderm may be specified to do so in response to mesoderm-inducing signals such as activin A (258, 283). One of the earliest behavioural responses noted in animal cap cells following activin A induction is the transition to a pro-migratory state on FN (283). Two major integrin-related questions arise from these studies. First, which integrin receptors are expressed around the time of gastrulation and, are their patterns of expression altered during mesoderm induction? Secondly, is differential 'position-specific' expression of integrins responsible for specifying the assembly of FN fibrils on the blastocoel roof and/or triggering mesoderm migration?

The search for integrins that participate in amphibian gastrulation resulted in the identification of subunits that are expressed maternally and others that are first expressed zygotically at the time of the mid-blastula transition (284–286). Both β1 and β3 integrins are expressed in the *Xenopus* egg. Interestingly, β1 integrins are present on the oocyte cell surface but are removed from the oolemma during maturation. At the time of fertilization, no β1 integrins are detected at the egg plasma membrane but are subsequently re-inserted in the new membrane that appears at each successive cleavage (287). This indicates that the proposed ADAM–integrin β1 interaction important in murine fertilization (Chapter 6) is not required for amphibian sperm–egg binding, but it remains unclear whether αVβ? or another integrin(s) is available at the egg surface to participate in this process. Integrin expression in *Xenopus* embryos falls into two general categories; those receptors that are maternally expressed and maintained in all cells through gastrulation/neurulation (84, 287), and those that are zygotically expressed and localized to specific cells and tissues (86, 87, 285, 286, 288). Both α5β1 (84) and an αv-containing integrin (85) are widely expressed at gastrulation in *Xenopus* (Hoffstrom and DeSimone, unpublished observations).

Taken together, current data suggest strongly that α5β1 alone is sufficient to mediate the major FN-dependent behaviours observed at gastrulation in *Xenopus* (Fig. 6). Cell attachment to the CCBD of FN is inhibited by anti-α5β1 blocking mAbs (289), and the switch from FN 'attached' behaviour to spread and migratory behaviour following mesoderm induction/involution involves the α5β1-dependent recognition of the synergy site in the CCBD (279). Synergy site recognition in these cells likely occurs via an inside-out signalling mechanism, which does not involve a change in integrin synthesis or expression at the cell surface (258). Antibodies that block *Xenopus* αv function have no apparent effect on mesodermal cell adhesive behaviour on FN and αv does not compensate for the loss of α5β1 functions following injection of α5β1 mAbs into the blastocoel (Hoffstrom, Dzamba, and DeSimone, in preparation, 289). In contrast, the *Pleurodeles* αv subunit, although widely expressed early on, is progressively lost from the surfaces of most cells during cleavage but is retained on the mesoderm (290), where it may mediate spreading and migration; the role of α5β1 in these embryos may be to initiate FN fibril assembly and promote mesoderm attachment to FN (291). Recent studies with *Xenopus* indicate that α5β1-dependent recognition of FN fibrils by blastocoel roof cells, is required for radial intercalation movements and epiboly (292). If fibril assembly is inhibited, the blastocoel roof fails to thin and superficial cell involution does not occur. Interestingly, α5β1 binding to FN appears to be required for regulating cell polarity in the blastocoel roof, which in turn may be necessary to direct radial intercalation movements (292).

Thus, the *Xenopus* gastrula provides a striking example of a vertebrate developmental system in which multiple adhesive behaviours can be mediated by a single integrin. Precisely how α5β1 function is regulated at gastrulation is unclear, but is likely to involve changes in receptor activation state (affinity modulation), signal transduction, and cytoskeletal interactions in response to inductive signals. Tissue-specific expression of integrins is also important in later morphogenetic behaviours and this first becomes apparent at the end of gastrulation with the appearance of α6β1. *Xenopus* α6β1 is expressed in the neural plate ectoderm. Antisense inhibition of α6 expression in these cells disrupts adhesion to laminin *in vitro* and prevents neural tube closure *in vivo* (86). The inability to progress through neural morphogenesis in the absence of α6β1 appears to arise from a mechanical failure of neurectoderm adhesion to the basal lamina and is not an indirect consequence of alterations in general neural induction and patterning (293).

5.3 Analyses of integrin function in mice

Of the 24 known mammalian integrin subunits, 13 α and seven β subunit genes have thus far been targeted by homologous recombination in mouse embryonic stem cells. Table 3 provides a brief description of phenotypes for each targeted deletion and relevant source citations. The reader is referred to the primary references for a detailed discussion of these mice and to several excellent reviews for additional perspectives (294–296). The following discussion focuses on select examples from these studies but emphasizes the general lessons and concepts that emerge from an overview of this invaluable data set.

Mouse genetics, and knock-out technologies in particular, offer an unquestionably powerful strategy for elucidating specific integrin functions *in vivo*. However, analyses of these mutant mice are often complicated by several factors worth keeping in mind. First, because the integrin family is comprised of multiple receptors with overlapping ligand binding specificities, the phenotypes that result from a targeted deletion of a single subunit gene (and the resultant failure to express a specific functional hetero-dimer) may be 'obscured' by the compensatory functions of another receptor(s). This 'rescue' can be affected by an integrin that is able to substitute for the loss of function in the knock-out but that may not participate in this process under normal circumstances. For example, FN matrix assembly is typically regulated by α5β1, but αvβ3 is able to assemble FN fibrils in embryos lacking α5 (186). Secondly, many receptors share a common subunit and a targeted deletion of a shared subunit is typically more severe than the loss of a single receptor. Thirdly, while knock-out mice have confirmed many of the roles attributed to specific integrin receptors based on earlier non-genetic studies, they have also provided investigators with many significant surprises. These include the elucidation of additional functions and demonstrations that functions thought to be essential for certain processes are, in fact, dispensable.

The inactivation of the gene encoding the β1 subunit (205, 297) results in the loss of at least 11 different heterodimers. As predicted, this results in early embryonic lethality but, notably, fertilization occurs and the embryos implant. Even progression

Table 3 Targeted deletions of integrin subunit genes

Subunit	Phenotype[a]	Reference
α1	Viable, no obvious phenotype	317
α2	NA	
α3	Neonatal lethality, basement membrane defects in kidney and lung	318
α4	Defects in placentation, heart development, haematopoesis, and leukocyte trafficking	304, 319
α5	Embryonic lethal, posterior mesodermal defects	186
α6	Perinatal lethality, epithelial detachment	320
α7	Myopathy, disruption of myotendinous junction	321
α8	Perinatal lethality, defects in kidney morphogenesis	322
α9	Die within 6–10 days of birth, bilateral chylothorax	323
αv	80% embryonic lethal—placental defects, 20% perinatal lethal—intracerebral and intestinal haemorrhages, cleft palates	307
αIIb	Thrombasthenia	324
αD	NA	
αE	Viable, reduced numbers of intraepithelial lymphocytes in gut and vagina	325
αL	Viable, defect in peripheral immune response	326, 327
αM	Viable, increase in neutrophils due to defect in apoptosis, reduced numbers of mast cells	328, 329
αX	NA	
β1	Peri-implantation lethality, ICM deterioration	205, 297
β2	Mild granulocytosis, impaired inflammatory response, skin lesions, defects in T cell proliferation	330, 331
β3	Viable, prolonged bleeding times, platelet aggregation defects, osteosclerosis due to osteoclast dysfunction	308, 332
β4	Perinatal lethality, lack of hemidesmosomes, reduction in epithelial adhesion	333
β5	Viable, no obvious phenotype	334
β6	Viable, inflammation of airway and skin epithelia	335
β7	Viable, gut associated lymphocytes reduced	336
β8	NA	

[a] NA, not available.

to this stage is somewhat surprising given that members of the β1 integrin family have been implicated in both fertilization (298) and implantation (299). However, it is important to note that maternal contributions of β1 mRNA and/or protein could enable these early events to proceed normally, in addition to the possibility of functional compensation by another receptor(s). In the case of implantation, it is possible that αv integrins, possibly up-regulated in the knock-out, facilitate tropho-blast cell attachment to the uterine epithelium. Interestingly, cells deficient in β1 integrins were shown to populate multiple adult organs (with the exception of liver, spleen, bone marrow, thymus, and blood) in chimeric animals composed of wild-type and β1 null cells (297, 300). Aside from indicating a clear role for β1 integrins in aspects of haematopoiesis, these results further suggest that embryonic lethality in the β1 knock-out mice is due to critical β1 functions at very early stages as opposed to later developmental events such as organogenesis.

In contrast to the β1 knock-out, targeted deletions of genes that encode α subunits that pair with β1 vary widely in terms of the timing of observed defects, and their severity and functional consequences. Consider the case of integrins that bind FN; up to 12 different receptors are reported to fall into this category (Table 2), yet the ablation of each individual subunit involved yields a distinct set of phenotypes (Table 3). Notably, the phenotype of the FN null mouse is more severe than the loss of any single integrin heterodimer that recognizes FN (301). This tells us several things. First, each integrin involved clearly mediates a subset of integrin–FN functions. Secondly, FN null mice lack notochords and somites and die at day 8–8.5 of embryogenesis, whereas α5 null mice survive until day 9.5–10 and lack these structures only posteriorly. Given that the α5 knock-out is the most severe of the FN receptors in terms of early embryonic lethality, the extended viability of the α5 null could be explained by the up-regulation of an additional FN binding integrin able to compensate for the loss of α5 in these mice. Alternatively, some critical FN functions in early embryogenesis may be non-integrin dependent (i.e. interactions with syndecans, other ECM molecules, etc.). Thirdly, lack of integrin receptor binding to FN cannot account for the early embryonic lethality of the β1 knock-out, therefore, non-FN receptor integrins play a critical role at these stages.

In order to investigate the possibility of overlapping integrin functions, mice have been generated that are deficient in more than one integrin α subunit. These animals demonstrate that there is often little synergy among different receptors; the phenotypes of double knock-outs can frequently be interpreted as simply 'additive'. To continue with the example of FN receptors, mice deficient in both α5 and α3 integrins exhibit the same phenotypes as those lacking only α5, probably because the loss of α5 causes lethality before α3 is required. The loss of both α5 and α4 results in mice with a summation of the α5 and α4 phenotypes, i.e. they have a defect in placentation in addition to the α5 null defects (302). Because these embryos die early in development, critical overlapping functions that may normally occur later in development would be missed in this analysis. A good example of receptors with unexpected overlapping functions comes from the double knock-out of the laminin binding integrins α3 and α6. These embryos exhibit limb abnormalities and a failure of neural tube closure, which are not seen in either single mutant animal (303). Thus, it is likely that α3 and α6 have compensatory functions that can substitute for the loss of each individual subunit. Comparisons of phenotypes obtained from knock-outs of individual integrin receptors and their ligands can also be instructive. For example, integrin α4β1 acts as both a cell–matrix and cell–cell adhesion molecule by binding either FN or VCAM-1. Prior to the availability of α4 null mice it was thought that the primary function of α4β1 binding to VCAM-1 was to mediate adherence of blood cells to endothelium and that α4β1 adhesion to the V-region of fibronectin would be critical in embryogenesis. However, defects in allantois and chorion fusion, and abnormalities in epicardium and coronary vessels, were observed in the α4 knock-out mouse (304). Very similar phenotypes are seen in VCAM-1 null animals (305, 306), indicating that the early defects in the α4 nulls are not due to a failure of α4β1–FN interactions at these stages. Mouse embryos deficient in both α5 and αv live to day 7.5 of gestation, have gastrulation defects, and generate only two germ layers (302). Therefore, the α5/αv double knock-out is significantly more severe than the phenotypes observed in either integrin subunit-null alone, or in the FN knock-out. This suggests that αv- and α5-dependent interactions with another ligand(s) is required for normal development at the earlier stages.

Antibody inhibition of αvβ3 function is reported to result in both vasculogenesis and angiogenesis defects (269, 270), yet 20% of αv null mouse embryos survive embryogenesis only to die after birth (307). While vasculogenesis clearly proceeds in the absence of αv integrins (80% of embryos that do not survive embryogenesis still exhibit extensive vasculogenesis) angiogenesis is not entirely normal because perinatal lethality is accompanied by evidence of cerebral haemorrhage. Similarly β3 null mice undergo extensive vasculogenesis and even postnatal retinal neovascularization (308). The recurring theme of compensatory mechanisms may again apply in this case and help explain the apparent dispensability αv/β3 integrin functions. Alternatively, in contrast to earlier indications, these integrins may not be critical players in vascular development but may instead be involved in the maintenance of mature vascular tissues and/or postnatal angiogenic functions. In any event, these surprising results underscore the underlying complexities associated with analyses of specific integrin functions *in vivo* and the need for cautious interpretation of data obtained using more traditional 'cellular' approaches.

The complexity of the integrin family with its inherent functional overlap makes elucidation of the precise roles of these receptors in development (and beyond) a particularly challenging problem. The study of organisms such as *C. elegans* and *Drosophila* simplifies analyses of integrin functions considerably, yet ultimately, understanding the specific roles of the 24 or more integrin heterodimers found in mammals will require increasing levels of sophistication. Current techniques being applied to the problem include chimera analyses, multiple knock-outs, and conditional and tissue-specific knock-outs (e.g. Cre/lox system). The application of these strategies will also allow investigators to address the roles of integrins in processes such as organogenesis, which may occur after the point in development at which a 'traditional' knock-out mutation is lethal.

6. Integrins in disease

There are several human hereditary diseases that are caused by mutations in specific integrin subunits. Not surprisingly, the phenotypes bear strong resemblance to similar mutations derived by gene targeting in mice. The correct balance of integrin-mediated adhesive function is often critical to human health; either 'too much or too little' integrin activity can lead to problems. For example, as discussed in Section 3.2, Glanzmann's thrombasthenia is a bleeding disorder characterized by the failure of platelets to aggregate and form clots. The disease arises from mutations in the platelet integrin subunits αIIb or β3. While Glanzmann's patients would benefit from increased αIIbβ3 activity, antagonists of αIIbβ3 are being developed and clinically tested as inhibitors of thrombus formation in patients with coronary artery disease. These agents include mAbs (e.g. Fab fragments), synthetic peptides (e.g. RGD-based), and inorganic inhibitors of αIIbβ3 ligand binding site (309). Because αvβ3 has been implicated in promoting the survival of proliferating endothelial cells during angiogenesis (310) there is also considerable interest in using antagonists of αvβ3 to disrupt the sprouting of blood vessels, which nourish growing tumours and provide routes for metastasis.

Various mutations in β2 integrins lead to the condition known as leukocyte adhesion deficiency or LAD. Neutrophils of LAD patients can roll along activated endothelia in response to inflammatory signals but are unable to arrest and extravasate through the endothelial wall. The result is immunodeficiency and severe, recurrent bacterial infections with a characteristic absence of pus (i.e. no neutrophils). Current treatment modalities include haematopoetic stem cell transplantation but clinical trials are also underway, which involve gene transfer strategies to deliver the wild-type β2 subunit gene into LAD patients (311). Conversely, agents are also being developed that inhibit β2 function and hold promise in the treatment of chronic inflammatory diseases.

Other human diseases involving defects in integrin function include epidermolysis bullosa with pyloric atresia, which is caused by mutations in either subunit of the α6β4 hemidesmosomal integrin and characterized by epithelial blistering (312, 313). Mutations in the laminin receptor α7(β1) have been correlated with congenital myopathy (314). Not surprisingly, based on the likelihood of embryonic lethality, there are no human diseases attributed to mutations in the β1 subunit.

7. Emerging concepts and perspectives

A review of the integrins is necessarily limited due to the sheer complexity and scope of the topic, but a number of key concepts and likely directions for future studies emerge from an overview of research done in this area during the past 15 years. Perhaps chief among these is the realization that integrins mediate not only the mechanical adhesion of cells to their environments but also that they are key players in the bidirectional flow of information, which is critical for directing many types of cellular responses and behaviours. While a great deal is known about the adhesive ligands and other protein complexes that interact with integrins on both sides of the cell membrane, high resolution structures of intact heterodimers and their associated complexes await clarification. Nevertheless, a variety of empirical methods continue to be used to glean information about integrin binding sites and other key functional domains. This work is likely to lead to the availability of additional reagents that will serve as effective antagonists or, in some cases, activators of integrin functions. Aside from helping dissect the functions of integrins in multiple systems, these tools may also be used to treat a wide range of pathological conditions that involve inflammation and immune response, angiogenesis, wound healing, and haemostasis. Future studies will also focus on how integrins fit into the inevitable cross-talk between multiple cell signalling pathways and the other cell adhesion mechanisms available to cells. Together, this information will be key to understanding how cellular responses are dictated by the synergistic effects of multiple adhesive and signalling inputs.

References

1. Hynes, R. O. (1987). Integrins: a family of cell surface receptors. *Cell*, **48**, 549.
2. Hynes, R. O. (1990). *Fibronectins*. New York, Springer–Verlag.
3. Hynes, R. O. (1974). Role of surface alterations in cell transformation: The importance of proteases and surface proteins. *Cell*, **1**, 147.
4. Gahmberg, C. G. and Hakomori, S. (1974). Organization of glycolipids and glycoproteins in surface membranes: Dependency on cell cycle and on transformation. *Biochem. Biophys. Res. Commun.*, **59**, 283.
5. Yamada, K. M., Yamada, S. S., and Pastan, I. (1976). Cell surface protein partially restores morphology, adhesiveness,and contact inhibition of movement to transformed fibroblasts. *Proc. Natl. Acad. Sci. USA*, **73**, 1217.
6. Ali, I. U., Mautner, V. M., Lanza, R. P., and Hynes, R. O. (1977). Restoration of normal morphology, adhesion and cytoskeleton in transformed cells by addition of a transformation-sensitive surface protein. *Cell*, **11**, 115.
7. Hynes, R. O. and Destree, A. T. (1978). Relationships between fibronectin (LETS protein) and actin. *Cell*, **15**, 875.
8. Singer, II. (1979). The fibronexus: a transmembrane association of fibronectin-containing fibers and bundles of 5 nm microfilaments in hamster and human fibroblasts. *Cell*, **16**, 675.
9. Neff, N. T., Lowrey, C., Decker, C., Tovar, A., Damsky, C., Buck, C., *et al.* (1982). A monoclonal antibody detaches embryonic skeletal muscle from extracellular matrices. *J. Cell Biol.*, **95**, 654.
10. Greve, J. M. and Gottlieb, D. I. (1982). Monoclonal antibodies which alter the morphology of cultured chick myogenic cells. *J. Cell. Biochem.*, **18**, 221.
11. Chen, W. T., Hasegawa, E., Hasegawa, T., Weinstock, C., and Yamada, K. M. (1985). Development of cell surface linkage complexes in cultured fibroblasts. *J. Cell Biol.*, **100**, 1103.
12. Damsky, C. H., Knudsen, K. A., Bradley, D., Buck, C. A., and Horwitz, A. F. (1985). Distribution of the cell-substratum attachment (CSAT) antigen on myogenic and fibroblastic cells in culture. *J. Cell Biol.*, **100**, 1528.
13. Brown, P. J. and Juliano, R. L. (1985). Selective inhibition of fibronectin-mediated cell adhesion by monoclonal antibodies to a cell-surface glycoprotein. *Science*, **228**, 1448.
14. Pierschbacher, M. D. and Ruoslahti, E. (1984). Cell attachment activity of fibronectin can be duplicated by small synthetic fragments of the molecule. *Nature*, **309**, 30.
15. Pytela, R., Pierschbacher, M. D., and Ruoslahti, E. (1985). Identification and isolation of a 140 kd cell surface glycoprotein with properties expected of a fibronectin receptor. *Cell*, **40**, 191.
16. Patel, V. P. and Lodish, H. F. (1986). The fibronectin receptor on mammalian erythroid precursor cells: characterization and developmental regulation. *J. Cell Biol.*, **102**, 449.
17. Johansson, S., Forsberg, E., and Lundgren, B. (1987). Comparison of fibronectin receptors from rat hepatocytes and fibroblasts. *J. Biol. Chem.*, **262**, 7819.
18. Pytela, R., Pierschbacher, M. D., and Ruoslahti, E. (1985). A 125/115-kDa cell surface receptor specific for vitronectin interacts with the arginine-glycine-aspartic acid adhesion sequence derived from fibronectin. *Proc. Natl. Acad. Sci. USA*, **82**, 5766.
19. Pytela, R., Pierschbacher, M. D., Ginsberg, M. H., Plow, E. F., and Ruoslahti, E. (1986). Platelet membrane glycoprotein IIb/IIIa: member of a family of Arg-Gly-Asp-specific adhesion receptors. *Science*, **231**, 1559.

20. Dedhar, S., Ruoslahti, E., and Pierschbacher, M. D. (1987). A cell surface receptor complex for collagen type I recognizes the Arg-Gly-Asp sequence. *J. Cell Biol.*, **104**, 585.

21. Knudsen, K. A., Horwitz, A. F., and Buck, C. A. (1985). A monoclonal antibody identifies a glycoprotein complex involved in cell-substratum adhesion. *Exp. Cell Res.*, **157**, 218.

22. Akiyama, S. K., Yamada, S. S., and Yamada, K. M. (1986). Characterization of a 140-kD avian cell surface antigen as a fibronectin-binding molecule. *J. Cell Biol.*, **102**, 442.

23. Hynes, R. O., Marcantonio, E. E., Stepp, M. A., Urry, L. A., and Yee, G. H. (1989). Integrin heterodimer and receptor complexity in avian and mammalian cells. *J. Cell Biol.*, **109**, 409.

24. Buck, C. A. and Horwitz, A. F. (1987). Cell surface receptors for extracellular matrix molecules. *Annu. Rev. Cell Biol.*, **3**, 179.

25. Ruoslahti, E. and Pierschbacher, M. D. (1987). New perspectives in cell adhesion: RGD and integrins. *Science*, **238**, 491.

26. Jennings, L. K. and Phillips, D. R. (1982). Purification of glycoproteins IIb and III from human platelet plasma membranes and characterization of a calcium-dependent glycoprotein IIb-III complex. *J. Biol. Chem.*, **257**, 10458.

27. Hemler, M. E., Huang, C., and Schwarz, L. (1987). The VLA protein family. Characterization of five distinct cell surface heterodimers each with a common 130,000 molecular weight beta subunit. *J. Biol. Chem.*, **262**, 3300.

28. Harris, E. S., McIntyre, T. M., Prescott, S. M., and Zimmerman, G. A. (2000). The leukocyte integrins. *J. Biol. Chem.*, **275**, 23409.

29. Leptin, M., Aebersold, R., and Wilcox, M. (1987). *Drosophila* position-specific antigens resemble the vertebrate fibronectin-receptor family. *EMBO J.*, **6**, 1037.

30. Tamkun, J. W., DeSimone, D. W., Fonda, D., Patel, R. S., Buck, C., Horwitz, A. F., *et al.* (1986). Structure of integrin, a glycoprotein involved in the transmembrane linkage between fibronectin and actin. *Cell*, **46**, 271.

31. Bogaert, T., Brown, N., and Wilcox, M. (1987). The *Drosophila* PS2 antigen is an invertebrate integrin that, like the fibronectin receptor, becomes localized to muscle attachments. *Cell*, **51**, 929.

32. Fitzgerald, L. A., Steiner, B., Rall, S. C. J., Lo, S. S., and Phillips, D. R. (1987). Protein sequence of endothelial glycoprotein IIIa derived from a cDNA clone. Identity with platelet glycoprotein IIIa and similarity to 'integrin'. *J. Biol. Chem.*, **262**, 3936.

33. Kishimoto, T. K., O'Connor, K., Lee, A., Roberts, T. M., and Springer, T. A. (1987). Cloning of the beta subunit of the leukocyte adhesion proteins: homology to an extracellular matrix receptor defines a novel supergene family. *Cell*, **48**, 681.

34. Nermut, M. V., Green, N. M., Eason, P., Yamada, S. S., and Yamada, K. M. (1988). Electron microscopy and structural model of human fibronectin receptor. *EMBO J.*, **7**, 4093.

35. Qu, A. and Leahy, D. J. (1995). Crystal structure of the I-domain from the CD11a/CD18 (LFA-1, alpha L beta 2) integrin. *Proc. Natl. Acad. Sci. USA*, **92**, 10277.

36. Lee, J. O., Rieu, P., Arnaout, M. A., and Liddington, R. (1995). Crystal structure of the A domain from the alpha subunit of integrin CR3 (CD11b/CD18). *Cell*, **80**, 631.

37. Emsley, J., King, S. L., Bergelson, J. M., and Liddington, R. C. (1997). Crystal structure of the I domain from integrin alpha2beta1. *J. Biol. Chem.*, **272**, 28512.

38. Loftus, J. C., O'Toole, T. E., Plow, E. F., Glass, A., Frelinger, A. L. D., and Ginsberg, M. H. (1990). A beta 3 integrin mutation abolishes ligand binding and alters divalent cation-dependent conformation. *Science*, **249**, 915.

39. Weisel, J. W., Nagaswami, C., Vilaire, G., and Bennett, J. S. (1992). Examination of the platelet membrane glycoprotein IIb-IIIa complex and its interaction with fibrinogen and other ligands by electron microscopy. *J. Biol. Chem.*, **267**, 16637.

40. Springer, T. A. (1997). Folding of the N-terminal, ligand-binding region of integrin alpha-subunits into a beta-propeller domain. *Proc. Natl. Acad. Sci. USA*, **94**, 65.

41. Burke, R. D. (1999). Invertebrate integrins: structure, function, and evolution. *Int. Rev. Cytol.*, **191**, 257.

42. Brower, D. L., Brower, S. M., Hayward, D. C., and Ball, E. E. (1997). Molecular evolution of integrins: genes encoding integrin beta subunits from a coral and a sponge. *Proc. Natl. Acad. Sci. USA*, **94**, 9182.

43. Pancer, Z., Kruse, M., Muller, I., and Muller, W. E. (1997). On the origin of Metazoan adhesion receptors: cloning of integrin alpha subunit from the sponge Geodia cydonium. *Mol. Biol. Evol.*, **14**, 391.

44. Randi, A. M. and Hogg, N. (1994). I domain of beta 2 integrin lymphocyte function-associated antigen-1 contains a binding site for ligand intercellular adhesion molecule-1. *J. Biol. Chem.*, **269**, 12395.

45. Ueda, T., Rieu, P., Brayer, J., and Arnaout, M. A. (1994). Identification of the complement iC3b binding site in the beta 2 integrin CR3 (CD11b/CD18). *Proc. Natl. Acad. Sci. USA*, **91**, 10680.

46. Tuckwell, D., Calderwood, D. A., Green, L. J., and Humphries, M. J. (1995). Integrin alpha 2 I-domain is a binding site for collagens. *J. Cell Sci.*, **108**, 1629.

47. Stanley, P. and Hogg, N. (1998). The I domain of integrin LFA-1 interacts with ICAM-1 domain 1 at residue Glu-34 but not Gln-73. *J. Biol. Chem.*, **273**, 3358.

48. Kamata, T., Puzon, W., and Takada, Y. (1994). Identification of putative ligand binding sites within I domain of integrin alpha 2 beta 1 (VLA-2, CD49b/CD29). *J. Biol. Chem.*, **269**, 9659.

49. Rieu, P., Ueda, T., Haruta, I., Sharma, C. P., and Arnaout, M. A. (1994). The A-domain of beta 2 integrin CR3 (CD11b/CD18) is a receptor for the hookworm-derived neutrophil adhesion inhibitor NIF. *J. Cell Biol.*, **127**, 2081.

50. Forsyth, C. B., Plow, E. F., and Zhang, L. (1998). Interaction of the fungal pathogen *Candida albicans* with integrin CD11b/CD18: recognition by the I domain is modulated by the lectin-like domain and the CD18 subunit. *J. Immunol.*, **161**, 6198.

51. Ivaska, J., Kapyla, J., Pentikainen, O., Hoffren, A. M., Hermonen, J., Huttunen, P., *et al.* (1999). A peptide inhibiting the collagen binding function of integrin alpha2I domain. *J. Biol. Chem.*, **274**, 3513.

52. Huang, C., Lu, C., and Springer, T. A. (1997). Folding of the conserved domain but not of flanking regions in the integrin beta2 subunit requires association with the alpha subunit. *Proc. Natl. Acad. Sci. USA*, **94**, 3156.

53. Smith, J. W. and Cheresh, D. A. (1988). The Arg-Gly-Asp binding domain of the vitronectin receptor. Photoaffinity cross-linking implicates amino acid residues 61–203 of the beta subunit. *J. Biol. Chem.*, **263**, 18726.

54. D'Souza, S. E., Ginsberg, M. H., Burke, T. A., Lam, S. C., and Plow, E. F. (1988). Localization of an Arg-Gly-Asp recognition site within an integrin adhesion receptor. *Science*, **242**, 91.

55. Pasqualini, R., Koivunen, E., and Ruoslahti, E. (1995). A peptide isolated from phage display libraries is a structural and functional mimic of an RGD-binding site on integrins. *J. Cell Biol.*, **130**, 1189.

56. Takada, Y., Ylanne, J., Mandelman, D., Puzon, W., and Ginsberg, M. H. (1992). A point mutation of integrin beta 1 subunit blocks binding of alpha 5 beta 1 to fibronectin and invasin but not recruitment to adhesion plaques. *J. Cell Biol.*, **119**, 913.

57. Puzon-McLaughlin, W. and Takada, Y. (1996). Critical residues for ligand binding in an I domain-like structure of the integrin beta1 subunit. *J. Biol. Chem.*, **271**, 20438.

58. Bajt, M. L., Goodman, T., and McGuire, S. L. (1995). Beta 2 (CD18) mutations abolish ligand recognition by I domain integrins LFA-1 (alpha L beta 2, CD11a/CD18) and MAC-1 (alpha M beta 2, CD11b/CD18). *J. Biol. Chem.*, **270**, 94.

59. Dransfield, I., Cabanas, C., Craig, A., and Hogg, N. (1992). Divalent cation regulation of the function of the leukocyte integrin LFA-1. *J. Cell Biol.*, **116**, 219.

60. Bazzoni, G., Shih, D. T., Buck, C. A., and Hemler, M. E. (1995). Monoclonal antibody 9EG7 defines a novel beta 1 integrin epitope induced by soluble ligand and manganese, but inhibited by calcium. *J. Biol. Chem.*, **270**, 25570.

61. Du, X. P., Plow, E. F., Frelinger, A. L., 3rd, O'Toole, T. E., Loftus, J. C., and Ginsberg, M. H. (1991). Ligands 'activate' integrin alpha IIb beta 3 (platelet GPIIb-IIIa). *Cell*, **65**, 409.

62. Bazzoni, G. and Hemler, M. E. (1998). Are changes in integrin affinity and conformation overemphasized? *Trends Biochem. Sci.*, **23**, 30.

63. Liu, S., Calderwood, D. A., and Ginsberg, M. H. (2000). Integrin cytoplasmic domain-binding proteins. *J. Cell Sci.*, **113**, 3563.

64. Schaller, M. D., Otey, C. A., Hildebrand, J. D., and Parsons, J. T. (1995). Focal adhesion kinase and paxillin bind to peptides mimicking beta integrin cytoplasmic domains. *J. Cell Biol.*, **130**, 1181.

65. Hannigan, G. E., Leung-Hagesteijn, C., Fitz-Gibbon, L., Coppolino, M. G., Radeva, G., Filmus, J., *et al.* (1996). Regulation of cell adhesion and anchorage-dependent growth by a new beta 1-integrin-linked protein kinase. *Nature*, **379**, 91.

66. Kolanus, W., Nagel, W., Schiller, B., Zeitlmann, L., Godar, S., Stockinger, H., *et al.* (1996). Alpha L beta 2 integrin/LFA-1 binding to ICAM-1 induced by cytohesin-1, a cytoplasmic regulatory molecule. *Cell*, **86**, 233.

67. Liu, S. and Ginsberg, M. H. (2000). Paxillin binding to a conserved sequence motif in the alpha 4 integrin cytoplasmic domain. *J. Biol. Chem.*, **275**, 22736.

68. Wary, K. K., Mainiero, F., Isakoff, S. J., Marcantonio, E. E., and Giancotti, F. G. (1996). The adaptor protein Shc couples a class of integrins to the control of cell cycle progression. *Cell*, **87**, 733.

69. Horwitz, A., Duggan, K., Buck, C., Beckerle, M. C., and Burridge, K. (1986). Interaction of plasma membrane fibronectin receptor with talin–a transmembrane linkage. *Nature*, **320**, 531.

70. Otey, C. A., Pavalko, F. M., and Burridge, K. (1990). An interaction between alpha-actinin and the beta 1 integrin subunit *in vitro*. *J. Cell Biol.*, **111**, 721.

71. Sharma, C. P., Ezzell, R. M., and Arnaout, M. A. (1995). Direct interaction of filamin (ABP-280) with the beta 2-integrin subunit CD18. *J. Immunol.*, **154**, 3461.

72. Solowska, J., Guan, J. L., Marcantonio, E. E., Trevithick, J. E., Buck, C. A., and Hynes, R. O. (1989). Expression of normal and mutant avian integrin subunits in rodent cells. *J. Cell Biol.*, **109**, 853.

73. Hayashi, Y., Haimovich, B., Reszka, A., Boettiger, D., and Horwitz, A. (1990). Expression and function of chicken integrin beta 1 subunit and its cytoplasmic domain mutants in mouse NIH 3T3 cells. *J. Cell Biol.*, **110**, 175.

74. LaFlamme, S. E., Akiyama, S. K., and Yamada, K. M. (1992). Regulation of fibronectin receptor distribution. *J. Cell Biol.*, **117**, 437.

75. LaFlamme, S. E., Thomas, L. A., Yamada, S. S., and Yamada, K. M. (1994). Single subunit chimeric integrins as mimics and inhibitors of endogenous integrin functions in receptor localization, cell spreading and migration, and matrix assembly. *J. Cell Biol.*, **126**, 1287.

76. Akiyama, S. K., Yamada, S. S., Yamada, K. M., and LaFlamme, S. E. (1994). Transmembrane signal transduction by integrin cytoplasmic domains expressed in single-subunit chimeras. *J. Biol. Chem.*, **269**, 15961.

77. Lukashev, M. E., Sheppard, D., and Pytela, R. (1994). Disruption of integrin function and induction of tyrosine phosphorylation by the autonomously expressed beta 1 integrin cytoplasmic domain. *J. Biol. Chem.*, **269**, 18311.

78. Hogervorst, F., Kuikman, I., von dem Borne, A. E., and Sonnenberg, A. (1990). Cloning and sequence analysis of beta-4 cDNA: an integrin subunit that contains a unique 118 kd cytoplasmic domain. *EMBO J.*, **9**, 765.

79. Suzuki, S. and Naitoh, Y. (1990). Amino acid sequence of a novel integrin beta 4 subunit and primary expression of the mRNA in epithelial cells. *EMBO J.*, **9**, 757.

80. Tamura, R. N., Rozzo, C., Starr, L., Chambers, J., Reichardt, L. F., Cooper, H. M., *et al.* (1990). Epithelial integrin alpha 6 beta 4: complete primary structure of alpha 6 and variant forms of beta 4. *J. Cell Biol.*, **111**, 1593.

81. Stepp, M. A., Spurr-Michaud, S., Tisdale, A., Elwell, J., and Gipson, I. K. (1990). Alpha 6 beta 4 integrin heterodimer is a component of hemidesmosomes. *Proc. Natl. Acad. Sci. USA*, **87**, 8970.

82. Spinardi, L., Ren, Y. L., Sanders, R., and Giancotti, F. G. (1993). The beta 4 subunit cytoplasmic domain mediates the interaction of alpha 6 beta 4 integrin with the cytoskeleton of hemidesmosomes. *Mol. Biol. Cell*, **4**, 871.

83. Whittaker, C. A. and DeSimone, D. W. (1998). Molecular cloning and developmental expression of the *Xenopus* homolog of integrin α4. *Ann. NY Acad. Sci.*, **857**, 56.

84. Joos, T. O., Whittaker, C. A., Meng, F., DeSimone, D. W., Gnau, V., and Hausen, P. (1995). Integrin alpha 5 during early development of *Xenopus laevis*. *Mech. Dev.*, **50**, 187.

85. Joos, T. O., Reintsch, W. E., Brinker, A., Klein, C., and Hausen, P. (1998). Cloning of the *Xenopus* integrin alpha(v) subunit and analysis of its distribution during early development. *Int. J. Dev. Biol.*, **42**, 171.

86. Lallier, T., Whittaker, C., and DeSimone, D. (1996). Integrin alpha6 expression is required for early nervous system development in *Xenopus laevis*. *Development*, **122**, 2539.

87. Meng, F., Whittaker, C. A., Ransom, D. G., and DeSimone, D. W. (1997). Cloning and characterization of cDNAs encoding the integrin alpha2 and alpha3 subunits from *Xenopus laevis*. *Mech. Dev.*, **67**, 141.

88. Rojiani, M. V., Finlay, B. B., Gray, V., and Dedhar, S. (1991). *In vitro* interaction of a polypeptide homologous to human Ro/SS-A antigen (calreticulin) with a highly conserved amino acid sequence in the cytoplasmic domain of integrin alpha subunits. *Biochemistry*, **30**, 9859.

89. Wixler, V., Laplantine, E., Geerts, D., Sonnenberg, A., Petersohn, D., Eckes, B., *et al.* (1999). Identification of novel interaction partners for the conserved membrane proximal region of alpha-integrin cytoplasmic domains. *FEBS Lett.*, **445**, 351.

90. Naik, U. P., Patel, P. M., and Parise, L. V. (1997). Identification of a novel calcium-binding protein that interacts with the integrin alphaIIb cytoplasmic domain. *J. Biol. Chem.*, **272**, 4651.

91. Wixler, V., Geerts, D., Laplantine, E., Westhoff, D., Smyth, N., Aumailley, M., *et al.* (2000). The LIM-only protein DRAL/FHL2 binds to the cytoplasmic domain of several alpha and beta integrin chains and is recruited to adhesion complexes. *J. Biol. Chem.*, **275**, 33669.

92. Coppolino, M. G., Woodside, M. J., Demaurex, N., Grinstein, S., St-Arnaud, R., and Dedhar, S. (1997). Calreticulin is essential for integrin-mediated calcium signalling and cell adhesion. *Nature*, **386**, 843.

93. Wary, K. K., Mariotti, A., Zurzolo, C., and Giancotti, F. G. (1998). A requirement for caveolin-1 and associated kinase Fyn in integrin signalling and anchorage-dependent cell growth. *Cell*, **94**, 625.

94. Liu, S., Thomas, S. M., Woodside, D. G., Rose, D. M., Kiosses, W. B., Pfaff, M., *et al.* (1999). Binding of paxillin to alpha4 integrins modifies integrin-dependent biological responses. *Nature*, **402**, 676.

95. Chan, B. M., Kassner, P. D., Schiro, J. A., Byers, H. R., Kupper, T. S., and Hemler, M. E. (1992). Distinct cellular functions mediated by different VLA integrin alpha subunit cytoplasmic domains. *Cell*, **68**, 1051.

96. Kassner, P. D., Alon, R., Springer, T. A., and Hemler, M. E. (1995). Specialized functional properties of the integrin alpha 4 cytoplasmic domain. *Mol. Biol. Cell*, **6**, 661.

97. Weber, K. S., Klickstein, L. B., and Weber, C. (1999). Specific activation of leukocyte beta2 integrins lymphocyte function-associated antigen-1 and Mac-1 by chemokines mediated by distinct pathways via the alpha subunit cytoplasmic domains. *Mol. Biol. Cell*, **10**, 861.

98. Keely, P. J., Rusyn, E. V., Cox, A. D., and Parise, L. V. (1999). R-Ras signals through specific integrin alpha cytoplasmic domains to promote migration and invasion of breast epithelial cells. *J. Cell Biol.*, **145**, 1077.

99. Fornaro, M. and Languino, L. R. (1997). Alternatively spliced variants: a new view of the integrin cytoplasmic domain. *Matrix Biol.*, **16**, 185.

100. Belkin, A. M., Zhidkova, N. I., Balzac, F., Altruda, F., Tomatis, D., Maier, A., *et al.* (1996). Beta 1D integrin displaces the beta 1A isoform in striated muscles: localization at junctional structures and signalling potential in nonmuscle cells. *J. Cell Biol.*, **132**, 211.

101. Balzac, F., Belkin, A. M., Koteliansky, V. E., Balabanov, Y. V., Altruda, F., Silengo, L., *et al.* (1993). Expression and functional analysis of a cytoplasmic domain variant of the beta 1 integrin subunit. *J. Cell Biol.*, **121**, 171.

102. Balzac, F., Retta, S. F., Albini, A., Melchiorri, A., Koteliansky, V. E., Geuna, M., *et al.* (1994). Expression of beta 1B integrin isoform in CHO cells results in a dominant negative effect on cell adhesion and motility. *J. Cell Biol.*, **127**, 557.

103. Meredith, J., Jr., Takada, Y., Fornaro, M., Languino, L. R., and Schwartz, M. A. (1995). Inhibition of cell cycle progression by the alternatively spliced integrin beta 1C. *Science*, **269**, 1570.

104. Shaw, L. M. and Mercurio, A. M. (1994). Regulation of cellular interactions with laminin by integrin cytoplasmic domains: the A and B structural variants of the alpha 6 beta 1 integrin differentially modulate the adhesive strength, morphology, and migration of macrophages. *Mol. Biol. Cell*, **5**, 679.

105. Sastry, S. K., Lakonishok, M., Thomas, D. A., Muschler, J., and Horwitz, A. F. (1996). Integrin alpha subunit ratios, cytoplasmic domains, and growth factor synergy regulate muscle proliferation and differentiation. *J. Cell Biol.*, **133**, 169.

106. Fornaro, M., Manzotti, M., Tallini, G., Slear, A. E., Bosari, S., Ruoslahti, E., *et al.* (1998). Beta1C integrin in epithelial cells correlates with a nonproliferative phenotype: forced expression of beta1C inhibits prostate epithelial cell proliferation. *Am. J. Pathol.*, **153**, 1079.

107. Giancotti, F. G. and Ruoslahti, E. (1999). Integrin signalling. *Science*, **285**, 1028.

108. Hemler, M. E. (1998). Integrin associated proteins. *Curr. Opin. Cell Biol.*, **10**, 578.

109. Woods, A. and Couchman, J. R. (2000). Integrin modulation by lateral association. *J. Biol. Chem.*, **275**, 24233.

110. Lindberg, F. P., Gresham, H. D., Schwarz, E., and Brown, E. J. (1993). Molecular cloning of integrin-associated protein: an immunoglobulin family member with multiple

membrane-spanning domains implicated in alpha v beta 3-dependent ligand binding. *J. Cell Biol.*, **123**, 485.

111. Lindberg, F. P., Bullard, D. C., Caver, T. E., Gresham, H. D., Beaudet, A. L., and Brown, E. J. (1996). Decreased resistance to bacterial infection and granulocyte defects in IAP-deficient mice. *Science*, **274**, 795.

112. Lindberg, F. P., Gresham, H. D., Reinhold, M. I., and Brown, E. J. (1996). Integrin-associated protein immunoglobulin domain is necessary for efficient vitronectin bead binding. *J. Cell Biol.*, **134**, 1313.

113. Zhang, X. A., Bontrager, A. L., and Hemler, M. E. (2001). TM4SF proteins associate with activated PKC and Link PKC to specific beta1 integrins. *J. Biol. Chem.*, **26**, 26.

114. Yauch, R. L. and Hemler, M. E. (2000). Specific interactions among transmembrane 4 superfamily (TM4SF) proteins and phosphoinositide 4-kinase. *Biochem. J.*, **351** Pt 3, 629.

115. Chen, M. S., Tung, K. S., Coonrod, S. A., Takahashi, Y., Bigler, D., Chang, A., *et al.* (1999). Role of the integrin-associated protein CD9 in binding between sperm ADAM 2 and the egg integrin alpha6beta1: implications for murine fertilization. *Proc. Natl. Acad. Sci. USA*, **96**, 11830.

116. Schwartz, M. A., Schaller, M. D., and Ginsberg, M. H. (1995). Integrins: emerging paradigms of signal transduction. *Annu. Rev. Cell Dev. Biol.*, **11**, 549.

117. Schlaepfer, D. D. and Hunter, T. (1998). Integrin signalling and tyrosine phosphorylation: just the FAKs? *Trends Cell Biol.*, **8**, 151.

118. Howe, A., Aplin, A. E., Alahari, S. K., and Juliano, R. L. (1998). Integrin signalling and cell growth control. *Curr. Opin. Cell Biol.*, **10**, 220.

119. Coppolino, M. G. and Dedhar, S. (2000). Bi-directional signal transduction by integrin receptors. *Int. J. Biochem. Cell Biol.*, **32**, 171.

120. Ingber, D. E., Prusty, D., Frangioni, J. V., Cragoe, E. J., Jr., Lechene, C., and Schwartz, M. A. (1990). Control of intracellular pH and growth by fibronectin in capillary endothelial cells. *J. Cell Biol.*, **110**, 1803.

121. Schwartz, M. A., Ingber, D. E., Lawrence, M., Springer, T. A., and Lechene, C. (1991). Multiple integrins share the ability to induce elevation of intracellular pH. *Exp. Cell Res.*, **195**, 533.

122. Schwartz, M. A., Lechene, C., and Ingber, D. E. (1991). Insoluble fibronectin activates the Na/H antiporter by clustering and immobilizing integrin alpha 5 beta 1, independent of cell shape. *Proc. Natl. Acad. Sci. USA*, **88**, 7849.

123. Fuortes, M., Jin, W. W., and Nathan, C. (1993). Adhesion-dependent protein tyrosine phosphorylation in neutrophils treated with tumor necrosis factor. *J. Cell Biol.*, **120**, 777.

124. Nathan, C. and Sanchez, E. (1990). Tumor necrosis factor and CD11/CD18 (beta 2) integrins act synergistically to lower cAMP in human neutrophils. *J. Cell Biol.*, **111**, 2171.

125. Ng-Sikorski, J., Andersson, R., Patarroyo, M., and Andersson, T. (1991). Calcium signalling capacity of the CD11b/CD18 integrin on human neutrophils. *Exp. Cell Res.*, **195**, 504.

126. Jaconi, M. E., Theler, J. M., Schlegel, W., Appel, R. D., Wright, S. D., and Lew, P. D. (1991). Multiple elevations of cytosolic-free Ca^{2+} in human neutrophils: initiation by adherence receptors of the integrin family. *J. Cell Biol.*, **112**, 1249.

127. Zhu, X. and Assoian, R. K. (1995). Integrin-dependent activation of MAP kinase: a link to shape-dependent cell proliferation. *Mol. Biol. Cell*, **6**, 273.

128. Guan, J. L., Trevithick, J. E., and Hynes, R. O. (1991). Fibronectin/integrin interaction induces tyrosine phosphorylation of a 120-kDa protein. *Cell. Regul.*, **2**, 951.

129. Kornberg, L., Earp, H. S., Parsons, J. T., Schaller, M., and Juliano, R. L. (1992). Cell adhesion or integrin clustering increases phosphorylation of a focal adhesion-associated tyrosine kinase. *J. Biol. Chem.*, **267**, 23439.

130. Burridge, K., Turner, C. E., and Romer, L. H. (1992). Tyrosine phosphorylation of paxillin and pp125FAK accompanies cell adhesion to extracellular matrix: a role in cytoskeletal assembly. *J. Cell Biol.*, **119**, 893.

131. Guan, J. L. and Shalloway, D. (1992). Regulation of focal adhesion-associated protein tyrosine kinase by both cellular adhesion and oncogenic transformation. *Nature*, **358**, 690.

132. Miyamoto, S., Akiyama, S. K., and Yamada, K. M. (1995). Synergistic roles for receptor occupancy and aggregation in integrin transmembrane function. *Science*, **267**, 883.

133. Schaller, M. D., Hildebrand, J. D., Shannon, J. D., Fox, J. W., Vines, R. R., and Parsons, J. T. (1994). Autophosphorylation of the focal adhesion kinase, pp125FAK, directs SH2-dependent binding of pp60src. *Mol. Cell. Biol.*, **14**, 1680.

134. Cobb, B. S., Schaller, M. D., Leu, T. H., and Parsons, J. T. (1994). Stable association of pp60src and pp59fyn with the focal adhesion- associated protein tyrosine kinase, pp125FAK. *Mol. Cell. Biol.*, **14**, 147.

135. Schlaepfer, D. D., Hanks, S. K., Hunter, T., and van der Geer, P. (1994). Integrin-mediated signal transduction linked to Ras pathway by GRB2 binding to focal adhesion kinase. *Nature*, **372**, 786.

136. Hildebrand, J. D., Schaller, M. D., and Parsons, J. T. (1995). Paxillin, a tyrosine phosphorylated focal adhesion-associated protein binds to the carboxyl terminal domain of focal adhesion kinase. *Mol. Biol. Cell*, **6**, 637.

137. Schlaepfer, D. D., Broome, M. A., and Hunter, T. (1997). Fibronectin-stimulated signalling from a focal adhesion kinase-c-Src complex: involvement of the Grb2, p130cas, and Nck adaptor proteins. *Mol. Cell. Biol.*, **17**, 1702.

138. Klemke, R. L., Cai, S., Giannini, A. L., Gallagher, P. J., de Lanerolle, P., and Cheresh, D. A. (1997). Regulation of cell motility by mitogen-activated protein kinase. *J. Cell Biol.*, **137**, 481.

139. Cary, L. A., Han, D. C., Polte, T. R., Hanks, S. K., and Guan, J. L. (1998). Identification of p130Cas as a mediator of focal adhesion kinase- promoted cell migration. *J. Cell Biol.*, **140**, 211.

140. Barry, S. T., Flinn, H. M., Humphries, M. J., Critchley, D. R., and Ridley, A. J. (1997). Requirement for Rho in integrin signalling. *Cell Adhes. Commun.*, **4**, 387.

141. Clark, E. A., King, W. G., Brugge, J. S., Symons, M., and Hynes, R. O. (1998). Integrin-mediated signals regulated by members of the rho family of GTPases. *J. Cell Biol.*, **142**, 573.

142. Price, L. S., Leng, J., Schwartz, M. A., and Bokoch, G. M. (1998). Activation of Rac and Cdc42 by integrins mediates cell spreading. *Mol. Biol. Cell*, **9**, 1863.

143. Klemke, R. L., Leng, J., Molander, R., Brooks, P. C., Vuori, K., and Cheresh, D. A. (1998). CAS/Crk coupling serves as a 'molecular switch' for induction of cell migration. *J. Cell Biol.*, **140**, 961.

144. Plopper, G. E., McNamee, H. P., Dike, L. E., Bojanowski, K., and Ingber, D. E. (1995). Convergence of integrin and growth factor receptor signalling pathways within the focal adhesion complex. *Mol. Biol. Cell*, **6**, 1349.

145. Miyamoto, S., Teramoto, H., Coso, O. A., Gutkind, J. S., Burbelo, P. D., Akiyama, S. K., *et al.* (1995). Integrin function: molecular hierarchies of cytoskeletal and signalling molecules. *J. Cell Biol.*, **131**, 791.

146. Miyamoto, S., Teramoto, H., Gutkind, J. S., and Yamada, K. M. (1996). Integrins can collaborate with growth factors for phosphorylation of receptor tyrosine kinases and

MAP kinase activation: roles of integrin aggregation and occupancy of receptors. *J. Cell Biol.*, **135**, 1633.

147. Klinghoffer, R. A., Sachsenmaier, C., Cooper, J. A., and Soriano, P. (1999). Src family kinases are required for integrin but not PDGFR signal transduction. *EMBO J.*, **18**, 2459.

148. Schwartz, M. A. and Ingber, D. E. (1994). Integrating with integrins. *Mol. Biol. Cell*, **5**, 389.

149. McNamee, H. P., Ingber, D. E., and Schwartz, M. A. (1993). Adhesion to fibronectin stimulates inositol lipid synthesis and enhances PDGF-induced inositol lipid breakdown. *J. Cell Biol.*, **121**, 673.

150. Rankin, S. and Rozengurt, E. (1994). Platelet-derived growth factor modulation of focal adhesion kinase (p125FAK) and paxillin tyrosine phosphorylation in Swiss 3T3 cells. Bell-shaped dose response and cross-talk with bombesin. *J. Biol. Chem.*, **269**, 704.

151. Vuori, K. and Ruoslahti, E. (1994). Association of insulin receptor substrate-1 with integrins. *Science*, **266**, 1576.

152. Meredith, J. E., Jr., Fazeli, B., and Schwartz, M. A. (1993). The extracellular matrix as a cell survival factor. *Mol. Biol. Cell*, **4**, 953.

153. Frisch, S. M. and Francis, H. (1994). Disruption of epithelial cell-matrix interactions induces apoptosis. *J. Cell Biol.*, **124**, 619.

154. Zhang, Z., Vuori, K., Reed, J. C., and Ruoslahti, E. (1995). The alpha 5 beta 1 integrin supports survival of cells on fibronectin and up-regulates Bcl-2 expression. *Proc. Natl. Acad. Sci. USA*, **92**, 6161.

155. Khwaja, A. and Downward, J. (1997). Lack of correlation between activation of Jun-NH2-terminal kinase and induction of apoptosis after detachment of epithelial cells. *J. Cell Biol.*, **139**, 1017.

156. O'Toole, T. E., Katagiri, Y., Faull, R. J., Peter, K., Tamura, R., Quaranta, V., *et al.* (1994). Integrin cytoplasmic domains mediate inside-out signal transduction. *J. Cell Biol.*, **124**, 1047.

157. O'Toole, T. E., Mandelman, D., Forsyth, J., Shattil, S. J., Plow, E. F., and Ginsberg, M. H. (1991). Modulation of the affinity of integrin alpha IIb beta 3 (GPIIb-IIIa) by the cytoplasmic domain of alpha IIb. *Science*, **254**, 845.

158. Hughes, P. E., Diaz-Gonzalez, F., Leong, L., Wu, C., McDonald, J. A., Shattil, S. J., *et al.* (1996). Breaking the integrin hinge. A defined structural constraint regulates integrin signalling. *J. Biol. Chem.*, **271**, 6571.

159. Chen, Y. P., O'Toole, T. E., Shipley, T., Forsyth, J., LaFlamme, S. E., Yamada, K. M., *et al.* (1994). 'Inside-out' signal transduction inhibited by isolated integrin cytoplasmic domains. *J. Biol. Chem.*, **269**, 18307.

160. Loftus, J. C. and Liddington, R. C. (1997). New insights into integrin-ligand interaction. *J. Clin. Invest.*, **100**, S77.

161. Kashiwagi, H., Schwartz, M. A., Eigenthaler, M., Davis, K. A., Ginsberg, M. H., and Shattil, S. J. (1997). Affinity modulation of platelet integrin alphaIIbbeta3 by beta3-endonexin, a selective binding partner of the beta3 integrin cytoplasmic tail. *J. Cell Biol.*, **137**, 1433.

162. Chung, J., Gao, A. G., and Frazier, W. A. (1997). Thrombspondin acts via integrin-associated protein to activate the platelet integrin alphaIIbbeta3. *J. Biol. Chem.*, **272**, 14740.

163. Zhang, Z., Vuori, K., Wang, H., Reed, J. C., and Ruoslahti, E. (1996). Integrin activation by R-ras. *Cell*, **85**, 61.

164. Hughes, P. E., Renshaw, M. W., Pfaff, M., Forsyth, J., Keivens, V. M., Schwartz, M. A., *et al.* (1997). Suppression of integrin activation: a novel function of a Ras/Raf- initiated MAP kinase pathway. *Cell*, **88**, 521.

165. Sethi, T., Ginsberg, M. H., Downward, J., and Hughes, P. E. (1999). The small GTP-binding protein R-Ras can influence integrin activation by antagonizing a Ras/Raf-initiated integrin suppression pathway. *Mol. Biol. Cell*, **10**, 1799.

166. Hato, T., Pampori, N., and Shattil, S. J. (1998). Complementary roles for receptor clustering and conformational change in the adhesive and signalling functions of integrin alphaIIb beta3. *J. Cell Biol.*, **141**, 1685.

167. O'Toole, T. E., Ylanne, J., and Culley, B. M. (1995). Regulation of integrin affinity states through an NPXY motif in the beta subunit cytoplasmic domain. *J. Biol. Chem.*, **270**, 8553.

168. Vignoud, L., Albiges-Rizo, C., Frachet, P., and Block, M. R. (1997). NPXY motifs control the recruitment of the alpha5beta1 integrin in focal adhesions independently of the association of talin with the beta1 chain. *J. Cell Sci.*, **110**, 1421.

169. Lallier, T. E., Hens, D. W., and DeSimone, D. W. (1994). Integrins in development. In *Integrins: molecular and biological responses to the extracellular matrix* (ed. D. A. Cheresh and R. P. Mecham), p. 111. Academic Press, San Diego.

170. Horwitz, A. R. and Parsons, J. T. (1999). Cell migration–movin' on. *Science*, **286**, 1102.

171. Palecek, S. P., Loftus, J. C., Ginsberg, M. H., Lauffenburger, D. A., and Horwitz, A. F. (1997). Integrin-ligand binding properties govern cell migration speed through cell-substratum adhesiveness. *Nature*, **385**, 537.

172. Ren, X. D., Kiosses, W. B., and Schwartz, M. A. (1999). Regulation of the small GTP-binding protein Rho by cell adhesion and the cytoskeleton. *EMBO J.*, **18**, 578.

173. Chrzanowska-Wodnicka, M. and Burridge, K. (1996). Rho-stimulated contractility drives the formation of stress fibers and focal adhesions. *J. Cell Biol.*, **133**, 1403.

174. Choquet, D., Felsenfeld, D. P., and Sheetz, M. P. (1997). Extracellular matrix rigidity causes strengthening of integrin-cytoskeleton linkages. *Cell*, **88**, 39.

175. Smilenov, L. B., Mikhailov, A., Pelham, R. J., Marcantonio, E. E., and Gundersen, G. G. (1999). Focal adhesion motility revealed in stationary fibroblasts. *Science*, **286**, 1172.

176. Ilic, D., Furuta, Y., Kanazawa, S., Takeda, N., Sobue, K., Nakatsuji, N., *et al.* (1995). Reduced cell motility and enhanced focal adhesion contact formation in cells from FAK-deficient mice. *Nature*, **377**, 539.

177. Ilic, D., Kanazawa, S., Furuta, Y., Yamamoto, T., and Aizawa, S. (1996). Impairment of mobility in endodermal cells by FAK deficiency. *Exp. Cell Res.*, **222**, 298.

178. Lee, J., Ishihara, A., Oxford, G., Johnson, B., and Jacobson, K. (1999). Regulation of cell movement is mediated by stretch-activated calcium channels. *Nature*, **400**, 382.

179. Huttenlocher, A., Sandborg, R. R., and Horwitz, A. F. (1995). Adhesion in cell migration. *Curr. Opin. Cell Biol.*, **7**, 697.

180. Hendey, B., Klee, C. B., and Maxfield, F. R. (1992). Inhibition of neutrophil chemokinesis on vitronectin by inhibitors of calcineurin. *Science*, **258**, 296.

181. Sechler, J. L., Cumiskey, A. M., Gazzola, D. M., and Schwarzbauer, J. E. (2000). A novel RGD-independent fibronectin assembly pathway initiated by alpha4beta1 integrin binding to the alternatively spliced V region. *J. Cell Sci.*, **113**, 1491.

182. McDonald, J. A., Quade, B. J., Broekelmann, T. J., LaChance, R., Forsman, K., Hasegawa, E., *et al.* (1987). Fibronectin's cell-adhesive domain and an amino-terminal matrix assembly domain participate in its assembly into fibroblast pericellular matrix. *J. Biol. Chem.*, **262**, 2957.

183. Akiyama, S. K., Yamada, S. S., Chen, W. T., and Yamada, K. M. (1989). Analysis of fibronectin receptor function with monoclonal antibodies: roles in cell adhesion, migration, matrix assembly, and cytoskeletal organization. *J. Cell Biol.*, **109**, 863.

184. Fogerty, F. J., Akiyama, S. K., Yamada, K. M., and Mosher, D. F. (1990). Inhibition of binding of fibronectin to matrix assembly sites by anti-integrin (alpha 5 beta 1) antibodies. *J. Cell Biol.*, **111**, 699.

185. Wu, C., Bauer, J. S., Juliano, R. L., and McDonald, J. A. (1993). The alpha 5 beta 1 integrin fibronectin receptor, but not the alpha 5 cytoplasmic domain, functions in an early and essential step in fibronectin matrix assembly. *J. Biol. Chem.*, **268**, 21883.

186. Yang, J. T., Rayburn, H., and Hynes, R. O. (1993). Embryonic mesodermal defects in alpha 5 integrin-deficient mice. *Development*, **119**, 1093.

187. Wennerberg, K., Lohikangas, L., Gullberg, D., Pfaff, M., Johansson, S., and Fassler, R. (1996). Beta 1 integrin-dependent and -independent polymerization of fibronectin. *J. Cell Biol.*, **132**, 227.

188. Wu, C., Hughes, P. E., Ginsberg, M. H., and McDonald, J. A. (1996). Identification of a new biological function for the integrin alpha v beta 3: initiation of fibronectin matrix assembly. *Cell Adhes. Commun.*, **4**, 149.

189. Wu, C., Keivens, V. M., O'Toole, T. E., McDonald, J. A., and Ginsberg, M. H. (1995). Integrin activation and cytoskeletal interaction are essential for the assembly of a fibronectin matrix. *Cell*, **83**, 715.

190. Sechler, J. L., Corbett, S. A., and Schwarzbauer, J. E. (1997). Modulatory roles for integrin activation and the synergy site of fibronectin during matrix assembly. *Mol. Biol. Cell*, **8**, 2563.

191. Hocking, D. C., Sottile, J., and McKeown-Longo, P. J. (1994). Fibronectin's III-1 module contains a conformation-dependent binding site for the amino-terminal region of fibronectin. *J. Biol. Chem.*, **269**, 19183.

192. Hocking, D. C., Smith, R. K., and McKeown-Longo, P. J. (1996). A novel role for the integrin-binding III-10 module in fibronectin matrix assembly. *J. Cell Biol.*, **133**, 431.

193. Morla, A. and Ruoslahti, E. (1992). A fibronectin self-assembly site involved in fibronectin matrix assembly: reconstruction in a synthetic peptide. *J. Cell Biol.*, **118**, 421.

194. Sottile, J. and Mosher, D. F. (1997). N-terminal type I modules required for fibronectin binding to fibroblasts and to fibronectin's III1 module. *Biochem. J.*, **323**, 51.

195. Bultmann, H., Santas, A. J., and Peters, D. M. (1998). Fibronectin fibrillogenesis involves the heparin II binding domain of fibronectin. *J. Biol. Chem.*, **273**, 2601.

196. Ingham, K. C., Brew, S. A., Huff, S., and Litvinovich, S. V. (1997). Cryptic self-association sites in type III modules of fibronectin. *J. Biol. Chem.*, **272**, 1718.

197. Ali, I. U. and Hynes, R. O. (1977). Effects of cytochalasin B and colchicine on attachment of a major surface protein of fibroblasts. *Biochim. Biophys. Acta*, **471**, 16.

198. Halliday, N. L. and Tomasek, J. J. (1995). Mechanical properties of the extracellular matrix influence fibronectin fibril assembly *in vitro*. *Exp. Cell Res.*, **217**, 109.

199. Ohashi, T., Kiehart, D. P., and Erickson, H. P. (1999). Dynamics and elasticity of the fibronectin matrix in living cell culture visualized by fibronectin-green fluorescent protein. *Proc. Natl. Acad. Sci. USA*, **96**, 2153.

200. Zhang, Q., Magnusson, M. K., and Mosher, D. F. (1997). Lysophosphatidic acid and microtubule-destabilizing agents stimulate fibronectin matrix assembly through Rho-dependent actin stress fiber formation and cell contraction. *Mol. Biol. Cell*, **8**, 1415.

201. Zhong, C., Chrzanowska-Wodnicka, M., Brown, J., Shaub, A., Belkin, A. M., and Burridge, K. (1998). Rho-mediated contractility exposes a cryptic site in fibronectin and induces fibronectin matrix assembly. *J. Cell Biol.*, **141**, 539.

202. Pankov, R., Cukierman, E., Katz, B. Z., Matsumoto, K., Lin, D. C., Lin, S., *et al.* (2000). Integrin dynamics and matrix assembly: tensin-dependent translocation of alpha(5) beta(1) integrins promotes early fibronectin fibrillogenesis. *J. Cell Biol.*, **148**, 1075.

203. Thomas, T. and Dziadek, M. (1993). Genes coding for basement membrane glycoproteins laminin, nidogen, and collagen IV are differentially expressed in the nervous system and by epithelial, endothelial, and mesenchymal cells of the mouse embryo. *Exp. Cell Res.*, **208**, 54.

204. Graham, P. L., Johnson, J. J., Wang, S., Sibley, M. H., Gupta, M. C., and Kramer, J. M. (1997). Type IV collagen is detectable in most, but not all, basement membranes of *Caenorhabditis elegans* and assembles on tissues that do not express it. *J. Cell Biol.*, **137**, 1171.

205. Stephens, L. E., Sutherland, A. E., Klimanskaya, I. V., Andrieux, A., Meneses, J., Pedersen, R. A., *et al.* (1995). Deletion of beta 1 integrins in mice results in inner cell mass failure and peri-implantation lethality. *Genes Dev.*, **9**, 1883.

206. Sasaki, T., Forsberg, E., Bloch, W., Addicks, K., Fassler, R., and Timpl, R. (1998). Deficiency of beta 1 integrins in teratoma interferes with basement membrane assembly and laminin-1 expression. *Exp. Cell Res.*, **238**, 70.

207. Raghavan, S., Bauer, C., Mundschau, G., Li, Q., and Fuchs, E. (2000). Conditional ablation of beta1 integrin in skin. Severe defects in epidermal proliferation, basement membrane formation, and hair follicle invagination. *J. Cell Biol.*, **150**, 1149.

208. Brakebusch, C., Grose, R., Quondamatteo, F., Ramirez, A., Jorcano, J. L., Pirro, A., *et al.* (2000). Skin and hair follicle integrity is crucially dependent on beta 1 integrin expression on keratinocytes. *EMBO J.*, **19**, 3990.

209. Henry, M. D. and Campbell, K. P. (1998). A role for dystroglycan in basement membrane assembly. *Cell*, **95**, 859.

210. Ingber, D. E., Dike, L., Hansen, L., Karp, S., Liley, H., Maniotis, A., *et al.* (1994). Cellular tensegrity: exploring how mechanical changes in the cytoskeleton regulate cell growth, migration, and tissue pattern during morphogenesis. *Int. Rev. Cytol.*, **150**, 173.

211. Chen, Q., Kinch, M. S., Lin, T. H., Burridge, K., and Juliano, R. L. (1994). Integrin-mediated cell adhesion activates mitogen-activated protein kinases. *J. Biol. Chem.*, **269**, 26602.

212. Morino, N., Mimura, T., Hamasaki, K., Tobe, K., Ueki, K., Kikuchi, K., *et al.* (1995). Matrix/integrin interaction activates the mitogen-activated protein kinase, p44erk-1 and p42erk-2. *J. Biol. Chem.*, **270**, 269.

213. Schaller, M. D., Borgman, C. A., Cobb, B. S., Vines, R. R., Reynolds, A. B., and Parsons, J. T. (1992). pp125FAK a structurally distinctive protein-tyrosine kinase associated with focal adhesions. *Proc. Natl. Acad. Sci. USA*, **89**, 5192.

214. Nobes, C. D. and Hall, A. (1995). Rho, rac and cdc42 GTPases: regulators of actin structures, cell adhesion and motility. *Biochem. Soc. Trans.*, **23**, 456.

215. Roskelley, C. D. and Bissell, M. J. (1995). Dynamic reciprocity revisited: a continuous, bidirectional flow of information between cells and the extracellular matrix regulates mammary epithelial cell function. *Biochem. Cell Biol.*, **73**, 391.

216. Barcellos-Hoff, M. H., Aggeler, J., Ram, T. G., and Bissell, M. J. (1989). Functional differentiation and alveolar morphogenesis of primary mammary cultures on reconstituted basement membrane. *Development*, **105**, 223.

217. Roskelley, C. D., Desprez, P. Y., and Bissell, M. J. (1994). Extracellular matrix-dependent tissue-specific gene expression in mammary epithelial cells requires both physical and biochemical signal transduction. *Proc. Natl. Acad. Sci. USA*, **91**, 12378.

218. Streuli, C. H., Schmidhauser, C., Bailey, N., Yurchenco, P., Skubitz, A. P., Roskelley, C., *et al.* (1995). Laminin mediates tissue-specific gene expression in mammary epithelia. *J. Cell Biol.*, **129**, 591.

219. Schmidhauser, C., Casperson, G. F., Myers, C. A., Sanzo, K. T., Bolten, S., and Bissell, M. J. (1992). A novel transcriptional enhancer is involved in the prolactin- and extracellular matrix-dependent regulation of beta-casein gene expression. *Mol. Biol. Cell*, **3**, 699.

220. Edwards, G. M., Wilford, F. H., Liu, X., Hennighausen, L., Djiane, J., and Streuli, C. H. (1998). Regulation of mammary differentiation by extracellular matrix involves protein-tyrosine phosphatases. *J. Biol. Chem.*, **273**, 9495.

221. Muschler, J., Lochter, A., Roskelley, C. D., Yurchenco, P., and Bissell, M. J. (1999). Division of labor among the alpha6beta4 integrin, beta1 integrins, and an E3 laminin receptor to signal morphogenesis and beta-casein expression in mammary epithelial cells. *Mol. Biol. Cell*, **10**, 2817.

222. Okura, M., Shirasuna, K., Hiranuma, T., Yoshioka, H., Nakahara, H., Aikawa, T., *et al.* (1993). Characterization of growth and differentiation of normal human submandibular gland epithelial cells in a serum-free medium. *Differentiation*, **54**, 143.

223. Royce, L. S., Kibbey, M. C., Mertz, P., Kleinman, H. K., and Baum, B. J. (1993). Human neoplastic submandibular intercalated duct cells express an acinar phenotype when cultured on a basement membrane matrix. *Differentiation*, **52**, 247.

224. Hoffman, M. P., Kibbey, M. C., Letterio, J. J., and Kleinman, H. K. (1996). Role of laminin-1 and TGF-beta 3 in acinar differentiation of a human submandibular gland cell line (HSG). *J. Cell Sci.*, **109**, 2013.

225. Lafrenie, R. M., Bernier, S. M., and Yamada, K. M. (1998). Adhesion to fibronectin or collagen I gel induces rapid, extensive, biosynthetic alterations in epithelial cells. *J. Cell. Physiol.*, **175**, 163.

226. Werb, Z., Tremble, P. M., Behrendtsen, O., Crowley, E., and Damsky, C. H. (1989). Signal transduction through the fibronectin receptor induces collagenase and stromelysin gene expression. *J. Cell Biol.*, **109**, 877.

227. Tremble, P., Chiquet-Ehrismann, R., and Werb, Z. (1994). The extracellular matrix ligands fibronectin and tenascin collaborate in regulating collagenase gene expression in fibroblasts. *Mol. Biol. Cell*, **5**, 439.

228. Huhtala, P., Humphries, M. J., McCarthy, J. B., Tremble, P. M., Werb, Z., and Damsky, C. H. (1995). Cooperative signalling by alpha 5 beta 1 and alpha 4 beta 1 integrins regulates metalloproteinase gene expression in fibroblasts adhering to fibronectin. *J. Cell Biol.*, **129**, 867.

229. Tremble, P., Damsky, C. H., and Werb, Z. (1995). Components of the nuclear signalling cascade that regulate collagenase gene expression in response to integrin-derived signals. *J. Cell Biol.*, **129**, 1707.

230. Kheradmand, F., Werner, E., Tremble, P., Symons, M., and Werb, Z. (1998). Role of Rac1 and oxygen radicals in collagenase-1 expression induced by cell shape change. *Science*, **280**, 898.

231. Adams, J. C. and Watt, F. M. (1993). Regulation of development and differentiation by the extracellular matrix. *Development*, **117**, 1183.

232. Jensen, U. B., Lowell, S., and Watt, F. M. (1999). The spatial relationship between stem cells and their progeny in the basal layer of human epidermis: a new view based on whole-mount labelling and lineage analysis. *Development*, **126**, 2409.

233. Zhu, A. J., Haase, I., and Watt, F. M. (1999). Signalling via beta1 integrins and mitogen-activated protein kinase determines human epidermal stem cell fate *in vitro*. *Proc. Natl. Acad. Sci. USA*, **96**, 6728.

234. Roovers, K. and Assoian, R. K. (2000). Integrating the MAP kinase signal into the G1 phase cell cycle machinery. *Bioessays*, **22**, 818.

235. Assoian, R. K. and Schwartz, M. A. (2001). Coordinate signalling by integrins and receptor tyrosine kinases in the regulation of G1 phase cell-cycle progression. *Curr. Opin. Genet. Dev.*, **11**, 48.

236. Adams, J. C. and Watt, F. M. (1990). Changes in keratinocyte adhesion during terminal differentiation: reduction in fibronectin binding precedes alpha 5 beta 1 integrin loss from the cell surface. *Cell*, **19**, 425.

237. Nicholson, L. J. and Watt, F. M. (1991). Decreased expression of fibronectin and the alpha 5 beta 1 integrin during terminal differentiation of human keratinocytes. *J. Cell Sci.*, **98**, 225.

238. Hotchin, N. A. and Watt, F. M. (1992). Transcriptional and post-translational regulation of beta 1 integrin expression during keratinocyte terminal differentiation. *J. Biol. Chem.*, **267**, 14852.

239. Adams, J. C. and Watt, F. M. (1989). Fibronectin inhibits the terminal differentiation of human keratinocytes. *Nature*, **340**, 307.

240. Marsden, M. and Burke, R. D. (1997). Cloning and characterization of novel beta integrin subunits from a sea urchin. *Dev. Biol.*, **181**, 234.

241. Marsden, M. and Burke, R. D. (1998). The betaL integrin subunit is necessary for gastrulation in sea urchin embryos. *Dev. Biol.*, **203**, 134.

242. Murray, G., Reed, C., Marsden, M., Rise, M., Wang, D., and Burke, R. D. (2000). The alphaBbetaC integrin is expressed on the surface of the sea urchin egg and removed at fertilization. *Dev. Biol.*, **227**, 633.

243. Hertzler, P. L. and McClay, D. R. (1999). alphaSU2, an epithelial integrin that binds laminin in the sea urchin embryo. *Dev. Biol.*, **207**, 1.

244. Gettner, S. N., Kenyon, C., and Reichardt, L. F. (1995). Characterization of beta pat-3 heterodimers, a family of essential integrin receptors in *C. elegans*. *J. Cell Biol.*, **129**, 1127.

245. Baum, P. D. and Garriga, G. (1997). Neuronal migrations and axon fasciculation are disrupted in ina-1 integrin mutants. *Neuron*, **19**, 51.

246. Tomaselli, K. J., Neugebauer, K. M., Bixby, J. L., Lilien, J., and Reichardt, L. F. (1988). N-cadherin and integrins: two receptor systems that mediate neuronal process outgrowth on astrocyte surfaces. *Neuron*, **1**, 33.

247. Tomaselli, K. J., Doherty, P., Emmett, C. J., Damsky, C. H., Walsh, F. S., and Reichardt, L. F. (1993). Expression of beta 1 integrins in sensory neurons of the dorsal root ganglion and their functions in neurite outgrowth on two laminin isoforms. *J. Neurosci.*, **13**, 4880.

248. Rubin, G. M., Yandell, M. D., Wortman, J. R., Gabor Miklos, G. L., Nelson, C. R., Hariharan, I. K., *et al.* (2000). Comparative genomics of the eukaryotes. *Science*, **287**, 2204.

249. Brower, D. L., Wilcox, M., Piovant, M., Smith, R. J., and Reger, L. A. (1984). Related cell-surface antigens expressed with positional specificity in *Drosophila* imaginal discs. *Proc. Natl. Acad. Sci. USA*, **81**, 7485.

250. Wilcox, M., Brown, N., Piovant, M., Smith, R. J., and White, R. A. (1984). The *Drosophila* position-specific antigens are a family of cell surface glycoprotein complexes. *EMBO J.*, **3**, 2307.

251. Stark, K. A., Yee, G. H., Roote, C. E., Williams, E. L., Zusman, S., and Hynes, R. O. (1997). A novel alpha integrin subunit associates with betaPS and functions in tissue morphogenesis and movement during *Drosophila* development. *Development*, **124**, 4583.

252. Yee, G. H. and Hynes, R. O. (1993). A novel, tissue-specific integrin subunit, beta nu, expressed in the midgut of *Drosophila melanogaster*. *Development*, **118**, 845.

253. Wright, T. R. F. (1960). The phenogenetics of the embryonic mutant, lethal myospheroid, in *Drosophila melanogaster*. *J. Exp. Zool.*, **143**, 77.

254. Brabant, M. C. and Brower, D. L. (1993). PS2 integrin requirements in *Drosophila* embryo and wing morphogenesis. *Dev. Biol.*, **157**, 49.

255. Brabant, M. C., Fristrom, D., Bunch, T. A., and Brower, D. L. (1996). Distinct spatial and temporal functions for PS integrins during *Drosophila* wing morphogenesis. *Development*, **122**, 3307.

256. Brower, D. L., Bunch, T. A., Mukai, L., Adamson, T. E., Wehrli, M., Lam, S., *et al.* (1995). Nonequivalent requirements for PS1 and PS2 integrin at cell attachments in *Drosophila*: genetic analysis of the alpha PS1 integrin subunit. *Development*, **121**, 1311.

257. Grotewiel, M. S., Beck, C. D., Wu, K. H., Zhu, X. R., and Davis, R. L. (1998). Integrin-mediated short-term memory in *Drosophila*. *Nature*, **391**, 455.

258. Ramos, J. W., Whittaker, C. A., and DeSimone, D. W. (1996). Integrin-dependent adhesive activity is spatially controlled by inductive signals at gastrulation. *Development*, **122**, 2873.

259. Martin-Bermudo, M. D., Dunin-Borkowski, O. M., and Brown, N. H. (1998). Modulation of integrin activity is vital for morphogenesis. *J. Cell Biol.*, **141**, 1073.

260. Martin-Bermudo, M. D. and Brown, N. H. (1999). Uncoupling integrin adhesion and signalling: the betaPS cytoplasmic domain is sufficient to regulate gene expression in the *Drosophila* embryo. *Genes Dev.*, **13**, 729.

261. Drake, C. J. and Little, C. D. (1991). Integrins play an essential role in somite adhesion to the embryonic axis. *Dev. Biol.*, **143**, 418.

262. Drake, C. J., Davis, L. A., Hungerford, J. E., and Little, C. D. (1992). Perturbation of beta 1 integrin-mediated adhesions results in altered somite cell shape and behavior. *Dev. Biol.*, **149**, 327.

263. Jaffredo, T., Horwitz, A. F., Buck, C. A., Rong, P. M., and Dieterlen-Lievre, F. (1988). Myoblast migration specifically inhibited in the chick embryo by grafted CSAT hybridoma cells secreting an anti-integrin antibody. *Development*, **103**, 431.

264. Menko, A. S. and Boettiger, D. (1987). Occupation of the extracellular matrix receptor, integrin, is a control point for myogenic differentiation. *Cell*, **51**, 51.

265. Boucaut, J. C., Darribere, T., Poole, T. J., Aoyama, H., Yamada, K. M., and Thiery, J. P. (1984). Biologically active synthetic peptides as probes of embryonic development: a competitive peptide inhibitor of fibronectin function inhibits gastrulation in amphibian embryos and neural crest cell migration in avian embryos. *J. Cell Biol.*, **99**, 1822.

266. Poole, T. J. and Thiery, J. P. (1986). Antibodies and a synthetic peptide that block cell-fibronectin adhesion arrest neural crest cell migration *in vivo*. *Prog. Clin. Biol. Res.*, **217B**, 235.

267. Lallier, T. and Bronner-Fraser, M. (1993). Inhibition of neural crest cell attachment by integrin antisense oligonucleotides. *Science*, **259**, 692.

268. Kil, S. H., Lallier, T., and Bronner-Fraser, M. (1996). Inhibition of cranial neural crest adhesion *in vitro* and migration *in vivo* using integrin antisense oligonucleotides. *Dev. Biol.*, **179**, 91.

269. Brooks, P. C., Montgomery, A. M., Rosenfeld, M., Reisfeld, R. A., Hu, T., Klier, G., *et al.* (1994). Integrin alpha v beta 3 antagonists promote tumor regression by inducing apoptosis of angiogenic blood vessels. *Cell*, **79**, 1157.

270. Drake, C. J., Cheresh, D. A., and Little, C. D. (1995). An antagonist of integrin alpha v beta 3 prevents maturation of blood vessels during embryonic neovascularization. *J. Cell Sci.*, **108**, 2655.

271. Bronner-Fraser, M., Artinger, M., Muschler, J., and Horwitz, A. F. (1992). Developmentally regulated expression of alpha 6 integrin in avian embryos. *Development*, **115**, 197.

272. Boettiger, D., Enomoto-Iwamoto, M., Yoon, H. Y., Hofer, U., Menko, A. S., and Chiquet-Ehrismann, R. (1995). Regulation of integrin alpha 5 beta 1 affinity during myogenic differentiation. *Dev. Biol.*, **169**, 261.

273. Keller, R. (1991). Early embryonic development of *Xenopus laevis*. *Methods Cell Biol.*, **36**, 61.

274. Lee, G., Hynes, R., and Kirschner, M. (1984). Temporal and spatial regulation of fibronectin in early *Xenopus* development. *Cell*, **36**, 729.

275. DeSimone, D. W., Norton, P. A., and Hynes, R. O. (1992). Identification and characterization of alternatively spliced fibronectin mRNAs expressed in early *Xenopus* embryos. *Dev. Biol.*, **149**, 357.

276. Nakatsuji, N., Smolira, M. A., and Wylie, C. C. (1985). Fibronectin visualized by scanning electron microscopy immunocytochemistry on the substratum for cell migration in *Xenopus laevis* gastrulae. *Dev. Biol.*, **107**, 264.

277. Boucaut, J. C., Darribere, T., Boulekbache, H., and Thiery, J. P. (1984). Prevention of gastrulation but not neurulation by antibodies to fibronectin in amphibian embryos. *Nature*, **307**, 364.

278. Howard, J. E., Hirst, E. M., and Smith, J. C. (1992). Are beta 1 integrins involved in *Xenopus* gastrulation? *Mech. Dev.*, **38**, 109.

279. Ramos, J. W. and DeSimone, D. W. (1996). *Xenopus* embryonic cell adhesion to fibronectin: position-specific activation of RGD/synergy site-dependent migratory behavior at gastrulation. *J. Cell Biol.*, **134**, 227.

280. Winklbauer, R. (1990). Mesodermal cell migration during *Xenopus* gastrulation. *Dev. Biol.*, **142**, 155.

281. Winklbauer, R., Nagel, M., Selchow, A., and Wacker, S. (1996). Mesoderm migration in the *Xenopus* gastrula. *Int. J. Dev. Biol.*, **40**, 305.

282. Darribere, T., Yamada, K. M., Johnson, K. E., and Boucaut, J. C. (1988). The 140-kDa fibronectin receptor complex is required for mesodermal cell adhesion during gastrulation in the amphibian Pleurodeles waltlii. *Dev. Biol.*, **126**, 182.

283. Smith, J. C., Symes, K., Hynes, R. O., and DeSimone, D. (1990). Mesoderm induction and the control of gastrulation in *Xenopus laevis*: the roles of fibronectin and integrins. *Development*, **108**, 229.

284. DeSimone, D. W. and Hynes, R. O. (1988). *Xenopus laevis* integrins. Structural conservation and evolutionary divergence of integrin beta subunits. *J. Biol. Chem.*, **263**, 5333.

285. Ransom, D. G., Hens, M. D., and DeSimone, D. W. (1993). Integrin expression in early amphibian embryos: cDNA cloning and characterization of *Xenopus* beta 1, beta 2, beta 3, and beta 6 subunits. *Dev. Biol.*, **160**, 265.

286. Whittaker, C. A. and DeSimone, D. W. (1993). Integrin alpha subunit mRNAs are differentially expressed in early *Xenopus* embryos. *Development*, **117**, 1239.

287. Gawantka, V., Ellinger-Ziegelbauer, H., and Hausen, P. (1992). Beta 1-integrin is a maternal protein that is inserted into all newly formed plasma membranes during early *Xenopus* embryogenesis. *Development*, **115**, 595.

288. Gawantka, V., Joos, T. O., and Hausen, P. (1994). A beta 1-integrin associated alpha-chain is differentially expressed during *Xenopus* embryogenesis. *Mech. Dev.*, **47**, 199.

289. Davidson, L. A., Hoffstrom, B. G., Keller, R. E., and DeSimone, D. W. (2001). Mesendoderm migration and mantle closure in *Xenopus laevis* gastrulation: combined roles for integrin α5β1, fibronectin and tissue geometry, submitted.

290. Alfandari, D., Whittaker, C. A., DeSimone, D. W., and Darribere, T. (1995). Integrin alpha v subunit is expressed on mesodermal cell surfaces during amphibian gastrulation. *Dev. Biol.*, **170**, 249.

291. Skalski, M., Alfandari, D., and Darribere, T. (1998). A key function for alphav containing integrins in mesodermal cell migration during Pleurodeles waltl gastrulation. *Dev. Biol.*, **195**, 158.

292. Marsden, M. and DeSimone, D. W. (2001). Regulation of cell polarity, radial intercalation and epiboly in *Xenopus laevis*: novel roles for integrin and fibronectin development, 128, *in press*.

293. Lallier, T. E. and DeSimone, D. W. (2000). Separation of neural induction and neurulation in *Xenopus*. *Dev. Biol.*, **225**, 135.

294. Hynes, R. O. (1996). Targeted mutations in cell adhesion genes: what have we learned from them? *Dev. Biol.*, **180**, 402.

295. Hynes, R. O. and Bader, B. L. (1997). Targeted mutations in integrins and their ligands: their implications for vascular biology. *Thromb. Haemost.*, **78**, 83.

296. Sheppard, D. (2000). *In vivo* functions of integrins: lessons from null mutations in mice. *Matrix Biol.*, **19**, 203.

297. Fassler, R. and Meyer, M. (1995). Consequences of lack of beta 1 integrin gene expression in mice. *Genes Dev.*, **9**, 1896.

298. Almeida, E. A., Huovila, A. P., Sutherland, A. E., Stephens, L. E., Calarco, P. G., Shaw, L. M., *et al.* (1995). Mouse egg integrin alpha 6 beta 1 functions as a sperm receptor. *Cell*, **81**, 1095.

299. Sutherland, A. E., Calarco, P. G., and Damsky, C. H. (1988). Expression and function of cell surface extracellular matrix receptors in mouse blastocyst attachment and outgrowth. *J. Cell Biol.*, **106**, 1331.

300. Hirsch, E., Iglesias, A., Potocnik, A. J., Hartmann, U., and Fassler, R. (1996). Impaired migration but not differentiation of haematopoietic stem cells in the absence of beta1 integrins. *Nature*, **380**, 171.

301. George, E. L., Georges-Labouesse, E. N., Patel-King, R. S., Rayburn, H., and Hynes, R. O. (1993). Defects in mesoderm, neural tube and vascular development in mouse embryos lacking fibronectin. *Development*, **119**, 1079.

302. Yang, J. T., Bader, B. L., Kreidberg, J. A., Ullman-Cullere, M., Trevithick, J. E., and Hynes, R. O. (1999). Overlapping and independent functions of fibronectin receptor integrins in early mesodermal development. *Dev. Biol.*, **215**, 264.

303. De Arcangelis, A., Mark, M., Kreidberg, J., Sorokin, L., and Georges-Labouesse, E. (1999). Synergistic activities of alpha3 and alpha6 integrins are required during apical ectodermal ridge formation and organogenesis in the mouse. *Development*, **126**, 3957.

304. Yang, J. T., Rayburn, H., and Hynes, R. O. (1995). Cell adhesion events mediated by alpha 4 integrins are essential in placental and cardiac development. *Development*, **121**, 549.

305. Kwee, L., Baldwin, H. S., Shen, H. M., Stewart, C. L., Buck, C., Buck, C. A., *et al.* (1995). Defective development of the embryonic and extraembryonic circulatory systems in vascular cell adhesion molecule (VCAM-1) deficient mice. *Development*, **121**, 489.

306. Gurtner, G. C., Davis, V., Li, H., McCoy, M. J., Sharpe, A., and Cybulsky, M. I. (1995). Targeted disruption of the murine VCAM1 gene: essential role of VCAM-1 in chorioallantoic fusion and placentation. *Genes Dev.*, **9**, 1.

307. Bader, B. L., Rayburn, H., Crowley, D., and Hynes, R. O. (1998). Extensive vasculogenesis, angiogenesis, and organogenesis precede lethality in mice lacking all alpha v integrins. *Cell*, **95**, 507.

308. Hodivala-Dilke, K. M., McHugh, K. P., Tsakiris, D. A., Rayburn, H., Crowley, D., Ullman-Cullere, M., *et al.* (1999). Beta3-integrin-deficient mice are a model for Glanzmann thrombasthenia showing placental defects and reduced survival. *J. Clin. Invest.*, **103**, 229.

309. Bennett, J. S. (2001). Novel platelet inhibitors. *Annu. Rev. Med.*, **52**, 161.

310. Brooks, P. C., Clark, R. A., and Cheresh, D. A. (1994). Requirement of vascular integrin alpha v beta 3 for angiogenesis. *Science*, **264**, 569.

311. Bauer, T. R., Jr. and Hickstein, D. D. (2000). Gene therapy for leukocyte adhesion deficiency. *Curr. Opin. Mol. Ther.*, **2**, 383.

312. Vidal, F., Aberdam, D., Miquel, C., Christiano, A. M., Pulkkinen, L., Uitto, J., *et al.* (1995). Integrin beta 4 mutations associated with junctional epidermolysis bullosa with pyloric atresia. *Nature Genet.*, **10**, 229.

313. Ruzzi, L., Gagnoux-Palacios, L., Pinola, M., Belli, S., Meneguzzi, G., D'Alessio, M., *et al.* (1997). A homozygous mutation in the integrin alpha6 gene in junctional epidermolysis bullosa with pyloric atresia. *J. Clin. Invest.*, **99**, 2826.

314. Hayashi, Y. K., Chou, F. L., Engvall, E., Ogawa, M., Matsuda, C., Hirabayashi, S., *et al.* (1998). Mutations in the integrin alpha7 gene cause congenital myopathy. *Nature Genet.*, **19**, 94.

315. Hynes, R. O. (1992). Integrins: versatility, modulation, and signalling in cell adhesion. *Cell*, **69**, 11.

316. Charo, I. F., Nannizzi, L., Phillips, D. R., Hsu, M. A., and Scarborough, R. M. (1991). Inhibition of fibrinogen binding to GP IIb-IIIa by a GP IIIa peptide. *J. Biol. Chem.*, **266**, 1415.

317. Gardner, H., Kreidberg, J., Koteliansky, V., and Jaenisch, R. (1996). Deletion of integrin alpha 1 by homologous recombination permits normal murine development but gives rise to a specific deficit in cell adhesion. *Dev. Biol.*, **175**, 301.

318. Kreidberg, J. A., Donovan, M. J., Goldstein, S. L., Rennke, H., Shepherd, K., Jones, R. C., *et al.* (1996). Alpha 3 beta 1 integrin has a crucial role in kidney and lung organogenesis. *Development*, **122**, 3537.

319. Arroyo, A. G., Yang, J. T., Rayburn, H., and Hynes, R. O. (1996). Differential requirements for alpha4 integrins during fetal and adult hematopoiesis. *Cell*, **85**, 997.

320. Georges-Labouesse, E., Messaddeq, N., Yehia, G., Cadalbert, L., Dierich, A., and Le Meur, M. (1996). Absence of integrin alpha 6 leads to epidermolysis bullosa and neonatal death in mice. *Nature Genet.*, **13**, 370.

321. Mayer, U., Saher, G., Fassler, R., Bornemann, A., Echtermeyer, F., von der Mark, H., *et al.* (1997). Absence of integrin alpha 7 causes a novel form of muscular dystrophy. *Nature Genet.*, **17**, 318.

322. Muller, U., Wang, D., Denda, S., Meneses, J. J., Pedersen, R. A., and Reichardt, L. F. (1997). Integrin alpha8beta1 is critically important for epithelial-mesenchymal interactions during kidney morphogenesis. *Cell*, **88**, 603.

323. Huang, X. Z., Wu, J. F., Ferrando, R., Lee, J. H., Wang, Y. L., Farese, R. V., Jr., *et al.* (2000). Fatal bilateral chylothorax in mice lacking the integrin alpha9beta1. *Mol. Cell. Biol.*, **20**, 5208.

324. Tronik-Le Roux, D., Roullot, V., Poujol, C., Kortulewski, T., Nurden, P., and Marguerie, G. (2000). Thrombasthenic mice generated by replacement of the integrin alpha(IIb) gene: demonstration that transcriptional activation of this megakaryocytic locus precedes lineage commitment. *Blood*, **96**, 1399.

325. Schon, M. P., Arya, A., Murphy, E. A., Adams, C. M., Strauch, U. G., Agace, W. W., *et al.* (1999). Mucosal T lymphocyte numbers are selectively reduced in integrin alpha E (CD103)-deficient mice. *J. Immunol.*, **162**, 6641.

326. Schmits, R., Kundig, T. M., Baker, D. M., Shumaker, G., Simard, J. J., Duncan, G., *et al.* (1996). LFA-1-deficient mice show normal CTL responses to virus but fail to reject immunogenic tumor. *J. Exp. Med.*, **183**, 1415.

327. Shier, P., Otulakowski, G., Ngo, K., Panakos, J., Chourmouzis, E., Christjansen, L., *et al.* (1996). Impaired immune responses toward alloantigens and tumor cells but normal thymic selection in mice deficient in the beta2 integrin leukocyte function-associated antigen-1. *J. Immunol.*, **157**, 5375.

328. Coxon, A., Rieu, P., Barkalow, F. J., Askari, S., Sharpe, A. H., von Andrian, U. H., *et al.* (1996). A novel role for the beta 2 integrin CD11b/CD18 in neutrophil apoptosis: a homeostatic mechanism in inflammation. *Immunity*, **5**, 653.

329. Rosenkranz, A. R., Coxon, A., Maurer, M., Gurish, M. F., Austen, K. F., Friend, D. S., *et al.* (1998). Impaired mast cell development and innate immunity in Mac-1 (CD11b/CD18, CR3)-deficient mice. *J. Immunol.*, **161**, 6463.

330. Wilson, R. W., Ballantyne, C. M., Smith, C. W., Montgomery, C., Bradley, A., O'Brien, W. E., *et al.* (1993). Gene targeting yields a CD18-mutant mouse for study of inflammation. *J. Immunol.*, **151**, 1571.

331. Scharffetter-Kochanek, K., Lu, H., Norman, K., van Nood, N., Munoz, F., Grabbe, S., *et al.* (1998). Spontaneous skin ulceration and defective T cell function in CD18 null mice. *J. Exp. Med.*, **188**, 119.

332. McHugh, K. P., Hodivala-Dilke, K., Zheng, M. H., Namba, N., Lam, J., Novack, D., *et al.* (2000). Mice lacking beta3 integrins are osteosclerotic because of dysfunctional osteoclasts. *J. Clin. Invest.*, **105**, 433.

333. Dowling, J., Yu, Q. C., and Fuchs, E. (1996). Beta4 integrin is required for hemi-desmosome formation, cell adhesion and cell survival. *J. Cell Biol.*, **134**, 559.

334. Huang, X., Griffiths, M., Wu, J., Farese, R. V., Jr., and Sheppard, D. (2000). Normal development, wound healing, and adenovirus susceptibility in beta5-deficient mice. *Mol. Cell. Biol.*, **20**, 755.

335. Huang, X. Z., Wu, J. F., Cass, D., Erle, D. J., Corry, D., Young, S. G., *et al.* (1996). Inactivation of the integrin beta 6 subunit gene reveals a role of epithelial integrins in regulating inflammation in the lung and skin. *J. Cell Biol.*, **133**, 921.

336. Wagner, N., Lohler, J., Kunkel, E. J., Ley, K., Leung, E., Krissansen, G., *et al.* (1996). Critical role for beta7 integrins in formation of the gut-associated lymphoid tissue. *Nature*, **382**, 366.

5 | Cell surface heparan sulfate proteoglycans

OFER REIZES, PYONG WOO PARK, and MERTON BERNFIELD

1. Introduction

Cell adhesion plays a central role in a wide variety of biological processes. Depending on how cells adhere to different substrates with their adhesion receptors, engagement of a cell to the extracellular matrix (ECM) or another cell can lead to proliferation, differentiation, aggregation, or even apoptosis. In addition to the major adhesion receptors, such as integrins, cadherins, the Ig superfamily proteins, and selectins (reviewed in this book), it is now known that cell surface heparan sulfate proteoglycans (HSPGs) also function as important cell–cell and cell–matrix adhesion receptors. Binding of cell surface HSPGs to insoluble ligands, such as ECM components or cell adhesion molecules, immobilizes the proteoglycan (PG) in the plasma membrane, enabling the cytoplasmic domain of the PG to interact with signalling components and the actin cytoskeleton. Cell surface HSPGs are co-receptors with other adhesion receptors in many of these interactions, potentially generating a second signal upon formation of the signalling complex.

The ECM was originally thought of as an inert structure functioning solely to maintain the architecture of tissues and organs (1). Studies within the last 20 years have revealed that ECMs also have active roles in mediating both normal and pathological events. The major components of the ECM have been defined, and one of them, the PGs are ubiquitous components of most, if not all, ECMs (2). Proteoglycans have extended structures in solution, owing to their relatively stiff and highly anionic glycosaminoglycan (GAG) component, and occupy very large hydrodynamic volumes relative to their molecular mass, indicating that they retain water and produce a swelling pressure in the tissue. During the 1960s, studies of cartilage PGs revealed that these molecules are reversibly compressible. Upon application of a compressive force, the water molecules are displaced and intramolecular interactions between the GAG chains increase. When the load is removed, the GAG chains of the PGs once again can retain water and expand. This property is critical for the normal function of cartilage, which must act as cushions for variable, compressive loads.

Once proposed to be specific to cartilage, PGs are emerging as 'global regulators' of biological processes. PGs are now known to exist in many tissues and found on cell surfaces, within intracellular compartments, in addition to the ECMs (2). At these sites, PGs mediate a plethora of biological events including, among others, cell–cell and cell–matrix adhesion, growth factor signalling, ligand internalization, microbial invasion, regulation of macromolecule permeability, and assembly and maintenance of the basement membranes. These molecular interactions mediated by PGs frequently dictate the outcome of pathophysiological manifestations such as morphogenesis, thrombosis, angiogenesis, microbial infection, and obesity (3).

PGs are glycoconjugates comprised of GAG chains covalently attached to a core protein. The GAG chains are linear polysaccharides of repeating disaccharide units, and the existing GAGs are classified as heparan sulfate (HS)/heparin, chondroitin sulfates (includes dermatan sulfate), keratan sulfate and hyaluronic acid, based on the composition and linkage of the monosaccharides in the repeating disaccharide unit. With the exception of hyaluronic acid, the GAG chains are covalently bound to core proteins. HS chains are the most structurally heterogeneous GAGs due to their extensive modifications (see below), and are conjugated to a variety of HSPG core proteins, such as the ECM components perlecan and agrin, the intracellular HSPG serglycin, and the cell surface HSPGs syndecans and glypicans (3). Until recently, the physiological functions of HSPGs remained largely speculative (4). For the cell surface HSPGs, it has long been postulated that these molecules are critical for fundamental cellular activities based on their wide (on all adherent cell type) and abundant ($\sim 10^6$ syndecan-1/epithelial cell) expression pattern (5). Recent studies, demonstrating that cell surface HSPGs act as co-receptors or receptors for many ligands, have provided some evidence in support of this hypothesis. In order to provide a concise review of HSPGs, this chapter will not re-examine the early work of HSPG interactions with collagen and fibronectin, since these have been reviewed previously (5–9). The objective of this chapter is to review the current understanding of the structural and functional aspects of cell surface HSPGs, focusing on cell adhesion events, and to discuss the emerging pathophysiological roles of these exciting molecules evident from *Drosophila* and mouse mutants.

2. A primer on heparan sulfate

2.1 Fine structure and biosynthesis

HS is the most structurally complex GAG and is qualitatively identical in structure to the pharmaceutical product heparin (10). Biosynthesis is initiated by covalent linkage between a serine residue on the core proteins to the reducing end of a xylose residue. Figure 1 depicts the biosynthetic machinery involved in the generation of HS chains. To the non-reducing end of xylose, two galactose and a single glucuronic acid (GlcUA) residue are then added forming a linkage region, which can also serve to initiate chondroitin sulfate chain synthesis. Initiation of HS chain synthesis occurs by addition of a *N*-acetyl glucosamine (GlcNAc) residue by the GlcUA/GlcNAc co-

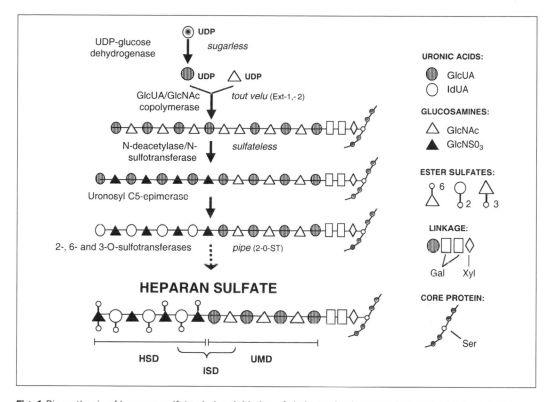

Fig. 1 Biosynthesis of heparan sulfate chains. Initiation of chain synthesis occurs by covalent linkage between a xylose and a specific serine residue on the core protein followed by addition of two galactosyl and a glucuonosyl residue. Chain elongation and modification reactions are carried out sequentially, yet reactions not going to completion. As a result each HS chain is likely unique. HS chain sizes depicted here are arbitrary. Highly sulfated heparin-like domains (HSD) alternate with relatively unmodified domains (UMD) along the chain with sites of ligand binding likely occurring along the HSDs as well as at the intermediate sulfated domains (ISDs) that lie at the interface between the HSDs and UMDs. HSDs and UMDs are distinguished by their susceptibility to HNO_2 at pH 1.5 as well as to heparin lyase I (heparinase) and heparin lyase III (heparitinase). Also indicated are mutants in the biosynthetic machinery identified by genetic screens, transgenic, and gene mapping approaches in *Drosophila*, mouse, and humans, respectively; these include the UDP-glucose dehydrogenase (*sgl*), several sulfotransferases (*sfl*, *pipe*, and the mouse HS 2-*O*-sulfotransferase), and the HS co-polymerases (*ttv*, and human EXT1 and EXT2).

polymerase. The polymer is extended by the same enzyme as repeating disaccharide units of βGlcUA-1,4-αGlcNAc residues. The substrates for the chain initiation and polymerization reactions are UDP-sugars generated by a UDP-glucose dehydrogenase.

The HS chain then may undergo a set of modifications including random *N*-deacetylation and *N*-sulfation of the GlcNAc residue to $GlcNSO_3$ by an *N*-deacetylase *N*-sulfotransferase (NDST) and C5 epimerization of the GlcUA to iduronic acid (IdUA). Both sugars in the repeating disaccharide may be *O*-sulfated on the 2-O position of GlcUA or IdUA, and on the 3-O and 6-O position of the $GlcNSO_3$ sugars by the multiple 2-*O*-, 3-*O*-, and 6-*O*-sulfotransferase isoforms, resulting in synthesis of at least 18 distinct disaccharide units (11). In the case of heparin chains, $> 80\%$ are composed of IdUA and $GlcNSO_3$ disaccharides, which contain $>$ two SO_4 groups/unit. In contrast, HS chains tend to be more complex because of these multiple modification reactions with many not go to completion. Thus, no two HS chains are identical. Despite its complexity, HS is widely distributed in the animal kingdom, identified in primitive invertebrates, as early as Cnidaria, to humans (12).

2.2 Domain structure

The HS at cell surfaces has been shown to have a novel macroscopic structure; highly sulfated IdUA-rich, *c.* 3 kDa heparin-like domains (highly sulfated domains, HSDs) alternate with two- to three-fold larger, GlcUA-rich, poorly sulfated domains (unmodified domains, UMDs) (13). HS chain complexity derives from the combination of these HSD and UMDs as well as the unique interface sequences between them, referred to as intermediate sulfated domains (ISDs). The macroscopic structure of the HS chains, including chain length (*c.* 20–150 disaccharides) and number and size of HSDs and UMDs, varies reproducibly between cell types (13, 14), despite the conservation in the HS biosynthetic machinery among different cell types and organisms. The complexity in HS domain structure may, in part, be due to the multiple isoforms of the modification enzymes in different cell types (15–17). These differences in macroscopic structures lead to alteration in the interaction between HS chains and their protein ligands.

The shape of HS chains can be modified by its ligands, analogous to the induced fit model of enzyme–substrate interactions. Ligand binding affinity is enhanced by the IdUA residues, which impart conformational flexibility to the chains (18). The same characteristics that encourage ligand binding also enable the HS to be a selective filter, as seen in the pulmonary alveolar and renal glomerular basal laminae. Importantly, HS chains on the same core protein can vary in size, sequence, and extent of modifications between cell types (13). Thus, the nature of the HS chains is a differentiated characteristic specific to cells of the same type, suggesting that certain structural aspects of the HS chains may be informational. It is unclear how HS chain structures can be stable and heritable because there is no pre-existing template.

HS chains are designed for binding ligands and this is the likely basis for their maintenance throughout evolution. HS binding ligands have highly variable characteristics. While a large number of proteins bind at physiological concentrations and ionic conditions, the amino acid sequences responsible for this binding vary widely (19). The sequences are generally rich in the basic amino acids, lysine and arginine, but no unique or invariant sequences can be predicted with certainty to bind to HS. Indeed, the key amino acids may not be in a contiguous sequence. Nevertheless, the binding is relatively high affinity, with K_ds ranging from 1–100 nM. Cell surface HSPGs provide cells with a mechanism to snare a wide variety of physiological effectors without requiring that evolution generate multiple novel binding proteins. The interaction of fibroblast growth factor-2 (FGF-2) with HS is a well-studied example: this growth factor binds at nM affinities to HS, which is more abundant than the signal transducing

FGF receptors at the cell surface. Once formed, the FGF-HS complex forms a higher affinity ternary complex with the FGF receptor which, when occupied, initiates a signalling cascade within the cell (20, 21). This interaction provides the cell with a way to regulate receptor signalling by means other than receptor expression and occupancy. Analogous co-receptor interactions occur with other extracellular effectors. For example, fibronectin initiates stress fibre formation only when it engages the cell surface by both its HS and integrin binding domains (22).

3. Cell surface heparan sulfate proteoglycan core proteins

The cell surface PG protein backbone, or core protein, directs the intracellular and extracellular trafficking of HS to its proper site and circumstance for ligand binding and signalling (3, 23–25). Once bound, the ligand can then be stored in a stable ECM compartment, interact with signal transducing or endocytic receptors at the cell surface, activate or inhibit enzymes, and associate with adhesion proteins for cell attachment or with a variety of other protein ligands (3). Where studied, the number of HS chains on most cell surfaces far exceeds the number of specific receptors. Thus, although HS binding is not highly specific, it is sufficiently high in affinity and specificity to be physiologically relevant.

Syndecans and glypicans comprise the most abundant HSPGs at the cell surface, while betaglycan and CD44E are referred to as 'part-time PGs', because they can also exist as non-glycanated forms and have HS-independent biological functions (3). Due to space limitations, betaglycan and CD44E will not be discussed further. The syndecans are single membrane spanning proteins with an apparently extended extracellular domain that bears HS chains near the N-terminus distal from the plasma membrane (5, 26, 27). The glypicans are disulfide bonded, probably globular proteins, that likely bear HS chains near their C-terminus, and are linked to the plasma membrane via a glycosyl phosphatidylinositol (GPI) linkage (23, 25, 28). Both families are conserved from *Drosophila* to humans. To date, four syndecan and six glypican genes have been found in vertebrates (3). Where studied, the gene products differ in tissue and developmental expression. While they appear functionally homologous in some assays (29), recent *in vitro* and *in vivo* evidence suggests that glypicans and syndecans have distinct functions (30–32).

3.1 Syndecans

The syndecan family of transmembrane PGs is the major source of cell surface HS. The syndecans have been known since syndecan-1 was molecularly cloned in 1989 (33). The four gene products, syndecan-1, -2, -3, and -4, are expressed on the surface of most adherent cells. The identification of only one *Drosophila* gene suggests that the four mammalian syndecans arose through gene duplication and divergent evolution from a single ancestor. Their extracellular domains, except for their GAG attachment regions, differ substantially and syndecan-1 and -4 may contain chondroitin sulfate in addition to HS (34). However, their transmembrane domains are highly conserved and their short (33 to 36 amino acids) cytoplasmic domains are quite homologous among the family members and across species; the significant differences reside in sequences between two of the three invariant cytoplasmic domain tyrosines. On the basis of similarities in core protein size and sequence, and reciprocity in developmental expression, syndecan-1 and -3 are considered to be in one subfamily and syndecan-2 and -4 in another (5).

Syndecan expression is highly regulated during embryonic development and wound repair (Table 1). In the mouse, syndecan-1 appears earliest, at the four cell

Table 1 Characteristics of vertebrate glypicans and syndecans

| Proteoglycan | Molecular weight (kDa) | | Expression | | |
	Predicted[a]	PAGE mobility	Embryo	Adult	Feature
Glypican-1 (Glypican)	62	64	CNS	Ubiquitous, prevalent in CNS	–
Glypican-2 (Cerebroglycan)	63	57	CNS	None	Early post-mitotic neurons and axon tracts
Glypican-3 (OCI-5)	66	69	Ubiquitous	Markedly reduced in CNS	Simpson–Golabi–Behmel syndrome (SGBS)
Glypican-4 (K-glypican)	62	60	Kidney, CNS, adrenal	Reduced in CNS	Replicating neural precursors
Glypican-5	64	65	Kidney, limb, CNS	–	Late post-mitotic neurons
Glypican-6	63	UN	–	–	–
Syndecan-1 (Syndecan)	33	80	Pre-implantation germ layers, organogenesis	Epithelia, plasma cells	Basolateral epithelial surface
Syndecan-2 (Fibroglycan)	22	48	Organogenesis	Endothelia, fibroblasts	–
Syndecan-3 (N-syndecan)	46	120	Neuroectoderm, neural crest, limb	Neural crest derivatives, CNS	Neural fibre tracts
Syndecan-4 (Ryudocan)	22	35	Mature tissues	Ubiquitous on adherent cells	Focal adhesions

[a] Includes signal peptide, molecular weight is based on mouse cDNAs for syndecans and human cDNAs for glypicans. UN, unknown.

stage, it is expressed intracellularly, and is subsequently found at cell surfaces in the morula. Expression thereafter coincides with the cells fated to become the embryo proper, including the inner cell mass and embryonic ectoderm, and then the definitive endoderm and ectoderm and their epithelial derivatives (35). Expression is transiently induced in mesodermal derivatives when the associated epithelia change shape. Thus, syndecan-1 expression is early, widespread, and follows morphogenetic rather than histological boundaries. Syndecan-2 is expressed later, appearing during organogenesis and predominates in mesodermal derivatives. Syndecan-3 also arises during late embryogenesis, primarily in the neural crest, neural crest-derived cranial mesenchyme, and in the limb bud mesenchyme. Finally, syndecan-4 appears at the end of embryogenesis and is found with a widespread distribution in neonate and adult tissues (5, 26).

3.1.1 Extracellular domains

The extracellular domains (ectodomains) contain the signal peptide, HS attachment, and proteolytic cleavage (shedding) sites, as well as putative sites for cell interaction, and oligomerization (Fig. 3). There are two regions for GAG attachment. An N-terminal site that bears only HS chains, and a juxtamembrane site, which may attach chondroitin sulfate to syndecan-1 and -4 core proteins (34). The attachment sites generally contain two to three consecutive Ser–Gly residues with flanking acidic amino acids and C-terminal hydrophobic residues (36, 37). GAG-independent core protein interaction with various cell types has been documented with syndecan-4 and weakly with syndecan-1 (38). The cell interaction domain on syndecan-4 is specific, shows high affinity (2 nM), and has been mapped to a region distal to the N-terminal GAG attachment sites, though the cell binding proteins have not been identified. Syndecans may also form oligomers, a suggestion based on their anomalous migration in SDS–polyacrylamide gels (33). Oligomerization of the ectodomain does not require the transmembrane and cytoplasmic domains since SDS-resistant dimers and oligomers form even when the ectodomain is expressed in the absence of the transmembrane and cytoplasmic domains. Despite the large molecular volume of the HS chains, which may themselves self-associate, dimer or oligomer formation would serve to enlarge their interaction surface both extracellularly and with intracellular components such as the actin cytoskeleton (39).

Syndecans can be shed from cell surfaces by cleavage of the core protein at a juxtamembrane site (40, 41). Regulated shedding of syndecan-1 and -4 occurs via a tissue inhibitor of membrane metalloprotease-3 sensitive zinc metalloprotease at a site within 10 amino acids from the plasma membrane (42). Conversion of cell surface components into soluble molecules is a post-translational regulatory event for a diverse group of membrane-anchored proteins. When these proteins are shed from the cell surface, their functional domains are often released as soluble paracrine effectors. In several instances the physiological role of the cell surface protein differs from that of its soluble counterpart, e.g. for CSF-1, TNF-α, TGF-α, and HB-EGF (see refs 43 and 44 for reviews). Cleavage converts the syndecan from a cell surface receptor or co-receptor to a soluble effector. Because both the soluble and cell surface syndecans compete for the same ligands, the soluble syndecan may function as an inhibitor, similar to the soluble forms of FGF-R and VEGF-R (45, 46).

3.1.2 Transmembrane domains

Syndecan transmembrane domains may act in localizing these PGs to unique plasma membrane compartments and may also facilitate dimer formation. In contrast to the ectodomains, these domains are highly conserved among and between family members. Localization of syndecan-1 and -4 to the basolateral membrane and to focal contacts, respectively, is mediated by their transmembrane domains (39, 47, 48). Finally, formation of syndecan-3 oligomers requires glycine residues in the trans-membrane domain, which along with residues in the ectodomain result in formation of the SDS-resistant oligomers (49).

3.1.3 Cytoplasmic domains

In contrast to the highly divergent extracellular domain of the syndecans, the short cytoplasmic tail (33–36 amino acids) is the most evolutionarily conserved region among the syndecans (5). Within this region there are three distinct domains with interaction sites for cytoskeletal proteins and signalling molecules, as well as phosphorylation sites. The three domains are referred to as C1, V, and C2; where C1 and C2 are highly conserved and V is quite variable (see Fig. 2 and Section 5). While the cytoplasmic tail participates in multiple intracellular interactions, its function in mediating cell adhesion events remains under intense investigation. For example, while formation of focal contacts relies on the presence of the syndecan-4 cyto-plasmic tail, this does not hold true for syndecan-1-mediated cell spreading of lymphoblasts on thrombospondin and fibronectin matrices. Thus, integration of dif-ferent cytoplasmic signalling events mediated by the syndecan cytoplasmic tails will likely depend on the particular co-receptor that is engaged for specific cell adhesion events.

3.2 Glypicans

The glypicans are 50–60 kDa HSPGs composed of an amino terminal 50 kDa cysteine-rich domain followed by the HS attachment motifs, and are anchored to the plasma membrane via a GPI linkage to their C-terminus (23) (Fig. 2). The cysteine-rich domain contains 14 conserved cysteines, which likely form a compact disulfide linked globular domain and may be involved in protein–protein interactions. This domain is highly conserved among family members and throughout evolution based on homology with the *Drosophila* glypican homologue *dally* (31). Like the syndecans the HS attachment sequences are highly conserved but the adjacent regions are less well conserved. Based on available sequence data, their HS chains are located near the plasma membrane in contrast to the syndecans where the HS attachment occurs at sites distal from the cell surface.

Glypicans associate with the plasma membrane via a GPI anchor (50). These anchors associate with membrane microdomains or 'rafts' that are rich in glyco-

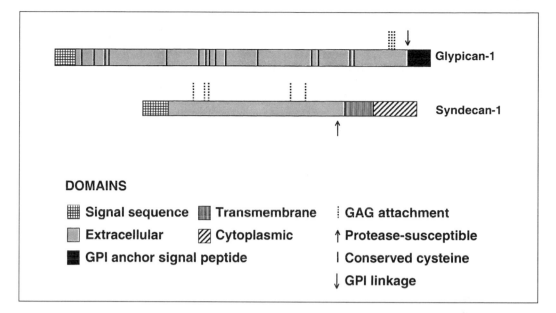

Fig. 2 Syndecan and glypican core protein design. Syndecan and glypican core proteins can be subdivided into distinct domains. Starting at the N-terminus, the syndecan core protein contains the secretory signal peptide, extracellular (ectodomain), transmembrane, and cytoplasmic domains. There are several GAG attachment sites, and the figure shows that of syndecan-1 where HS chains are attached distal to the plasma membrane whereas CS chains are generally attached proximal to the plasma membrane. The syndecan core protein is thought to assume the structure of a linear extended polypeptide. The glypican core protein, starting at the N-terminus, contains the signal peptide for secretion, globular extracellular domain with 14 conserved Cys residues for intra-molecular disulfide bonding, and the C-terminal signal peptide domain for GPI linkage. The three HS attachment sites are located proximal to the plasma membrane.

sphingolipid, which were originally thought to be caveolae. In contrast to other GPI-anchored proteins, which localize to apical surfaces, glypicans associate with basolateral surfaces. Deletion of the HS attachment sites reverts their localization to the apical surface, suggesting that the HS chains or their ligands dictate the glypican subcellular localization (51). The GPI anchor also dictates the rapid turnover and metabolism of linked proteins and their ligands. This endocytic pathway may mediate the rapid turnover of HS binding ligands like anti-thrombin III, follistatin, among others (52–55).

Soluble glypicans have been detected in conditioned media of cultured cells, yet neither regulation nor *in vivo* evidence glypican shedding or secretion has been documented (28). Based on their structure, there are two likely mechanisms to release the glypicans from the cell surface, either via proteolytic cleavage at a juxtamembrane site or hydrolysis mediated by a phosphoinositol-specific phospholipase C (56). It remains to be determined whether either mechanism physiologically regulates glypican shedding/release from the cell surface.

Glypicans are developmentally regulated and widely expressed in the adult and all are expressed in the central nervous system (Table 1). The glypican family likely form three subfamilies, glypicans-1 and -2, glypicans-3 and -5, and glypicans-4 and -6, based on sequence conservation and exon organization (3). Surprisingly, glypicans-3 and -4 and glypicans-5 and -6 are closely linked, mapping to human chromosomal loci Xq26 and 13q31–32, respectively. Mutants in the *Drosophila* and human glypican genes have been identified and give rise to unique developmental phenotypes (see Section 6).

4. Cell adhesion mediated by cell surface heparan sulfate proteoglycans

Cell surface HSPGs form signalling complexes with various extracellular proteins and modulate cell adhesion events including cell–cell, cell–matrix, and cell–microbe adhesion (Fig. 3). Upon ligand binding, cell surface HSPGs are immobilized in the plane of the membrane leading to interaction with the actin cytoskeleton. These HSPGs, particularly syndecans, have been shown to interact with signalling components associated with the cytoskeleton and may even generate a signal upon complex formation independent of other adhesion receptors.

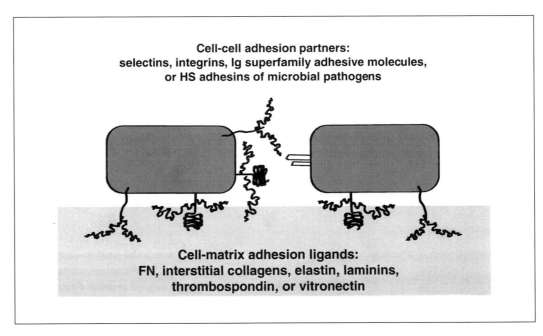

Fig. 3 Schematic representation of adhesive events mediated by cell surface HSPGs. Cell surface HSPGs can mediate cell–cell adhesion by interacting with HS binding cell adhesion molecules such as selectins, integrins, and Ig superfamily proteins, and also with HS adhesins of microbial pathogens. In cell–matrix adhesion, cell surface HSPGs via their HS chains, interact with ECM components including fibronectin, interstitial collagens, laminins, thrombospondin, elastin, and vitronectin. These cell adhesive events mediated by cell surface HSPGs play important roles in morphogenesis, tissue repair, feeding behaviour, and host defence.

4.1 Cell–cell adhesion

4.1.1 Cell surface HSPGs in intercellular adhesions

Initial evidence suggesting that cell surface HSPGs function in cell–cell adhesion came from the observation that syndecans localize at adherens junctions (47) and that several cell adhesion molecules bind HS (see Table 2). Subsequently, loss of syndecan-1 expression in myeloma cells has been found to correlate with the inability of these cells to aggregate (57), and stable transfection of the syndecan-1 or -4, but not betaglycan, cDNAs has been demonstrated to restore cellular aggregation (58), suggesting that these syndecans mediate cell–cell adhesion in this cell type. Syndecan-1 and -4 have also been implicated in cell–cell adhesion of fibroblasts (38, 59). Cells adhere to the extracellular core protein domain of these syndecans and this adhesion can be specifically blocked by the corresponding recombinant extracellular domains or the shed ectodomains. These findings were unique, since they clearly demonstrated that the core protein, and not the HS chains, of cell surface HSPGs can act in cell–cell adhesion. Furthermore, the observation that the shed syndecan ectodomains can promote adhesion suggests that shed ectodomains bound to the ECM may mediate cell–matrix adhesion of fibroblasts (59).

Although these findings indicate that cell surface HSPGs are active components of the cell–cell adhesion complex, the counter-receptors for the PGs have not been defined. Similarly, despite reports of HS binding activity by various cell adhesion molecules, their counter cell surface HSPG receptors are unknown. It is not clear whether these cell adhesion receptors specifically interact with syndecans, glypicans, or part-time HSPGs, or with any available cell surface HSPG. For example, HIP

Table 2 Partial list of adhesive molecules/microbes bound by heparin/HS

Cell–cell	Cell–microbe
L-selectin	**Viruses**
MAC-1	Adeno-associated virus
N-CAM	Cytomegalovirus
PECAM-1	Dengue virus
DCC	Foot-and-mouth disease virus
HIP	Herpes simplex virus
Cell–ECM	**Bacteria**
Elastin	*Bordetella pertussis*
Fibrin	*Borrelia burgdorferia*
Fibronectin	*Chlamydia trachomatis*
Interstitial collagens	*Listeria monocytogenes*
Laminins	*Neisseria gonorrhoeae*
Pleiotropin (HB-GAM)	*Staphylococcus aureus*
Tenascin	
Thrombospondin	**Parasites**
Vitronectin	*Leishmania* spp.
	Plasmodium spp.
	Trypanosoma cruzi

(HS/heparin interacting protein) (60, 61), L-selectin (62), the integrin Mac-1 (63), and the adhesion molecules of the immunoglobulin superfamily PECAM-1 (64), N-CAM (65), and DCC (Deleted in Colorectal Cancer) (66), have all been demonstrated to bind HS, but the identity of their cell surface HSPG receptors has not been defined and the physiological relevance of these interactions is not clear. The Mac-1 interaction with HS by itself is too weak to mediate leukocyte–endothelial cell adhesion under flow conditions, and the primary mode of cell–cell adhesion mediated by PECAM-1 and NCAM is homophilic. These studies raise several intriguing questions, most importantly whether there is a counter-receptor for these cell–cell adhesion events that are mediated by cell surface HSPGs. The composite data, however, do suggest that cell–cell adhesion events mediated by cell surface HSPGs are heterophilic in nature. Thus, not only is the molecular mechanism of cell surface HSPG-mediated cell–cell adhesion largely unknown, but the physiological role of these interactions is speculative.

4.1.2 Effects on E-cadherin expression and epithelial phenotype

Cell surface HSPGs can also modulate cell–cell adhesion by affecting the expression of other adhesion molecules. Syndecan-1 affects the organization of the actin cytoskeleton and the expression of E-cadherin, the molecule necessary for epithelial intercellular adhesions (67). Reduced expression of syndecan-1 and E-cadherin and subsequent cytoskeleton disorganization correlate with transformation of mouse midline epithelial to mesenchymal transformation (68), as well as when squamous cell carcinoma cells become poorly differentiated (69). When endogenous expression of syndecan-1 in mouse mammary gland epithelial cells is down-regulated by transfecting antisense syndecan-1 cDNA, expression of E-cadherin is also reduced. This is accompanied by morphological and functional changes of the epithelia; cells lose their cuboidal shape and palisaded organization, become fusiform, and gain the ability to invade and migrate within collagen gels and to grow in an anchorage-independent manner (67). Similarly, cell surface syndecan-1 is reduced when expression of E-cadherin is suppressed in these cells (70). However, epithelial morphology is not restored when syndecan-1 is induced or overexpressed in mesenchymal cells (which lack E-cadherin) (71, 72). These findings suggest that syndecan-1 and E-cadherin communicate to influence each other's expression, which regulates the intercellular adhesive phenotype of epithelial cells. However, the mechanism underlying this correlation is unknown.

4.2 Cell–extracellular matrix adhesion

4.2.1 Evidence for cell surface HSPGs as cell–matrix adhesion receptors

The ECM is comprised of many large multidomain components of which most bind, via discrete domains, to both integrin and cell surface HSPGs (see Table 2). The ECM components bind to HS with relatively low affinities; for example, HS binds fibronectin with a K_A of approximately 2 μM. However, it is important to note that the behaviour of the soluble or isolated components may differ from the native molecule incorporated in the matrix. Furthermore, in the physiological milieu of the insoluble ECM amalgam, HSPGs will have multiple, relatively low affinity inter-actions (73, 74). Moreover, cell surface HSPG-ECM interactions are usually specific in that intact heparin competes in cell adhesion assays, whereas chemically modified heparin, undersulfated HS, and chondroitin sulfate do not compete. These findings suggest that specific HS sequences may mediate cell surface HSPG-dependent cell–matrix adhesions. Although glypicans can bind to interstitial collagens and fibronectin, their role in cell–matrix adhesion is still speculative. In contrast, syndecans clearly act as co-receptors in cell–matrix interactions, acting to modify the cytoskeletal organization and the adhesive phenotype (27).

The expression of syndecans is consistent with a cell–matrix receptor role. For example, syndecan-1 polarizes to the basolateral surfaces of cultured epithelial cells (39) and of simple epithelial sheets (47) where they interact with the basement membrane ECM components. It also localizes to the initial site of ECM accumulation in early mouse embryos (75). Syndecan-1 and -3 co-localize with tenascin during tooth (76) and limb (77) development, respectively, and bind to tenascin. Syndecan-1 is expressed on developing B cells only when they are in contact with ECM (78) and binds B cells to type I collagen (57). Finally, syndecan-4 and under certain conditions syndecan-1 (4) localize to focal adhesions (48, 79).

4.2.2 Syndecans in the formation of focal adhesions

Focal adhesions are multicomponent signalling organelles where actin stress fibres insert into the plasma membrane at sites of close contact (10–15 nm) with the substratum (80, 81); see Chapter 9. The assembly of focal adhesions and actin stress fibres by cells plated on fibronectin require both integrin- and cell surface HSPG-mediated signals. Experimental evidence indicating that cell surface HSPGs are necessary is compelling: focal adhesions are not formed by HS-deficient CHO cells (82) or by fibroblasts plated on fibronectin lacking its heparin binding domain unless a heparin binding domain peptide is added (22). In contrast, expression of exogenous syndecan-1 cDNA in rat Schwann cells increases the number of focal adhesions (83). Cell surface HSPGs and integrins act co-operatively in generating the signals necessary for the assembly of focal adhesions. Antibody-mediated clustering of syndecan-4 at the cell surface induces focal adhesions in cells pre-spread on sub-strates coated with the integrin binding domain of fibronectin (27). Thus, cell surface HSPGs immobilized in the membrane by binding to the heparin binding domain of fibronectin provide a second signal that is required for formation of stress fibres and maturation of focal adhesions in mesenchymal cells (cf. ref. 27). This second signal requires the syndecan-4 cytoplasmic domain, is Rho-dependent (84) and presumably involves the interactions with activated PKC-α (79).

4.2.3 Syndecans in cell spreading

Human lymphoblastoid cells transfected with syndecan-1 spread on immobilized thrombospondin or fibronectin. Surprisingly, the cytoplasmic domain is apparently not required for the binding and spreading (85), and neither this domain nor the transmembrane domain is needed for the formation of actin-containing dendritic processes by antibody-mediated ligation of syndecan-4 on activated B cells (86). Thus, it has been proposed that cell spreading is likely mediated by interactions of the syndecan core protein with signalling elements associated with membrane lipid rafts (87). Alternatively, syndecans along with other membrane proteins may form a multimolecular signalling complex at the cell surface that signals through cytoskeletal reorganization, similar to integrin signalling (80). However, the recent identification of syndesmos, a novel cytoplasmic protein that binds specifically to the cytoplasmic domain of syndecan-4 suggests that the cytoplasmic domain of syndecans may mediate cell spreading under certain circumstances (84). Overexpression of syndesmos in NIH 3T3 cells has been found to induce cell spreading and actin cytoskeletal reorganization. Thus, while syndecans appear to mediate cell spreading events, the signalling mechanisms remain unclear, specifically in regards to the role of the extracellular and cytoplasmic domains.

4.3 Cell–microbe adhesion

4.3.1 *In vitro* evidence

The initial binding of microbial pathogens to host tissue components is a prerequisite for a successful infection. Numerous bacteria, protozoa, and viruses have been found to bind specifically to cell surface HSPGs (88) (see Table 2). The list of microorganisms binding to HSPGs is extensive, and includes both extracellular and intracellular pathogens such as *Bordetella pertussis*, *Neisseria gonorrhoeae*, *Chlamydia trachomatis*, and Herpes simplex virus (HSV), among others (see Table 2). Except for *Chlamydia*, which has been reported to express a heparin-like molecule (89), none of the identified HSPG binding microbes express an endogenous HS ligand for their HSPG binding proteins (adhesins). Furthermore, the majority of the listed intracellular microbes can no longer infect target cells *in vitro* when expression of the HSPGs has been abrogated by chemical or mutagenesis methods, or when their HSPG binding is inhibited by addition of exogenous heparin or HS (88). Based on these findings, the interaction with cell surface HSPGs has been proposed to be an important determinant of microbial attachment and invasion into host cells (88).

The various binding interactions between cell surface HSPGs and microbial pathogens were established by measuring binding properties such as kinetics, protease susceptibility and affinity of the interaction, and also by studying the microbial specificity for GAGs. Results from these studies show that specific HSPG adhesins of pathogens recognize the HS chains, and not the core protein component of cell surface HSPGs. Identified microbial HSPG binding proteins do not show significant sequence homology with each other. Where studied, the HS binding domains in the HSPG adhesins are comprised of a region rich in basic residues, flanked by hydrophobic domains (88). Since information is scarce in this area of research, it is not known if a similar binding domain is utilized by other HSPG binding microbial pathogens.

4.3.2 Relevance to microbial pathogenesis

HSPGs are ubiquitous at the cell surface. If the HS interaction is important for pathogenesis, it remains to be elucidated as to how some HSPG binding microbial pathogens maintain tissue tropism. The specificity for certain HS macrostructures has been defined in some cases, and these properties may account for the pathogen's *in vivo* tropism (90–92). Although it remains to be clarified whether HS chains at different tissue sites are indeed structurally and functionally distinct (13), microbial pathogens appear to clearly prefer certain HSPGs for their adhesion. For example, the aetiologic agent of malaria, *Plasmodium falciparum*, binds preferentially to the sinusoidal side of target hepatocytes via its HSPG interaction. These results suggest that syndecan-1, which is targeted to this site, is this pathogen's cell surface HSPG determinant (93).

Despite a plethora of biochemical information, the exact role of cell–microbe adhesion in pathogenesis is still not clear. For some viruses, the HSPG interaction has been suggested to be physiologically non-functional, since HSPG binding has been found to be non-essential for infection (94) or because HSPG-dependent pathogenesis was determined to be an *in vitro* phenomenon caused by adaptation of the virus to cell culture conditions (95, 96). Available data, however, indicate that the interaction with cell surface HSPGs is indeed an important component of pathogenesis for certain pathogens. The composite data suggest that cell surface HSPG is a co-receptor of a two receptor mechanism utilized by the intracellular pathogens for their infection. In this model, one host receptor provides a scaffold for attachment whereas the other mediates entry of the pathogen into host cells. Cell surface HSPGs appear to play a role as an attachment site. For example, the initial binding of HSV and other alpha herpes viruses to host cells is mediated by the interaction between viral glycoprotein C and cell surface HSPG of the host (97). This binding then allows the surface bound virions to fuse with the host cell membrane for internalization using secondary interactions between viral glycoproteins B, D, and L, and host determinants such as the poliovirus receptor-related protein 1 (98). Similarly, it has been proposed that the initial interaction between Dengue and foot-and-mouth viruses and cell surface HSPG serves to concentrate virus particles at the cell surface for subsequent binding to host integrin receptors (99, 100). These properties of the invasion mechanism also suggest that the expression pattern of the second receptor may dictate the observed microbial tissue tropism.

The signalling cascade mediating internalization of *Neisseria gonorrhoeae* has been found to involve cell surface HSPGs. Adhesion of *Neisseria* to a cell surface syndecan-like molecule triggers an intracellular signalling cascade in the host cell involving generation of active phosphatidylcholine-specific phospholipase C, diacylglycerol, acidic sphingomyelinase, and membrane sphingomyelin-derived ceramide (101). Importantly, generation of ceramide is essential for *Neisseria* invasion. The mechanism as to how adhesion of *Neisseria* to syndecans activates this internalization signalling in host cells, however, remains to be delineated.

Available data clearly show that cell surface HSPGs are important host determinants for the attachment and invasion of several intracellular pathogens. However, the physiological significance of HSPG binding is still controversial for most microbial pathogens. Furthermore, none of the studies have thus far defined the molecular identity of cell surface HSPGs involved. Syndecans have been implicated in some interactions, but this hypothesis needs experimental verification. Other questions also remain to be answered. For example, do microbial pathogens recognize specific or multiple cell surface HSPGs? Does binding of an intracellular pathogen to cell surface HSPG trigger a common internalization signal in the target host cell? Does microbial binding to cell surface HSPGs benefit only the pathogens, or in some cases, can it be a host defence mechanism promoting clearance of the pathogen? These questions are significant public health concerns since effective prophylactic or therapeutic measures are either losing their potency or currently not available for many microbial pathogens that bind to HSPGs. With the generation of specific cell surface HSPG mutants, it has become feasible to test directly the role of these interactions in microbial pathogenesis.

5. Cellular consequences of HSPG-mediated adhesion

Core protein interactions involving the cytoplasmic domains of syndecans have been well studied. The cytoplasmic domain is subdivided into three regions: the highly conserved C1 and C2 regions and the variable V region (see Fig. 4). Starting at the N-terminus of the cytoplasmic domain, the C1 region is proximal to the plasma

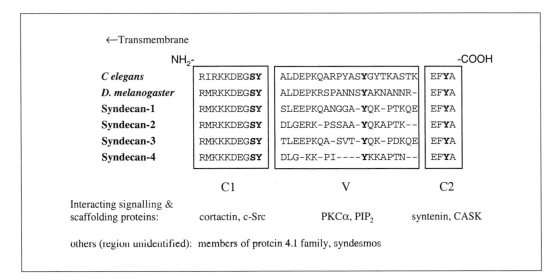

Fig. 4 Domain structure of syndecan cytoplasmic regions. Syndecan cytoplasmic tails are 33–36 amino acids in length and have been further subdivided into three domains named C1, C2, and V. As shown these domains contain sites for interactions with multiple intracellular targets including cell scaffolding proteins, actin cytoskeleton, and signalling molecules. The C1 and C2 domains are highly conserved among all the syndecans and thus it is likely that all bind to the same intracellular targets. In contrast, the V domain is not conserved among the syndecans, thus may contain sites for unique protein interactions with the individual syndecans.

membrane and consists of a stretch of 10 highly homologous amino acids, R(M/I) (R/K)KKDEGSY. The V region (c. 16–22 amino acids) follows the C1 region, and is homologous between species, but varies among the different syndecan family members. The C2 region is at the C-terminus of the protein and distal from the plasma membrane. This region is comprised of four amino acids, EFYA, completely conserved between the four mammalian, *C. elegans,* and *D. melanogaster* syndecans. These findings suggest that the conserved C1 and C2 regions mediate core protein interactions common to syndecan function, whereas, the variable V region mediates interactions specific for each syndecan type. Furthermore, presence of highly conserved serine and tyrosine residues in the cytoplasmic domain suggests regulation of syndecan activity by Ser/Thr and protein tyrosine kinases (PTKs). These hypotheses have been tested experimentally, and the results have provided potential mechanisms as to how cell surface HSPGs transduce extracellular signals and link the ECM with the intracellular cytoskeleton.

5.1 Core protein interactions

5.1.1 Interactions with the actin cytoskeleton and cell scaffolding proteins

There is ample evidence indicating that the cytoplasmic domain of syndecans interacts with the actin cytoskeleton during cell adhesion. For example, syndecan-1 and F-actin co-localize at the basal surfaces of polarized mammary epithelial cells (102), actin filaments are reorganized upon engagement of the syndecans with an ECM ligand (103) or upon depletion of syndecan-1 from epithelial cells (67), and syndecan 1 binds to a crude preparation of F-actin filaments (104). Moreover, Schwann cells transfected with syndecan-1, not normally expressed in these cells, show actin cytoskeleton reorganization and transient co-localization of cell surface syndecan-1 and F-actin during cell spreading (105). Co-localization of syndecan-1 and F-actin and depends on the cytoplasmic domain (106) and is disrupted by cytochalasin treatment (107). Finally, clustering of cell surface syndecan-1 by anti-syndecan-1 antibodies, or by FGF-2- or collagen I-coated beads also leads to recruitment of actin microfilaments (108).

These interactions with the actin microfilaments are thought to be mediated by the C2 region of the syndecan cytoplasmic domain. The highly conserved EFYA motif can interact with the actin cytoskeleton and cell scaffolding components via binding to PDZ domain-containing proteins (109). PDZ domain proteins (named for *P*SD-95, *D*iscs-large and *Z*onula occludens-1 proteins) bind specific C-terminal sequences, organize and assemble protein complexes at the inner face of the plasma membrane, and link membrane components to the underlying actin-containing cytoskeleton. Using the yeast two-hybrid system, several PDZ domain-containing proteins that interact with the C2 region of syndecans have been identified (110–112). Grootjans *et al.* (110) identified binding of syntenin to the syndecan C-terminus using a yeast two-hybrid screen. Syntenin is a widely distributed 33 kDa protein containing two class II PDZ domains. Syntenin co-localizes with mouse syndecan-2 at the plasma membrane and in intracellular vesicles of CHO cells transfected with a GFP–syntenin fusion protein. Furthermore, syntenin is likely involved in cytoskeleton organization because high levels of syntenin expression results in the formation of cell surface projections and a flattened cell morphology (110). A similar function has been ascribed to another PDZ domain protein, CASK, which also binds to the EFYA motif of syndecans (111). CASK (the mammalian homologue of LIN2A from *C. elegans*) is a 120 kDa protein belonging to the family of *m*embrane-*a*ssociated *gu*anyate *k*inase (MAGUK) PDZ domain proteins. CASK is a 909 aa protein composed of several distinct domains, including a Ca^{2+}/*Ca*lmodulin-dependent protein *k*inase domain, a *S*rc homology type 3 domain, and an apparently inactive guanylate kinase domain (113). It was identified in a yeast two-hybrid screen for proteins which interact with the cytoplasmic domain of neurexins, a highly polymorphic gene family expressed in the peripheral and central nervous system that share high homology with the syndecan cytoplasmic domains. MAGUK proteins often localize to adherens junctions, and are likely scaffolding proteins acting to recruit and organize cytoskeletal components and signalling molecules in adhesion complexes at the cell surface (114). CASK co-localizes with syndecan-1 in a variety of epithelial tissues, co-localizes with syndecan-2 in rat brain synapses, and can be co-immunoprecipitated with rat syndecan-2 from COS-7 cells expressing both proteins (112). CASK can bind to protein 4.1 family members, including protein 4.1, ERM (*E*zrin, *R*adixin, *M*oesin) proteins and talin, each known to associate with the actin cytoskeleton. Additionally, protein 4.1 can also bind directly to syndecan cytoplasmic domains, though the specific binding site on the syndecan cytoplasmic domain is not known (24). Thus, protein 4.1 family members can provide indirect links between syndecans and the actin microfilament. Taken together, the findings indicate that cell adhesion events mobilize syndecan cytoplasmic domains into an organized complex of sub-plasmalemmal scaffolding proteins. Further delineation of the mechanism awaits identification of additional components.

5.1.2 Interactions with signalling molecules

Phosphorylation of cell adhesion receptors is a common mechanism regulating cellular adhesive behaviours. The syndecan-4 cytoplasmic domain can bind to activated PKC-α via a syndecan-4 specific amino acid sequence, LGKKPIYKK, located in the V region (115). This interaction is thought to regulate the recruitment of both syndecan-4 and PKC-α to focal adhesion sites and lead to formation of syndecan-4 oligomers (24).

The C1 region of syndecan-3 cytoplasmic domain has also been reported to form a complex with Src and Fyn, member of the Src PTKs. Because the C1 region is highly conserved among different syndecans and across species, these results suggest that other syndecans also bind Src family PTKs via their C1 region. Moreover, the presence of one invariant serine and three invariant tyrosine residues in the syndecan cytoplasmic domain suggests regulation of syndecan activity by phosphorylation. Indeed, available evidence indicates that each of these amino acids can be phosphorylated. When syndecan-3 was expressed in bacteria harbouring Elk kinase, all three tyrosine residues were phosphorylated (116). The invariant serine and tyrosine residue of syndecan-4 are also phosphorylated endogenously (117, 118). Interestingly, the serine appears to be dephosphorylated in response to FGF-2 (119). Cell adhesion to ECM modifies the action of growth factors including FGF-2 (120). Thus, one possible mechanism for this effect could be that the adhesion induces an interaction between the V region and phosphatidyl inositol-4,5 bisphosphate (PtdIns-4,5-P_2), leading to syndecan oligomerization and subsequent phosphorylation/ dephosphorylation events. In this manner, syndecans act both to assist in focal adhesion formation and recruitment of signalling molecules.

6. Consequences of altered heparan sulfate proteoglycan cell adhesions

HS linked to the syndecans and glypicans interacts with multiple soluble and insoluble ligands at cell surfaces and helps co-ordinate diverse processes. Furthermore, cell surface HSPGs also participate in intracellular complexes and perhaps signalling. While these biochemical and cellular activities implicate HSPGs in multiple processes these must be evaluated in the whole animal in the context of physiology and pathophysiology. Identification of human disease loci as well as gene deletion in mice and genetic screens in *Drosophila* are unravelling novel roles for these cell surface HSPGs. Interestingly, though the mechanisms underlying these phenotypes have not been fully explored, these phenotypes are not obviously related to the identified cellular roles of HSPGs in cell adhesion events.

6.1 Neoplastic transformation

Lesions in GlcUA/GlcNAc co-polymerase (HS co-polymerases) and glypican-3 lead to human bone and generalized overgrowth syndromes, respectively, and may lead to neoplasia. The underlying mechanism is unknown though as described earlier in this chapter multiple ligands with roles in cell proliferation and adhesion interact with cell surface HSPGs. Furthermore, the recent cloning of a mammalian heparanase and its ability to transform low metastatic T lymphoma and melanoma cells to a highly metastatic phenotype further implicate HSPGs on the cell surface and/or in the ECM as tumour suppressors.

6.1.1 HS co-polymerase and benign tumours

EXT1 and EXT2 code for HS co-polymerases and mutations in these genes are the cause of the autosomal dominant hereditary multiple exostoses syndrome in humans (121, 122). This disorder is characterized by development of multiple benign bone tumours, derived from the growth plate, also called exostoses, which may transform to malignant chondrosarcomas and osteosarcomas (123). A *Drosophila* homologue of EXT1, *tout velu* (*ttv*), has been identified, and mutations in this gene lead to segment polarity defects likely as a result of inappropriate diffusion of the Hedgehog signalling molecule (124). Both EXT1 and EXT2 have been shown to have D-glucoronyl (GlcA) and *N*-acetyl-D-glucosaminyl (GlcNAc) transferase activity (121). Despite the homology between *ttv* and EXT1, transferase activity has not been demonstrated for the *Drosophila* gene. The mechanism(s) underlying the phenotypes of the HS co-polymerase mutants in both humans and fly remains unknown. Furthermore, the EXTs appear to function as tumour suppressors because in their absence multiple bone tumours develop, yet it is not clear how the reduction or absence of HS would lead to uncontrolled cell proliferation (125).

6.1.2 Glypican-3 and overgrowth syndrome

Mutations in the human glypican-3 gene result in the X-linked disorder Simpson–Golabi–Behmel (SGB) syndrome (32). SGB syndrome is characterized by pre- and postnatal generalized somatic overgrowth, macroglossia, unique facial appearance, and increased incidence of embryonal tumours and a spectrum of anomalies that include heart defects, umbilical hernias, renal defects, vertebral and rib defects, polydactyly, imperforate anus and genito-urinary anomalies (126). The observed phenotypes are likely due to reduced cell death or increased cell proliferation. *In vitro*, glypican-3 has been shown to regulate the entry of cells into programmed cell death (apoptosis), though this function does not require the presence of its HS chains (127). Interestingly, the *Drosophila* glypican homologue was identified in a genetic screen for cell division mutants (31). Identification of glypican-3 as the mutated gene in SGB syndrome is only the beginning of elucidating the pathophysiology of this disorder. It was originally proposed, based on clinical similarity to the Beckwith–Wiedmann syndrome, that glypican-3 binds to insulin-like growth factor-2 (IGF-2) and acts as a negative modulator of IGF-2 (32). Beckwith–Wiedmann syndrome is thought to be due to overproduction of IGF-2. Yet, to date no interaction between glypican-3 and IGF-2 has been documented and the mechanism underlying this diverse syndrome remains unclear. The recent generation of glypican-3 null mice, which give rise to a similar overgrowth phenotype, will likely be useful in gaining a better understanding of this complex syndrome (128).

6.1.3 Mammalian heparanases and metastasis

Multiple studies have shown an association between either decreased HS or increased heparanase activity with increased metastatic potential in tumours (reviewed in refs 129–131). Additionally, adhesive and invasive properties of tumour cell lines correlate with the presence of syndecan-1, whereas loss of syndecan-1 is associated with acquisition of a malignant phenotype (30, 132). Furthermore, in squamous cell carcinomas, presence of syndecan-1 is associated with a more favourable outcome (69). The recent cloning of the mammalian heparanase opens new avenues for understanding mechanisms involved in metastases as well as inflammation (133). Active heparanase is a 50 kDa endoglucuronidase that cleaves HS chains preferentially in the UMDs and leaving the HSDs intact (134–136). The released HSDs likely have ligands, including growth factors, associated with them have been identified in human acute wound fluids (137).

6.2 *Drosophila* mutants

Drosophila development follows a series of morphogenetic events, some occur sequentially while others are simultaneous (138). As a genetic organism easily amenable to genetic and transgenic manipulation, *Drosophila* mutants have been useful in defining at the molecular level the events underlying its development (139, 140). Unique mutants have been identified and cloned, which regulate the anterior–posterior axis, dorso–ventral patterning, gastrulation, and organogenesis (138). These studies have identified diverse signalling pathways regulated by receptor tyrosine kinases, G protein-coupled receptors, cell adhesion, and ECM molecules to name a few.

Using screens, genetic interactions, transgenic, and RNA interference (RNAi) methods, the function of HS and its core proteins are being elucidated in *Drosophila* (141). Importantly, these studies provide *in vivo* validation for a role of HS as co-receptor in the action of fibroblast growth factors (142), evidence for developmental roles for HSPGs, and identification of new receptor ligand encounters modulated by HS, e.g. *wingless* (*wg*), *decapentaplegic* (*dpp*), *hedgehog* (*Hh*), and *spaetzle* (124, 143, 144). These studies also point to the specificity of the interactions between HSPGs and their ligands. It is commonly perceived that HS interaction with ligands and receptors is non-specific, generally functioning to sequester ligands or protect them from proteolysis. Finally, it is becoming apparent that HSPGs play a specific role in modulating interactions between ligands and their signal transducing receptors.

6.2.1 HSPG biosynthetic enzymes

Expression of HS and HSPGs is detected early in the developing *Drosophila* embryo. HS expression is detected upon cellularization and at the site of initial gastrulation, the ventral epithelium. Subsequently it is found widespread in all three germ layers (Lincecum and Bernfield, unpublished observation). Importantly, mutations in HS biosynthetic enzymes modify signalling mediated via secreted morphogens such as *spaetzle*, *wingless*, and *hedgehog* (124, 143, 144). These include the UDP-glucose dehydrogenase (*sugarless*, *sgl*), N-deacetylase N-sulfotransferase (NDST, *sulfateless*, *sfl*), HS co-polymerase (*tout velu*, *ttv*), and the HS 2-*O*-sulfotransferase (HS2ST, *pipe*). Finally, these studies implicate HSPGs as modulators of diffusible ligands and proteolytic enzymes involved in dorso–ventral patterning and segmentation (Table 3).

Drosophila dorso–ventral patterning is dependent on signals initially established in the pre-fertilized egg. There are 12 genes involved in this patterning event, including Toll, a transmembrane receptor homologous to the interleukin-1 receptor, its ligand, spaetzle, and several immediate targets of this signalling pathway (145, 146). *Spaetzle* is activated by several proteases resulting in a ventralizing gradient that induces the nuclear localization of transcription factor Dorsal (147). The products of the *pipe* locus encode homologues of the vertebrate HS2ST. *Pipe* expression in the ovarian follicular epithelium is required for the ventral specific activation of *spaetzle* (143). Thus, activation of this proteolytic cascade is modulated by a specific HS sequence with requirement for 2-*O* sulfated IdUA residues. These studies are similar to the action of the syndecan ectodomains in activating proteolytic cascades in wounds by protecting proteinases from the action of serpins (148).

Drosophila segmentation establishes the future adult organs. Segmentation is set up by the gap genes whose localization is determined by the maternal genes Bicoid and *Nanos*, which initially establish anterior–posterior axis. These segments are further subdivided into anterior and posterior parasegments, which are governed by the action of the secreted morphogens *wg* and *Hh*, and both modulated by HSPGs (144, 149). The boundaries set up by the parasegments will lead to the deposition of the imaginal discs, sacs of cells that develop into the adult organs during pupal metamorphosis. The action of *wg* is likely modulated specifically by the glypican homologue *dally* (150, 151). Importantly, in the absence of the GAG chains, *wg* signalling is not properly transduced because mutants in HS biosynthesis (*sgl* and *sfl*) recapitulate the *wg* cuticle phenotype (144, 152, 153). Importantly, UDP-glucose dehydrogenase activity has been assayed in the wild-type *sgl* allele (154). Additionally, *Hh* long distance signalling is also modified by HS since mutants in the HS co-polymerase, *ttv*, show decreased diffusion of this diffusible morphogen (124, 155, 156). Yet, it is presently not known whether the phenotype is due to loss of a cell surface or soluble HSPG. Finally, a limitation of the *sfl* and *ttv* biosynthetic mutants is that they are based on homology to the vertebrate enzymes. To date no direct activity has been documented for these *Drosophila* enzymes. Although HS is reduced in these mutants based on limited biochemical and immunohistological analyses, chondroitin sulfate is not affected (156). It is interesting to note that while the *sgl*, *sfl*,

Table 3 Mutations in PG biosynthetic enzymes and PGs affect morphogenesis and growth

Gene	Mutation[a]		Phenotype	
	Drosophila	**Mammal**	**Drosophila**[b]	**Mammal**
UDPG dehydrogenase	*sugarless*		*Wingless Heartless, Breathless, Branchless*	
HS polymerase	*tout velu*	Null mutation	*Hedgehog* diffusion	Bone tumours exostoses 1 and 2
HS *N*-deacetylase/ *N*-sulfotransferase	*sulfateless*	Null mutation	*Wingless, Heartless, Breathless, Branchless*	Mast cell defects[c]
HS 2-*O*-sulfotransferase	*pipe*	Null mutation	*Spaetzle*	No kidneys
Glypican	*Dally*	Glypican-3 null	*Wingless* phenocopy	Overgrowth syndrome (Simpson–Golabi– Behmel syndrome)

[a] See text for details.

[b] Humphries, D. E., Wong, G. W., Friend, D. S., Gurish, M. F., Qiu, W. T., Huang, C., *et al.* (1999). Heparin is essential for the storage of specific granule proteases in mast cells. *Nature*, **400**, 769; and Forsberg, E., Pejler, G., Ringwall, M., Lunderius, C., Tomasini-Johansson, B., Kusche-Gullberg, M., *et al.* (1999). Abnormal mast cells in mice deficient in a heparin-synthesizing enzyme. *Nature*, **400**, 773.

[c] *Wingless*, segmental patterning; *Heartless*, mesoderm cell migration and cardiac defects; *Breathless*, abnormal tracheal branching; *Branchless*, abnormal tracheal branching; *Hedgehog*, defective wing anterior–posterior patterning; *Spaetzle*, defective dorso–ventral axis.

and *ttv* mutants all have reduced HS their effects are not necessarily overlapping. For example, *ttv* show specificity for the *Hh* signalling pathway with no effect on FGF or *Wg* signalling (155), and in a similar manner *sfl* and *sgl* do not affect *Hh* signalling. This lack of overlap may argue for specificity in HS chains produced by the different enzymes or may in fact indicate that there are additional unidentified *Drosophila* HS biosynthetic enzymes or related enzymes that compensate for the loss of these enzymes.

Finally, the *in vivo* role of HSPGs as FGF co-receptors was elegantly demonstrated in the HS biosynthetic mutants *sfl* and *sgl* (142). As discussed in this chapter, HSPGs act as co-receptors by modulating ligand receptor encounters originally demonstrated for the FGF-FGFR pathway. These studies indicate that HS catalyses the molecular interaction similar to the action of enzymes. Thus, the effect of the HS can be overcome by increasing the concentration of ligand, essentially nudging the reactants to interact. In *Drosophila*, FGF is implicated in branching morphogenesis and mesodermal patterning. There are two FGF receptors known in *Drosophila* (157, 158), which give rise to distinct phenotypes when mutated; *Heartless* (*Htl*) (159) that lacks visceral mesoderm and heart, and *Breathless* (*Btl*) (160) that lack tracheal branching. The ligand for *Btl* is *Branchless* (*Bnl*), the *Drosophila* FGF, and *Bnl* mutants share many characteristics with *Btl* (161). Strikingly, the *sgl* and *sfl* mutant embryos have phenotypes similar to *Htl* and *Btl* (142). As predicted from *in vitro* experiments, constitutively active forms of *Htl* partially rescues both of these HS biosynthetic mutants. Furthermore, overexpression of the ligand, *Bnl*, can partially overcome the requirement of either *sgl* and *sfl* on *Btl* activity. Thus, either the FGF receptor, *Htl*, or the FGF ligand, *Bnl*, can overcome the modulatory role of the HS co-receptor.

6.2.2 Core proteins

Currently, only two HS core PGs have been identified, D-syndecan and *dally*, the glypican homologue. Expression of D-syndecan transcripts is observed in the epidermis, the abdominal and thoracic segments, the differentiating CNS, the ventral furrow, and the ventral nerve cord of the early embryo (162). As the embryo matures, expression is detected in tissues such as the lymph glands, central and peripheral nervous system, and in the gut epithelium. Hypomorphic alleles in the D-syndecan gene do not give rise to early or late embryonic phenotypes, apparently due to maternal zygotic contribution, but result in third instar larval lethality.

The glypican homologue was identified in a screen for mutants in cell division patterning in visual system development, and named *dally* for division abnormally delayed (31). *Dally* like the vertebrate glypicans is a GPI-anchored HSPG (150, 151). Using genetic interaction and RNAi experiments, *dally* was shown to regulate the activity of *wg* in cuticle patterning (150, 151). *Dally* expression has not been extensively studied, though it is expressed in a segmented pattern in cells adjacent to *wg* expressing, correlating with its function in *wg* signalling (151). *Dally* has also been implicated in *dpp* signalling in imaginal tissue development including the genitalia. In fact, *dally* mutants have a highly penetrant genitalia defect. Using the same genetic approaches, *dally* was shown to potentiate the action of *dpp* signalling on genitalia development (151). These *in vivo* studies strongly support the co-receptor model of HS action.

6.3 Mouse mutants

Both HS biosynthetic enzymes and the core proteins are expressed early in mouse development. In fact, syndecan-1 appears as early as the four cell blastocyst. Yet, mice with a null allele for syndecan-1 as well as with mutations in the HS 2-*O*-sulfotransferase do not show early embryonic phenotype. Lack of phenotypes may indicate selectivity in unique HS structures that are required for early morphogenetic events. In contrast, the apparently normal development in syndecan-1 null mice suggests compensation by other HSPGs.

6.3.1 HS 2-*O*-sulfotransferase null mice

Mice deficient for the HS2ST, created by a gene trap transgenic approach, exhibit late embryonic defects and die of a syndrome that includes bilateral renal agenesis, bilateral coloboma of the iris, skeletal fusions, and ectopic ossifications (163). The enzyme catalyses the transfer of sulfate to the 2 position of the IdUA in HS chains and has at least two isoforms (163, 164). The kidney defect observed in these mice is failure of the uteric bud to branch and lack of mesenchyme condensation surrounding it. It is likely that this enzyme generates a unique HS sequence that interacts with a specific ligand involved in organogenesis of the kidney. Clearer understanding of this process awaits identification of the cognate ligand that binds to this unique sulfated IdUA sequence.

6.3.2 Syndecan-1 null mice

Syndecan-1 null mice are phenotypically identical to their wild-type littermates in viability, development, and gross anatomy (Hinkes, Gibson, and Bernfield, unpublished observation). This lack of phenotype was quite unexpected because syndecan-1 is expressed on the cell surface of the morula, has been implicated in multiple developmental processes including the epithelial–mesenchymal interactions of organogenesis, and is highly expressed in adult epithelia and plasma cells (52, 165). Nonetheless, syndecan-1 null mice show a delayed repair of skin and corneal wounds (unpublished), likely due to a defect in keratinocytes or keratocytes in restoring their stable cell–cell and cell–matrix contacts at a normal rate (41, 137). Thus, one function of syndecan-1 may be in the regulation of normal epithelial response to challenges such as during wound repair or microbial invasion.

7. Future prospects

We have reviewed the available evidence implicating cell surface HSPGs as major cell adhesion receptors. These PGs bind to numerous cell–cell and cell–matrix adhesive ligands and transduce extracellular signals to the intracellular signalling machinery. The details of the mechanisms, however, have not been completely delineated. How do cell surface HSPGs engage adhesive ligands? Are specific sequences of the HS chains involved? If so, does this imply that there are biologically relevant microstructural differences in the HS chains among different cell surface HSPGs? These are only some of the questions that remain unanswered.

At the cellular level it remains to be determined whether expression and function of cell surface HSPGs are regulated to maintain the specificity of cell adhesion events. The finding that syndecan-1 and E-cadherin affect each other's expression pattern suggests that cell surface HSPGs communicate with other adhesion receptors. Our preliminary data showing that fragments of ECM components, such as hydrolysed elastin peptides and fibrinogen peptides derived from clot supernatant, induce syndecan-1 expression (Park and Bernfield, unpublished data) suggest that matrix components may also regulate expression in the context of wound healing. In fact, it is known that syndecan expression and shedding are tightly regulated during tissue repair (41, 72). However, it remains to be examined whether cell surface HSPGs can modulate the expression and function of the ECM (e.g. matrix assembly).

The recent generation of various mutant animal models has allowed us to appreciate the physiological significance of cell surface HSPGs. These valuable tools should also enable us to better understand how the molecular mechanisms of cell adhesions mediated by cell surface HSPGs translate into pathophysiological manifestations.

Acknowledgements

The authors would like to thank members of the Bernfield Laboratory for their input and critical evaluation of the work. We would also like to thank Mary Todd for editing and assistance with generation of figures and tables. Pyong Woo Park thanks the Parker B. Francis Foundation for their generous support. Ofer Reizes was supported by an individual NRSA grant from the NIH. Work in Dr Bernfield's laboratory has been supported by grants from the NIH (HD 06763, CA 28735).

References

1. Trelstad, R. L. (1984). The role of extracellular matrix in development. In *The Proceedings of the 42nd Annual Symposium of the Society of Developmental Biology*. A. R. Liss, New York.
2. Wight, T. N. and Mecham, R. P. (eds.) (1987). *Biology of proteoglycans*. Academic Press, Orlando.
3. Bernfield, M., Götte, M., Park, P. W., Reizes, O., Fitzgerald, M. L., Lincecum, J., *et al.* (1999). Functions of cell surface heparan sulfate proteoglycans. *Annu. Rev. Biochem.*, **68**, 729.
4. Yanagishita, M. and Hascall, V. C. (1992). Cell surface heparan sulfate proteoglycans. *J. Biol. Chem.*, **267**, 9451.
5. Bernfield, M., Kokenyesi, R., Kato, M., Hinkes, M. T., Spring, J., Gallo, R. L., *et al.* (1992). Biology of the syndecans: a family of transmembrane heparan sulfate proteoglycans. *Annu. Rev. Cell Biol.*, **8**, 365.
6. Hay, E. D. (1985). Matrix-cytoskeletal interactions in the developing eye. *J. Cell. Biochem.*, **27**, 143.
7. Jalkanen, M. (1987). Biology of cell surface heparan sulfate proteoglycans. *Med. Biol.*, **65**, 41.

8. Woods, A. and Couchman, J. R. (1988). Focal adhesions and cell-matrix interactions. *Coll. Relat. Res.*, **8**, 155.

9. Couchman, J. R., Austria, M. R., and Woods, A. (1990). Fibronectin-cell interactions. *J. Invest. Dermatol.*, **94**, 7S.

10. Lindahl, U. (1989). Biosynthesis of heparin and related polysaccharides. In *Heparin, chemical and biological properties, clinical applications* (ed. D. A. Lane and U. Lindahl), p. 159. CRC Press, Inc., Boca Raton.

11. Conrad, H. E. (1998). *Heparin-binding proteins*. Academic Press, San Diego.

12. Nader, H. B., Ferreira, T. M. P. C., Toma, L., Chavante, S. F., Dietrich, C. P., Basu, B., *et al.* (1988). Maintenance of heparan sulfate structure throughout evolution: chemical and enzymic degradation, and ^{13}C-n.m.r.-spectral evidence. *Carbohydr. Res.*, **184**, 292.

13. Kato, M., Wang, H., Bernfield, M., Gallagher, J. T., and Turnbull, J. E. (1994). Cell surface syndecan-1 on distinct cell types differs in fine structure and ligand binding of its heparan sulfate chains. *J. Biol. Chem.*, **269**, 18881.

14. Sanderson, R. D., Turnbull, J. E., Gallagher, J. T., and Lander, A. D. (1994). Fine structure of heparan sulfate regulates syndecan-1 function and cell behavior. *J. Biol. Chem.*, **269**, 13100.

15. Aikawa, J. and Esko, J. D. (1999). Molecular cloning and expression of a third member of the heparan sulfate/heparin GlcNAc N-deacetylase/ N-sulfotransferase family. *J. Biol. Chem.*, **274**, 2690.

16. Liu, J., Shworak, N. W., Sinay, P., Schwartz, J. J., Zhang, L., Fritze, L. M., *et al.* (1999). Expression of heparan sulfate D-glucosaminyl 3-O-sulfotransferase isoforms reveals novel substrate specificities. *J. Biol. Chem.*, **274**, 5185.

17. Shworak, N. W., Liu, J., Petros, L. M., Zhang, L., Kobayashi, M., Copeland, N. G., *et al.* (1999). Multiple isoforms of heparan sulfate D-glucosaminyl 3-O-sulfotransferase. Isolation, characterization, and expression of human cDNAs and identification of distinct genomic loci. *J. Biol. Chem.*, **274**, 5170.

18. Casu, B., Petitou, M., Provasoli, M., and Sinay, P. (1988). Conformational flexibility: a new concept for explaining binding and biological properties of iduronic acid-containing glycosaminoglycans. *Trends Biochem. Sci.*, **13**, 221.

19. Cardin, A. D., Demeter, D. A., Weintraub, H. J., and Jackson, R. L. (1991). Molecular design and modeling of protein-heparin interactions. In *Methods in enzymology: Molecular design and modeling: Concepts and applications, part B: Antibodies and antigens, nucleic acids, polysaccharides, and drugs*, Langone, J. J. (ed.), Academic Press, San Diego, Vol. 203, p. 556.

20. Rapraeger, A. C. (1995). In the clutches of proteoglycans: how does heparan sulfate regulate FGF binding? *Chem. Biol.*, **2**, 645.

21. Gallagher, J. T. and Turnbull, J. E. (1992). Heparan sulphate in the binding and activation of basic fibroblast growth factor. *Glycobiology*, **2**, 523.

22. Woods, A., McCarthy, J. B., Furcht, L. T., and Couchman, J. R. (1993). A synthetic peptide from the COOH-terminal heparin-binding domain of fibronectin promotes focal adhesion formation. *Mol. Biol. Cell*, **4**, 605.

23. Lander, A. D., Stipp, C. S., and Ivins, J. K. (1996). The glypican family of heparan sulfate proteoglycans: major cell surface proteoglycans of the developing nervous system. *Perspect. Dev. Neurobiol.*, **3**, 347.

24. Rapraeger, A. C. and Ott, V. L. (1998). Molecular interactions of the syndecan core proteins. *Curr. Opin. Cell Biol.*, **10**, 620.

25. Veugelers, M. and David, G. (1998). The glypicans: a family of GPI-anchored heparan sulfate proteoglycans with a potential role in the control of cell division. *Trends Glycosci. Glycotechnol.*, **10**, 145.

26. Salmivirta, M. and Jalkanen, M. (1995). Syndecan family of cell surface proteoglycans: developmentally regulated receptors for extracellular effector molecules. *Experientia*, **51**, 863.

27. Woods, A. and Couchman, J. R. (1998). Syndecans: synergistic activators of cell adhesion. *Trends Cell Biol.*, **8**, 189.

28. David, G., Lories, V., Decock, B., Marynen, P., Cassiman, J.-J., and Van den Berghe, H. (1990). Molecular cloning of a phosphatidylinositol-anchored membrane heparan sulfate proteoglycan from human lung fibroblasts. *J. Cell Biol.*, **111**, 3165.

29. Steinfeld, R., Van Den Berghe, H., and David, G. (1996). Stimulation of fibroblast growth factor receptor-1 occupancy and signaling by cell surface-associated syndecans and glypican. *J. Cell Biol.*, **133**, 405.

30. Liu, W., Litwack, E. D., Stanley, M. J., Langford, J. K., Lander, A. D., and Sanderson, R. D. (1998). Heparan sulfate proteoglycans as adhesive and anti-invasive molecules. Syndecans and glypican have distinct functions. *J. Biol. Chem.*, **273**, 22825.

31. Nakato, H., Futch, T. A., and Selleck, S. B. (1995). The division abnormally delayed (*dally*) gene: a putative integral membrane proteoglycan required for cell division patterning during postembryonic development of the nervous system in *Drosophila*. *Development*, **121**, 3687.

32. Pilia, G., Hughes-Benzie, R. M., MacKenzie, A., Baybayan, P., Chen, E. Y., Huber, R., *et al.* (1996). Mutations in GPC3, a glypican gene, cause the Simpson–Golabi–Behmel overgrowth syndrome. *Nature Genet.*, **12**, 241.

33. Saunders, S., Jalkanen, M., O'Farrell, S., and Bernfield, M. (1989). Molecular cloning of syndecan, an integral membrane proteoglycan. *J. Cell Biol.*, **108**, 1547.

34. Kokenyesi, R. and Bernfield, M. (1994). Core protein structure and sequence determine the site and presence of heparan sulfate and chondroitin sulfate on syndecan-1. *J. Biol. Chem.*, **269**, 12304.

35. Trautman, M. S., Kimelman, J., and Bernfield, M. (1991). Developmental expression of syndecan, an integral membrane proteoglycan, correlates with cell differentiation. *Development*, **111**, 213.

36. Zhang, L., David, G., and Esko, J. D. (1995). Repetitive Ser-Gly sequences enhance heparan sulfate assembly in proteoglycans. *J. Biol. Chem.*, **270**, 27127.

37. Zhang, L. and Esko, J. D. (1994). Amino acid determinants that drive heparan sulfate assembly in a proteoglycan. *J. Biol. Chem.*, **269**, 19295.

38. McFall, A. J. and Rapraeger, A. C. (1997). Identification of an adhesion site within the syndecan-4 extracellular protein domain. *J. Biol. Chem.*, **272**, 12901.

39. Rapraeger, A., Jalkanen, M., and Bernfield, M. (1986). Cell surface proteoglycan associates with the cytoskeleton at the basolateral cell surface of mouse mammary epithelial cells. *J. Cell Biol.*, **103**, 2683.

40. Jalkanen, M., Nguyen, H., Rapraeger, A., Kurn, N., and Bernfield, M. (1985). Heparan sulfate proteoglycans from mouse mammary epithelial cells: localization on the cell surface with a monoclonal antibody. *J. Cell Biol.*, **101**, 976.

41. Subramanian, S. V., Fitzgerald, M. L., and Bernfield, M. (1997). Regulated shedding of syndecan-1 and -4 ectodomains by thrombin and growth factor activation. *J. Biol. Chem.*, **272**, 14713.

42. Fitzgerald, M. L., Wang, A., Park, P. W., Murphy, G., and Bernfield, M. (2000). Shedding of syndecan-1 and -4 ectodomains is regulated by multiple signaling pathways and mediated by a TIMP-3-sensitive metalloproteinase. *J. Cell Biol.*, **148**, 811.

43. Ehlers, M. R. and Riordan, J. F. (1991). Membrane proteins with soluble counterparts: role of proteolysis in the release of transmembrane proteins. *Biochemistry*, **30**, 10065.

44. Massagué, J. and Pandiella, A. (1993). Membrane-anchored growth factors. *Annu. Rev. Biochem.*, **62**, 515.
45. Hanneken, A., Maher, P. A., and Baird, A. (1995). High affinity immunoreactive FGF receptors in the extracellular matrix of vascular endothelial cells–implications for the modulation of FGF-2. *J. Cell Biol.*, **128**, 1221.
46. Kendall, R. L. and Thomas, K. A. (1993). Inhibition of vascular endothelial cell growth factor activity by an endogenously encoded soluble receptor. *Proc. Natl. Acad. Sci. USA*, **90**, 10705.
47. Hayashi, K., Hayashi, M., Jalkanen, M., Firestone, J., Trelstad, R. L., and Bernfield, M. (1987). Immunocytochemistry of cell surface heparan sulfate proteoglycan in mouse tissues. A light and electron microscopic study. *J. Histochem. Cytochem.*, **35**, 1079.
48. Woods, A. and Couchman, J. R. (1994). Syndecan 4 heparan sulfate proteoglycan is a selectively enriched and widespread focal adhesion component. *Mol. Biol. Cell*, **5**, 183.
49. Asundi, V. K. and Carey, D. J. (1995). Self-association of *N*-syndecan (syndecan-3) core protein is mediated by a novel structural motif in the transmembrane domain and ectodomain flanking region. *J. Biol. Chem.*, **270**, 26404.
50. Brown, D. A. and London, E. (1998). Functions of lipid rafts in biological membranes. *Annu. Rev. Cell Dev. Biol.*, **14**, 111.
51. Mertens, G., Van der Schueren, B., van den Berghe, H., and David, G. (1996). Heparan sulfate expression in polarized epithelial cells: the apical sorting of glypican (GPI-anchored proteoglycan) is inversely related to its heparan sulfate content. *J. Cell Biol.*, **132**, 487.
52. Mertens, G., Cassiman, J. J., Van den Berghe, H., Vermylen, J., and David, G. (1992). Cell surface heparan sulfate proteoglycans from human vascular endothelial cells. Core protein characterization and antithrombin III binding properties. *J. Biol. Chem.*, **267**, 20435.
53. Mast, A. E., Higuchi, D. A., Huang, Z. F., Warshawsky, I., Schwartz, A. L., and Broze, G. J. J. (1997). Glypican-3 is a binding protein on the HepG2 cell surface for tissue factor pathway inhibitor. *Biochem. J.*, **327**, 577.
54. Tibell, L. A., Sethson, I., and Buevich, A. V. (1997). Characterization of the heparin-binding domain of human extracellular superoxide dismutase. *Biochim. Biophys. Acta*, **1340**, 21.
55. Hashimoto, O., Nakamura, T., Shoji, H., Shimasaki, S., Hayashi, Y., and Sugino, H. (1997). A novel role of follistatin, an activin-binding protein, in the inhibition of activin action in rat pituitary cells. Endocytotic degradation of activin and its acceleration by follistatin associated with cell-surface heparan sulfate. *J. Biol. Chem.*, **272**, 13835.
56. Carey, D. J., Stahl, R. C., Asundi, V. K., and Tucker, B. (1993). Processing and subcellular distribution of the Schwann cell lipid-anchored heparan sulfate proteoglycan and identification as glypican. *Exp. Cell Res.*, **208**, 10.
57. Sanderson, R. D., Sneed, T., Young, L., Sullivan, G., and Lander, A. (1992). Adhesion of B lymphoid (MPC-11) cells to type I collagen is mediated by the integral membrane proteoglycan, syndecan. *J. Immunol.*, **148**, 3902.
58. Stanley, M. J., Liebersbach, B. F., Liu, W., Anhalt, D. J., and Sanderson, R. D. (1995). Heparan sulfate-mediated cell aggregation. Syndecans-1 and -4 mediate intercellular adhesion following their transfection into human B lymphoid cells. *J. Biol. Chem.*, **270**, 5077.
59. McFall, A. J. and Rapraeger, A. C. (1998). Characterization of the high affinity cell-binding domain in the cell surface proteoglycan syndecan-4. *J. Biol. Chem.*, **273**, 28270.

60. Liu, S., Hoke, D., Julian, J., and Carson, D. D. (1997). Heparin/heparan sulfate (HP/HS) interacting protein (HIP) supports cell attachment and selective, high affinity binding of HP/HS. *J. Biol. Chem.*, **272**, 25856.

61. Hoke, D. E., Regisford, E. G., Julian, J., Amin, A., Begue-Kirn, C., and Carson, D. D. (1998). Murine HIP/L29 is a heparin-binding protein with a restricted pattern of expression in adult tissues. *J. Biol. Chem.*, **273**, 25148.

62. Giuffre, L., Cordey, A. S., Monai, N., Tardy, Y., Schapira, M., and Spertini, O. (1997). Monocyte adhesion to activated aortic endothelium: role of L-selectin and heparan sulfate proteoglycans. *J. Cell Biol.*, **136**, 945.

63. Diamond, M. S., Alon, R., Parkos, C. A., Quinn, M. T., and Springer, T. A. (1995). Heparin is an adhesive ligand for the leukocyte integrin Mac-1 (CD11b/CD1). *J. Cell Biol.*, **130**, 1473.

64. Albelda, S. M., Smith, C. W., and Ward, P. A. (1994). Adhesion molecules and inflammatory injury. *FASEB J.*, **8**, 504.

65. Reyes, A. A., Akeson, R., Brezina, L., and Cole, G. J. (1990). Structural requirements for neural cell adhesion molecule-heparin interaction. *Cell Regul.*, **1**, 567.

66. Bennett, K. L., Bradshaw, J., Youngman, T., Rodgers, J., Greenfield, B., Aruffo, A., *et al.* (1997). Deleted in colorectal carcinoma (DCC) binds heparin via its fifth fibronectin type III domain. *J. Biol. Chem.*, **272**, 26940.

67. Kato, M., Saunders, S., Nguyen, H., and Bernfield, M. (1995). Loss of cell surface syndecan-1 causes epithelia to transform into anchorage-independent mesenchyme-like cells. *Mol. Biol. Cell*, **6**, 559.

68. Sun, D., Mcalmon, K. R., Davies, J. A., Bernfield, M., and Hay, E. D. (1998). Simultaneous loss of expression of syndecan-1 and E-cadherin in the embryonic palate during epithelial-mesenchymal transformation. *Int. J. Dev. Biol.*, **42**, 733.

69. Inki, P. and Jalkanen, M. (1996). The role of syndecan-1 in malignancies. *Ann. Med.*, **28**, 63.

70. Leppa, S., Vleminckx, K., Van Roy, F., and Jalkanen, M. (1996). Syndecan-1 expression in mammary epithelial tumor cells is E-cadherin-dependent. *J. Cell Sci.*, **109**, 1393.

71. Elenius, K., Maatta, A., Salmivirta, M., and Jalkanen, M. (1992). Growth factors induce 3T3 cells to express bFGF-binding syndecan. *J. Biol. Chem.*, **267**, 6435.

72. Gallo, R. G., Ono, M., Povsic, T., Page, C., Eriksson, E., Klagsbrun, M., *et al.* (1994). Syndecans, cell surface heparan sulfate proteoglycans, are induced by a proline-rich antimicrobial peptide from wounds. *Proc. Natl. Acad. Sci. USA*, **91**, 11035.

73. Lyon, M. and Gallagher, J. T. (1998). Bio-specific sequences and domains in heparan sulfate and the regulation of cell growth and adhesion. *Matrix Biol.*, **17**, 485.

74. Sweeney, S. M., Guy, C. A., Fields, G. B., and Antonio, J. D. (1998). Defining the domains of type I collagen involved in heparin-binding and endothelial tube formation. *Proc. Natl. Acad. Sci. USA*, **95**, 7275.

75. Sutherland, A. E., Sanderson, R. D., Mayes, M., Siebert, M., Calarco, P. G., Bernfield, M., *et al.* (1991). Expression of syndecan, a putative low affinity fibroblast growth factor receptor, in the early mouse embryo. *Development*, **113**, 339.

76. Vainio, S., Lehtonen, E., Jalkanen, M., Bernfield, M., and Saxen, L. (1989). Epithelial-mesenchymal interactions regulate the stage-specific expression of a cell surface proteoglycan, syndecan, in the developing kidney. *Dev. Biol.*, **134**, 382.

77. Koyama, E., Shimazu, A., Leatherman, J. L., Golden, E. B., Nah, H. D., and Pacifici, M. (1996). Expression of syndecan-3 and tenascin-C: possible involvement in periosteum development. *J. Orthop. Res.*, **14**, 403.

78. Sanderson, R. D., Lalor, P., and Bernfield, M. (1989). B lymphocytes express and lose syndecan at specific stages of differentiation. *Cell Regul.*, **1**, 27.
79. Baciu, P. C. and Goetinck, P. F. (1995). Protein kinase C regulates the recruitment of syndecan-4 into focal contacts. *Mol. Biol. Cell*, **6**, 1503.
80. Thomas, S. M. and Brugge, J. S. (1997). Cellular functions regulated by SRC family kinases. *Annu. Rev. Cell Dev. Biol.*, **13**, 513.
81. Burridge, K. and Chrzanowska-Wodnicka, M. (1996). Focal adhesions, contractility and signaling. *Annu. Rev. Cell Dev. Biol.*, **12**, 463.
82. LeBaron, R. G., Esko, J. D., Woods, A., Johnsson, S., and Hook, M. (1988). Adhesion of glycosaminoglycan-deficient Chinese hamster ovary cell mutants to fibronectin substrata. *J. Cell Biol.*, **106**, 945.
83. Hansen, C. A., Schroering, A. G., Carey, D. J., and Robishaw, J. D. (1994). Localization of a heterotrimeric G protein gamma subunit to focal adhesions and associated stress fibers. *J. Cell Biol.*, **126**, 811.
84. Saoncella, S., Echtermeyer, F., Denhez, F., Nowlen, J. K., Mosher, D. F., Robinson, S. D., *et al.* (1999). Syndecan-4 signals cooperatively with integrins in a Rho-dependent manner in the assembly of focal adhesions and actin stress fibers. *Proc. Natl. Acad. Sci. USA*, **96**, 2805.
85. Lebakken, C. S. and Rapraeger, A. C. (1996). Syndecan-1 mediates cell spreading in transfected human lymphoblastoid (Raji) cells. *J. Cell Biol.*, **132**, 1209.
86. Yamashita, Y., Oritani, K., Miyoshi, E. K., Wall, R., Bernfield, M., and Kincade, P. W. (1999). Syndecan-4 is expressed by B lineage lymphocytes and can transmit a signal for formation of dendritic processes. *J. Immunol.*, **162**, 5940.
87. Martin, T. F. J. (1998). Phosphoinositide lipids as signaling molecules: common themes for signal transduction, cytoskeletal regulation, and membrane trafficking. *Annu. Rev. Cell Dev. Biol.*, **14**, 231.
88. Rostand, K. S. and Esko, J. D. (1997). Microbial adherence to and invasion through proteoglycans. *Infect. Immun.*, **65**, 1.
89. Zhang, J. P. and Stephens, R. S. (1992). Mechanism of *C. trachomatis* attachment to eukaryotic host cells. *Cell*, **69**, 861.
90. Lycke, E., Johansson, M., Svennerholm, B., and Lindahl, U. (1991). Binding of herpes simplex virus to cellular heparan sulphate, an initial step in the adsorption process. *J. Gen. Virol.*, **72**, 1131.
91. Chen, J. C. R., Zhang, J. P., and Stephens, R. S. (1996). Structural requirements of heparin binding to *Chlamydia trachomatis*. *J. Biol. Chem.*, **271**, 11134.
92. Chen, Y., Maguire, T., Hileman, R. E., Fromm, J. R., and Esko, J. D. (1997). Dengue virus infectivity depends on envelope protein binding to target cell heparan sulfate. *Nature Med.*, **3**, 866.
93. Frevert, U., Sinnis, P., Cerami, C., Shreffler, W., Takacs, B., and Nussenzweig, V. (1993). Malaria circumsporozoite protein binds to heparan sulfate proteoglycans associated with the surface membrane of hepatocytes. *J. Exp. Med.*, **177**, 1287.
94. Karger, A., Saalmuller, A., Tufaro, F., Banfield, B. W., and Mettenleiter, T. C. (1995). Cell surface proteoglycans are not essential for infection by pseudorabies virus. *J. Virol.*, **69**, 3482.
95. Klimstra, W. B., Ryman, K. D., and Johnston, R. E. (1998). Adaptation of Sindbis virus to BHK cells selects for use of heparan sulfate as an attachment receptor. *J. Virol.*, **72**, 7357.
96. Neff, S., Sa-Carvalho, D., Rieder, E., Mason, P. W., Blystone, S. D., Brown, E. J., *et al.* (1998). Foot-and-mouth disease virus virulent for cattle utilizes integrin alpha(v)beta3 as its receptor. *J. Virol.*, **72**, 3587.

97. Feyzi, E., Trybala, E., Bergström, T., Lindhal, U., and Spillman, D. (1997). Structural requirement of heparan sulfate for interaction with Herpes simplex virus type I virions and isolated glycoprotein C. *J. Biol. Chem.*, **272**, 24850.

98. Geraghty, R. J., Krummenacher, C., Cohen, G. H., Eisenberg, R. J., and Spear, P. G. (1998). Entry of alphaherpesviruses mediated by poliovirus receptor-related protein 1 and poliovirus receptor. *Science*, **280**, 1618.

99. Jackson, T. (1996). Efficient infection of cells in culture by type O foot-and-mouth disease requires binding to cell surface heparan sulfate. *J. Virol.*, **70**, 5282.

100. Putnak, J. R., Kanesa-Thasan, N., and Innis, B. L. (1997). A putative cellular receptor for Dengue viruses. *Nature Med.*, **3**, 828.

101. Grassmé, H., Gulbins, E., Brenner, B., Ferlinz, K., and Sandhoff, K. (1997). Acidic sphingomyelinase mediates entry of *N. gonorrhoeae* into nonphagocytic cells. *Cell*, **91**, 605.

102. Rapraeger, A., Jalkanen, M., and Bernfield, M. (1987). Integral membrane proteoglycans as matrix receptors: role in cytoskeleton and matrix assembly at the epithelial cell surface. In *Biology of extracellular matrix, biology of proteoglycans* (ed. R. N. Wight and R. P. Mecham), Vol. II, p. 129. Marcel Dekker, Inc., Academic Press, New York.

103. Woods, A., Couchman, J. R., Johansson, S., and Höök, M. (1986). Adhesion and cytoskeletal organization of fibroblasts in response to fibronectin fragments. *EMBO J.*, **5**, 665.

104. Rapraeger, A. C. and Bernfield, M. (1982). An integral membrane proteoglycan is capable of binding components of the cytoskeleton and the extracellular matrix. In *Extracellular matrix* (ed. S. Hawkes and J. Wang), p. 265. Academic Press, New York.

105. Carey, D. J., Stahl, R. C., Cizmeci-Smith, G., and Asundi, V. K. (1994). Syndecan-1 expressed in Schwann cells causes morphological transformation and cytoskeletal reorganization and associates with actin during cell spreading. *J. Cell Biol.*, **124**, 161.

106. Carey, D. J., Bendt, K. M., and Stahl, R. C. (1996). The cytoplasmic domain of syndecan-1 is required for cytoskeleton association but not detergent insolubility. Identification of essential cytoplasmic domain residues. *J. Biol. Chem.*, **271**, 15253.

107. Carey, D. J., Stahl, R. C., Tucker, B., Bendt, K. A., and Cizmeci-Smith, G. (1994). Aggregation-induced association of syndecan-1 with microfilaments mediated by the cytoplasmic domain. *Exp. Cell Res.*, **214**, 12.

108. Carey, D. J. (1997). Syndecans: multifunctional cell-surface co-receptors. *Biochem. J.*, **327**, 1.

109. Fanning, A. S. and Anderson, J. M. (1996). Protein-protein interactions: PDZ domain networks. *Curr. Biol.*, **6**, 1385.

110. Grootjans, J. J., Zimmermann, P., Reekmans, G., Smets, A., Degeest, G., Durr, J., *et al.* (1997). Syntenin, a PD2 protein that binds syndecan cytoplasmic domains. *Proc. Natl. Acad. Sci. USA*, **94**, 13683.

111. Cohen, A. R., Woods, D. F., Marfatia, S. M., Walther, Z., Chishti, A. H., Anderson, J. M., *et al.* (1998). Human CASK/LIN-2 binds syndecan-2 and protein4.1 and localized to the basolateral membrane of epithelial cells. *J. Cell Biol.*, **142**, 129.

112. Hsueh, Y. P., Yang, F. C., Kharazia, V., Naisbitt, S., Cohen, A. R., Weinberg, R. J., *et al.* (1998). Direct interaction of CASK/LIN-2 and syndecan heparan sulfate proteoglycan and their overlapping distribution in neuronal synapses. *J. Cell Biol.*, **142**, 139.

113. Hata, Y., Butz, S., and Südhof, T. C. (1996). CASK: a novel dlg/PSD95 homolog with an N-terminal calmodulin-dependent protein kinase domain identified by interaction with neurexins. *J. Neurosci.*, **16**, 2488.

114. Kim, S. K. (1995). Tight junctions, membrane-associated guanylate kinases and cell signaling. *Curr. Opin. Cell Biol.*, **7**, 641.

115. Oh, E.-S., Woods, A., Lim, S.-T., Thiebert, A. W., and Couchman, J. R. (1998). Syndecan-4 cytoplasmic domain and phosphatidylinositol 4,5-biphosphate coordinately regulate protein kinase C activity. *J. Biol. Chem.*, **273**, 10624.

116. Asundi, V. K. and Carey, D. J. (1997). Phosphorylation of recombinant N-syndecan (syndecan 3) core protein. *Biochem. Biophys. Res. Commun.*, **240**, 502.

117. Ott, V. L. and Rapraeger, A. C. (1998). Tyrosine phosphorylation of syndecan-1 and -4 cytoplasmic domains in adherent B82 fibroblasts. *J. Biol. Chem.*, **273**, 35291.

118. Horowitz, A. and Simons, M. (1998). Phosphorylation of the cytoplasmic tail of syndecan-4 regulates activation of protein kinase Cα. *J. Biol. Chem.*, **273**, 25548.

119. Horowitz, A. and Simons, M. (1998). Regulation of syndecan-4 phosphorylation *in vivo*. *J. Biol. Chem.*, **273**, 10914.

120. Ingber, D. E. and Folkman, J. (1989). Mechanochemical switching between growth and differentiation during fibroblast growth factor-stimulated angiogenesis *in vitro*: role of extracellular matrix. *J. Cell Biol.*, **109**, 317.

121. Lind, T., Tufaro, F., McMormick, C., Lindahl, U., and Lidholt, K. (1998). The putative tumor suppressors EXT1 and EXT2 are glycosyltransferases required for the bio-synthesis of heparan sulfate. *J. Biol. Chem.*, **273**, 26265.

122. McCormick, C., Leduc, Y., Martindale, D., Mattison, K., Esford, L. E., Dyer, A. P., *et al.* (1998). The putative tumor suppressor *EXT1* alters the expression of cell-surface heparan sulfate. *Nature Genet.*, **19**, 158.

123. Stickens, D., Clines, G., Burbee, D., Ramos, P., Thomas, S., Hogue, D., *et al.* (1996). The EXT2 multiple exostoses gene defines a family of putative tumor suppressor genes. *Nature Genet.*, **14**, 25.

124. Bellaiche, Y., The, I., and Perrimon, N. (1998). *Tout-velu* is a *Drosophila* homologue of the putative tumor suppressor *EXT-1* and is needed for Hh diffusion. *Nature*, **394**, 85.

125. Hecht, J. T., Hogue, D., Strong, L. C., Hansen, M. F., Blanton, S. H., and Wagner, M. (1995). Hereditary multiple exostosis and chondrosarcoma: linkage to chromosome II and loss of heterozygosity for EXT-linked markers on chromosomes II and 8. *Am. J. Hum. Genet.*, **56**, 1125.

126. Neri, G., Gurrieri, F., Zanni, G., and Lin, A. (1998). Clinical and molecular aspects of the Simpson–Golabi–Behmel syndrome. *Am. J. Med. Genet.*, **79**, 279.

127. Gonzalez, A. D., Kaya, M., Shi, W., Song, H., Testa, J. R., Penn, L. Z., *et al.* (1998). OCI-5/GPC3, a glypican encoded by a gene that is mutated in the Simpson–Golabi–Behmel overgrowth syndrome, induces apoptosis in a cell line-specific manner. *J. Cell Biol.*, **141**, 1407.

128. Cano-Gauci, D. F., Song, H. H., Yang, H., McKerlie, C., Choo, B., Shi, W., *et al.* (1999). Glypican-3-deficient mice exhibit developmental overgrowth and some of the abnormalities typical of Simpson–Golabi–Behmel Syndrome. *J. Cell Biol.*, **146**, 255.

129. Turley, E. A. (1984). Proteoglycans and cell adhesion. Their putative role during tumorigenesis. *Cancer Metastasis Rev.*, **3**, 325.

130. Nakajima, M., Irimura, T., and Nicolson, G. L. (1988). Heparanases and tumor metastasis. *J. Cell. Biochem.*, **36**, 157.

131. Vlodavsky, I., Eldor, A., Haimovitz-Friedman, A., Matzner, Y., Ishai-Michaeli, R., Lider, O., *et al.* (1992). Expression of heparanase by platelets and circulating cells of the immune system: possible involvement in diapedesis and extravasation. *Invasion Metastasis*, **12**, 112.

132. Dhodapkar, M. V., Kelly, T., Theus, A., Athota, A. B., Barlogie, B., and Sanderson, R. D. (1997). Elevated levels of shed syndecan-1 correlate with tumour mass and decreased

matrix metalloproteinase-9 activity in the serum of patients with multiple myeloma. *Br. J. Haematol.*, **99**, 368.

133. Eccles, S. A. (1999). Heparanase: breaking down barriers in tumors. *Nature Med.*, **5**, 735.

134. Vlodavsky, I., Friedmann, Y., Elkin, M., Aingorn, H., Atzmon, R., Ishai-Michaeli, R., *et al.* (1999). Mammalian heparanase: gene cloning, expression and function in tumor progression and metastasis. *Nature Med.*, **5**, 793.

135. Hulett, M. D., Freeman, C., Hamdorf, B. J., Baker, R. T., Harris, M. J., and Parish, C. R. (1999). Cloning of mammalian heparanase, an important enzyme in tumor invasion and metastasis. *Nature Med.*, **5**, 803.

136. Toyoshima, M. and Nakajima, M. (1999). Human heparanase. Purification, characterization, cloning, and expression. *J. Biol. Chem.*, **274**, 24153.

137. Kato, M., Wang, H., Kainulainen, V., Fitzgerald, M. F., Ledbetter, S., Ornitz, D. M., *et al.* (1998). Physiological degradation converts the soluble syndecan-1 ectodomain from an inhibitor to a potent activator of FGF-2. *Nature Med.*, **4**, 691.

138. Lawrence, P. A. (1992). *The making of a fly: the genetics of animal design*. Blackwell Science, Oxford.

139. Bate, M. and Martinez Arias, A. (ed.) (1993). *The development of* Drosophila melanogaster. Cold Spring Harbor Laboratory Press, Plainview.

140. Campos-Ortega, J. A. and Hartenstein, V. (1997). *The embryonic development of* Drosophila melanogaster. Springer, Berlin.

141. Selleck, S. (1998). Genetic analysis of functions for cell surface proteoglycans. *Matrix Biol.*, **17**, 473.

142. Lin, X., Buff, E. M., Perrimon, N., and Michelson, A. M. (1999). Heparan sulfate proteoglycans are essential for FGF receptor signaling during *Drosophila* embryonic development. *Development*, **126**, 3715.

143. Sen, J., Goltz, J. S., Stevens, L., and Stein, D. (1998). Spatially restricted expression of *pipe* in the *Drosophila* egg chamber defines embryonic dorsal-ventral polarity. *Cell*, **95**, 471.

144. Häcker, U., Lin, X., and Perrimon, N. (1997). The *Drosophila sugarless* gene modulates *Wingless* signaling and encodes an enzyme involved in polysaccharide biosynthesis. *Development*, **124**, 3565.

145. Lipshitz, H. D. (1991). Axis specification in the *Drosophila* embryo. *Curr. Opin. Cell Biol.*, **3**, 966.

146. Keith, F. J. and Gay, N. J. (1990). The *Drosophila* membrane receptor Toll can function to promote cellular adhesion. *EMBO J.*, **9**, 4299.

147. Anderson, K. V. (1998). Pinning down positional information: dorsal-ventral polarity in the *Drosophila* embryo. *Cell*, **95**, 439.

148. Kainulainen, V., Wang, H., Schick, C., and Bernfield, M. (1998). Syndecans, heparan sulfate proteoglycans, maintain the proteolytic balance of acute wound fluids. *J. Biol. Chem.*, **273**, 11563.

149. Haerry, T. E., Heslip, T. R., Marsh, J. L., and O'Conner, M. B. (1997). Defects in glucuronate biosynthesis disrupt *Wingless* signaling in *Drosophila*. *Development*, **124**, 3055.

150. Lin, X. and Perrimon, N. (1999). Dally cooperates with *Drosophila* Frizzled 2 to transduce Wingless signalling. *Nature*, **400**, 281.

151. Tsuba, M., Kamimura, K., Nakato, H., Archer, M., Staatz, W., Fox, B., *et al.* (1999). The cell-surface proteoglycan Dally regulates Wingless signalling in *Drosophila*. *Nature*, **400**, 276.

152. Binari, R. C., Staveley, B. E., Johnson, W. A., Godavarti, R., Sasisekharan, R., and Manoukian, A. S. (1997). Genetic evidence that heparin-like glycosaminoglycans are involved in *wingless* signaling. *Development*, **124**, 2623.

153. Cumberledge, S. and Reichsman, F. (1997). Glycosaminoglycans and WNTs: just a spoonful of sugar helps the signal go down. *Trends Genet.*, **13**, 421.

154. Benevolenskaya, E. V., Frolov, M. V., and Birchler, J. A. (1998). The sugarless mutation affects the expression of the white eye color gene in *Drosophila melanogaster*. *Mol. Gen. Genet.*, **260**, 131.

155. The, I., Bellaiche, Y., and Perrimon, N. (1999). Hedgehog movement is regulated through tout velu-dependent synthesis of a heparan sulfate proteoglycan. *Mol. Cell*, **4**, 633.

156. Toyoda, H., Kinoshita-Toyoda, A., and Selleck, S. B. (2000). Structural analysis of glycosaminoglycans in *Drosophila* and *Caenorhabditis elegans* and demonstration that tout-velu, a *Drosophila* gene related to EXT tumor suppressors, affects heparan sulfate *in vivo*. *J. Biol. Chem.*, **275**, 2269.

157. Glazer, L. and Shilo, B. Z. (1991). The *Drosophila* FGF-R homolog is expressed in the embryonic tracheal system and appears to be required for directed tracheal cell extension. *Genes Dev.*, **5**, 697.

158. Shishido, E., Higashijima, S., Emori, Y., and Saigo, K. (1993). Two FGF-receptor homologues of *Drosophila*: one is expressed in mesodermal primordium in early embryos. *Development*, **117**, 751.

159. Beiman, M., Shilo, B. Z., and Volk, T. (1996). Heartless, a *Drosophila* FGF receptor homolog, is essential for cell migration and establishment of several mesodermal lineages. *Genes Dev.*, **10**, 2993.

160. Klambt, C., Glazer, L., and Shilo, B. Z. (1992). breathless, a *Drosophila* FGF receptor homolog, is essential for migration of tracheal and specific midline glial cells. *Genes Dev.*, **6**, 1668.

161. Sutherland, D., Samakovlis, C., and Krasnow, M. A. (1996). branchless encodes a *Drosophila* FGF homolog that controls tracheal cell migration and the pattern of branching. *Cell*, **87**, 1091.

162. Kopczynski, C. C., Noordermeer, J. N., Serano, T. L., Chen, W. Y., Pendleton, J. D., Lewis, S., *et al.* (1998). A high throughput screen to identify secreted and transmembrane proteins involved in *Drosophila* embryogenesis. *Proc. Natl. Acad. Sci. USA*, **95**, 9973.

163. Bullock, S. L., Fletcher, J. M., Beddington, R. S. P., and Wilson, V. A. (1998). Renal agenesis in mice homozygous for a gene trap mutation in the gene encoding heparan sulfate 2-sulfotransferase. *Genes Dev.*, **12**, 1894.

164. Wlad, H., Maccarana, M., Eriksson, I., Kjellen, L., and Lindahl, U. (1994). Biosynthesis of heparin. Different molecular forms of O-sulfotransferases. *J. Biol. Chem.*, **269**, 24538.

165. Bernfield, M., Hinkes, M. T., and Gallo, R. L. (1993). Developmental expression of the syndecans: possible function and regulation. *Development*, **Supplement**, 205.

6 | ADAMs

JUDITH M. WHITE, DORA BIGLER, MICHELLEE CHEN, YUJI
TAKAHASHI, and TYRA G. WOLFSBERG

1. Introduction

ADAMs are cell surface proteins that contain *A Disintegrin* and *A* Metalloprotease domain. They are unique in that they can display both a cell adhesion domain (a disintegrin domain) and a protease domain to the cell exterior. In addition to the disintegrin and metalloprotease domains, ADAMs have other domains that may be involved in additional adhesive events as well as in cell–cell fusion and cell signalling (Fig. 1). ADAMs have been implicated in a wide array of functions including fertilization, neurogenesis, and myogenesis (Table 1) and they have been implicated in several disease states. Collectively the ADAMs constitute a large, widely expressed, and fascinating group of cell interactive proteins with multiple potential functions in development and disease.

ADAMs 1 and 2 (also referred to as fertilin α and β) are sperm surface glycoproteins (1, 2). They were pursued as molecular entities because they had been implicated in the important process of sperm–egg fusion (3, 4). A subsequent PCR analysis led to the identification of ADAMs 3–6 (5).[1] The gene family was constituted in 1995 (5, 6) with ADAMs 1 to 6 and five additional cDNAs (ADAMs 7 to 11) that had been cloned for disparate reasons. Several short sequences as well as several expressed sequence tags encoding ADAMs were also available in 1995 (5–7). Currently, there are 31 distinct full (or nearly full)-length ADAM sequences in public databases. An updated Table of the ADAMs can be found at

http://www.people.Virginia.EDU/~jag6n/Table_of_the_ADAMs.html.[2]

Although members of the gene family have been referred to by other names, for example metalloprotease-disintegrins or MDCs (8) (metalloprotease disintegrin cysteine-rich proteins), we use the name ADAM since it is an appropriate acronym (5, 6) and because ADAM is the stem symbol that has been assigned to the gene

[1] Mouse ADAM 3 was independently cloned and identified as a cysteine-rich testis-specific protein (128).

[2] The website Table lists ADAMs up to and including ADAM 33. Collectively the Table includes 31 distinct ADAMs. The discrepancy in the numbering is because two ADAMs (nos 27 and 31) are literature aliases. (See the website Table for details.)

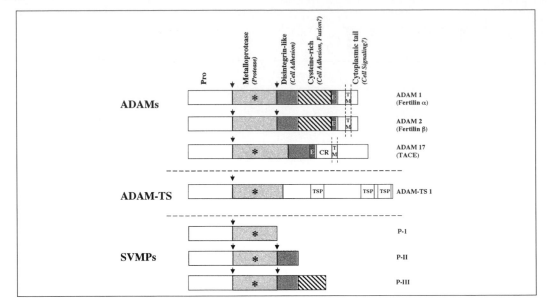

Fig. 1 Domain structure of ADAMs. Following a signal sequence, most ADAMs contain Pro (Pro), metalloprotease, disintegrin, and cysteine-rich domains followed by an EGF-like repeat (E), a spacer region, a transmembrane domain (TM), and a cytoplasmic tail. Some ADAMs (e.g. ADAMs 1 and 17) have a catalytic active site sequence (*) whereas others (e.g. ADAM 2) do not. ADAM 17 differs from most ADAMs in that following its disintegrin domain, it has an EGF-like repeat and a crambin domain (CR) before its transmembrane domain and cytoplasmic tail. The cytoplasmic tails of the ADAMs vary in length and truncated forms lacking transmembrane and cytoplasmic tail domains or with alternate cytoplasmic tails have been observed. ADAM-TS proteins are secreted proteins: following Pro and metalloprotease domains, they contain long C-terminal domains with thrombospondin (TSP) motifs. ADAMs are most closely related to PIII SVMPs. PIV SVMPs (not shown) have lectin-like domains following their cysteine-rich domains. *Arrowheads* indicate sites of interdomain proteolytic processing.

Table 1 (facing page)

[a] 'm, d, f, and s' indicate, respectively, the presence of a catalytically active metalloprotease signature sequence, a disintegrin domain, a fusion peptide, and SH3 binding sites in the cytoplasmic tail. 'm' and 'f' indicate that all known orthologues (except pseudogenes) have the requisite sequences. An 's' indicates that at least one orthologue has at least one SH3 binding site in its cytoplasmic tail. We allow this latitude since several ADAMs appear to have alternatively spliced cytoplasmic tails. If there is a minor sequencing error, ADAM 21 would encode a metalloprotease active site; hence the designation, '(m)'. Candidate fusion peptide assignments should be considered tentative. Where sequence is only available for one orthologue, the assignment is '(f)'. Given the limited information currently available, we presently consider it possible that all disintegrin domains are functional.

[b] na, not applicable. These ADAMs do not have a signature sequence for a catalytically active metalloprotease.

[c] The disintegrin domains of ADAMs 2, 3, 9, 12, 15, and 23 have been shown to be functional based on studies with peptide analogues or antibodies to the disintegrin loop and/or with recombinant disintegrin domains. The assignments marked (yes) denote early evidence that the disintegrin domains of *Xenopus* ADAMs 13, and 16 may be functional based on the ability of disintegrin loop peptides to inhibit *Xenopus* fertilization *in vitro*.

[d] KNOWN roles are based on functional studies. Predicted roles are based largely on tissue distributions. For example, ADAMs 2, 3, 5, 6, 16, 18, 20, 21, 24, 25, 26, 29, and 30 appear to be (largely) testis specific and are therefore predicted to play a role in reproduction (in the male).

[e] Rat ADAM 32, the only ADAM 32 sequence available at this time, does not have a HEXGH box. As the available sequence is only partial, we cannot yet assess whether it encodes potential disintegrin, fusion, or signalling functions. Information is not yet available on the expression pattern or functions of ADAM 32.

[f] Human ADAM 33, the only ADAM 33 sequence available at this time, has a HEXGH box, a disintegrin loop, and a candidate fusion peptide; these motifs are, however, not in the same open reading frame. It is not yet clear whether human ADAM 33 is a pseudogene or whether there are errors in the available sequence. Information is not yet available on the expression pattern or functions of ADAM 33.

Table 1 ADAMs: domain, functions, and roles in development

ADAM no.	Alias(es)	Predicted functional domains[a]	Protease activity shown[b]	Disintegrin activity shown[c]	KNOWN or predicted roles[d]
ADAM 1	Fertilin α	m, d, f, s	No	No	spermatogenesis, fertilization
ADAM 2	Fertilin β	d	na	Yes	FERTILIZATION
ADAM 3	Cyritestin, tMDC I	d	na	Yes	FERTILIZATION
ADAM 4	tMDC V	d	na	No	
ADAM 5	tMDC II	d	na	No	reproduction (male)
ADAM 6	tMDC IV	d, s	na	No	reproduction (male)
ADAM 7	EAP I	d, s	na	No	reproduction (male)
ADAM 8	MS2, CD156	m, d, s	No	No	immune function
ADAM 9	MDC9, meltrin γ	m, d, f, s	Yes	Yes	myogenesis, bone/joint biology
ADAM 10	MADM, kuzbanian	m, d, s	Yes	No	NEUROGENESIS, AXON EXTENSION, AXON REPULSION
ADAM 11	MDC	d, f	na	No	tumour suppresser, neurogenesis
ADAM 12	Meltrin α	m, d, f, s	Yes	Yes	MYOGENESIS, osteogenesis, osteoclast formation
ADAM 13		m, d, s	Yes	(Yes)	NEURAL CREST CELL MIGRATION, somitogenesis
ADAM 14	adm-1	d, (f)	na	No	cell fusion, other
ADAM 15	Metargidin, MDC 15	m, d, s	No	Yes	vascular function, bone/joint biology
ADAM 16	MDC 16	m, d, (f)	No	(Yes)	*Xenopus* fertilization
ADAM 17	TACE	m, d, s	Yes	No	SHEDDING of growth factors, cytokines, etc.
ADAM 18	tMDC III	d, s	na	No	reproduction (male)
ADAM 19	Meltrin β	m, d, s	No	No	myogenesis, osteogenesis, neurogenesis
ADAM 20		m, d	No	No	reproduction (male)
ADAM 21		(m), d, (f)	No	No	reproduction (male)
ADAM 22	MDC 2	d, f	na	No	brain function
ADAM 23	MDC 3	d, (f)	na	Yes	brain function
ADAM 24	Testase-1	m, d	No	No	reproduction (male)
ADAM 25	Testase-2	m, d, s	No	No	reproduction (male)
ADAM 26	Testase-3	m, d	No	No	reproduction (male)
ADAM 27	See ADAM 18				
ADAM 28	eMDCII, MDC-L	m, d, f, s	No	No	reproduction (male), immune function
ADAM 29		d,	No	No	reproduction (male)
ADAM 30		m, d,	No	No	reproduction (male)
ADAM 31	See ADAM 21				
ADAM 32		Footnote [e]	na	No	Footnote [e]
ADAM 33		Footnote [f]	No	No	Footnote [f]

family by the human genome nomenclature committee (http://www.gene.ucl.ac.uk/users/hester/metallo.html).[3] During the past few years there have been several exciting reports on the ADAM gene family. For recent reviews with general focuses on either the roles of ADAMs in development and disease or as proteases, the reader is referred to refs 9–18. The focus of this chapter is on the known and potential adhesive activities of the ADAMs.

2. Molecular structure of the ADAMs

2.1 Overview of ADAM domains and functions

All ADAM proteins exhibit, from their N-terminal ends, a Pro domain, a metalloprotease domain, a disintegrin domain,[4] a cysteine-rich domain, a spacer region, a transmembrane domain, and a cytoplasmic tail. ADAMs can be proteolytically processed at interdomain boundaries to yield proteins that lack Pro or Pro and metalloprotease domains (Fig. 1). The relationships of the ADAMs to their closest relatives, the PIII snake venom metalloproteases (SVMPs) and to the newly described ADAM-TS proteins—ADAM-like proteins with thrombospondin (TS) motifs—are also shown in Fig. 1. Several ADAMs are known to function as metalloproteases or as cell adhesion molecules, and some ADAMs have been postulated to participate in cell fusion and cell signalling events (Fig. 1). It is already clear, however, that not all ADAMs will display all four functions. One of the major current challenges is to decipher which ADAMs display which functions and in what developmental or physiological contexts.

Among the 31 distinct ADAMs, 18 are predicted to have Zn-dependent metalloprotease activity based on the presence of a signature catalytic site sequence, HEXGHXXGXXHD, in their metalloprotease domains. Of these, six (ADAMs 9, 10, 12, 13, 17, and 28) have been shown experimentally to possess protease activity. The protease activities of ADAMs 10 and 17 have been shown to be biologically relevant based on functional studies. ADAM 10 (kuzbanian) has been implicated in neurogenesis by virtue of its participation in the Notch signalling pathway (19–23) as well as in axon extension (19) and axon repulsion (24). ADAM 17 (TACE) plays a pivotal role in shedding several ligands from cell surface-anchored precursors, notably the potent cytokine, tumour necrosis factor (25–27). The other ADAMs with documented protease activity are ADAM 9 (meltrin γ), ADAM 12 (meltrin α), ADAM 13, and

---

[3] The website http://www.gene.ucl.ac.uk/users/hester/metallo.html maintains a table that focuses on human ADAMs. The website http://www.people.Virginia.EDU/~jag6n/Table_of_the_ADAMs.html maintains a table of all known ADAMs. For more information see http://www.ncbi.nlm.nih.gov/LocusLink/list.cgi?Q=adam*&V=0.

[4] For ease of discussion we will refer to ADAM disintegrin-like and metalloprotease-like domains as, respectively, disintegrin and metalloprotease domains. ADAM disintegrin domains are, more accurately, disintegrin-like in contrast to the classic disintegrins from PII SVMPs (67). As a group, ADAM metalloprotease domains are, more accurately, metalloprotease-like because only about half encode active site signature sequences.

ADAM 28 (28–31, 137). The expression patterns of ADAMs 9 and 12, as well as several cell culture-based studies, have suggested that they participate in myogenesis, osteogenesis, and osteoclast formation (32–36). Candidate substrates for ADAMs 9 and 12 have been described (12). ADAM 13 has been implicated in neural crest cell migration (37). And, since ADAM 28 is highly expressed in epididymis, it may play a role in spermatogenesis (28).

Six ADAMs that will be discussed below have been implicated as cell adhesion molecules by virtue of either their disintegrin or their cysteine-rich domains. They are ADAMs 2, 3, 9, 12, 15, and 23 (1, 38–49).

Six of the known ADAMs have candidate fusion peptides in the cysteine-rich domains of all known orthologues (see Table 1). Among these, three have been implicated in cell fusion events: ADAM 1 (fertilin α) in fertilization,[5] ADAM 12 (meltrin α) and ADAM 9 (meltrin γ) in myoblast fusion, and ADAM 12 (meltrin α) in osteoclast formation (1, 32, 33, 36, 50). Even though several studies have shown that synthetic analogues of the candidate fusion peptide of ADAM 1 (fertilin α) have properties consistent with fusion peptide function (51–54), there is, as yet, no formal proof that any ADAM functions as a cell–cell fusion protein in a manner analogous to the glycoproteins that promote virus–cell fusion (55).

And, of the 31 known ADAMs, 14 have SH3 binding sites (56) in their cytoplasmic tails[6] (in at least one orthologue). The cytoplasmic tails of ADAMs 9, 12, 13, 15, and 17 have been shown to interact with signalling molecules, notably ones containing SH3 domains (29, 57–61) and/or with cytoskeletal associated proteins (62). Future studies will address the role of these cytoplasmic tail interactions in cellular and developmental processes.

2.2 ADAM dimers

The mature forms of ADAMs 1 and 2 (fertilin α and β) are found as tight hetero-dimers on the surface of guinea pig (3, 4) and bovine (63) sperm. Other ADAMs have, however, not yet been found as dimer pairs.[7] Future experiments are necessary to elucidate which ADAMs exist as dimers, if dimerization (or higher order oligomerization) is developmentally regulated, if dimerization varies among species, if, as is seen for integrins (see Chapter 4), different ADAMs can pair with each other, and if dimerization has any particular functional significance. The issue if whether an

---

[5] The notion that fertilin α participates in sperm–egg fusion has been questioned because we (17) and others (129) have only found a pseudogene for human ADAM 1. In the case of the human, ADAM 21, which has a candidate fusion peptide, may serve the analogous purpose (50). Other ADAM pseudogenes have been found.

[6] Many ADAM cytoplasmic tails also have sites for serine, threonine, or tyrosine phosphorylation.

[7] Although evidence has been presented that the precursor forms of mouse pro-fertilin α and pro-fertilin β exist as a dimer pair on testicular sperm (39, 130), a dimer of mature mouse fertilin β (i.e. the form lacking Pro and metalloprotease domains) has not been detected on mature fusion competent mouse sperm (39, 130, 131).

ADAM exists as a dimer is relevant to cases where ADAM mutants have been reported to act in a dominant-negative fashion. With respect to the possibility of ADAM dimers, it is interesting that whereas most SVMP disintegrins are monomeric, several homodimeric and heterodimeric snake venom disintegrins have recently been described (64–66).

3.   Disintegrin domains

3.1   Snake venom disintegrin domains

Disintegrins are small proteins that are excised from the polyproteins of PII SVMPs (Fig. 1). Snake venom disintegrins are notable for their contribution to the haemorrhagic response in snakebite victims and because of their potential use as anticoagulants (67). The disintegrin domains of several PII SVMPs have been shown to bind to integrins (67), heterodimeric cell surface glycoproteins that are well known to bind to extracellular matrix (ECM) molecules such as laminin and fibronectin (Fig. 2A) as well as to certain cell surface proteins. For a review of the integrins, see Chapter 4. By binding to integrins, snake venom disintegrins prevent integrins from

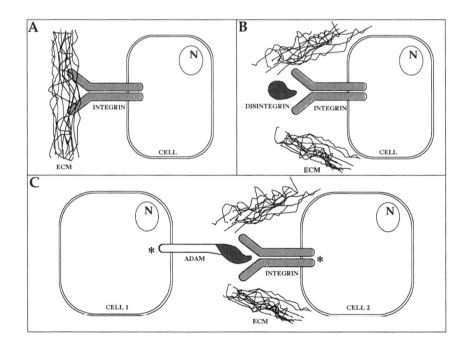

Fig. 2 Model for ADAM–integrin interaction. (A) Integrins bind ECM ligands. (B) Snake disintegrins can bind to integrins, thereby disrupting or preventing integrin–ECM interactions. (C) ADAMs (at least ADAMs 2, 3, 9,15, and 23) bind to integrins through their disintegrin domains. This may foster a positive cell–cell interaction and/or disrupt an integrin–ECM interaction. Because integrins are known to transmit intracellular signals and because many ADAM cytoplasmic tails have signalling motifs, there is a possibility for bidirectional signalling (*) through ADAM–integrin pairs.

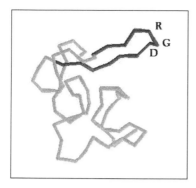

Fig. 3 NMR structure of kistrin. The NMR structure of kistrin (68) is available as file kst1 from the Research Collaboratory for Structural Bioinformatics (133); the database can be accessed at http://www.rcsb.org/pdb/. The disintegrin loop with its central RGD sequence has been highlighted.

binding to their ECM ligands (Fig. 2B). The result, in the case of platelets, is prevention of fibrinogen binding and, hence, prevention of platelet aggregation and blood coagulation.

NMR structures have been determined for four small (PII) disintegrins (for review, see ref. 67). The NMR structure of kistrin (68) is shown in Fig. 3. Kistrin binds to the platelet integrin through an RGD sequence found at the centre of a 13 amino acid protruding loop. Although PIII disintegrin domains (from PIII SVMPs) differ in disulfide bonding pattern from their PII counterparts, although they possess a downstream cysteine-rich domain not seen in PII disintegrins (Fig. 1), although they contain a 14 amino acid disintegrin loop with a cysteine near the centre, and although said cysteine appears to be in a disulfide bond, at least in the cases of atrolysin A (69) and catrocollastatin (70), there may be important structural similarities between the PII disintegrins and the disintegrin-like domains of PIII SVMPs. A modelling study indicates that the disintegrin-like domain of the PIII SVMP catrocollastatin conforms well to the known (Fig. 3) structure of kistrin (J. W. Fox, personal communication). By deduction, if the disintegrin-like domains of PIII SVMPs share general structural features with their PII counterparts (e.g. Fig. 3), then ADAM disintegrin domains may as well. Structural similarities between disintegrin domains of ADAMs and PII/PIII SVMPs may parallel the recently observed structural similarities (with a few notable exceptions) between the metalloprotease domain of ADAM 17 and that of the SVMP adamylysin (9, 71).

PII snake venom disintegrins including kistrin, echistatin, and bitistatin have been shown to bind to αIIbβ3, αvβ3, and α5β1 (67), integrins that are well known to bind RGD-containing ligands such as fibronectin (see Chapter 4). The aforementioned disintegrins have sequences such as RGD, KGD, MGD, MVD, or MLD at the centre of their disintegrin loops. Different PII disintegrins bind with different affinities and selectivities to different RGD binding integrins, and sequences flanking the 'RGD' tripeptide influence binding specificity (67).

Several PIII SVMPs have been shown to inhibit collagen-induced platelet aggregation. In the most comprehensive study, Fox and co-workers have ascribed the platelet aggregation inhibitory activity, at least in part, to the disintegrin-like domain of atrolysin A. Both a recombinant protein expressing the disintegrin-like and cysteine-rich domains of atrolysin A, as well as a 14 residue peptide analogue of the atrolysin A disintegrin loop, inhibit ADP- and collagen-induced platelet aggregation (69). Although these inhibitory activities are likely due to binding to the α2β1 integrin, a formal bimolecular interaction has not yet been shown.

3.2 ADAM disintegrin domains

Since ADAMs are the first known cell surface proteins to possess disintegrin domains, we proposed (1) that they might foster cell–cell adhesive interactions via integrin co-receptors (Fig. 2C). Recent data suggest that at least six ADAM disintegrin domains can bind to integrins (see below). Whether the primary reason for an ADAM–integrin interaction is to foster a positive (perhaps signalling) cell–cell interaction (Fig. 2C) or to disrupt or preclude a cell–ECM interaction, akin to the action of a snake disintegrin (Fig. 2B), remains to be determined. The two possibilities are, of course, not mutually exclusive. In the model for the ADAM– integrin interaction depicted in Fig. 2C, we have shown a proteolytically processed 'mature' ADAM subunit in which the Pro and metalloprotease domains have been removed. Although some can (43, 72), it is not yet certain whether all ADAM disintegrin domains can be recognized in the context of their larger forms with metalloprotease and/or Pro domains intact (Fig. 1). Also, since it is well known that engaging an integrin can transmit a signal into a cell (star in Fig. 2C, right), and since many ADAM cytoplasmic tails have signalling motifs, it is possible that there may be bidirectional signalling through ADAM–integrin pairs (e.g. stars in Fig. 2C).

Since it appears that ADAMs species that contain metalloprotease domains can function as integrin co-receptors (43, 72; and see below), it will be interesting to determine if the disintegrin–integrin interaction influences the metalloprotease domain. We consider three scenarios (Fig. 4). In the first (Fig. 4A), the disintegrin– integrin interaction may bring the protease to an ECM ligand targeted for ADAM (protease)-mediated destruction. The targeted ECM ligand may even be one that was displaced by the disintegrin–integrin interaction. ADAM-mediated cleavage of the ECM ligand may, in turn, promote cell migration by physically degrading matrix or by sending a pro-migratory signal into the integrin-expressing cell (16, 73). This scenario (Fig. 4A) is reminiscent of proposals for interactions between other proteases (e.g. matrix metalloproteases) and integrins (73–75). In the second scenario (Fig. 4B), binding of the ADAM disintegrin domain to an integrin may position the metalloprotease so that it can cleave the integrin co-receptor. Support for this possibility is the observation that the PIII SVMP jararhagin can cleave the α2β1 integrin (76). In the third scenario (Fig. 4C), the disintegrin–integrin interaction may send a signal back into the ADAM-expressing cell (Fig. 2C, left, star) such that the ADAM metalloprotease activity is manifest or enhanced. In this context it is worth noting that most ADAMs that are expected to possess protease activity also have SH3 binding sites in their cytoplasmic tails (Table 1). Perhaps binding of an SH3-containing protein to an ADAM cytoplasmic tail changes the conformation, location, or clustering of the ADAM thereby influencing its protease activity. With respect to

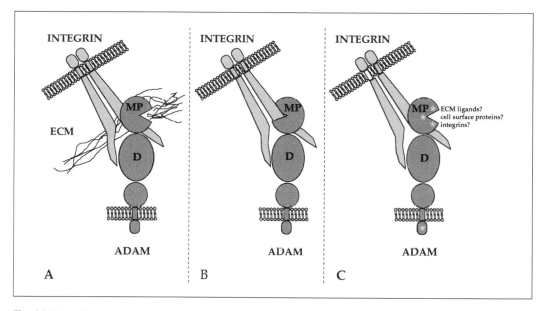

Fig. 4 Models for how the disintegrin domain of an ADAM might influence its metalloprotease domain. The ADAM disintegrin–integrin interaction may position the ADAM metalloprotease so as to (A) cleave an ECM ligand or (B) cleave the integrin co-receptor. Alternatively, (C) binding of an integrin to an ADAM disintegrin domain may send a signal into the ADAM-bearing cell such that the metalloprotease activity of the ADAM can be manifest or enhanced.

this possibility it is interesting that ADAM 9 is phosphorylated in response to treatment of cells with phorbol esters (31) and appears to be involved in a phorbol ester-stimulated ectodomain shedding event (29).

3.3 ADAM disintegrin loops

A compilation of a representative disintegrin loop from each of the known ADAMs is given in Fig. 5. A compilation of the disintegrin loop sequences from all known orthologues of those ADAMs for which the disintegrin domain of at least one orthologue has been shown to be functional is shown in Fig. 6A. From these analyses, the only conclusions that can be drawn at present are:

(a) There are many highly varied disintegrin loop sequences.

(b) While most ADAM disintegrin loops are 14 residues in length, some contain 15 and some contain 13 residues (Figs 5 and 6).

(c) The only residues that are absolutely conserved are the three cysteines at the beginning, middle, and end of the motif; these cysteines presumably play general structural roles.

(d) The additional residues that are absolutely conserved in all ADAM disintegrin loops, except those of ADAMs 10 and 17, are three charged residues: an R at position two, a D at position nine, and an E at position twelve (Fig. 6A).

ADAM #	Species	Disintegrin Loop Sequence
ADAM 1	Monkey	C R P T Q D . E | C | D L P E Y | C
ADAM 2	Human	C R P S F E . E | C | D L P E Y | C
ADAM 3	Human	C R K S V D . M | C | D F P E Y | C
ADAM 4	Mouse	C R D K N G . I | C | D L P E Y | C
ADAM 5	Monkey	C R K S V D V E | C | D F T E F | C
ADAM 6	Monkey	C R P I Q N . I | C | D L P E Y | C
ADAM 7	Human	C R P A K D . E | C | D F P E M | C
ADAM 8	Human	C R P K K D . M | C | D L E E F | C
ADAM 9	Human	C R G K T S . E | C | D V P E Y | C
ADAM 10	Human	C R D D . S . D | C | A R E G I | C
ADAM 11	Human	C R E A V N . E | C | D I A E T | C
ADAM 12	Human	C R D S S N . S | C | D L P E F | C
ADAM 13	Xenopus	C R E M A G . S | C | D L P E F | C
ADAM 14	C.elegans	C R S S K S . P | C | D V A E Q | C
ADAM 15	Human	C R P T R G . D | C | D L P E F | C
ADAM 16	Xenopus	C R M P K T . E | C | D L A E Y | C
ADAM 17	Human	C Q E A I N A T | C | K G V S Y | C
ADAM 18	Human	C R K S I D P E | C | D F T E Y | C
ADAM 19	Mouse	C R E Q V R . Q | C | D L P E F | C
ADAM 20	Human	C R Q Q V G . E | C | D L P E W | C
ADAM 21	Human	C R Q E V N . E | C | D L P E W | C
ADAM 22	Human	C R E A V N . D | C | D I R E T | C
ADAM 23	Human	C R D A V N . E | C | D I T E Y | C
ADAM 24	Mouse	C R A R E N . E | C | D L P E W | C
ADAM 25	Mouse	C R Q E V N . E | C | D L P E W | C
ADAM 26	Mouse	C R E E K N . E | C | D L P E W | C
ADAM 28	Human	C R P A K D . E | C | D L P E W | C
ADAM 29	Human	C R K E V N . E | C | D L P E W | C
ADAM 30	Human	C R Q E G N . E | C | D L A E Y | C

Fig. 5 Disintegrin loop sequences of the ADAMs. A representative sequence of each of the 31 ADAMs known at the time of this writing is shown. Where possible human or monkey sequences are shown. The only absolutely conserved residues are the three cysteines at the beginning, middle, and end of the loop.

(e) Additional blocks of identity are seen in the disintegrin loops of orthologues of a given ADAM. It is not known whether it is significant that the disintegrin loop sequences of all known mammalian orthologues of ADAMs 10 and 17 are, respectively, highly and absolutely conserved (Fig. 6B).

A study with several mutant disintegrin loop peptides from the SVMP atrolysin A (Fig. 6A) suggested that the negatively charged residues at positions seven and nine (e.g. flanking the central cysteine) are important for function, with that at position nine appearing especially important (69). Alanine scanning mutagenesis of the disintegrin loop within the mouse ADAM 2 disintegrin domain indicated that the aspartic acid at position nine is critically important (49, 77) (also see ref. 78). Although single mutations at aspartic acid position six and at glutamic acid position seven in mADAM 2 had small effects, the combined mutation (Asp6Ala/Glu7Ala) showed about 50% loss of biological activity (inhibition of sperm–egg binding) (77). In the case of mouse ADAM 3, alanine substitution of the aspartic acid at position nine had only a small effect, whereas substitution of the glutamine at position seven

		1	2	3	4	5	6	7	8	9	10	11	12	13	14	
A	**Atrolysin A**	C	R	P	A	R	S	E	C	D	I	A	E	S	C	
	ADAM 2															
	Mouse	C	R	L	A	Q	D	E	C	D	V	T	E	Y	C	
	Rat	C	R	P	A	N	Q	E	C	D	V	T	E	Y	C	
	Guinea pig	C	R	E	S	T	D	E	C	D	L	P	E	Y	C	
	Rabbit	C	R	P	P	V	G	E	C	D	L	F	E	Y	C	
	Bovine	C	R	G	S	T	D	E	C	D	L	H	E	Y	C	
	Human	C	R	P	S	F	E	E	C	D	L	P	E	Y	C	
	Monkey	C	R	P	S	F	D	E	C	D	L	P	E	Y	C	
	ADAM 3															
	Mouse	C	R	K	S	K	D	Q	C	D	F	P	E	F	C	
	Rat	C	R	K	S	T	D	Q	C	D	F	P	E	F	C	
	Human	C	R	K	S	V	D	M	C	D	F	P	E	Y	C	
	Monkey	C	R	K	S	I	D	M	C	D	F	P	E	Y	C	
	ADAM 15															
	Human	C	R	P	T	R	G	D	C	D	L	P	E	F	C	
	Mouse	C	R	P	P	T	D	D	C	D	L	P	E	F	C	
	Rat	C	R	L	P	T	D	D	C	D	L	P	E	F	C	
	ADAM 23															
	Human	C	R	D	A	V	N	E	C	D	I	T	E	Y	C	
	Mouse	C	R	D	A	V	N	S	C	D	I	T	E	Y	C	
B	**ADAM 10**															
	Bovine	C	R	D	D	.	S	.	D	C	A	K	E	G	I	C
	Rat	C	R	D	D	.	S	.	D	C	A	K	E	G	I	C
	Human	C	R	D	D	.	S	.	D	C	A	R	E	G	I	C
	Mouse	C	R	D	D	.	S	.	D	C	A	K	E	G	I	C
	Xenopus	C	R	E	E	.	S	.	D	C	A	K	M	G	T	C
	Drosophila	C	K	E	E	.	T	.	E	C	S	W	S	S	T	C
	C. elegans	C	R	Q	E	.	S	.	E	C	S	N	L	Q	T	C
	ADAM 17															
	Human	C	Q	E	A	I	N	A	T	C	K	G	V	S	Y	C
	Rat	C	Q	E	A	I	N	A	T	C	K	G	V	S	Y	C
	Mouse	C	Q	E	A	I	N	A	T	C	K	G	V	S	Y	C

Fig. 6 Disintegrin loop sequences of all known orthologues of ADAMs 2, 3, 9, 12, 15, and 23 (A), and ADAMs 10 and 17 (B). The disintegrin loops of atrolysin A and ADAMs that have been implicated in cell adhesive events are shown in (A). Residues that are absolutely conserved amongst them (in addition to the three absolutely conserved cysteines) are boxed. In (B), the boxes denote the absolutely conserved residues in the disintegrin loops of mammalian ADAMs 10 and 17.

(with alanine), significantly impaired the ability of the mouse ADAM 3 disintegrin domain to inhibit sperm–egg binding and fusion (49). In the case of human ADAM 15, which is unique in having an RGD sequence in the centre of its disintegrin loop (79), mutation of the RGD to SGA abolished binding to the αvβ3 integrin (45), tentatively implicating the negatively charged residue at position seven for integrin binding. In a study of the effects of peptides corresponding to the *Xenopus* ADAM 16 disintegrin loop (on fertilization), changing both the lysine at position five and the glutamic acid at position seven (to alanines) compromised activity (80). And, in the case of human ADAM 23, changing the glutamic acid at position seven of the loop (to alanine, the only substitution analysed to date) inhibited its ability to support cell adhesion (46). In addition, the cysteine in the middle of the disintegrin loops of mADAMs 2 and 3 appear to be important for optimal disintegrin function (49, 81).

Collectively, the data available to date indicate that residues near the middle of ADAM disintegrin loops (e.g. positions five to nine) are important for function.

Nonetheless, additional and more extensive mutagenesis studies are clearly needed to delineate the sequence requirements of ADAM disintegrin loops. It appears likely that there will be important differences in the sequence requirements of different ADAM disintegrin loops. It is also possible that sequences outside of the disintegrin loop influence disintegrin function (82).

3.4 Possibility of alternate disintegrin domains or loops

The ADAM genes analysed to date are large and are encoded by numerous exons. For example, mouse ADAM 2 (83) and human ADAM 11 (84) are each encoded by more than 20 exons. The intron–exon boundaries do not simply coincide with the interdomain boundaries of the ADAM proteins (Fig. 1). Several ADAMs appear to exhibit alternative splicing, for example to encode membrane-bound or secreted forms (33, 85) (see also GenBank AF137334 and AF137335) or forms with alternate cytoplasmic tails (86, 87). An interesting case of alternative splicing is seen for the PII SVMP, halystatin. In this case, the disintegrin domain is encoded by four exons, the third of which can be one of two alternatives. In both cases the resulting cDNAs encode disintegrin loops with a central RGD tripeptide. However, the two altern-ative splice forms encode different flanking sequences: CRMARGDDMDDYC or CRRARGDWNDNTC (GenBank accession no. D28871). It is not yet known whether the integrin specificity of the two alternate isoforms of halystatin differ. However, since data from studies on other PII disintegrins (67, 88) as well as on human ADAM 15 (45) suggest that sequences flanking the RGD affect integrin specificity, it seems plausible that alternative splicing, giving rise to different disintegrin loops (or domains) with altered integrin specificity, may occur for ADAMs.

4. ADAM family tree

A phylogenetic tree of the metalloprotease and disintegrin domain sequences of 29 full-length ADAMs is shown in Fig. 7. Six subfamilies stand out: ADAMs 12, 13, and 19; ADAMs 4 and 6; ADAMs 2, 3, 5, and 18; ADAMs 10 and 17; ADAMs 20, 21, 24, 25, 26, and 29; and ADAMs 11, 22, and 23. In an analysis of the metalloprotease domain by itself, the first group grows to include ADAMs 8, 15, and 28 in addition to 12, 13, and 19 (not shown; see also Fig. 2 in ref. 6). If the disintegrin domain is analysed separately, ADAMs 7 and 28 also group together. With one exception, all members of each subfamily either possess (asterisks) or lack the metalloprotease active site signature sequence. The exception is human ADAM 29, which has a histidine in place of the catalytic glutamic acid. The tree presented in Fig. 7 does not show evolutionary distances; however, pairwise alignments between ADAMs 10 and 17 and other ADAMs clearly show that these two ADAMs are distant from the rest.

The ADAM-TS proteins (see below) clearly fall on a separate branch of the family tree, with ADAM-TS1 grouping with ADAM-TS4, and ADAM-TS2 grouping with ADAM-TS3 (not shown). This distinct grouping of ADAM-TS proteins is not surprising as they do not have the same domain structure as the other ADAMs (Fig. 1).

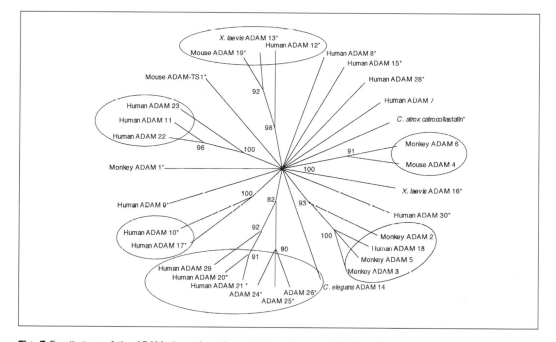

Fig. 7 Family tree of the ADAMs based on the metalloprotease and disintegrin domains. Phylogenetic analysis was performed on the metalloprotease and disintegrin domains of 31 sequences, which represent 29 published full-length ADAMs (Table 1), one PIII SVMP, catrocollastatin (GenBank U21003), and one ADAM-TS protein, ADAM-TS1. Asterisks denote ADAMs that are known or predicted metalloproteases. (See footnote [a] in the legend to Table 1 for an explanation of the asterisk by ADAM 21.) Ovals indicate clusters of related ADAMs. Numbers on the nodes indicate the bootstrap value, that is, the number of times that fork was generated in 100 separate analyses. Where available, human, monkey, or mouse sequences were used. The metalloprotease domain sequences were aligned using a structure based sequence alignment of ADAMs 10 and 17 with adamalysin (71). Other metalloprotease sequences, as well as all disintegrin sequences, were aligned by BLAST (134), with modifications performed by hand. Trees were constructed using algorithms in PHYLIP (Phylogeny Inference Package) version 3.5c (135). The sequences were subjected to 100 bootstrap replicates with SEQBOOT. PROTDIST was then used to calculate a distance matrix based on the Dayhoff PAM probability model. Phylogenies were estimated from this distance matrix using the FITCH program with global rearrangement and multiple jumbles (reordering the data set 25 times). The consensus trees were then determined by CONSENSE. Nodes with bootstrap values of 65 or less were discarded. The final tree diagram was generated with TREETOOL (136). A similar tree was obtained using PROTPARS, a different tree-building algorithm. This analysis was conducted before the sequences of ADAMs 32 and 33 were available.

All ADAM disintegrin domains except those of ADAMs 10 and 17 are 85–90 amino acids in length and include 15 cysteines[8] that align with each other and with the cysteines of the PIII SVMP disintegrin domains. The ADAM 10 disintegrin domains are somewhat larger (97–107 amino acids), but also contain 15 cysteines that align with those of the typical ADAMs. The ADAM 17 disintegrin domains are all the same length

---

[8] It has generally been thought that the disintegrin domains of most ADAMs contain 15 cysteines and that their downstream cysteine-rich domains contain 13 cysteines (see for example ref. 77). However, a recent analysis of the PIII SVMP, catrocollastatin C, suggests that ADAM disintegrin and cysteine-rich domains have, respectively, 16 and 12 cysteines (70).

(∼ 89 amino acids). They contain 13 cysteines that align with 13 of the 15 cysteines of the other ADAM disintegrin domains. And, as noted above, the ADAM 10 and 17 disintegrin loops vary in length from those of most other ADAMs (Figs 5 and 6). Given their importance as proteases in development (21, 27) it will be very interesting to determine whether the disintegrin domains of ADAMs 10 and 17 are functional. If the ADAM 10 and 17 disintegrin domains bind to integrins, it will be interesting to determine if integrin binding influences their proteolytic activity (Fig. 4).

4.1 ADAM-TS proteins and other relatives

ADAM and SVMP proteases are the founding members of the adamylysin/ reprolysin family of the metzincin superfamily of Zn-dependent metalloproteases. ADAM and SVMP proteases share an active site signature sequence and a mechanism for protease activation as well as other structural features (9, 14, 18). Recently a third group of the adamylysin/reprolysin family of metzincins has emerged. The proteins in the latter group are referred to as ADAM-TS proteins, the TS denoting the presence of thrombospondin motifs following Pro, metalloprotease, and 'disintegrin' domains (Fig. 1). What has been termed the 'disintegrin domain' of the ADAM-TS proteins is a cysteine-rich region that does not contain the characteristic number or spacing of cysteines seen in the disintegrin domains of the ADAMs and SVMPs. In fact it may encode a fourth (degenerate) thrombospondin motif. Hence, unlike SVMPs and (some) ADAMs, ADAM-TS proteins may not interact with integrins. Nonetheless, several ADAM-TS proteins appear to play important roles in development or disease. For example, ADAM-TS1 has been implicated in cancer cachexia and inflammatory responses (89), ADAM-TS4 and ADAM-TS5 (literature alias ADAM TS11; see http://www.gene.ucl.ac.uk/users/hester/adamts.html) have been implicated in osteoarthritis by virtue of their ability to cleave the cartilage protein aggrecan (90, 91). And, in *C. elegans*, an ADAM-TS protein is involved in organ shape determination (92) and an ADAM-TS-like protein is involved in directing the path of distal tip cell migration (93). Several other proteins (e.g. decysin; GenBank accession no. Y13323) bear some similarity to the ADAMs, and a sequence that might encode an ADAM-like protein has been observed in the genome of *S. pombe* (GenBank accession no. Z98849).

5. Integrins as co-receptors for ADAMs

A fundamental proposal (Fig. 2C) is that ADAMs bind to integrins via their disintegrin domains. Membrane-embedded ADAMs might therefore be co-receptors for integrins (and vice versa), and ADAM–integrin interactions may foster heterophilic cell–cell adhesive events. To date, four integrins have been implicated as ADAM co-receptors: α5β1, α6β1, α9β1, and αvβ3.

The integrin α6β1 (on the egg plasma membrane) was proposed as a receptor for sperm mouse ADAM 2 (mADAM 2) based on the ability of an anti-α6 mAb (GoH3) to inhibit sperm–egg binding, on the ability of somatic cells expressing α6β1 to bind sperm more avidly than cells that lack either α6 or β1, and on the ability of GoH3 to inhibit sperm binding to α6β1 expressing cells (38). Subsequent studies with a photoactivatable affinity analogue of the disintegrin loop of mADAM 2 (94), with a recombinant mADAM 2 disintegrin domain (77, 95), with a mature 57 kDa form of mADAM 2 purified from sperm (96), and with somatic cell adhesion assays to recombinant mADAM 2 (M. Tomczuk, Y. Takahashi, and J. M. White, unpublished results) have supported the notion that α6β1 can interact with mADAM 2. Recent studies also indicate that the disintegrin domain of mADAM 3 can interact with the α6β1 integrin (49) (M. Tomczuk, Y. Takahashi, and J. M. White, unpublished results). The recent finding that α6β1 is not essential for murine fertilization (97) has, however, raised the possibility that other β1 integrins, or other cell surface proteins, may be able to interact with the disintegrin domains of mADAMs 2 and 3. Intriguingly, recent studies have indicated that the disintegrin domain of ADAM 9 can interact with α6β1 on somatic cells (72).

Recombinant forms of the disintegrin domain or the full-length ectodomain of human ADAM 15 can bind to the αvβ3 and α5β1 integrins (43, 45).[9] These findings are consistent with the well-known ability of αvβ3 and α5β1 to bind RGD-containing ligands. Based on its up-regulated expression in diseased versus normal arterial cells, it has been proposed that human ADAM 15 may function in the development of atherosclerotic lesions (98). This proposal is consistent with an ADAM 15–αvβ3 interaction since αvβ3 is highly expressed in endothelial cells. ADAM 15 is also expressed in chondrocytes (99) as well as in osteoblasts and osteoclast-like cells (34) suggesting additional roles in bone (or joint) development or pathology. Human ADAM 15 is the only ADAM with the sequence RGD at the centre of its disintegrin loop. Mouse (100) and rat ADAM 15 (Fig. 6) have, instead, the sequence TDD at the equivalent position (Fig. 6A). Interestingly, the disintegrin domains of mouse and human ADAM 15 (as well as the disintegrin domain of human ADAM 12) have recently been shown to bind, in an RGD-independent manner, to the α9β1 integrin (47). Conversely, human ADAM 23, which does not have an RGD sequence in its disintegrin loop (Figs 5 and 6), has recently been shown to bind to the αvβ3 integrin (46).

5.1 Regulation of integrins for ADAM binding

We recently suggested that mADAM 2 binds to a different 'state' of the α6β1 integrin than does the classic α6β1 ECM ligand, laminin (49, 77, 96). For example, sperm, the 57 kDa form of mADAM 2, and the isolated mADAM 2 disintegrin domain (as well as the mADAM 3 disintegrin domain) all bind very well to eggs in media containing Ca^{2+}. And, binding of these entities (in Ca^{2+}-containing media) is inhibited by the anti-α6 mAb, GoH3. Conversely, murine eggs will not bind the classic α6β1 ligand,

[9] Takada and co-workers found that a recombinant disintegrin domain (GST chimera) of human ADAM 15 bound to cells expressing αvβ3 but not to cells expressing α5β1 (45). Murphy and co-workers employed an Fc–chimera of the entire extracellular domain of human ADAM 15 expressed in eukaryotic cells. With this protein they documented binding to both αvβ3 and α5β1 (43).

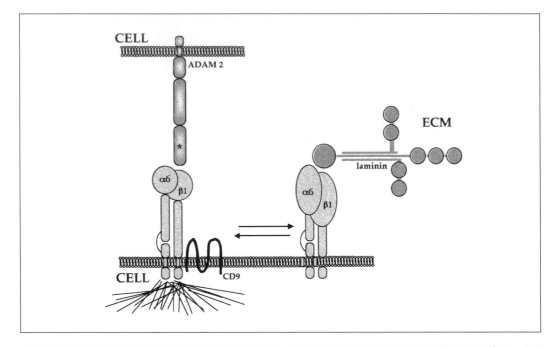

Fig. 8 Integrin affinity modulation: ADAM versus laminin. ADAM 2 appears to bind to a resting state of the α6β1 integrin on the egg. Laminin appears to bind only to an activated state of α6β1 on the egg. Recent work has shown that the α6β1 integrin-associated protein, CD9, is important for ADAMs 2 and 3 to bind to the egg and for murine fertilization. For ease of representation, the two alternate 'states' of the egg α6β1 are shown as having different conformations. In reality they may have the same conformation but be present in different geometric arrays or in different regions of the plasma membrane or they may be associated with different sets of proteins.

laminin, in Ca^{2+}-containing media; they will however, bind laminin (via α6β1) if placed in a Mn^{2+}-containing medium, an experimental treatment that activates α6β1 to bind to laminin on many somatic cells. For a more detailed discussion of integrin affinity states, see Chapter 4. Preliminary evidence suggests that the actin cytoskeleton may play a role in modulating the 'affinity state' of the egg α6β1 integrin (96, 101). Although the mechanism is far from clear, the findings suggest that integrins may be able to switch from cell–cell (integrin–ADAM) to cell–matrix (integrin–ECM ligand) binding modes (Fig. 8) and that such switches may be regulated during development or by physiological stimuli.

More recent data from our laboratory suggest that co-operation between a β1 integrin and the integrin-associated protein CD9 may be required for the disintegrin domains of mADAM 2 and mADAM 3 to interact with the egg surface (49, 77). These findings further suggest that there may be stringent cell type regulation of ADAM–integrin interactions dictated, for example, by the presence of a specific repertoire of integrin-associated proteins (102, 103). With respect to our observations that CD9 appears to be required for sperm–egg binding and fusion as well as for the binding of the ADAM 2 and 3 disintegrin domains to the egg (based on antibody inhibition studies), it is interesting that three recent studies have reported that the major defect in CD9 (–/–) mice is female infertility due to problems with sperm–egg fusion (104–106).

6. The cysteine-rich domain: possible roles in cell adhesion

All ADAMs have a 'cysteine-rich' domain adjacent to their disintegrin domains and preceding their transmembrane domains (Fig. 1). In the case of most ADAMs (e.g. ADAMs 1 and 2) the cysteine-rich domain is approximately 140 amino acids (with 13 cysteines)[8] and is followed by an approximately 30 residue EGF-like repeat (with six cysteines) and an approximately 30–40 residue spacer region before the transmembrane domain. The cysteine-rich regions of ADAMs 10 and 17 (and the ADAM-TS proteins) share no sequence similarity with those of the other ADAMs. The cysteine-rich region of ADAM 10 has, instead, approximately 120 amino acids (with 13 cysteines) between its disintegrin domain and its transmembrane domain. ADAM 17 has, instead, (immediately following its disintegrin domain) an EGF-like repeat (approximately 30 amino acids with six cysteines) followed by a crambin-like domain of approximately 70 amino acids which contains six cysteines (107). At present the functions of these cysteine-rich elements between the disintegrin and transmembrane domains are not clear.

One function of the cysteine-rich domains may be to affect the overall conformation of the ADAM so as to position the disintegrin domain for optimal interaction with an integrin on a neighbouring cell. Another possibility is that the cysteine-rich domain may itself play an adhesive role. Several recent observations support this possibility:

(a) The cysteine-rich domain of the SVMP atrolysin A has recently been shown to inhibit platelet aggregation (108).

(b) Recombinant constructs containing the cysteine-rich domain of human ADAM 12 support cell adhesion through syndecans (40, 48).

(c) Bacterial constructs encoding the cysteine-rich and EGF-like domains of fertilins α and β (ADAMs 1 and 2) inhibit fertilization *in vitro* (109).

(d) A peptide from the EGF-like repeat of mADAM 2 appears to enhance fertilization *in vitro* (110).

In view of the evidence for adhesive activity of ADAM cysteine-rich domains, an interesting possibility is that an ADAM can bind to both an integrin (through its disintegrin domain) and to a syndecan (through its cysteine-rich domain); integrins and syndecans are known to co-operate (111) (also see Chapters 4 and 5). Of possible relevance to this discussion is the recent observation that the C-terminal region of ADAM-TS1 (Fig. 1), which contains three thrombospondin motifs, interacts with sulfated glycosaminoglycans (112). Another possible role for the cysteine-rich domain is in dictating the target specificity of an ADAM protease domain. A recent study employing chimeras containing domains from ADAMs 10 and 17 supports the latter possibility (87). And, another recent study has indicated that the thrombospndin repeats of ADAM-TS4 are important for recognition of its substrate, aggrecan (113).

7. ADAMs in development

ADAMs (and ADAM-TS proteins) are emerging as important players in development. As displayed in Table 1, ADAMs have now been implicated, based on either functional data or on their expression patterns, in a variety of important events: spermatogenesis/fertilization (ADAMs 1–3, 16, 18, 20, 21, 24–26, 29, and 30), myogenesis (ADAMs 9, 12, 19), neurogenesis (ADAMs 10, 11, 19), neural crest cell migration (ADAM 13), bone (or joint) biology (ADAMs 9, 12, 15, and 19), immune function (ADAM 8), embryonic development and inflammatory reactions (ADAM 17), and vascular function (ADAM 15). Recent studies have indicated that 17 distinct ADAMs are expressed in the adult rodent brain (114). In the overview section we reviewed how specific domains of specific ADAMs are known or are thought to participate in these important developmental events. Here we focus on what is known about ADAM disintegrin domains in development, even though our knowledge on this subject is currently quite limited.

7.1 ADAM disintegrin domains in development

As discussed above, the disintegrin domains of mADAM 2, mADAM 3, and hADAM 9 have been shown to interact with the $\alpha6\beta1$ integrin (49, 72, 77, 95); the disintegrin domains of hADAM 15 and hADAM 23 have been shown to interact with the $\alpha v\beta3$ integrin (45, 46); the disintegrin domain of hADAM 15 has been shown to interact with the $\alpha5\beta1^9$ (43) integrin; and the disintegrin domains of hADAM 15, mADAM 15, and hADAM 12 have been shown to interact with the $\alpha9\beta1$ integrin (47). The disintegrin domains of mADAM 2 and mADAM 3 clearly play roles in development. *In vitro* studies with peptide analogues (38, 41, 42, 44, 94, 96, 115–117), with recombinant disintegrin domains (49, 77, 109, 118), as well as with a mature form of mADAM 2 purified from sperm (96) all support roles for the disintegrin domains of sperm mADAMs 2 and 3 in binding to the egg plasma membrane during fertilization. Most compellingly, male ADAM 2 'knock-out mice' are infertile (39). Male mice lacking ADAM 3 are also infertile (119).

Sperm from male ADAM 2 (–/–) mice are significantly compromised (albeit not abolished) in their ability to bind to the egg plasma membrane (39). An unexpected phenotype was that sperm from male ADAM 2 'knock-out mice' were compromised in their ability to cross a critical junction between the uterus and the oviduct. This phenotype suggested that mADAM 2 plays a role in sperm–epithelial cell interactions that are critical for the sperm's journey to reach the egg. Intriguingly both the $\alpha6$ integrin subunit as well as the integrin-associated protein CD9 (see above) are found on the apical surface of patches of epithelial cells at the critical uterine/oviduct junction (95).

8. ADAMs in disease

No ADAM gene has yet been mapped to a locus for a human genetic disease (see http://www.ncbi.nlm.nih.gov/LocusLink/list.cgi?Q=adam*&V=0).[3] Several ADAM proteases have, however, been implicated in disease states. Given the potential role of the disintegrin domain for protease function (Fig. 4), it is possible that ADAM adhesive functions (or lack thereof) may contribute to disease states. Enhanced expression or enhanced catalytic activity of ADAM 17, leading to exacerbated levels of soluble tumour necrosis factor, is involved in several inflammatory states including rheumatoid arthritis and Crone's disease (12, 120). By analogy with the effects of an ADAM 10 'dominant-negative' construct defective in protease activity (20), defects in ADAM 10 might cause defects in neurogenesis. Due to their ability to shed cell surface proteins, both ADAMs 10 and 17 have been suggested to play a role in the progression of Alzheimer's disease (121, 122). Because they have been shown to cleave a specific site in cartilage aggrecan, ADAM-TS4 and ADAM-TS5 (literature alias ADAM TS11; http://www.ncbi.nlm.nih.gov/LocusLink/LocRpt.cgi?l=11096) have been implicated in osteoarthritis (90, 91). ADAMs that are expressed in cartilage cells (ADAMs 9, 10, 15, and 17) may play roles in joint pathologies (99, 123).[10] ADAM 10 and ADAM 12 have been shown to be up-regulated in tumours from several tissues (35, 104), and ADAMs 9, 10, and 15 have been found in various leukaemic, lymphoma, and myeloma cells (124).[10] ADAMs have also been implicated in endometriosis (125), prostate cancer (126), and myocarditis (127). Lastly given that ADAMs can display both adhesive and proteolytic functions, they may play roles in tumour cell metastasis.

9. Remaining questions

Currently there are more questions than answers about the roles and molecular mechanisms of ADAMs in biology. We have recently enumerated a list of pressing questions regarding ADAM metalloprotease domains (9). Here we consider several key questions regarding ADAM disintegrin domains.

First, there is a great deal to learn about the structure and function of ADAM disintegrin domains. It will be very important to know whether all ADAM disintegrin domains interact with integrins. If not, it will be important to determine what defines those ADAMs that do bind to integrins. For example, are there specific sequences that define functional (integrin binding) disintegrin loops? Are there sequences outside of the disintegrin loop that contribute to or dictate the specificity of ADAM–integrin interactions? What are the relative affinities of different ADAM disintegrin domains for different integrins? Do ADAM disintegrin domains bind directly to integrins? Do they bind to complexes of integrins and integrin-associated proteins (for example tetraspanins such as CD9). Or, as some have proposed (97), do they only bind to non-integrin receptors? Ultimately it will be especially revealing to know what an ADAM

---

[10] Based on sequence analysis, the protein referred to as ADAM 12 in refs 99, 124, and 132 is ADAM 9.

disintegrin domain looks like in three-dimensions, both alone and in complex with its receptor.

Secondly, there is a great deal to learn about the cell and developmental biology of ADAM disintegrin domains. Some of the questions that are in pressing need of answers are: How many different integrins serve as co-receptors for ADAM disintegrin domains? What is the molecular basis for the apparent regulation of the integrin α6β1 on the egg for differential binding of ADAM 2 versus laminin (96)? What are the cellular factors (e.g. integrin-associated proteins, physiological conditions) that specify or facilitate binding of ADAM disintegrin domains to integrins? Are there developmentally or physiologically meaningful examples of ADAM–integrin (bidirectional) signalling? Which ADAM disintegrin domains play direct roles in development or disease?

Another major current challenge is to unravel the interplay between different ADAM domains (Fig. 1). A particularly important question in this regard is whether the disintegrin domain of any ADAM directs its upstream metalloprotease domain to its proteolytic substrate (e.g. an ECM component or a cell surface protein, for example its own integrin receptor; see Fig. 4). Similarly we would like to know if the disintegrin domains of any ADAMs influence either the cell signalling or the proposed cell fusion activity of any ADAMs? Conversely, it will be important to determine what influence other ADAM domains have on the function of the disintegrin domain. For example, do ADAM cysteine-rich domains co-operate with ADAM disintegrin domains to foster cell–cell interactions? Do interactions between ADAM cytoplasmic tails and adaptor proteins (e.g. SH3 binding proteins) influence the ability of an ADAM to bind to an integrin (or non-integrin receptor)?

There is clearly much work to be done to unravel the functions of the ADAMs, a large and fascinating group of multidomain proteins with multiple potential functions in cell biology, development, and pathology. The story of the ADAMs has just begun!

References

1. Blobel, C. P., Wolfsberg, T. G., Turck, C. W., Myles, D. G., Primakoff, P., and White, J. M. (1992). A potential fusion peptide and an integrin ligand domain in a protein active in sperm-egg fusion. *Nature*, **356**, 248.
2. Wolfsberg, T. G., Bazan, J. J., Blobel, C. P., Myles, D. G., Primakoff, P., and White, J. M. (1993). The precursor region of a protein active in sperm-egg fusion contains a metalloprotease and a disintegrin domain: Structural, functional and evolutionary implications. *Proc. Natl. Acad. Sci. USA*, **90**, 10783.
3. Blobel, C. P., Myles, D. G., Primakoff, P., and White, J. M. (1990). Proteolytic processing of a protein involved in sperm-egg fusion correlates with the acquisition of fertilization competence. *J. Cell Biol.*, **111**, 69.
4. Primakoff, P., Hyatt, H., and Tredick-Kline, J. (1987). Identification and purification of a sperm surface protein with a potential role in sperm-egg membrane fusion. *J. Cell Biol.*, **104**, 141.
5. Wolfsberg, T. G., Straight, P. D., Gerena, R. L., Huovila, A.-P. J., Primakoff, P., Myles, D. G., *et al.* (1995). ADAM, a widely distributed and developmentally regulated gene

family encoding membrane proteins with a disintegrin and metalloprotease domain. *Dev. Biol.*, **169**, 378.

6. Wolfsberg, T. G., Primakoff, P., Myles, D. G., and White, J. M. (1995). ADAM, a novel family of membrane proteins containing A Disintegrin And Metalloprotease Domain: multipotential functions in cell–cell and cell–matrix interactions. *J. Cell Biol.*, **131**, 275.

7. Weskamp, G. and Blobel, C. P. (1994). A family of cellular proteins related to snake venom disintegrins. *Proc. Natl. Acad. Sci. USA*, **91**, 2748.

8. Frayne, J., Jury, J. A., Barker, H. L., and Hall, L. (1998). The MDC family of proteins and their processing during epididymal transit. *J. Reprod. Fertil. Suppl.*, **53**, 149.

9. Black, R. A. and White, J. M. (1998). ADAMs: focus on the protease domain. *Curr. Opin. Cell Biol.*, **10**, 654.

10. Blobel, C. P. (1997). Metalloprotease-disintegrins: links to cell adhesion and cleavage of TNFα and notch. *Cell*, **90**, 589.

11. Fox, J. W. (1998). Introduction to the reprolysins. In *Handbook of proteolytic enzymes* (ed. A. J. Barrett, N. D. Rawlings, and J. F. Woessner), p. 1247. Academic Press, London.

12. Moss, M. L., White, J. M., Lambert, M. H., and Andrews, R. C. (2001). TACE and other ADAM proteases as targets for drug discovery. *Drug Disc. Therap.*, **6**, 417.

13. Primakoff, P. and Myles, D. G. (2000). The ADAM gene family: surface proteins with adhesion and protease activity. *Trends Genet.*, **16**, 83.

14. Stone, A. L., Kroeger, M., and Sang, Q. X. (1999). Structure–function analysis of the ADAM family of disintegrin-like and metalloproteinase-containing proteins. *J. Protein Chem.*, **18**, 447.

15. Turner, A. J. and Hooper, N. M. (1999). Role for ADAM-family proteinases as membrane protein secretases. *Biochem. Soc. Trans.*, **27**, 255.

16. Werb, Z. (1997). ECM and cell surface proteolysis: regulating cellular ecology. *Cell*, **91**, 439.

17. Wolfsberg, T. G. and White, J. M. (1996). ADAMs in fertilization and development. *Dev. Biol.*, **180**, 389.

18. Wolfsberg, T. G. and White, J. M. (1998). ADAM metalloproteases. In *Handbook of proteolytic enzymes* (ed. A. J. Barrett, N. D. Rawlings, and J. F. Woessner), p. 1310. Academic Press, London.

19. Fambrough, D., Pan, D., Rubin, G. M., and Goodman, C. S. (1996). The cell surface metalloprotease/disintegrin Kuzbanian is required for axonal extension in *Drosophila*. *Proc. Natl. Acad. Sci. USA*, **93**, 13233.

20. Pan, D. and Rubin, G. M. (1997). Kuzbanian controls proteolytic processing of notch and mediates lateral inhibition during *Drosophila* and vertebrate neurogenesis. *Cell*, **90**, 271.

21. Qi, H., Rand, M. D., Wu, X., Sestan, N., Wang, W., Rakic, P., *et al.* (1999). Processing of the notch ligand delta by the metalloprotease kuzbanian. *Science*, **283**, 91.

22. Rooke, J., Pan, D., Xu, T., and Rubin, G. M. (1996). KUZ, a conserved metalloprotease-disintegrin protein with two roles in *Drosophila* neurogenesis. *Science*, **273**, 1227.

23. Wen, C., Metzstein, M. M., and Greenwald, I. (1997). SUP-17, a *Caenorhabditis elegans* ADAM protein related to *Drosophila* kuzbanian, and its role in LIN-12/Notch signalling. *Development*, **124**, 4759.

24. Hattori, M., Osterfield, M., and Flanagan, J. G. (2000). Regulated cleavage of a contact-mediated axon repellent. *Science*, **289**, 1360.

25. Black, R. A., Rauch, C. T., Kozlosky, C. J., Peschon, J. J., Slack, J. L., Wolfson, M. F. *et al.* (1997). A metalloproteinase disintegrin that releases tumour-necrosis factor-alpha from cells. *Nature*, **385**, 729.

26. Moss, M. L., Jin, S. L., Milla, M. E., Burkhart, W., Carter, H. L., Chen, W. J. *et al.* (1997). Cloning of a disintegrin metalloproteinase that processes precursor tumour-necrosis factor-alpha. *Nature*, **385**, 733.

27. Peschon, J. J., Slack, J. L., Reddy, P., Stocking, K. L., Sunnarborg, B. J., Lee, D. C., *et al.* (1998). An essential role for ectodomain shedding in mammalian development. *Science*, **282**, 1281.

28. Howard, L., Maciewicz, R. A., and Blobel, C. P. (2000). Cloning and characterization of ADAM28: evidence for autocatalytic pro-domain removal and for cell surface localization of mature ADAM28. *Biochem. J.*, **348**, 21.

29. Izumi, Y., Hirata, M., Hasuwa, H., Iwamoto, R., Umata, T., Miyado, K., *et al.* (1998). A metalloprotease-disintegrin, MDC9/meltrin-γ/ADAM 9 and PKCδ are involved in TPA-induced ectodomain shedding of membrane anchored heparin-binding EGF-like growth factor. *EMBO J.*, **17**, 7260.

30. Loechel, F., Overgaard, M. T., Oxvig, C., Albrechtsen, R., and Wewer, U. M. (1999). Regulation of human ADAM 12 protease by the prodomain. Evidence for a functional cysteine switch. *J. Biol. Chem.*, **274**, 13427.

31. Roghani, M., Becherer, J. D., Moss, M. L., Atherton, R. E., Erdjument-Bromage, H., Arribas, J., *et al.* (1999). Metalloprotease-disintegtin MDC9: Intacellular maturation and catalytic activity. *J. Biol. Chem.*, **274**, 3531.

32. Abe, E., Mocharla, H., Yamate, T., Taguchi, Y., and Manolagas, S. C. (1999). Meltrin-alpha, a fusion protein involved in multinucleated giant cell and osteoclast formation. *Calcif. Tissue Int.*, **64**, 508.

33. Gilpin, B. J., Loechel, F., Mattei, M.-G., Engvall, E., Albrechtsen, R., and Wewer, U. M. (1998). A novel secreted form of human ADAM 12 (meltrin α) provokes myogenesis *in vivo. J. Biol. Chem.*, **273**, 157.

34. Inoue, D., Reid, M., Lum, L., Kratzschmar, J., Weskamp, G., Myung, Y. M., *et al.* (1998). Cloning and initial characterization of mouse meltrin β and analysis of the expression of four metalloprotease-disintegrins in bone cells. *J. Biol. Chem.*, **273**, 4180.

35. Kurisaki, T., Masuda, A., Osumi, N., Nabeshima, Y., and Fujisawa-Sehara, A. (1998). Spatially- and temporally-restricted expression of meltrin α (ADAM 12) and β (ADAM 19) in mouse embryo. *Mech. Dev.*, **73**, 211.

36. Yagami-Hiromasa, T., Sato, T., Kurisaki, T., Kamijo, K., Nabeshima, Y., and Fujisawa-Sehara, A. (1995). A metalloprotease-disintegrin participating in myoblast fusion. *Nature*, **377**, 652.

37. Alfandari, D., Wolfsberg, T. G., White, J. M., and DeSimone, D. W. (1997). ADAM 13: a novel ADAM expressed in somitic mesoderm and neural crest cells during *Xenopus laevis* development. *Dev. Biol.*, **182**, 314.

38. Almeida, E. A. C., Huovila, A.-P. J., Sutherland, A. E., Stephens, L. E., Calarco, P. G., Shaw, L. M., *et al.* (1995). Mouse egg integrin α6β1 functions as a sperm receptor. *Cell*, **81**, 1095.

39. Cho, C., O'Dell Bunch, D., Faure, J. E., Goulding, E. H., Eddy, E. M., Primakoff, P., *et al.* (1998). Fertilization defects in sperm from mice lacking fertilin β. *Science*, **281**, 1857.

40. Iba, K., Albrechtsen, R., Gilpin, B. J., Loechel, F., and Wewer, U. M. (1999). The cysteine-rich domain of human ADAM 12 (meltrin α) supports tumor cell adhesion. *Am. J. Pathol.*, **154**, 1489.

41. Linder, B. and Heinlein, U. A. O. (1997). Decreased *in vitro* fertilization efficiencies in the presence of specific cyritestin peptides. *Dev. Growth Differ.*, **39**, 243.

42. Myles, D. G., Kimmel, L. H., Blobel, C. P., White, J. M., and Primakoff, P. (1994). Identification of a binding site in the disintegrin domain of fertilin required for sperm-egg fusion. *Proc. Natl. Acad. Sci. USA*, **91**, 4195.

43. Nath, D., Slocombe, P. M., Stephens, P. E., Warn, A., Hutchinson, G. R., Yamada, K. M., *et al.* (1999). Interaction of metargidin (ADAM-15) with αvβ3 and α5β1 integrins on different haemopoietic cells. *J. Cell Sci.*, **112**, 579.

44. Yuan, R., Primakoff, P., and Myles, D. G. (1997). A role for the disintegrin domain of cyritestin, a sperm surface protein belonging to the ADAM family, in mouse sperm-egg plasma membrane adhesion and fusion. *J. Cell Biol.*, **137**, 105.

45. Zhang, X.-P., Kamata, T., Yokoyama, K., Puzon-McLaughlin, W., and Takada, Y. (1998). Specific interaction of the recombinant disintegrin-like domain of MDC-15 (Metargidin, ADAM-15) with integrin αvβ3. *J. Biol. Chem.*, **273**, 7345.

46. Cal, S., Freije, J. M., Lopez, J. M., Takada, Y., and Lopez-Otin, C. (2000). ADAM 23/MDC3, a human disintegrin that promotes cell adhesion via interaction with the αvβ3 integrin through an RGD-independent mechanism. *Mol. Biol. Cell*, **11**, 1457.

47. Eto, K., Puzon-McLaughlin, W., Sheppard, D., Sehara-Fujisawa, A., Zhang, X.-P., and Takada, Y. (2000). RGD-independent binding of integrin α9β1 to the ADAM 12 and -15 disintegrin domains mediates cell–cell interaction. *J. Biol. Chem.*, **275**, 34922.

48. Iba, K., Albrechtsen, R., Gilpin, B., Frohlich, C., Loechel, F., Zolkiewska, A., *et al.* (2000). The cysteine-rich domain of human ADAM 12 supports cell adhesion through syndecans and triggers signaling events that lead to beta1 integrin-dependent cell spreading. *J. Cell Biol.*, **149**, 1143.

49. Takahashi, Y., Bigler, D., Ito, Y., and White, J. M. (2001). Sequence-specific interaction between the disintegrin domain of mouse ADAM 3 and murine eggs: role of the β1 integrin associated proteins CD9, CD81 and CD98. *Mol. Biol. Cell*, **12**, 809.

50. Hooft van Huijsduijnen, R. (1998). ADAM 20 and 21: two novel human testis-specific membrane metalloproteases with similarity to fertilin-α. *Gene*, **206**, 273.

51. Martin, I., Epand, R. M., and Ruysschaert, J. (1998). Structural properties of the putative fusion peptide of fertilin, a protein active in sperm-egg fusion, upon interaction with the lipid bilayer. *Biochemistry*, **37**, 17030.

52. Muga, A., Neugebauer, W., Hirama, T., and Surewicz, W. K. (1994). Membrane interactive and conformational properties of the putative fusion peptide of PH-30, a protein active in sperm-egg fusion. *Biochemistry*, **33**, 4444.

53. Nidome, T., Kimura, M., Chiba, T., Ohmori, N., Mihara, H., and Aoyagi, H. (1997). Membrane interaction of synthetic peptides related to the putative fusogenic region of PH-30α, a protein in sperm-egg fusion. *J. Peptide Res.*, **49**, 563.

54. Wolfe, C. A., Cladera, J., Ladha, S., Senior, S., Jones, R., and O'Shea, P. (1999). Membrane interactions of the putative fusion peptide (MF αP) from fertilin-α, the mouse sperm protein complex involved in fertilization. *Mol. Membr. Biol.*, **16**, 257.

55. Hernandez, L. D., Hoffman, L. R., Wolfsberg, T. G., and White, J. M. (1996). Virus-cell and cell-cell fusion. *Annu. Rev. Cell Dev. Biol.*, **12**, 627.

56. Alexandropoulos, K., Cheng, G., and Baltimore, D. (1995). Proline-rich sequences that bind to Src homology 3 domains with individual specificities. *Proc. Natl. Acad. Sci. USA*, **92**, 3110.

57. Cousin, H., Gaultier, A., Bleux, C., Darribere, T., and Alfandari, D. (2000). PACSIN2 is a regulator of the metalloprotease/disintegrin ADAM13. *Dev. Biol.*, **227**, 197.

58. Howard, L., Nelson, K. K., Maciewicz, R. A., and Blobel, C. P. (1999). Interaction of the metalloprotease disintegrins MDC9 and MDC15 with two SH3 domain-containing proteins, endophilin I and SH3PX1. *J. Biol. Chem.*, **274**, 31693.

59. Kang, Q., Cao, Y., and Zolkiewska, A. (2000). Metalloprotease-disintegrin ADAM 12 binds to the SH3 domain of Src and activates Src tyrosine kinase in C2C12 cells. *Biochem. J.*, **352**, 883.

60. Nelson, K. K., Schlondorff, J., and Blobel, C. P. (1999). Evidence for an interaction of the metalloprotease-disintegrin tumour necrosis factor alpha convertase (TACE) with mitotic arrest deficient 2 (MAD2), and of the metalloprotease-disintegrin MDC9 with a novel MAD2-related protein, MAD2beta. *Biochem. J.*, **343**, 673.

61. Weskamp, G., Kratzschmar, J., Reid, M. S., and Blobel, C. P. (1996). MDC9, a widely expressed cellular disintegrin containing cytoplasmic SH3 ligand domains. *J. Cell Biol.*, **132**, 717.

62. Galliano, M. F., Huet, C., Frygelius, J., Polgren, A., Wewer, U. M., and Engvall, E. (2000). Binding of ADAM12, a marker of skeletal muscle regeneration, to the muscle-specific actin-binding protein, alpha -actinin-2, is required for myoblast fusion. *J. Biol. Chem.*, **275**, 13933.

63. Waters, S. I. and White, J. M. (1997). Biochemical and molecular characterization of bovine fertilin α and β (ADAM 1 and ADAM 2): a candidate sperm-egg binding/fusion complex. *Biol. Reprod.*, **56**, 1245.

64. Marcinkiewicz, C., Calvete, J. J., Marcinkiewicz, M. M., Raida, M., Vijay-Kumar, S., Huang, Z., *et al.* (1999). EC3, a novel heterodimeric disintegrin from *Echis carinatus* venom, inhibits α4 and α5 integrins in an RGD-independent manner. *J. Biol. Chem.*, **274**, 12468.

65. Marcinkiewicz, C., Calvete, J. J., Vijay-Kumar, S., Marcinkiewicz, M. M., Raida, M., Schick, P., *et al.* (1999). Structural and functional characterization of EMF10, a hetero-dimeric disintegrin from *Eristocophis macmahoni* venom that selectively inhibits alpha 5 beta 1 integrin. *Biochemistry*, **38**, 13302.

66. Ritter, M. R., Zhou, Q., and Markland, F. S. (2000). Contortrostatin, a snake venom disintegrin, induces alphavbeta3-mediated tyrosine phosphorylation of CAS and FAK in tumor cells. *J. Cell. Biochem.*, **79**, 28.

67. McLane, M. A., Marcinkiewicz, C., Vijay-Kumar, S., Wierzbicka-Patynowski, I., and Niewiarowski, S. (1998). Viper venom disintegrins and related molecules. *Proc. Soc. Exp. Biol. Med.*, **219**, 109.

68. Adler, M., Lazarus, R. A., Dennis, M. S., and Wagner, G. (1991). Solution structure of kistrin, a potent platelet aggregation inhibitor and GP IIb-IIIa antagonist. *Science*, **253**, 445.

69. Jia, L.-G., Wang, X.-M., Shannon, J. D., Bjarnason, J. B., and Fox, J. W. (1997). Function of disintegrin-like/cysteine-rich domains of atrolysin A. *J. Biol. Chem.*, **272**, 13094.

70. Calvete, J. J., Moreno-Murciano, M. P., Sanz, L., Jurgens, M., Schrader, M., Raida, M., *et al.* (2000). The disulfide bond pattern of catrocollastatin C, a disintegrin-like/cysteine-rich protein isolated from *Crotalus atrox* venom. A model for reprolysin structure. *Protein Sci.*, **7**, 1365.

71. Maskos, K., Fernandez-Catalan, C., Huber, R., Bourenkov, G. P., Bartunik, H., Ellestad, G. A., *et al.* (1998). Crystal structure of the catalytic domain of human tumor necrosis factor α-converting enzyme. *Proc. Natl. Acad. Sci. USA*, **95**, 3408.

72. Nath, D., Slocombe, P. M., Webster, A., Stephens, P. E., Docherty, A. J., and Murphy, G. (2000). Meltrin γ (ADAM-9) mediates cellular adhesion through α6β1 integrin, leading to a marked induction of fibroblast cell motility. *J. Cell Sci.*, **113**, 2319.

73. Giannelli, G., Falk-Marzillier, J., Schiraldi, O., Stetler-Stevenson, W. G., and Quaranta, V. (1997). Induction of cell migration by matrix metalloprotease-2 cleavage of laminin-5. *Science*, **277**, 225.

74. Brooks, P. C., Silletti, S., Von Schalscha, T. L., Friedlander, M., and Cheresh, D. A. (1998). Disruption of angiogenesis by PEX, a noncatalytic metalloproteinase fragment with integrin binding activity. *Cell*, **92**, 391.

75. Nakahara, H., Nomizu, M., Akiyama, S. K., Yamada, Y., Yeh, Y., and Chen, W. T. (1996). A mechanism for regulation of melanoma invasion. Ligation of alpha 6beta1 integrin by laminin G peptides. *J. Biol. Chem.*, **271**, 27221.

76. Kamiguti, A. S., Hay, C. R., and Zuzel, M. (1996). Inhibition of collagen-induced platelet aggregation as the result of cleavage of $\alpha2\beta1$-integrin by the snake venom metalloproteinase jararhagin. *Biochem. J.*, **320**, 635.

77. Bigler, D., Takahashi, Y., Chen, M. S., Almeida, E. A. C., Osbourne, L., and White, J. M. (2000). Sequence-specific interaction between the disintegrin domain of mouse ADAM 2 (fertilin β) and murine eggs: Role of the $\alpha6$ integrin subunit. *J. Biol. Chem.*, **275**, 11576.

78. Zhu, X., Bansal, N. P., and Evans, J. P. (2000). Identification of key functional amino acids of the mouse fertilin beta (ADAM2) disintegrin loop for cell-cell adhesion during fertilization. *J. Biol. Chem.*, **275**, 7677.

79. Kratzschmar, J., Lum, L., and Blobel, C. P. (1996). Metargidin, a membrane-anchored metalloprotease-disintegrin protein with an RGD integrin binding sequence. *J. Biol. Chem.*, **271**, 4593.

80. Shilling, F. M., Kratzschmar, J., Cai, H., Weskamp, G., Gayko, U., Leibow, J., et al. (1997). Identification of metalloprotease/disintegrins in *Xenopus laevis* testis with a potential role in fertilization. *Dev. Biol.*, **186**, 155.

81. Gupta, S., Li, H., and Sampson, N. S. (2000). Characterization of fertilin β-disintegrin binding specificity in sperm-egg adhesion. *Bioorg. Med. Chem.*, **8**, 723.

82. Marcinkiewicz, C., Vijayu-Kumar, S., McLane, M. A., and Nieiarowski, S. (1997). Significance of RGD Loop and C-terminal domain of echistatin for recognition of aIIbB3 and avB3 integrins and expression of ligand-induced binding site. *Blood*, **90**, 1565.

83. Cho, C., Turner, L., Primakoff, P., and Myles, D. G. (1997). Genomic organization of the mouse fertilin B gene that encodes an ADAM family protein active in sperm-egg fusion. *Dev. Genet.*, **20**, 320.

84. Katagiri, T., Harada, Y., Emi, M., and Nakamura, Y. (1995). Human metalloprotease/disintegrin-like (MDC) gene: exon-intron organization and alternative splicing. *Cytogenet. Cell. Genet.*, **68**, 39.

85. Emi, M., Katagiri, T., Harada, Y., Saito, H., Inazawa, J., Ito, I., et al. (1993). A novel metalloprotease/disintegrin-like gene at 17q21.3 is somatically rearranged in two primary breast cancers. *Nature Genet.*, **5**, 151.

86. Cerretti, D. P., DuBose, R. F., Black, R. A., and Nelson, N. (1999). Isolation of two novel metalloproteinase-disintegrin (ADAM) cDNAs that show testis-specific gene expression. *Biochem. Biophys. Res. Commun.*, **263**(3), 810.

87. Reddy, P., Slack, J. L., Davis, R., Cerretti, D. P., Kozlosky, C. J., Blanton, R. A., et al. (2000). Functional analysis of the domain structure of TNF-α converting enzyme. *J. Biol. Chem.*, **275**, 14608.

88. Wierzbicka-Patynowski, I., Niewiarowski, S., Marcinkiewicz, C., Calvete, J. J., Marcinkiewicz, M. M., and McLane, M. A. (1999). Structural requirements of echistatin for the recognition of $\alpha v\beta3$ and $\alpha5\beta1$ integrins. *J. Biol. Chem.*, **274**, 37809.

89. Kuno, K., Kanada, N., Nakashima, E., Fujiki, F., Ichimura, F., and Matsushima, K. (1997). Molecular cloning of a gene encoding a new type of metalloproteinase-disintegrin family protein with thrombospondin motifs as an inflammation associated gene. *J. Biol. Chem.*, **272**, 556.

90. Abbaszade, I., Liu, R. Q., Yang, F., Rosenfeld, S. A., Ross, O. H., Link, J. R. et al. (1999). Cloning and characterization of ADAMTS11, an aggrecanase from the ADAMTS family. *J. Biol. Chem.*, **274**(33), 23443.

91. Tortorella, M. D., Burn, T. C., Pratta, M. A., Abbaszade, I., Hollis, J. M., Liu, R. *et al.* (1999). Purification and cloning of aggrecanase-1: A member of ADAMTS family of proteins. *Science*, **284**, 1664.

92. Blelloch, R. and Kimble, J. (1999). Control of organ shape by a secreted metalloprotease in the nematode *Caenorhabditis elegans*. *Nature*, **399**, 586.

93. Nishiwaki, K., Hisamoto, N., and Matsumoto, K. (2000). A metalloprotease disintegrin that controls cell migration in *Caenorhabditis elegans*. *Science*, **288**, 2205.

94. Chen, H. and Sampson, N. S. (1999). Mediation of sperm-egg fusion: evidence that mouse egg α6β1 integrin is the receptor for sperm fertilin β. *Chem. Biol.*, **6**, 1.

95. Chen, M. S., Tung, K. S. K., Coonrod, S. A., Takahashi, Y., Bigler, D., Chang, A., *et al.* (1999). Role of the integrin-associated protein CD9 in binding between sperm ADAM 2 and the egg integrin α6β1: Implications for murine fertilization. *Proc. Natl. Acad. Sci. USA*, **96**, 11830.

96. Chen, M. S., Almeida, E. A. C., Huovila, A.-P. J., Takahashi, Y., Shaw, L. M., Mercurio, A. M., *et al.* (1999). Evidence that distinct 'states' of the integrin α6β1 interact with laminin and an ADAM. *J. Cell Biol.*, **144**, 549.

97. Miller, B. J., Georges-Labouesse, E., Primakoff, P., and Myles, D. G. (2000). Normal fertilization occurs with eggs lacking the integrin α6β1 and is CD9-dependent. *J. Cell Biol.*, **149**, 1289.

98. Herren, B., Raines, E. W., and Ross, R. (1997). Expression of a disintegrin-like protein in cultured human vascular cells and *in vivo*. *FASEB J.*, **11**, 173.

99. McKie, N., Edwards, T., Dallas, D. J., Houghton, A., Stringer, B., Graham, R., *et al.* (1997). Expression of members of a novel membrane linked metalloproteinase family (ADAM) in human articular chondrocytes. *Biochem. Biophys. Res. Commun.*, **230**, 335.

100. Lum, L., Reid, M. S., and Blobel, C. P. (1998). Intracellular maturation of the mouse metalloprotease disintegrin MDC15. *J. Biol. Chem.*, **273**, 26236.

101. Chen, M. S. (2000). Regulation of fertilin β (ADAM 2) binding to the α6β1 integrin. Unpublished Ph.D. thesis submitted to the Department of Microbiology, University of Virginia, Charlottesville, VA, 233.

102. Hemler, M. E. (1998). Integrin associated proteins. *Curr. Opin. Cell Biol.*, **10**, 578.

103. Porter, J. C. and Hogg, N. (1998). Integrins take partners: cross-talk between integrins and other membrane receptors. *Trends Cell Biol.*, **8**, 390.

104. Le Naour, F., Rubinstein, E., Jasmin, C., Prenant, M., and Boucheix, C. (2000). Severely reduced female fertility in CD9-deficient mice. *Science*, **287**, 319.

105. Kaji, K., Oda, S., Shikano, T., Ohnuki, T., Uematsu, Y., Sakagami, J., *et al.* (2000). The gamete fusion process is defective in eggs of CD9-deficient mice. *Nature Genet.*, **24**, 279.

106. Miyado, K., Yamada, G., Yamada, S., Hasuwa, H., Nakamura, Y., Ryu, F., *et al.* (2000). Requirement of CD9 on the egg plasma membrane for fertilization. *Science*, **287**, 321.

107. Black, R. (1998). TNF-α converting enzyme. In *Handbook of proteolytic enzymes* (ed. A. J. Barrett, N. D. Rawlings, and J. F. Woessner), p. 1315. Academic Press, London.

108. Jia, L. G., Wang, X.-M., Shannon, J. D., Bjarnason, J. B., and Fox, J. W. (2000). Inhibition of platelet aggregation by the recombinant cysteine-rich domain of the hemorrhagic snake venom metalloprotease, Atrolysin A. *Arch. Biochem. Biophys.*, **373**, 281.

109. Evans, J. P., Schultz, R. M., and Kopf, G. S. (1998). Roles of the disintegrin domains of mouse fertilins α and β in fertilization. *Biol. Reprod.*, **59**, 145.

110. Chen, H., Pyluck, A. L., Janik, M., and Sampson, N. S. (1998). Peptides corresponding to the epidermal growth factor-like domain of mouse fertilin: synthesis and biological activity. *Biopolymers*, **47**, 299.

111. Woods, A. and Couchman, J. (1998). Syndecans: synergistic activators of cell adhesion. *Trends Cell Biol.*, **8**, 189.

112. Kuno, K. and Matsushima, K. (1998). ADAMTS-1 protein anchors at the extracellular matrix through the thrombospondin type I motifs and its spacing region. *J. Biol. Chem.*, **273**, 13912.

113. Tortorella, M. D., Pratta, M., Liu, R. Q., Abbaszade, I., Ross, H., Burn, T., *et al.* (2000). The thrombospondin motif of aggrecanase -1 (ADAMTS-4)is critical for aggrecan substrate recognition and cleavage. *J. Biol. Chem.*, **275**, 25791.

114. Karkkainen, I., Rybnikova, E., Pelto-Huikko, M., and Huovila, A.-P. J. (2000). Metalloprotease-disintegrin (ADAM) genes are widely and differentially expressed in the adult CNS. *Mol. Cell. Neurosci.*, **6**, 547.

115. Bronson, R. A., Fusi, F. M., Calzi, F., Doldi, N., and Ferrari, A. (1999). Evidence that a functional fertilin-like ADAM plays a role in human sperm-oolemmal interactions. *Mol. Hum. Reprod.*, **5**, 433.

116. Evans, J. P., Schultz, R. M., and Kopf, G. S. (1995). Mouse sperm-egg plasma membrane interactions: analysis of roles of egg integrins and the mouse sperm homologue of PH-30 (fertilin) β. *J. Cell Sci.*, **108**, 3267.

117. Gichuhi, P. M., Ford, W. C. L., and Hall, L. (1997). Evidence that peptides derived from the disintegrin domain of primate fertilin and containing the ECD motif block the binding of human spermatozoa to the zona-free hamster oocyte. *Int. J. Androl.*, **20**, 165.

118. Evans, J. P., Kopf, G. S., and Schultz, R. M. (1997). Characterization of the binding of recombinant mouse sperm fertilin β subunit to mouse eggs: evidence for adhesive activity via an egg β1 integrin-mediated interaction. *Dev. Biol.*, **187**, 79.

119. Shamsadin, R., Adham, I. M., Nayernia, K., Heinlein, U. A., Oberwinkler, H., and Engel, W. (1999). Male mice deficient for germ-cell cyritestin are infertile. *Biol. Reprod.*, **61**, 1445.

120. Beutler, B. and Bazzoni, F. (1998). TNF, apoptosis and autoimmunity: a common thread? *Blood Cells, Molecules, Diseases*, **24**, 216.

121. Buxbaum, J. D., Liu, K. N., Luo, Y., Slack, J. L., Stocking, K. L., Peschon, J. J., *et al.* (1998). Evidence that tumor necrosis factor alpha converting enzyme is involved in regulated alpha-secretase clevage of the Alzheimer amyloid protein precursor. *J. Biol. Chem.*, **273**, 27765.

122. Lammich, S., Kojro, E., Postina, R., Gilbert, S., Pfeiffer, R., Jasionowski, M., *et al.* (1999). Constitutive and regulated alpha-secretase cleavage of Alzheimer's amyloid precursor protein by a disintegrin metalloprotease. *Proc. Natl. Acad. Sci. USA*, **96**, 3922.

123. Patel, I., Attur, M., Patel, R., Stuchin, S., Abagyan, R., Abramson, S., *et al.* (1998). TNF-α convertase enzyme from human arthritis-affected cartilage: isolation of cDNA by differential display, expression of the active enzyme, and regulation of TNF-α. *J. Immunol.*, **160**, 4570.

124. Wu, E., Croucher, P. I., and McKie, N. (1997). Expression of members of the novel membrane linked metalloproteinase family ADAM in cells derived from a range of haematological malignancies. *Biochem. Biophys. Res. Commun.*, **235**, 437.

125. Gottschalk, C., Malberg, K., Arndt, M., Schmitt, J., Roessner, A., Schultze, D., *et al.* (2000). Matrix metalloproteinases and TACE play a role in the pathogenesis of endometriosis. *Adv. Exp. Med. Biol.*, **477**, 483.

126. McCulloch, D. R., Harvey, M., and Herington, A. C. (2000). The expression of the ADAMs proteases in prostate cancer cell lines and their regulation by dihydrotestosterone. *Mol. Cell. Endocrinol.*, **167**, 11.

127. Satoh, M., Nakamura, M., Satoh, H., Saitoh, H., Segawa, I., and Hiramori, K. (2000). Expression of tumor necrosis factor-alpha—converting enzyme and tumor necrosis factor-alpha in human myocarditis. *J. Am. Coll. Cardiol.*, **36**, 1288.

128. Heinlein, U. A. O., Wallat, S., Senftleben, A., and Lemaire, L. (1994). Male germ cell-expressed mouse gene TAZ83 encodes a putative cysteine-rich transmembrane protein (cyritestin) sharing homologies with snake toxins and sperm-egg fusion proteins. *Dev. Growth Differ.*, **36**, 49.

129. Jury, J. A., Frayne, J., and Hall, L. (1997). The human fertilin α gene is non-functional: implications for its proposed role in fertilization. *Biochem. J.*, **321**, 577.

130. Cho, C., Ge, H., Branciforte, D., Primakoff, P., and Myles, D. G. (2000). Analysis of mouse fertilin in wild-type and fertilin beta(−/−) sperm: evidence for C-terminal modification, alpha/beta dimerization, and lack of essential role of fertilin alpha in sperm-egg fusion. *Dev. Biol.*, **222**, 289.

131. Rea, C. G. (1999). Biochemical properties of mouse fertilins α and β. Unpublished Master's thesis submitted to the Department of Cell Biology, University of Virginia, Charlottesville, VA, 110.

132. McKie, N., Dallas, D. J., Edwards, T., Apperley, J. F., Russell, R. G. G., and Croucher, P. I. (1996). Cloning of a novel membrane-linked metalloproteinase from human myeloma cells. *Biochem. J.*, **318**, 459.

133. Berman, H. M., Westbrook, J., Feng, Z., Gilliland, G., Bhat, T. N., Weissig, H., *et al.* (2000). The protein data bank. *Nucleic Acids Res.*, **28**, 235.

134. Altschul, S. F., Madden, T. L., Schaffer, A. A., Zhang, J., Zhang, Z., Miller, W., *et al.* (1997). Gapped BLAST and PSI-BLAST: a new generation of protein database search programs. *Nucleic Acids Res.*, **25**, 3389.

135. Feldenstein, J. (1993). PHYLIP (Phylogeny Inference Package), version 3.5c. Distributed by the author. Department of Genetics, University of Washington, Seattle.

136. Maciukenas, M. (1994). Treetool, version 2.0.1. Ribosomal RNA Database Project, University of Illinois.

137. Alfandari, D., Cousin, H., Gaultier, A., Smith, K., White, J. M., Darribere, T., and DeSimone, D. W. (2001). *Xenopus* ADAM 13 is a metalloprotease required for cranial neural crest-cell migration. *Curr. Biol.*, **11**, 918.

7 | Protein tyrosine phosphatases

SUSANN M. BRADY-KALNAY

1. Introduction

Cells use the reversible phosphorylation of the amino acid tyrosine as a molecular on/off switch. Tyrosine phosphorylation of proteins is an essential component of signal transduction pathways that regulate cell growth and death as well as differentiation. Cellular phosphotyrosine levels are controlled by both protein tyrosine kinases (PTKs) and protein tyrosine phosphatases (PTPs). An ever expanding family of receptor-like (RPTPs) and non-transmembrane PTPs have been isolated and characterized over the past decade. The diversity of the PTP family of enzymes resides in their non-catalytic sequences both within their intracellular and extracellular segments (for the RPTPs).

Even though the PTP field is approximately 13 years old, it is still in its infancy. There have been approximately 50 PTPs cloned of which 30 are cytosolic enzymes while 20 are receptor PTPs (RPTPs) (1). In this chapter, we will focus on the RPTPs. Of the 20 RPTPs that have been cloned, approximately 10 of them have been insufficiently characterized. The only information that has been published on these 10 PTPs consists of cloning data, gene expression studies, and demonstration of enzymatic activity. The reason for the RPTP field's infancy is that there is not a single RPTP where we know the ligands, substrates, and biological function of the enzyme. However, we have a greater understanding of a few prototypical RPTPs that will form the basis for this chapter. The most exciting feature of the RPTPs is that they have cell adhesion molecule-like extracellular segments directly linked to intracellular tyrosine phosphatase domains. This suggests that they are likely to transduce signals directly in response to cell adhesion. These types of molecules may be involved in the phenomenon of contact inhibition of growth and movement whose molecular mechanisms have remained elusive.

2. Molecular structure

2.1 Intracellular segments of RPTPs

The PTP enzymes share ~ 30% amino acid identity in the conserved catalytic domain of 240 residues. The catalytic domain has a unique sequence motif {(I/V)HCXAG

XXR(S/T)G} which forms a phosphate binding pocket where the cysteine residue is essential for catalysis (2). When the three-dimensional structure of the catalytic domain PTP1B was solved (3), it suggested precise mutations that could be made which would alter either the affinity for substrate (K_m) or the rate of catalysis (V_{max}). Studies of mutations made in PTP1B have led to three mutations that are useful for determining the function of these enzymes and in identification of their physiological substrates (4). Investigators are now using such mutants to analyse the physiological functions of the PTPs (5–7).

RPTPs are composed of one or two tandem phosphatase domains in their intracellular segment, a single transmembrane domain, and diverse extracellular segments. In the RPTPs that contain two PTP domains, the second phosphatase domain usually contains naturally occurring mutations that render it catalytically inactive. As discussed below, the second PTP domain may function to regulate the enzymatic activity of the first catalytic domain.

2.2 Extracellular segment of RPTPs

Some RPTPs have structural homology to cell–cell adhesion molecules such as the neural cell adhesion molecule, NCAM (Figs 1–3). NCAM is a cell–cell adhesion molecule that contains multiple immunoglobulin domains (Ig) and fibronectin type III (FNIII) repeats in its extracellular segment (8, 9). The intracellular domains of most adhesion molecules do not possess any enzymatic activity, suggesting that they must signal indirectly through associated proteins. Conversely, RPTPs have intracellular segments with PTP catalytic domains linked directly to adhesion molecule-like extracellular segments which suggests that they directly transduce signals in response to cell adhesion. There are currently eight subfamilies of RPTPs (see Figs 1–3) with quite diverse extracellular segments (10–12). The motifs found in the RPTPs described in this review include Ig domains, FNIII repeats, MAM domains, and carbonic anhydrase-like domains (Fig. 1).

The LAR-like (leukocyte antigen related) proteins (LAR, DLAR, PTPδ, and PTPσ) have multiple Ig domains and FNIII repeats (Fig. 1). The cell adhesion molecules and RPTPs contain C-2 type Ig domains which are disulfide bonded structures with key cysteine residues spaced approximately 60 residues apart. These domains contain the homophilic and heterophilic binding sites of many cell adhesion molecules (8). Ig domains are found in many RPTPs including LAR, DLAR, DPTP69D, PTPδ, PTPσ, *clr-1*, PTPμ, PTPκ, PCP-2, and PTPρ (10).

Most of the RPTPs contain FNIII repeats in their extracellular segments (Figs 1–2). FNIII repeats consist of 90–100 amino acids that are characterized by highly conserved hydrophobic residues (13) and were originally found in the extracellular matrix protein fibronectin. FNIII repeats have now been observed in many proteins (14). The FNIII repeat three-dimensional structure suggests that they fold with a similar topology to Ig C-type domains and cadherin repeats (9). In some proteins, FNIII repeats may be involved in protein–protein interactions, flexibility of the molecule, or determining the distance the protein projects from the plasma mem-

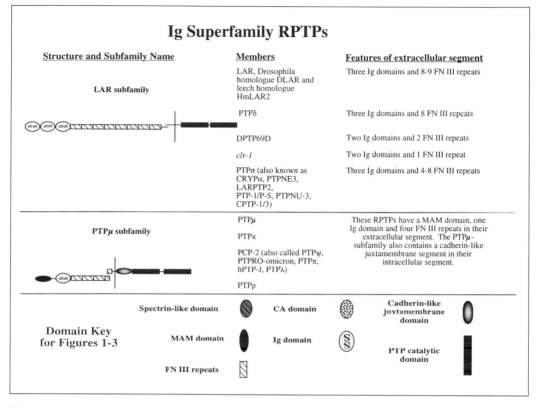

Fig. 1 Representative members of the family of receptor protein tyrosine phosphatases that have immunoglobulin domains are shown schematically. The types of domains found in receptor PTPs are shown at the bottom and include a spectrin-like domain, MAM domain, carbonic anhydrase domain (CA), Ig domain, fibronectin type III repeat (FNIII), cadherin-like juxtamembrane domain, and PTP catalytic domain.

brane. FNIII repeats are found in combination with Ig domains in many cell adhesion molecules (8). In addition, one subfamily of RPTPs has only FNIII repeats in their extracellular segments (Fig. 2). This subfamily includes DPTP10D, DPTP99A, DPTP4E, DEP-1, OST-PTP, SAP-1, GLEPP-1, and PTPβ (10).

The PTPμ-like subfamily (PTPμ, PTPκ, PCP-2, and PTPρ) contains a MAM domain, an Ig domain, and four FNIII repeats (Fig. 1). The MAM domain (meprins, A5, PTPMu) is a sequence motif identified in seven proteins: meprins A and B, neuropilin (A5), PTPμ, PTPκ, PCP-2, PTPρ (15–17). PCP-2 has been cloned by multiple groups and has many names (see Fig. 1) including PTPΨ, PTPRO (omicron), PTPπ, hPTP-J, PTPλ (16, 18–22). The MAM domain is comprised of 170 amino acids with four conserved cysteine residues and two conserved sequence motifs (*tChtFahhxxtt* and *ttGhhxhD-hxh* where h = hydrophobic, a = aromatic, and t = turn or polar residues) (15). The MAM domain appears to play a role in dimerization or oligomerization as well as proper protein folding and transport through the secretory pathway (23–25).

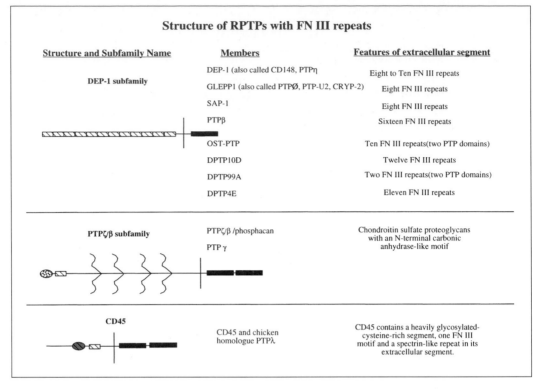

Fig. 2 RPTPs are shown which contain FNIII repeats. Three subfamilies are illustrated, the DEP-1 family, the RPTPζ/β family, and CD45.

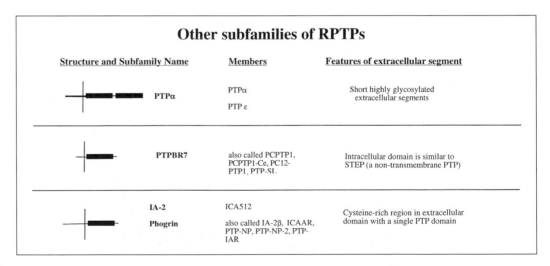

Fig. 3 RPTP subfamilies with short extracellular domains are shown schematically. These families include PTPα, PTPBR7, and IA-2/phogrin.

The RPTPζ/β-like enzymes (RPTPζ/β, PTPγ—Fig. 2) have carbonic anhydrase-like domains and a FNIII repeat in their extracellular segment (26). The carbonic anhydrase-like domain found in the RPTPζ/β-like phosphatases lacks key histidine residues required for hydration of CO_2 (27). Thus it has been proposed to function as a hydrophobic binding pocket for a heterophilic ligand. In the case of RPTPζ/β, the carbonic anhydrase-like domain binds to contactin/F11/F3 (28). As discussed below, RPTPζ/β binds to many different proteins and has a complex pattern of expression, which is regulated by alternative splicing. For other reviews on RPTPζ/β see refs 28, 29.

Other RPTP subfamilies described in this chapter include CD45, PTPα, PTPBR7, and the IA-2/phogrin family (Figs 2 and 3). The first RPTP identified, CD45 and its chick homologue PTPλ, have one FNIII repeat, a spectrin-like repeat, a serine/threonine/proline-rich region, and a cysteine-rich region in their extracellular segments (11). PTPα and PTPε have short highly glycosylated extracellular segments. PTPBR7 subfamily members also have short extracellular segments of unknown function. The IA-2/phogrin subfamily contains cysteine-rich extracellular domains. The function of these short extracellular segments is unclear.

3. Expression patterns

3.1 LAR-like RPTPs

LAR subfamily members include LAR, DLAR, PTPδ, and PTPσ. LAR is expressed in lung, liver, brain, heart, muscle, kidney, placenta, and pancreas (30, 31). PTPδ is expressed primarily in brain, kidney, placenta, and male germ cells (32–34). PTPσ is expressed primarily in brain, lung, heart, skin as well as at lower levels in kidney, pancreas, and intestine (31, 35–38). In addition, PTPσ is expressed in a variety of neuronal cell lines (36, 39). The chick homologue of PTPσ, CRYPα, is expressed in brain, spinal cord, retina, limb, breast muscle, heart, liver, gut, and lung (40). The wide distribution of these proteins suggests they will be important signalling molecules in many tissues.

The most extensive characterization of expression of the LAR subfamily members is in the nervous system. LAR and PTPσ are expressed by oligodendrocytes (41). PTPδ is expressed in the cerebellum, olfactory bulb, pyramidal cell layer of the hippocampus, dorsal root ganglia, thalamic reticular nucleus, pyramidal cell layer of the piriform cortex, and in the visual system (33, 42–44). LAR has been detected in subventricular layers, cortex, brainstem, cerebellum, olfactory bulb, spinal cord, geniculate ganglion, dorsal root ganglion, as well as glial cells and PC12 cells (30, 43). PTPσ is expressed in the cortex, trigeminal ganglia, dorsal root ganglia, telencephalon, diencephalon, hippocampus, midbrain, medulla, spinal cord, pituitary gland, retina, olfactory bulb, cerebellum, pituitary cells, neural crest derivatives, and thalamus (35, 39, 43, 45, 46). CRYPα is also broadly expressed in the nervous system during development, especially in the retina, optic tectum, hypothalamus, hippocampus, motor neurons in the spinal cord and in ganglia of the peripheral nervous system (40, 47, 48). These patterns of expression suggest that the LAR subfamily is likely to be important in many aspects of neural cell function.

Regulated expression of LAR and PTPσ has also been observed. For example, changes in LAR and PTPσ expression during sciatic nerve injury have been observed (49). In addition, changes in LAR expression were seen during NGF-induced differentiation of PC12 cells (30). However, a role for LAR in NGF-induced signalling by its receptor protein tyrosine kinase (RPTK) has not yet been demonstrated. LAR has been demonstrated to play a role in RPTK signalling in other systems (50, 51). Changes in protein expression are likely to be an important form of regulation of these enzymes.

Alternative splicing is commonly observed in RPTPs both in their extracellular and intracellular domains (52–55). A secreted form of LAR is expressed by alternative splicing (53, 54). Interestingly, an alternatively spliced region in the fifth FNIII repeat of LAR disrupts binding to the laminin–nidogen complex which could alter cell–matrix adhesion (56). The juxtamembrane domain of LAR contains an alternatively spliced 11 amino acid exon (LASE-a) that includes a serine phosphorylation site (53, 55). Inclusion of the LASE exon decreases during development (55). LASE-containing proteins were observed on cell bodies while forms of LAR lacking LASE were present on neurites (55). A similar alternatively spliced exon containing a phosphorylation site (RSLE) was found in the Ig superfamily adhesion molecule L1 and is differentially expressed in neurons and glia (57).

The essential functions of the LAR subfamily members in various tissues have recently been addressed by transgenic or knock-out experiments in mice. A transgenic mouse was engineered so that the neomycin gene was recombined into the LAR gene which resulted in the production of an RNA that lacked the PTP catalytic domains (58). These mice had defects in mammary gland development that resulted in an inability of females to deliver milk to their pups (58). A transgenic mouse containing a gene trap in the LAR gene resulted in a reduction of the size of basal forebrain cholinergic neurons and decreased innervation of the dentate gyrus (59). Knock-out mutations of the PTPσ gene resulted in death within a few weeks after birth due to severe developmental defects in the nervous system such as neuroendocrine dysplasia (37, 38). The PTPσ knock-out mice have stunted growth, spastic movements, tremor, ataxic gait, abnormal limb flexion, and defective proprioception at birth (37). These studies suggest that LAR subfamily members are very important in nervous system development (see Sections 7.1.2 and 7.1.3).

3.2 *Drosophila* RPTPs

All *Drosophila* RPTPs (Figs 1 and 2) are expressed in the nervous system and most of these RPTPs have mammalian counterparts (60). DLAR is structurally similar to mammalian LAR but it contains an extra FNIII repeat in its extracellular segment while DPTP69D contains two Ig-like domains and two FNIII repeats. Two other RPTPs contain only FNIII repeats in their extracellular domains, DPTP10D and DPTP99A (61). The *Drosophila* PTPs are named for their chromosomal location. The

DPTP69D protein is expressed in optic lobes where it is localized to neuropil of the lamina and medulla as well as transmedullary fibres of the developing lobula complex (62). In other areas, DPTP69D is expressed on subsets of neuronal processes in the brain, segmental ganglia, ventral nerve cord, and on photoreceptor axons (62). The DLAR, DPTP99A, and DPTP10D proteins are differentially expressed on axons in the ventral nerve cord (61). DPTP10D is expressed at the junction with the longitudinal tracts on the anterior commissure (61). This localization suggests a possible role for DPTP10D in choice point specification for migrating axons. The differential expression of the *Drosophila* RPTPs suggests that they are likely to have cell type-specific functions in migration or pathfinding during nervous system development (see Section 7.1.2).

3.3 PTPμ-like RPTPs

The PTPμ-like family shown in Fig. 1 includes PTPμ, PTPκ, PCP-2, and PTPρ. PTPκ and PCP-2 are the most closely related subfamily members as well as being the most broadly expressed. PTPρ is most similar to PTPμ (17). PTPμ is expressed predominately in heart, lung, and brain (63–65), however, it is expressed by endothelial cells both within the nervous system and throughout the body (66–69). Based on PCR amplification studies of the PTP family, PTPκ is ubiquitously expressed (D. Robinson and H. J. Kung, personal communication). PTPκ has specifically been shown to be expressed in brain, kidney, liver, spleen, prostate, and ovary (70, 71). PCP-2 is expressed in brain, liver, heart, skeletal muscle, lung, kidney, prostate, placenta, uterus, as well as pancreas and pancreatic islet tumour cells (16, 18–22, 72). PCP-2 is expressed in developing skeletal, epithelial, and neuronal structures (20). PTPρ is expressed predominately in the nervous system although it has also been found in pancreatic islet tumour cells (17, 44, 72, 73). Hormone and growth factor regulation of the expression of PTPμ subfamily members has already been reviewed (12). Various expression studies have been performed on this subfamily of enzymes; however, most work on the function of these enzymes has been in cell lines. The function of the PTPμ subfamily in most of these tissues remains to be addressed.

The PTPμ-like subfamily members are expressed at high levels in the developing nervous system (16, 48, 70). The following studies illustrate that the PTPμ-like proteins are expressed at critical periods during nervous system development and during times of neuronal migration and pathfinding. Expression studies of rat PTPκ indicate that it is found in the cerebral cortex and hippocampus, cerebellum, olfactory bulb, brainstem, and spinal cord (70). Interestingly, PTPκ expression is down-regulated in the adult cerebellum suggesting a possible role in granule cell migration during development (70). PCP-2 is expressed in the developing midbrain, choroid plexus and forebrain, in the roof plate, floor plate, telencephalon, and hindbrain (20, 67). In the adult, PCP-2 is expressed in the midbrain, septal area, basal ganglia, thalamus, substantia nigra and piriform cortex and endopiriform nucleus, amygdaloid nuclei, subiculum, and the hippocampus (20). PTPρ is expressed in the hippocampus and dentate gyrus, olfactory bulb, cerebral cortex, cerebellar cortex, and spinal cord (17, 73). PTPμ is abundant in many parts of the central nervous system including the retina (48, 63, 67, 74). Due to the broad distribution of the PTPμ-like phosphatases in the nervous system and overlapping patterns of expression, this suggests a possible role in axonal pathfinding and synaptic plasticity.

3.4   The carbonic anhydrase-like PTPs

There are three major forms of RPTPζ/β, two transmembrane and one soluble form that is called phosphacan (28). There is evidence that additional splice variants exist including an exon that encodes a seven amino acid insert in the juxtamembrane region (75). The expression of all isoforms of RPTPζ/β is restricted to nervous system as well as other cells of neural ectoderm origin (76–80). In early development, RPTPζ/β is expressed exclusively in the central nervous system (81). At later stages, RPTPζ/β is expressed in both neurons and glia of the central as well as the peripheral nervous system (81). Specifically, it is expressed in the ventricular and subventricular zones by radial glia, hippocampus, as well as in the outer layers of the cortex (77, 81). In addition, RPTPζ/β is expressed in the cerebellum by Purkinje cells as well as Müller glial cells of the retina (42, 77). Phosphacan is expressed in the brain, retina, spinal cord, dorsal root ganglia, as well as the cerebellum (82–84). Regulation of RPTPζ/β expression has been demonstrated. Up-regulation of all isoforms of RPTPζ/β has been observed in the distal nerve segment following sciatic nerve crush (75). RPTPζ/β expression by glial cells is also regulated during differentiation (77, 85). RPTPζ/β expression is likely to be highly regulated and quite complex during development of the nervous system and some studies suggest it is important for creating barriers to neuronal migration (see Section 7.1.4).

Multiple forms of PTPγ occur via alternative splicing of the intracellular and extracellular domains (86). PTPγ is expressed in brain, kidney, lung, and heart (86, 87). RPTPζ/β is found as a soluble form called phosphacan, which is expressed as a chondroitin sulfate proteoglycan (see Section 4.2). PTPγ is found in a soluble form similar to RPTPζ/β, however it is not expressed as a proteoglycan (86). It is likely that PTPγ will perform different physiological functions than RPTPζ/β based on the fact that they are not both proteoglycans and that they do not have similar ligands.

3.5   Other RPTPs

The other RPTPs that are reviewed in this chapter include PTPα and CD45. PTPα is expressed in lung, heart, liver, placenta, thymus, brain, kidney, Müller glial cells, and endothelial cells (42, 48, 88–90). PTPα is maternally expressed in *Xenopus* and its expression increases during gastrulation. It is primarily expressed in the brain and visceral arches at later stages in *Xenopus* development (91). CD45 is expressed in all nucleated cells of the haematopoietic lineage and plays an essential role in T cell signalling which is discussed in Section 6.4 (92).

4. Role in cell adhesion

4.1 Homophilic binding of Ig superfamily RPTPs (PTPμ, PTPκ, PCP-2, PTPδ)

The combination of Ig domains and FNIII repeats in the extracellular segment of RPTPs suggested that they might function in cell–cell adhesion. In fact, PTPμ induced the aggregation of non-adhesive insect cells when expressed by recombinant baculovirus infection (93, 94). The extracellular segment of PTPμ directly mediated aggregation, which was independent of both calcium and phosphatase activity. Binding between purified PTPμ-coated fluorescent beads and endogenously expressed PTPμ on lung cells has also been demonstrated (93). These studies suggest that PTPμ induced aggregation via a homophilic binding mechanism.

Other PTPμ subfamily proteins were also demonstrated to induce aggregation. PTPκ was shown to induce aggregation of *Drosophila* S2 cells (95). Similarly, the ability of PTPκ to induce aggregation was not dependent upon its proteolytic processing or the presence of the phosphatase domains (95). An *in vitro* approach was also used to demonstrate that PTPκ extracellular domain-coated beads could mediate aggregation or bind homophilically (95). In addition, when the extracellular domain of PCP-2 was covalenty coupled to fluorescent beads, it mediated bead aggregation via homophilic binding (20). It is clear that most members of the PTPμ subfamily mediate cell adhesion via homophilic binding.

The immunoglobulin domain of PTPμ is essential for promoting homophilic interactions (96). *In vitro* binding assays, where fragments of the extracellular segment of PTPμ were coated onto fluorescent beads, were used to demonstrate that only Ig domain-containing beads underwent aggregation or bound to PTPμ-coated surfaces (96). These results suggest that the Ig domain is necessary for homophilic binding. This study suggests that neither the MAM domain nor the FNIII repeats are directly responsible for this homophilic binding in *trans* (between two cells). Interestingly PTPδ, an immunoglobulin superfamily RPTP of the LAR subfamily, was recently shown to bind homophilically (97). This RPTP contains no MAM domain and is likely to mediate adhesion via its immunoglobulin domains.

Despite the fact that the MAM domain in PTPμ was not sufficient for homophilic binding, it is possible that the MAM or FNIII repeats are involved in heterophilic binding or lateral associations in the plane of the membrane (*cis* interactions). Interestingly, the MAM domain of PTPμ/PTPκ plays a role in cell–cell aggregation (98). PTPμ and PTPκ-expressing cells sort out into independent aggregates based on cell labelling experiments with lipophilic dyes (98). Deletion mutants, lacking the MAM domain, were used in aggregation experiments, and the results suggest that these mutant proteins were unable to induce aggregation. However, when a chimeric molecule was constructed in which the MAM domain in PTPκ was substituted with the MAM domain of PTPμ, expression of this chimera induced Sf9 cell aggregation. Mixing of PTPμ, PTPκ, or the chimera-expressing cells revealed that the chimera could self-associate but it did not bind to native PTPμ or PTPκ. Therefore, the MAM domain appears to be important for determining the specificity of homophilic binding or plays a role in adhesion by 'sorting' cells expressing closely related molecules during cell aggregation. Of interest, the MAM domain in meprins was demonstrated to induce dimerization (23). Neuropilin (previously called A5) is a cell surface protein that contains a MAM domain and binds semaphorins (99–101). The neuropilin MAM domain was recently shown to play a role in oligomerization (25). These data indicate that the MAM domain in the PTPμ-like proteins is probably functioning to promote *cis* oligomerization to generate a high affinity binding site for *trans* interactions, i.e. cell–cell adhesion via homophilic binding.

4.2   RPTPζ/β binds multiple cell adhesion molecules

RPTPζ/β is a chondroitin sulfate proteoglycan which contains a carbonic anhydrase-like (CAH) domain and an FNIII repeat in its extracellular segment (28). Interestingly, the CAH domain of RPTPζ/β specifically binds to the Ig superfamily member contactin/F11/F3 (28). Contactin contains six Ig domains, four FNIII repeats, and is linked to the cell membrane via a phosphatidyl inositol linkage (102, 103). Contactin also interacts in *cis* with a neurexin-like protein called Caspr which is probably involved in signalling via intracellular interactions with band 4.1-like proteins and SH3 domain-containing adapter proteins (28, 29). Recently, PTPα was also shown to associate with contactin within the plane of the membrane (*cis*) via the extracellular region of PTPα (104).

Phosphacan, a splice variant that contains only the RPTPζ/β extracellular segment and is expressed as a chondroitin sulfate proteoglycan, has been shown to bind heterophilically with NCAM, NgCAM, and tenascin (82, 83, 105, 106). The neural CAMs bind to the core glycoprotein of phosphacan (82, 83, 106). However, the presence of chondroitin sulfate is important for phosphacan binding to its ligand TAG-1/axonin-1, which is structurally similar to contactin (107). In addition, the heparin binding proteins, pleiotrophin/heparin binding growth-associated molecule (HB-GAM) and amphoterin, are also ligands for phosphacan (106, 108). The binding of pleiotrophin is also regulated by the presence of chondroitin sulfate (106, 108). The RPTPζ/β protein is likely to regulate multiple heterophilic binding interactions in both a carbohydrate-dependent and independent manner.

The various adhesion molecules and RPTPs can act both as soluble ligands and transmembrane receptors under different conditions. For example, transmembrane contactin could bind to its soluble ligand phosphacan. Alternatively, contactin can be secreted and act as a soluble ligand for the transmembrane RPTPζ/β. Glial cells expressing full-length RPTPζ/β extend cellular processes on NgCAM (79). However when soluble phosphacan is present in the medium, it inhibits glial adhesion to NgCAM (79). There is likely to be a complex hierarchy of interactions between soluble and transmembrane forms of these CAMs and RPTPζ/β-phosphacan to differentially regulate cell adhesion in a precise spatio-temporal fashion during development.

4.3   CD45 regulates adhesion in the haematopoietic system

CD45 controls cell adhesion by its involvement in both inside-out and outside-in signal transduction (11, 92). CD45 regulates integrin-dependent adhesion presumably by an inside-out signalling process. Specifically, CD45 regulates adhesion induced by the integrin LFA-1 binding to the Ig superfamily proteins ICAM-1 or ICAM-3 (109–112). Regulation of cell adhesion by CD45 appears to vary according to cell type. For example, incubation with antibodies to the extracellular segment of CD45 prevents LFA-1/ICAM-1 or LFA-1/ICAM-3-dependent adhesion and induces changes in tyrosine phosphorylation in T and B cell lines (113, 114). In contrast, in activated human T cells and thymocytes CD45 induces homotypic interactions (interactions between the same type of cells) (110, 111). Similarly, addition of anti-CD45 antibodies induced aggregation of B cells, natural killer cells, and mononuclear cells (109, 112). This aggregation was blocked by addition of EDTA and cytochalasin B, which also suggests a role in integrin-dependent adhesion as well as regulation of the actin cytoskeleton (109). Interestingly, CD45 has been reported to bind to fodrin, an actin-associated protein, via a region between PTP domain 1 and 2 (115). The stimulation of CD45-induced aggregation, through LFA-1/ICAM-1 interactions, somehow involves protein kinase A/G activation to increase cAMP levels in cells (116) although the precise mode of regulation is unknown. However, the positive and negative regulation of Src family tyrosine kinases by CD45 may indirectly regulate integrin-dependent adhesion (see below). CD45 has been shown to downregulate Src family kinase activity during integrin-mediated adhesion (117). CD45 may therefore change the tyrosine phosphorylation state of components of the cell adhesion or cytoskeletal machinery to regulate integrin-dependent adhesion.

Despite the fact that CD45 was the first RPTP cloned, the identity of its ligands remains elusive. However, various proteins that bind CD45 suggest a role in cell adhesion or alternatively, the induction of some type of outside-in signal transduction. CD22, a B cell surface protein, binds to CD45 and is involved in signal transduction by the T cell receptor (118). CD22 is a sialic acid-binding lectin and an Ig superfamily member, which interacts with the sialic acid moieties on CD45 (118). Sialic acid derivatives are important in selectin-dependent adhesion and leukocyte rolling (119, 120), raising the possibility that the CD22/CD45 interaction has the potential to play a role in lymphocyte adhesion. Interestingly, the leukocyte rolling response and subsequent strong adhesion involves co-ordination of sialic acid binding and integrin-dependent adhesion (121). The binding of CD22 to CD45 via sialic acid may be an early event that induces signals, which result in increased integrin-dependent adhesion.

Other ligands for CD45 have been isolated. Using a co-culture model system for the interaction of haematopoietic precursors with the bone marrow stroma, it has been shown that heparin sulfate on stromal cells serves as a ligand for CD45 (122). In addition, CD45 mediates heterotypic adhesion between a leukaemia/lymphoma cell line and bone marrow stromal cells (123), although the specific ligand has not yet been identified. The various isoforms of CD45 are differentially O-glycosylated, which may modulate the carbohydrate-dependent functions of CD45 thereby altering its ability to regulate adhesion. Interestingly, CD45 has recently been shown to interact with galectin-1 via a carbohydrate-dependent interaction (124). CD45 also interacts with the semaphorin CD100 on T cells (125). Semaphorins are proteins that mediate growth cone collapse and repulsion of neurons during migration in the nervous system (126–128). It is not clear what role CD45 binding to CD100 will play in lymphocyte adhesion, but semaphorins have been suggested to regulate cell–cell interactions in the immune system (129). Characterization of all ligands for CD45 will be required to precisely define its role in modulation of cell contact phenomena. However, it is likely that CD45 is involved in the regulation of cell adhesion-induced signals.

5. Regulation of the enzymatic function of RPTPs

The regulation of RPTP activity is shown schematically in Fig. 4 (see 130).

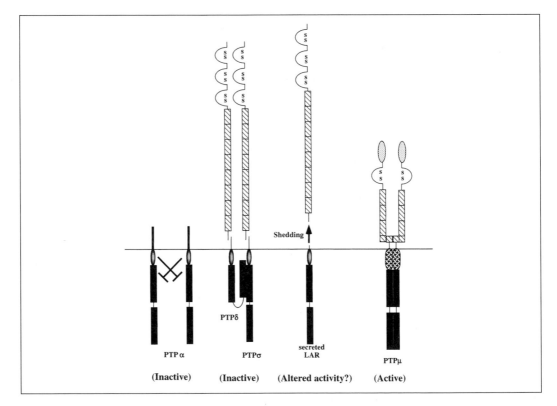

Fig. 4 Regulation of RPTP catalytic activity. There are at least four potential mechanisms of regulation: (1) dimerization that blocks the active site; (2) heterodimerization that negatively regulates catalytic activity; (3) cleavage of the extracellular segment at high density and shedding of the extracellular segment which may alter localization of the intracellular domain and thus modulate available substrates; (4) dimerization that does not block the active site.

5.1 Dimerization

5.1.1 The N-terminal wedge

When the crystal structure of PTPα was solved (131), it suggested that the first phosphatase domain of PTPα forms monomers, dimers, and oligomers. In contrast, they found that the second PTP domain only formed monomers. Full-length PTPα exists predominantly as a homodimer and dimerization is independently regulated by the extracellular domain, transmembrane domain, juxtamembrane region, and the second phosphatase domain (132). The authors propose that dimers are stabilized by multiple weak dimerization interfaces (132).

The PTPα data suggest that regulation of RPTP enzymatic activity occurs through dimerization and blockage of the active site (131). The authors demonstrated that blockage of the active site in an adjacent monomer occurs by insertion of a sequence that is just N-terminal to the PTP domain and is now termed the N-terminal wedge domain (Fig. 4). A role for the N-terminal wedge in the regulation of CD45 activity has also been demonstrated (133). Interestingly, an inactivating mutation within the N-terminal wedge (Glu613Arg), causes a dominant phenotype in transgenic mice resulting in lymphoproliferation, autoantibody production, autoimmune nephritis, and death (134).

The auto-inhibition of phosphatase activity by the N-terminal wedge may only occur in a subset of RPTPs. In the LAR crystal structure of the combined phosphatase domain 1 and 2 construct, it exists in a monomeric form and both catalytic domains are accessible to substrate (135). The crystal structure data from PTPμ suggest that it exists predominately as a dimer (Fig. 4), but that the active site is *not* blocked in the dimer by the N-terminal wedge domain (136). The N-terminal wedge region of PTPμ adopted a similar conformation to the wedge domain in PTPα but it did not insert into the active site of the adjacent catalytic domain of the PTPμ dimer (136). The wedge sequence is just adjacent to the catalytic domain of the RPTPs, which lies in the C-terminal region of what is currently termed the juxtamembrane domain. The juxtamembrane domain of the PTPμ subfamily proteins is approximately 70 amino acids longer than the same region in other RPTPs. The longer length of the PTPμ-like protein's juxtamembrane domain may result in different positioning or flexibility of the N-terminal wedge. A recent study suggests that the juxtamembrane domain of PTPμ forms an intramolecular interaction with both first and second catalytic domains of PTPμ (137). Furthermore, they find that the juxtamembrane domain of PTPμ is required for phosphatase activity of the first catalytic domain (137). Together, the crystal structure and functional studies suggest that RPTPs are likely to form homodimers and that they will fall into at least two categories where the dimers are either active or inactive based on the three-dimensional structure and position of the N-terminal wedge.

Interestingly, alternative splicing of the juxtamembrane domain near the N-terminal wedge region has been observed in PTPα, RPTPζ/β, and LAR (53, 55, 75, 89). PTPα is alternatively spliced in the N-terminal wedge region, which inserts 36 amino acids including seven serine/threonine residues, which are potential phosphorylation sites (89). Protein kinase C is known to phosphorylate serines in this region of PTPα (138, 139). One RPTPζ/β splice product inserts the amino acid sequence, TLKEFYQ, which adds two potential phosphorylation sites (threonine as well as tyrosine) to the second alpha helix of the N-terminal wedge (75). The juxtamembrane domain of LAR contains an alternatively spliced 11 amino acid exon called LASE-a (SSAPSCPNISS) that also includes serine phosphorylation sites (53, 55). Together, these studies suggest that the enzymatic activity of the RPTPs may be controlled by alternative splicing as well as serine/threonine or tyrosine phosphorylation of the N-terminal wedge domain.

5.1.2   PTP domain 2 alters PTP domain 1 function

The second catalytic domain in most RPTPs has naturally occurring mutations in the residues that are required for catalysing dephosphorylation. The one notable exception is PTPα where both phosphatase domains are active *in vitro* (140). Investigators have tried to convert an inactive domain 2 by point mutation to resemble an active phosphatase domain 1 (135, 140, 141). These studies have found that some phosphatase activity can be restored but substrate specificity is not restored (140, 141). It appears that domain 2 performs some function other than directly catalysing the dephosphorylation of proteins. Some investigators have observed that domain 2 mediates various protein–protein interactions which may control subcellular localization and/or regulate phosphatase activity (see LAR and CD45 below). The second phosphatase domain of PTPα binds to calmodulin, which inhibits its phosphatase activity (142). Alternatively, the second phosphatase domain may regulate dimerization and/or activity of the first phosphatase domain. Interestingly, a recent study suggests that domain 2 may be involved in RPTP heterodimerization (143). The first phosphatase domain of PTPσ (PTPσ-D1) specifically binds to PTPδ domain 2 (PTPδ-D2) (143) (Fig. 4). In addition, the binding of PTPδ-D2 resulted in reduced catalytic activity of PTPσ-D1. The interaction is mediated by the inhibitory wedge region of PTPσ-D1 binding to the pseudo-active site region of PTPδ-D2 (143). Domain 2 of PTPδ bound weakly to other subfamily members like PTPδ-D1 and LAR-D1 (143). Reciprocal interaction of PTPδ-D1 with PTPσ-D2 was not observed. Together, this study suggests that domain 2 may induce RPTP heterodimerization within a subfamily to negatively regulate the catalytic activity of the first phosphatase domain. In this regard, a recent study on PTPμ suggests that association between the juxtamembrane domain and catalytic domain 2 may regulate the activity of the first phosphatase domain keeping it in an inactive state (137). Therefore, the role of the second phosphatase domain may be to regulate phosphatase activity in either monomers, homodimers, or heterodimers.

5.2   Dissociation of cleaved extracellular and intracellular fragments

A post-translational modification that has been observed in both the RPTP family and in neural cell adhesion molecules is proteolytic cleavage (10, 12, 144). Full-length

LAR is cleaved into two non-covalently associated fragments, one (P-subunit) comprises the entire intracellular and transmembrane segments and a short stretch of extracellular sequence, the other (E-subunit) contains the rest of the extracellular segment (145, 146). The PTPμ-like subfamily members undergo a similar cleavage at a basic sequence in the fourth FNIII repeat which is catalysed by a subtilisin/kexin-like endoprotease, most likely the PC5 protease (20, 66, 68, 70, 96). The physiological significance of this cleavage is not well understood.

The cleavage of the extracellular segment might alter RPTP function in cell adhesion. However, mutation of this extracellular cleavage site in PTPκ, so that it could no longer be proteolytically processed, did not affect its ability to mediate adhesion (95). Interestingly, at high density the E-subunit of LAR is shed from the surface of cells following proteolytic cleavage at a different site closer to the transmembrane domain (147) (Fig. 4). The shedding of LAR and PTPσ is inducible by activating PKC via phorbol esters (147, 148). Shedding of the E-subunit resulted in internalization of the catalytically active P-subunit, which redistributed it away from cell–cell contacts (148). The shedding of the E-subunit of the PTPμ-like proteins would generate a fragment that is capable of homophilic binding and thus could antagonize cell surface interactions of the PTPμ-like proteins. In addition, internalization of the P-subunit of RPTPs is likely to change their substrate availability. Therefore, shedding could alter signal transduction by the RPTPs or indirectly regulate their enzymatic activity by changing their subcellular localization (Fig. 4). In this regard, enzymatic characterization of the full-length and intracellular forms of PTPμ demonstrated that the extracellular segment of RPTPs may exert a direct regulatory constraint on phosphatase activity (149). Therefore, it is possible that shedding of the E-subunit may relieve regulatory constraints on the P-subunit and therefore regulate enzymatic activity.

High cell density also affects the cell surface expression of the PTPμ protein, which increases with cell density (68). In addition, shedding of the extracellular segment at high cell density was observed (68). This density-dependent increase in protein was first observed for other RPTPs such as DEP-1 (150). The increase in PTPμ expression occurs post-transcriptionally but the mechanism is unclear (68). However, increased expression of cell adhesion molecules at high cell density is commonly observed, and may be a mechanism through which cell–cell adhesion is regulated.

6. Identification of RPTP-dependent signalling pathways

6.1 LAR subfamily members are localized to focal adhesions

Some members of the LAR subfamily are localized to focal adhesions and may regulate tyrosine phosphorylation during assembly or disassembly of these junctions (Fig. 5). Focal adhesions are dynamic junctions that transduce signals resulting from cell–extracellular matrix binding, and are remodelled by changes in tyrosine phosphorylation (151, 152). Assembly of focal adhesions induces tyrosine phos-

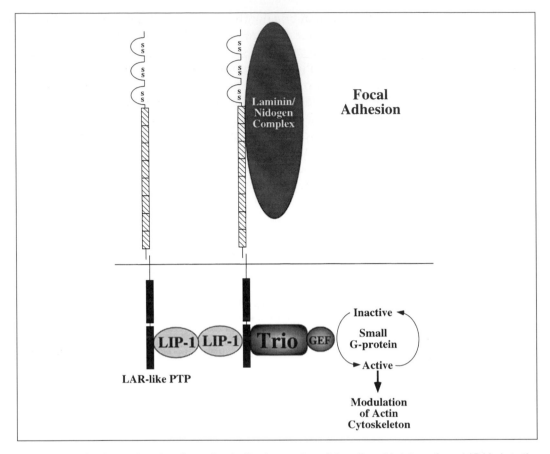

Fig. 5 LAR subfamily members localize to focal adhesions and regulate cell–matrix interactions. LAR binds to the laminin–nidogen complex extracellularly. LAR localizes to focal adhesions by binding to LIP-1 intracellularly. LAR also binds to Trio, which may regulate the small G proteins to alter the cytoskeleton in response to cell–extracellular matrix adhesion.

phorylation of the tyrosine kinase FAK and paxillin (151). Disassembly of FAs requires PTPs because the PTP inhibitor, pervanadate, increased the number of focal adhesions (153). Similarly, pervanadate treatment increased cell spreading via increased adhesion between integrins and their ligands in the extracellular matrix (154). Interestingly, a ligand for LAR was recently identified, LAR binds to the laminin–nidogen complex (56) which mediates cell–matrix adhesion (Fig. 5). The LAR PTP has been shown to localize to focal adhesions (155, 156). Specifically, LAR appears to bind to proximal ends of focal adhesions that are likely undergoing disassembly.

The localization of LAR subfamily members to focal adhesions is mediated by an interaction with a cytoplasmic protein called LAR-interacting protein 1 (LIP.1) (155, 156). LIP.1 binds to PTP domain 2 of LAR, PTPδ, and PTPσ (31, 155). LIP.1 contains coiled-coil alpha helical domains that are found in cytoskeletal proteins which self-associate into homodimers or oligomers (155). The self-association of LIP.1 promotes dimerization of the LAR subfamily members, which may regulate enzymatic activity (156). Together, these studies suggest that LAR subfamily members localize to focal adhesions and are regulated by LIP.1 association (Fig. 5). However, LAR and PTPσ have also been localized to cell–cell contacts such as adherens junctions and desmosomes (148). The LAR subfamily is likely to regulate multiple cell junctions.

Trio, another LAR-interacting protein, also binds to PTP domain 2 of LAR (157). LIP.1 and Trio binding may not be mutually exclusive since the structural data on RPTPs suggest that they exist in cells as dimers (Fig. 5). Trio contains four spectrin-like repeats, a rac-specific and rho-specific guanine nucleotide exchange factor (GEF) domain, two pleckstrin homology domains, an Ig-like domain, and a serine/threonine kinase domain. GTP binding proteins such as Ras and Rho family proteins are positively regulated by GEFs, which promote exchange of GDP for GTP, to regulate reorganization of the actin cytoskeleton. The serine/threonine kinase domain in Trio is most similar to the myosin light chain kinase (MLCK), suggesting a role in actin/myosin regulation. Trio is important in axonal growth and guidance in the nervous system (158). However, neither Trio nor LIP.1 is tyrosine phosphorylated indicating that they are not likely to be substrates of the LAR phosphatase (155, 157). Alternatively, Trio or LIP.1 binding may bring the LAR PTP to sites of actin reorganization and thus may be involved in regulating focal adhesion assembly or disassembly.

6.2   RPTPs and caspases

Programmed cell death or apoptosis is a part of normal development, however abnormal regulation of cell death is observed in various diseases. There has been recent progress on understanding the molecules involved in programmed cell death, which include TNF-α, caspases, and bcl-2 (159). However, all of the molecular components of the signals that induce cell death have not yet been identified. RPTKs and their soluble growth factor ligands are thought to be survival factors that signal to block apoptosis (159). RPTKs increase tyrosine phosphorylation of key cellular proteins, which are critical for cell survival. This raises the possibility that PTPs, through the dephosphorylation of key cellular proteins, may induce apoptosis. Interestingly, antisense suppression of LAR resulted in increased signalling of the insulin PTK, the EGF receptor, and the HGF receptor (50, 51). These data are consistent with the idea that LAR normally may down-regulate RPTK signalling to reduce cell survival. In that regard, LAR was recently shown to activate caspase-3 directly to induce p53-independent apoptosis (160). Since PTP inhibitors like pervanadate block apoptosis (161), it is interesting to note that when most RPTPs are overexpressed, they are cytotoxic presumably by inducing cell death. Most investigators have resorted to inducible or repressible expression systems to study RPTPs. This suggests that activation of apoptotic pathways may be a common feature of RPTPs.

6.3 PTPα affects integrin adhesion

PTPα regulates the Src family of protein tyrosine kinases (11). PTPα overexpression in rat embryo fibroblasts (162) or P19 cells (163) correlates with a dephosphorylation of the inhibitory Tyr 527 site and activation of Src (Fig. 6) (130, 164). The same dephosphorylation and activation event may lead to growth stimulation or differentiation depending upon the cell type. Overexpression of PTPα in fibroblasts resulted in cellular transformation (162). However, overexpression of PTPα in embryonal carcinoma cells resulted in neuronal differentiation (163). These data suggest that PTPα can ultimately affect cell growth as well as differentiation by regulating the Src phosphorylation state in cells. In addition, recent studies suggest the other Src family members such as p59[fyn] are also substrates for PTPα (165). The PTPα/contactin receptor complex associates with the Src family kinase p59[fyn] (104).

Src is localized to both focal adhesions and adherens junctions. Focal adhesions are sites of cell–matrix adhesion that contain many PTKs including Src and FAK that regulate the formation or disassembly of junctions during migration and cell growth (151, 152). A431 cells that overexpress PTPα showed increased cell–matrix adhesion, with coincident dephosphorylation and activation of Src (166). In contrast, a catalytically inactive form of PTPα (C433A) actually decreased the basal adhesion of A431 cells. In addition, PTPα expression led to an increase in Src association with FAK resulting in increased phosphorylation of paxillin. An increased association of paxillin, a cytoskeletal-associated protein found at focal adhesions, with the Csk tyrosine kinase was also observed (166). Together, these studies suggest that PTPα is

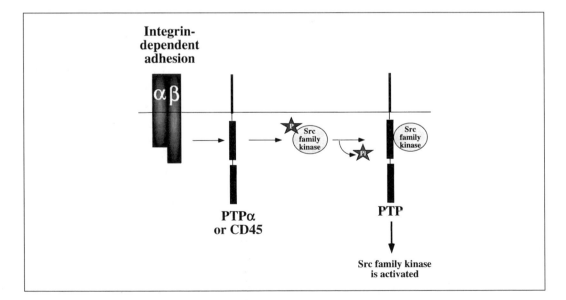

Fig. 6 Integrin-dependent adhesion to the extracellular matrix is regulated by RPTPs. Both CD45 and PTPα dephosphorylate an inhibitory tyrosine on Src family kinases thus stimulating their tyrosine kinase activity.

involved in tyrosine phosphorylation-dependent signals that modulate focal adhesions.

In growth cones, endogenous PTPα is abundant and its punctate pattern of expression is similar to β1 integrin (167). When neurons were plated on laminin, a ligand for β1 integrin, an increase in the amount of Src was associated with the cytoskeleton. The authors propose that integrin binding to laminin recruits PTPα to adhesion sites. The recruitment of PTPα results in dephosphorylation of Tyr 527 of Src thus stimulating its tyrosine kinase activity (167). Together, these studies suggest a role for PTPα in integrin-based adhesion and signalling.

6.4 CD45 interacts with Src family PTKs

CD45 is an essential component of antigen-induced signalling events in lymphocytes, through T and B cell receptors (92). CD45 exerts some of its effects through the dephosphorylation and activation of the Src family kinases (Fig. 6), Lck and Fyn, which are important in antigen-induced signalling (92, 130). Although, CD45 can also down-regulate the kinase activity of Src family kinases during integrin-mediated adhesion (117). CD45 has also recently been shown to dephosphorylate another cytosolic kinase (JAK) to negatively regulate cytokine receptor signalling (168). The second PTP domain of CD45 is critical for its interaction with the T cell receptor (TCR-zeta) and interleukin-2 production (169). Cells expressing CD45 lacking domain 2 exhibited hyper-phosphorylation of TCR-zeta and loss of ZAP-70 phosphorylation (169). CD45-deficient mice have impaired T cell maturation and lack cytotoxic T cell responses (170). Knock-out mice also revealed that CD45 regulates the negative and positive selection of certain subpopulations of B cells (171). This study implicates the strength of signalling through the T cell receptor as an important factor for B cell selection (171). In addition, it suggests that mutations in proteins like CD45 may result in a predisposition to autoimmune diseases (171). In fact, this hypothesis was recently confirmed when an inactivating mutation, within the N-terminal wedge (Glu613Arg) of CD45, resulted in lymphoproliferation, autoantibody production, autoimmune nephritis, and death (134). The regulation of Src family kinases by CD45 is likely to be important to T cell receptor signalling and immune responses.

6.5 Multiple PTPs associate with cadherins or catenins at adherens junctions

6.5.1 Protein–protein interactions of PTPs with cadherins and/or catenins

Adherens junctions are points of cell–cell contact that are linked to the actin cytoskeleton (see Chapter 8). The cadherins are the integral membrane proteins that induce cell–cell adhesion to form adherens junctions (see Chapter 3). The cadherins bind intracellular proteins called catenins which allow the complex to associate with the actin cytoskeleton (Fig. 7). The catenins are a family of proteins including, α, β,

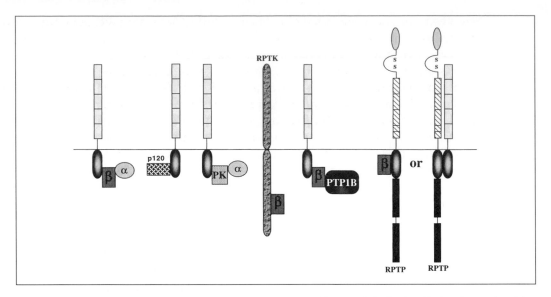

Fig. 7 The cadherins and some of their associated proteins are illustrated. Cadherins bind to β-catenin, plakoglobin, p120, and PTPμ. The PTKs, some PTPs and α-catenin associate indirectly by binding to β-catenin or plakoglobin. These complexes are dynamic and are likely to play different roles in signal transduction and cytoskeletal regulation.

γ/plakoglobin, *armadillo* and p120^{ctn}. α-Catenin is homologous to the cytoskeletal-associated protein vinculin. β-Catenin is an 'ARM repeat' protein and is homologous to plakoglobin, p120^{ctn}, and *armadillo*. γ-Catenin and plakoglobin are identical (172, 173). Various protein tyrosine kinases (PTKs) including *src*, EGF receptor, and HGF/*met* receptor (Fig. 7) phosphorylate components of the cadherin/catenin complex (174–176). Tyrosine phosphorylation of key components of the cadherin/catenin complex destabilizes adherens junctions and reduces cadherin-mediated adhesion (10, 12, 177).

A few PTPs have been shown to bind to the cadherin–catenin complex of proteins (Fig. 7). A LAR-like RPTP was found to associate with the cadherin–catenin complex in PC12 cells via binding directly to the amino terminal domain of β-catenin (178). The LAR-like PTP/β-catenin association is negatively regulated by nerve growth factor (NGF)-induced tyrosine phosphorylation of the LAR-like PTP. The intracellular domain of LAR was shown to bind directly to β-catenin and plakoglobin both *in vitro* (148) and *in vivo* (179). The interaction between LAR and β-catenin controls the tyrosine phosphorylation state of β-catenin as well as epithelial cell migration (179). Recently, pleiotrophin binding to the RPTPζ/β was shown to inhibit its phosphatase activity, which resulted in increased tyrosine phosphorylation of β-catenin (180).

The PTP1B cytoplasmic phosphatase interacts with the N-cadherin/catenin complex and induces the dephosphorylation of β-catenin (181, 182). They showed that β-catenin dephosphorylation is required for N-cadherin-dependent cell adhesion and association of the complex with the actin cytoskeleton. It was also demonstrated that an association between N-cadherin and a cell surface galacto-saminyl phosphotransferase regulates the tyrosine phosphorylation state of PTP1B. In addition, PTP1B had to be tyrosine phosphorylated in order to interact with the N-cadherin/β-catenin complex (181, 182). The binding of PTP1B to the EGF receptor (4), may also regulate the phosphorylation state of β-catenin and ultimately regulate cadherin-dependent adhesion.

The juxtamembrane domain of PTPμ-like subfamily members contains a region of homology to the conserved intracellular domain of the cadherins (Fig. 7) (144). This segment of the cadherins is essential for their adhesive function (183) because it binds to catenins, which associate with actin (184). Therefore, the PTPμ-like proteins might also interact with catenins based on their homology to the cytoplasmic domain of cadherins. PTPκ interacts with β- and γ-catenin via its juxtamembrane domain (185). PCP-2 was also shown to associate with β-catenin, which required the juxta-membrane segment of the phosphatase (20). PTPμ associates with a complex containing classical cadherins, α-catenin, and β-catenin (64). The intracellular segment of PTPμ was demonstrated to bind directly to the intracellular domain of E-cadherin but not to α-catenin or β-catenin *in vitro* (64, 65). Another study confirmed that PTPμ does not bind to β-catenin (186). PTPμ interacts with N-cadherin, E-cadherin, and R-cadherin (65). The binding of PTPμ directly to classical cadherins has also been confirmed by another group (187, 188). However, another group found that PTPμ binds to the cadherin-associated protein p120[ctn] (186). Therefore, it is highly likely that regulation of cadherins by the PTPμ-like RPTPs will be an important regulatory mechanism.

The C-terminal 38 amino acids of E-cadherin, including the catenin binding domain, were required for the interaction with PTPμ (65). The C-terminal 38 aa are currently divided into two domains (Fig. 8), the catenin binding domain and a domain encompassing the C-terminal 19 aa (189). β-Catenin and plakoglobin bind directly to the catenin binding domain (underlined in Fig. 8) (190, 191). The binding partner of the C-terminal 19 aa of E-cadherin is not currently known. However, it has recently been shown that deletion of the C-terminal 19 amino acids of the cadherin cytoplasmic domain had little effect on catenin binding but had a dramatic effect on cadherin-dependent adhesion and association with the cytoskeleton (189). These data indicate that the cytoskeletal stabilization required for cadherin adhesion is not entirely dependent on catenin binding (189). Since the proteins that bind to this site are unknown, it is possible that PTPμ may bind to this site. The association of PTPμ with the C-terminal domain of E-cadherin suggests that PTPμ may be the protein that regulates cytoskeletal association and cadherin-dependent adhesion.

6.5.2 Reversible tyrosine phosphorylation regulates cell–cell adhesion

Tyrosine phosphorylation by PTKs can negatively regulate cadherin-dependent adhesion (10, 12, 192, 193). However, the mechanism by which tyrosine phos-phorylation mediates this decrease in adhesion is not understood. The effect of phosphorylation on the composition of the PTPμ/cadherin complex was recently

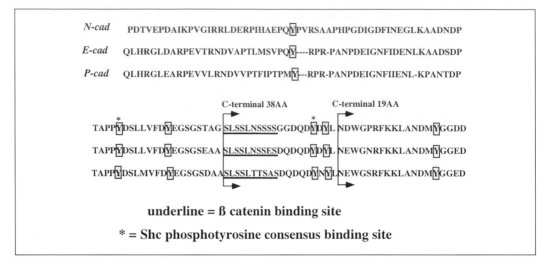

N-cad   PDTVEPDAIKPVGIRRLDERPIHAEPQYPVRSAAPHPGDIGDFINEGLKAADNDP

E-cad   QLHRGLDARPEVTRNDVAPTLMSVPQY----RPR-PANPDEIGNFIDENLKAADSDP

P-cad   QLHRGLEARPEVVLRNDVVPTFIPTPMY---RPR-PANPDEIGNFIIENL-KPANTDP

C-terminal 38AA   C-terminal 19AA

TAPPYDSLLVFDYEGSGSTAG SLSSLNSSSSGGDQDYDYL NDWGPRFKKLANDMYGGDD

TAPPYDSLLVFDYEGSGSEAA SLSSLNSSESDQDQDYDYL NEWGNRFKKLANDMYGGED

TAPPYDSLMVFDYEGSGSDAA SLSSLTTSASDQDQDYNYL NEWGSRFKKLANDMYGGED

underline = ß catenin binding site

* = Shc phosphotyrosine consensus binding site

Fig. 8 The C-terminal 38 aa are currently divided into two domains, the catenin binding domain and a domain encompassing the C-terminal 19 aa. β-Catenin and plakoglobin bind directly to the catenin binding domain (underlined). The binding partner of the C-terminal 19 aa of E-cadherin is not currently known. However, it has recently been shown that deletion of the C-terminal 19 amino acids of the cadherin cytoplasmic domain had little effect on catenin binding but had a dramatic effect on cadherin-dependent adhesion and association with the cytoskeleton. Tyrosine phosphorylation regulates cadherin function and the potential tyrosine phosphorylation sites are shown (shaded boxes). The SH2 domain of Shc only associates with the intracellular domain of cadherin when it is tyrosine phosphorylated. Tyr 851 and Tyr 883 in the intracellular domain of cadherins are consensus sites for recognition by the Shc SH2 domain (marked by asterisks).

determined and the data suggest that increased tyrosine phosphorylation of E-cadherin results in decreased association with PTPμ (65). In addition, preliminary experiments indicate that the major site of tyrosine phosphorylation is in the C-terminal 37 aa of E-cadherin (S. B. K. unpublished data). These data indicate that tyrosine phosphorylation may change the affinity of E-cadherin and PTPμ for one another, since the domain in E-cadherin that is tyrosine phosphorylated contains the PTPμ binding site (65). Therefore, reversible tyrosine phosphorylation may regulate the adhesive function of cadherins by altering the presence of PTPs, which maintain cadherin in the functional dephosphorylated state.

Recent studies implicate components of the PTK signalling pathways in the regulation of the cadherin/catenin complex. The cytoplasmic domain of N-cadherin was recently cloned as a Shc interacting protein using a modified two-hybrid screen (194). The SH2 domain of Shc associated with the intracellular domain of cadherin *in vitro* only when it was tyrosine phosphorylated (194). Tyr 851 and Tyr 883 in the intracellular domain of cadherins are consensus sites for recognition by the Shc SH2 domain (marked by asterisks in Fig. 8). At present, it is not known whether Shc binds to either of these two sites *in vivo*. Both of the tyrosine residues in the Shc consensus binding site are conserved in E-, N-, R-, P-cadherin, and cadherin 5 in all species (195). Pervanadate treatment, which inhibits tyrosine phosphatase activity, increased the level of Shc–cadherin association (194). This result implies that reversible tyrosine phosphorylation of the cytoplasmic domain of cadherins regulates Shc binding (194).

The C-terminal 37 aa are crucial for cadherin-dependent adhesion and cytoskeletal association (183, 189). Deletion of the C-terminal 37 aa of E-cadherin may result in changes in the tyrosine phosphorylation state of E-cadherin by the Src PTK or alter the association between PTPμ/E-cadherin or Shc/E-cadherin. Ultimately, the regulation of protein–protein interactions by tyrosine phosphorylation is likely to be a component of a signal transduction cascade that regulates cadherin-dependent cell adhesion and cytoskeletal association. The study of the regulation of adhesion molecules and tyrosine phosphorylation will be important for understanding both the loss of adhesion and loss of growth control that occur in malignant transformation.

6.5.3 Binding of PTPμ to the scaffolding protein RACK1

We have recently found that PTPμ binds to a scaffolding protein called RACK1 (196). The signalling pathway downstream of PTPμ was unknown; therefore, we used a yeast two-hybrid screen to identify additional PTPμ interacting proteins. The membrane proximal catalytic domain of PTPμ was used as bait and two interacting clones were identified as the scaffolding protein RACK1 (*r*eceptor for *a*ctivated protein *C k*inase) (196).

While RACK1 was originally identified as a receptor for activated PKC, more recent data suggest that RACK1 is a scaffolding protein that recruits a number of signalling molecules into a complex. RACK1 is a member of the Gβ family of seven propeller proteins (197) and therefore could theoretically bind up to seven different proteins. RACK1 binds PKC, PLCγ, the *src* cytoplasmic protein tyrosine kinase (PTK), cAMP phosphodiesterase-4, the β subunit of integrins, and the β chain of IL-5R (an interleukin receptor) (198–202). RACK1 also binds to some pleckstrin homology (PH) domains *in vitro* including dynamin, β-spectrin, Ras GRF, and oxysterol binding protein (OSBP) (203). Some of the interactions between RACK1 and the proteins listed above are mutually exclusive (199) and only some of these interactions depend upon PKC stimulation (201). These studies suggest that RACK1 may form distinct complexes in response to various cellular signals.

We set out to characterize the PTPμ/RACK1 interaction and demonstrated that RACK1 interacted with PTPμ when co-expressed in a recombinant baculovirus expression system. RACK1 was known to bind to the *src* protein tyrosine kinase (PTK). Our study demonstrated that PTPμ association with RACK1 was disrupted by the presence of constituitively active *src* (196). RACK1 is thought to be a scaffolding protein that recruits proteins to the plasma membrane via an unknown mechanism. We also demonstrated that the association of endogenous PTPμ and RACK1 in a lung cell line increases at high cell density. The recruitment of RACK1 to both the plasma membrane and cell–cell contact sites was dependent upon the presence of the PTPμ protein in these cells (196). Therefore, PTPμ may be one of the proteins that recruits RACK1 to points of cell–cell contact which may be important for PTPμ-dependent signalling in response to cell–cell adhesion.

Other PTPs may also interact with RACK1 since RACK1 binds to the conserved PTP catalytic domain of PTPμ. Interestingly, many PTPs regulate the *src* cytoplasmic PTK (see above). Since RACK1 is known to bind to *src*, our data suggest that one of the links between PTPs and *src* PTK signalling may be via the RACK1 protein. The PTPμ association with RACK1 is altered in the presence of *src* suggesting that there may be mutually exclusive interactions of *src* and PTPμ with RACK1. The PTP versus PTK competition for binding to scaffolding proteins like RACK1 may be a common form of regulation to control tyrosine phosphorylation of various substrate proteins via their mutually exclusive interactions between a scaffolding protein and a tyrosine kinase or phosphatase. The ability of *src* and PTPμ (or other PTPs) to compete for RACK1 binding may be an important mechanism for regulation of cell adhesion and signal transduction.

7. Physiological functions of RPTPs

7.1 Nervous system development and axon pathfinding

The role of RPTPs in the nervous system has been reviewed previously (12, 164, 204–206).

7.1.1 MAM domain containing proteins play a role in neurite outgrowth

PTPμ is expressed at high levels in the developing nervous system (63, 64, 67). The expression of PTPμ is developmentally regulated in the retina and increases in a pattern similar to N-cadherin expression (74). The retina is comprised of a number of different cell types, and the interactions of cell adhesion molecules are known to control retinal histogenesis and axonal pathfinding. In the retina, PTPμ is primarily expressed on retinal ganglion cells (RGC) who communicate with the brain via their neuronal processes. RGC axons migrate along the surfaces of other neuronal and glial cells during development, thus utilizing cell surface adhesion molecules as a 'substrate' for neuronal migration. PTPμ was shown to promote neurite outgrowth from RGC neurons, as well as the migration of bipolar neurons and Müller glia from embryonic day 8 chick retinal explants (74). Addition of antibodies directed against PTPμ blocked neurite outgrowth on a PTPμ substrate, indicating that the neurite outgrowth activity was specific to PTPμ. PTPμ and N-cadherin form a complex in retinal tissue and in RGC neurites, suggesting that PTPμ may regulate N-cadherin-dependent neurite outgrowth (74). PTPμ expression was down-regulated by anti-sense techniques which resulted in a decreased neurite outgrowth on a N-cadherin substrate, but did not affect neurite outgrowth on laminin or L1 substrates (74). A catalytically inactive form of PTPμ was overexpressed which also inhibited N-cadherin-mediated neurite outgrowth. This is the strongest evidence to date that a component of the N-cadherin/catenin complex is a PTPμ substrate (74). Therefore, PTPμ is capable of both promoting neurite outgrowth as well as regulating neurite outgrowth mediated by N-cadherin. Interestingly, PTPκ has recently been shown to promote neurite outgrowth of cerebellar neurons (207). The PTPμ subfamily is likely to regulate neurite outgrowth in many different types of neurons.

It is interesting to note that neuropilin-1 (previously named A5) has a MAM domain (15). Neuropilin-1 was demonstrated to promote neurite outgrowth of retinal explants and trigeminal ganglia neurons (208). Neuropilin-1 is a receptor for the group III semaphorins (99–101). However, semaphorins usually mediate growth cone collapse and repulsive cues in many systems. There may be other ligands for neuropilin-1 that mediate the neurite outgrowth promoting signals. The role of the MAM domain in neuropilin-1-mediated neurite outgrowth or binding to the sema- phorins is to promote oligomerization (25). The role of PTPμ's MAM domain in neurite outgrowth is currently under investigation.

7.1.2   *Drosophila* RPTPs in motor neuron pathfinding

The role of *Drosophila* RPTPs in motor neuron pathfinding has been reviewed previously (209). The expression of DLAR and DPTP69D is critical for neurons to innervate appropriate target muscles (210, 211). Knocking out the function of DLAR in *Drosophila* causes embryos to die at the late instar larvae stage due to defects in motor neuron pathfinding (211). DPTP69D and DPTP99A are also expressed on motor axons and growth cones. In embryos where DPTP69D was knocked out, motor neuron growth cones stopped growing before reaching their target muscles (210), resulting in death by the pupal stage. The single mutation of DPTP99A had no effect on the embryos. However, a more severe phenotype occurred when DPTP69D and DPTP99A mutant flies were crossed (210), suggesting a synergy or overlapping function of the two PTPs. The phenotypes exhibited suggest that these PTPs control the ability of the neurons to recognize guidance cues as well as regulate defasci- culation, which helps neurons extend axons to their appropriate target muscles.

Recent studies confirm that DLAR, DPTP69D, and DPTP99A phosphatases are required for proper neuronal guidance of two particular motor neurons (ISN and SNb) in *Drosophila* (209). Mutation of any one of the phosphatases still allows the ISN nerve to progress to an intermediate target. However, the presence of DLAR and DPTP69D is required for the neurons to progress beyond a second intermediate target. Only the presence of DLAR is required for a terminal nerve arbor to form at the neuromuscular junction (209). In DLAR mutants, the SNb axons bypass their target muscle. However, this defect can be suppressed by mutating DPTP99A. These data suggest that DLAR and DPTP99A co-operate in neurons growing along one nerve (ISN) and act in opposition to one another along another nerve (SNb). Interestingly, DPTP69D has also been shown to be important in retinal photoreceptor axon targeting (212). Together, these studies suggest that these RPTPs may transiently down-regulate adhesion of neurons at the choice point to allow defasci- culation and steering toward the target or possibly play a role in target recognition (209, 212).

DLAR associates with Abl and its substrate, Enabled (Ena), and this complex is involved in axonal pathfinding (213). It appears that profilin is downstream of Ena and directly regulates actin polymerization (158). DLAR binding to Trio (see above) may also be a component of this signalling pathway. Trio contains two Dbl homology domains that regulate Rho family GTPases which regulate the actin cytoskeleton (158). Recent studies suggest that the Abl tyrosine kinase antagonizes the function of DLAR in the ISNb nerve (213). Mutations of Trio and Abl lead to a similar phenotype in the ISNb nerve. In fact, mutation of Trio, Abl, Rac, Ena, and profilin result in overlapping phenotypes in neurons (158). This DLAR signalling pathway is likely to be involved in the regulation of axonal growth and guidance in multiple parts of the nervous system (158).

DPTP10D is expressed in developing *Drosophila* central nervous system on axons (60). The localization of DPTP10D suggests that it may also be involved in neuronal pathfinding. Although the ligand of DPTP10D is not known, a substrate called gp150 was identified that suggests a role in cell adhesion (214). gp150 is a transmembrane protein, which contains leucine-rich repeats that are found in proteoglycans involved in cell–matrix adhesion (biglycan and decorin) as well as homophilic cell–cell adhesion molecules (chaoptin and connectin). DPTP10D may alter the adhesive function of gp150 by regulating the phosphorylation state of gp150 or other associated proteins (214). These data suggest that DPTP10D may regulate cell–matrix adhesion to control neuronal pathfinding or guidance.

7.1.3 LAR-like PTPs and neurite outgrowth

The leech homologue of LAR (HmLAR2) is concentrated in growth cones during a time when neurons undergo rapid outgrowth (215). When HmLAR2 antibodies generated against the extracellular domain were used to perturb protein function, neurons had abnormal projections and their outgrowth was shorter than normal (215). Recent studies suggest that HmLAR2 may function in growth cone regulation and mutual repulsion via a homophilic binding mechanism (216). Interestingly, the HmLAR2 protein is predominately sequestered in an intracellular pool, which may be available for rapid release to the growth cone surface (215). This phenomenon may be a regulated protein trafficking event as has been demonstrated for the L1 cell adhesion molecule (57).

In addition, both PTPδ and CRYPα have been shown to promote neurite outgrowth. PTPδ promotes neurite outgrowth of forebrain and some cerebellar neurons (97). Interestingly, a local gradient of soluble PTPδ acted as a chemoattractant in embryonic forebrain neurons and the attraction depended upon tyrosine phosphatase activity (217). CRYPα regulates interactions between neurons and glial cells in the retina as well as in retinotectal pathfinding during development (218, 219). Perturbation of CRYPα affects growth cone morphology in retinal ganglion cell axons (220). The LAR subfamily PTPs are likely to control neuronal pathfinding or guidance in both invertebrate and vertebrate systems.

7.1.4 Phosphacan as a regulator of neuronal migration

RPTPζ/β or phosphacan binds to both adhesive (CAMs) and repulsive (tenascin) molecules (28). RPTPζ/β is expressed on radial glial cells in central nervous system at embryonic day 14–18 (77) where radial glia form barriers to neuronal migration.

Phosphacan, the soluble splice variant of RPTPζ/β, promoted neurite outgrowth of cortical neurons, however it has also been reported to be repulsive to the adhesion of cortical and thalamic neurons (78). RPTPζ/β either promoted neurite outgrowth and/or mediated cell adhesion via the NrCAM/contactin complex depending upon which fragments of the extracellular domain were used (80). Neurite outgrowth of primary tectal neurons was promoted by the CAH domain of RPTPζ/β and this outgrowth was blocked by antibodies to contactin (28). Interestingly, addition of phosphacan antibodies inhibited pleiotrophin-induced neurite outgrowth of cortical neurons (108). HB-GAM, a receptor for phosphacan, promotes neurite outgrowth of cortical neurons, but the addition of antibodies to phosphacan or soluble phosphacan protein inhibited neurite outgrowth on HB-GAM (221). This study suggests that HB-GAM promotes neurite outgrowth by binding to RPTPζ/β which can be competitively inhibited by phosphacan or regulated by chondroitin sulfate (221). Soluble forms of contactin, axonin-1, and RPTPζ/β (phosphacan) are expressed in the nervous system. These soluble forms may compete with one another or the transmembrane forms for binding during development. The regulation of neuronal migration by RPTPζ/β or phosphacan in the nervous system is likely to be very complex due to the diversity of ligands with which they can interact.

7.2 *clr-1* plays a role in FGFR signalling in *C. elegans*

The role of *clr-1* in FGF receptor signalling has been reviewed previously (222). Mutation of the FGF receptor (*egl-15*) in *C. elegans* results in a defect in sex myoblast migration. Null alleles for *egl-15* result in a larval lethality whereas hypomorphic mutations result in the Scrawny phenotype (223). A genetic screen was performed to identify suppressors of the Scrawny phenotype (223). One gene that was isolated encoded the *clr-1* RPTP protein (223). The *clr-1* RPTP consists of two Ig domains and one FNIII repeat in its extracellular segment as well as two phosphatase domains in its intracellular segment (223). When the *clr-1* gene is deleted, some animals die in the larval stage due to fluid accumulation in the pseudocoelom. The first phosphatase domain of *clr-1* was essential for the negative regulation of FGFR signalling. The authors mention preliminary data to suggest that *clr-1* may also affect EGFR signalling. Together, their data suggest a role for *clr-1* phosphatase activity in the negative regulation of receptor tyrosine kinase signalling (223).

7.3 Disease

7.3.1 Cancer

The role of RPTPs in cancer is an understudied area of this field. However, changes in gene expression of some PTPs have been observed during tumour progression. Increased expression of PTPα is observed in late stage colon cancers (54). PTPγ is found in a region of chromosome 3 (3p21) that is frequently deleted in renal and lung cancers (87). The transmembrane form of RPTPζ/β is lost during progression from low to high-grade gliomas (224). RPTPζ/β expression as well as its ligand, pleiotrophin, were reduced in colorectal cancers while expression of PTPγ was not altered (225). Protein expression of the LAR subfamily is altered during cancer progression, which plays a role in negative growth regulation. PTPδ expression was down-regulated in hepatomas (226). LAR is localized to human chromosome 1p32–34.1, which is a region that is altered in tumours of neuroectodermal origin (30). In addition, mutated forms of LAR were found in colon and breast cancers (11). LAR is able to negatively regulate proliferation of transformed breast cancer cells and suppress tumour growth in nude mice (227). Overexpression of LAR inhibits epithelial cell migration and tumour formation in nude mice (179). It is possible that changes in LAR expression during cancer progression contribute to tumour cell growth or metastasis.

7.3.2 Diabetes

A subfamily of RPTPs, IA-2/phogrin, may act as autoantigens in insulin-dependent diabetes mellitus (228, 229). These two PTPs are called IA-2 (also called ICA512) or IA-2β/phogrin (also called PTP-NP, PTP-NP-2, PTP-IAR) see Fig. 3 (228, 229). These proteins have a cysteine-rich region in their extracellular segments, a transmembrane domain, and a single PTP domain. The phosphatase domain of the IA-2 protein lacks key residues required to catalyse dephosphorylation (229). The PTP domain of IA-2 is similar to domain 2 of RPTPs (229). Proteins like IA-2 that have a PTP-like domain that cannot catalyse dephosphorylation are termed STYX domain-containing proteins (229). They may function to attenuate phosphatase activity in cells (229). Even though the presence of autoantibodies to IA-2/phogrin in diabetes mellitus correlates with the onset of disease, a pathogenic role for the antibodies has not been demonstrated.

8. Emerging concepts

There are still many unanswered questions for RPTPs. For example, does RPTP-dependent adhesion trigger a specific signalling pathway? Does adhesion mediated by the RPTPs regulate enzymatic activity? If the RPTPs naturally exist as dimers does adhesion induce oligomerization and how does oligomerization regulate enzymatic activity? These questions represent areas of current research by many investigators. General concepts that are emerging for the function of RPTPs in various organisms are discussed.

One emerging concept highlighted by this chapter is that RPTPs play key roles in regulating neuronal migration and axonal pathfinding in many species from insects to mammals. The roles that RPTPs could be playing in nervous system development are outlined. RPTPs may act as cell adhesion molecules that are involved in maintenance of nervous system integrity. This could occur through homophilic or heterophilic binding of RPTPs to promote axon fasciculation, a process required for nerve formation. A similar role has been suggested for other CAMs such as L1 and NCAM. Alternatively, RPTPs may act as permissive or restrictive molecules for axonal growth. A contact attraction or repulsive role for RPTPs is another possibility, such that RPTPs actively guide axons during pathfinding. For example, RPTP expression at specific choice points where axons must choose the appropriate pathway has already been observed. Neurons encounter many choice points during outgrowth to their target. The mechanisms regulating stereotypical innervation of targets is only partly understood, but involves tyrosine phosphorylation in many cases. In addition, RPTPs could function as sensor molecules. For example, changes in the adhesive state of the extracellular environment may be transmitted through regulation of an RPTP catalytic domain to directly control the phosphorylation state of a number of cytosolic proteins. The *in vitro* promotion of neurite outgrowth by certain RPTPs could be mimicking the ability of neurons to respond to signalling events initiated by that RPTP *in vivo*. Together, all the data summarized in this chapter on the role of RPTPs in the nervous system suggest that they will be important for axonal migration and pathfinding.

Another new area that deserves further investigation is the idea that the RPTPs are likely to be involved in unidirectional and bidirectional signalling. Specifically, there are likely to be both unidirectional (signals within a single cell) and bidirectional (signalling in both apposing cells) signals that result from RPTP-dependent adhesion. For heterophilic interactions, the bidirectional signals are not necessarily the same in both of the apposing cells. Bidirectional signalling is important for Eph tyrosine kinases which have heterophilic binding partners called ephrins (230). Further identification of ligands and substrates will allow the nature of RPTP-induced signals to be studied both *in vitro* and *in vivo*.

Acknowledgements

I would like to thank Dr. Susan Burden-Gulley for helpful comments on the manuscript. Dr. Brady-Kalnay is supported by a NIH grant (1RO1-EY12251) and an Army Prostate Cancer Grant (DAMD17-98-1-8586).

References

1. Hooft van Huijsduijnen, R. (1998). Protein tyrosine phosphatases: counting the trees in the forest. *Gene*, **225**, 1.
2. Denu, J. M., Stuckey, J. A., Saper, M. A., and Dixon, J. E. (1996). Form and function in protein dephosphorylation. *Cell*, **87**, 361.
3. Barford, D., Flint, A. J., and Tonks, N. K. (1994). Crystal structure of human protein tyrosine phosphatase 1B. *Science*, **263**, 1397.
4. Flint, A. J., Tiganis, T., Barford, D., and Tonks, N. K. (1997). Development of 'substrate-trapping' mutants to identify physiological substrates of protein tyrosine phosphatases. *Proc. Natl. Acad. Sci. USA*, **94**, 1680.
5. Garton, A. J., Flint, A. J., and Tonks, N. K. (1996). Identification of p130[cas] as a substrate for the cytosolic protein tyrosine phosphatase PTP-PEST. *Mol. Cell. Biol.*, **16**, 6408.
6. LaMontagne, K. R., Flint, A. J., Franza, B. R., Pendergast, A. M., and Tonks, N. K. (1998). Protein tyrosine phosphatase 1B antagonizes signalling by oncoprotein tyrosine kinase p210 bcr-abl *in vivo*. *Mol. Cell. Biol.*, **18**, 2965.

7. Tiganis, T., Bennett, A. M., Ravichandran, K. S., and Tonks, N. K. (1998). Epidermal growth factor receptor and the adaptor protein p52Shc are specific substrates of T-cell protein tyrosine phosphatase. *Mol. Cell. Biol.*, **18**, 1622.

8. Brummendorf, T. and Rathjen, F. G. (1998). Molecular interactions involving immunoglobulin superfamily adhesion proteins. In *Ig superfamily molecules in the nervous system* (ed. P. Sonderegger), p. 23. Harwood Academic Publishers, Amsterdam.

9. Rader, C. and Sonderegger, P. (1998). Structural features of neural immunoglobulin superfamily adhesion molecules. In *Ig superfamily molecules in the nervous system* (ed. P. Sonderegger), p. 1. Harwood Academic Publishers, Amsterdam.

10. Brady-Kalnay, S. M. and Tonks, N. K. (1995). Protein tyrosine phosphatases as adhesion receptors. *Curr. Opin. Cell Biol.*, **7**, 650.

11. Schaapveld, R., Wieringa, B., and Hendriks, W. (1997). Receptor-like protein tyrosine phosphatases: alike and yet so different. *Mol. Biol. Rep.*, **24**, 247.

12. Brady-Kalnay, S. (1998). Ig-superfamily phosphatases. In *Immunoglobulin superfamily adhesion molecules in neural development, regeneration, and disease* (ed. P. Sonderegger), p. 133. Harwood Academic Publishers, Amsterdam.

13. Potts, J. R. and Campbell, I. D. (1994). Fibronectin structure and assembly. *Curr. Opin. Cell Biol.*, **6**, 648.

14. Bork, P. and Doolittle, R. F. (1992). Proposed acquisition of an animal protein domain by bacteria. *Proc. Natl. Acad. Sci. USA*, **89**, 8990.

15. Beckmann, G. and Bork, P. (1993). An adhesive domain detected in functionally diverse receptors. *Trends Biochem. Sci.*, **18**, 40.

16. Wang, H., Lian, Z., Lerch, M. M., Chen, Z., Xie, W., and Ulrich, A. (1996). Characterization of PCP-2, a novel receptor protein tyrosine phosphatase of the MAM domain family. *Oncogene*, **12**, 2555.

17. McAndrew, P. E., Frostholm, A., White, R. A., Rotter, A., and Burghes, A. H. M. (1998). Identification and characterization of RPTPρ, a novel RPTPμ/κ-like receptor protein tyrosine phosphatase whose expression is restricted to the central nervous system. *Mol. Brain Res.*, **56**, 9.

18. Crossland, S., Smith, P. D., and Crompton, M. R. (1996). Molecular cloning and characterization of PTPπ, a novel receptor-like protein-tyrosine phosphatase. *Biochem. J.*, **319**, 249.

19. Wang, B., Kishihara, K., Zhang, D., Hara, H., and Nomoto, K. (1997). Molecular cloning and characterization of a novel human receptor protein tyrosine phosphatase gene, hPTP-J: down-regulation of gene expression by PMA and calcium ionophore in Jurkat T lymphoma cells. *Biochem. Biophys. Res. Commun.*, **231**, 77.

20. Cheng, J., Wu, K., Armanini, M., O'Rourke, N., Dowbenko, D., and Lasky, L. A. (1997). A novel protein-tyrosine phosphatase related to the homotypically adhering κ and μ receptors. *J. Biol. Chem.*, **272**, 7264.

21. Avraham, S., London, R., Tulloch, G. A., Ellis, M., Fu, Y., Jiang, S., *et al.* (1997). Characterization and chromosomal localization of PTPRO, a novel receptor protein tyrosine phosphatase, expressed in hematopoietic stem cells. *Gene*, **204**, 5.

22. Yoneya, T., Yamada, Y., Kakeda, M., Osawa, M., Arai, E., Hayashi, K., *et al.* (1997). Molecular cloning of a novel receptor-type protein tyrosine phosphatase from murine fetal liver. *Gene*, **194**, 241.

23. Marchand, P., Volkmann, M., and Bond, J. S. (1996). Cysteine mutations in the MAM domain result in monomeric meprin and alter stability and activity of the proteinase. *J. Biol. Chem.*, **271**, 24236.

24. Tsukuba, T. and Bond, J. S. (1998). Role of the COOH-terminal domains of meprin A in folding, secretion, and activity of the metalloendopeptidase. *J. Biol. Chem.*, **273**, 35260.

25. Chen, H., He, Z., Bagri, A., and Tessier-Lavigne, M. (1998). Semaphorin-neuropilin interactions underlying sympathetic axon responses to class III semaphorins. *Neuron*, **21**, 1283.

26. Barnea, G., Grumet, M., Sap, J., Margolis, R. U., and Schlessinger, J. (1993). Close similarity between receptor-linked tyrosine phosphatase and rat brain proteoglycan. *Cell*, **76**, 205.

27. Krueger, N. X. and Saito, H. (1992). A human transmembrane protein-tyrosine-phosphatase, PTPζ, is expressed in brain and has an N-terminal receptor domain homologous to carbonic anhydrases. *Proc. Natl. Acad. Sci. USA*, **89**, 7417.

28. Peles, E., Schlessinger, J., and Grumet, M. (1998). Multi-ligand interactions with receptor-like protein tyrosine phosphatase beta: implications for intercellular signaling. *Trends Biochem. Sci.*, **23**, 121.

29. Holland, S. J., Peles, E., Pawson, T., and Schlessinger, J. (1998). Cell-contact-dependent signalling in axon growth and guidance: Eph receptor tyrosine kinases and receptor protein tyrosine phosphatase beta. *Curr. Opin. Neurobiol.*, **8**, 117.

30. Longo, F. M., Martignetti, J. A., Le Beau, J. M., Zhang, J. S., Barness, J. P., and Brosius, J. (1993). Leukocyte common antigen-related receptor-linked tyrosine phosphatase. *J. Biol. Chem.*, **268**, 26503.

31. Pulido, R., Serra-Pages, C., Tang, M., and Streuli, M. (1995). The LAR/PTPδ/PTPσ subfamily of transmembrane protein-tyrosine-phosphatases: Multiple human LAR, PTPδ, and PTPσ isoforms are expressed in a tissue-specific manner and associate with the LAR-interacting protein LIP.1. *Proc. Natl. Acad. Sci. USA*, **92**, 11686.

32. Kaneko, Y., Takano, S., Okumura, K., Takenawa, J., Higashituji, H., Fukumoto, M., *et al.* (1993). Identification of protein tyrosine phosphatases expressed in murine male germ cells. *Biochem. Biophys. Res. Commun.*, **197**, 625.

33. Mizuno, K., Hasegawa, K., Katagiri, T., Ogimoto, M., Ichikawa, T., and Yakura, H. (1993). MPTPδ, a putative murine homolog of HPTPδ, is expressed in specialized regions of the brain and in the B-cell lineage. *Mol. Cell. Biol.*, **13**, 55130.

34. Pulido, R., Krueger, N. X., Serra-Pages, C., Saito, H., and Streuli, M. (1995). Molecular characterization of the human transmembrane protein-tyrosine phosphatase δ. *J. Biol. Chem.*, **270**, 6722.

35. Yan, H., Grossman, A., Wang, H., D'Eustachio, P., Mossie, K., Musacchio, J. M., *et al.* (1993). A novel receptor tyrosine phosphatase-σ that is highly expressed in the nervous system. *J. Biol. Chem.*, **268**, 24880.

36. Wagner, J., Boerboom, D., and Tremblay, M. L. (1994). Molecular cloning and tissue-specific RNA processing of a murine receptor-type protein tyrosine phosphatase. *Eur. J. Biochem.*, **226**, 773.

37. Wallace, M. J., Batt, J., Fladd, C. A., Henderson, J. T., Skarnes, W., and Rotin, D. (1999). Neuronal defects and posterior pituitary hypoplasia in mice lacking the receptor tyrosine phosphatase PTPσ. *Nature Genet.*, **21**, 334.

38. Elchebly, M., Wagner, J., Kennedy, T. E., Lanctot, C., Michaliszyn, E., Itie, A., *et al.* (1999). Neuroendocrine dysplasia in mice lacking protein tyrosine phosphatase σ. *Nature Genet.*, **21**, 330.

39. Pan, M. G., Rim, C., Lu, K. P., Florio, T., and Stork, P. J. (1993). Cloning and expression of two structurally distinct receptor-linked protein-tyrosine phosphatases generated by RNA processing from a single gene. *J. Biol. Chem.*, **268**, 19284.

40. Stoker, A. W. (1994). Isoforms of a novel cell adhesion molecule-like protein tyrosine phosphatase are implicated in neural development. *Mech. Dev.*, **46**, 201.

41. Ranjan, M. and Hudson, L. D. (1996). Regulation of tyrosine phosphorylation and protein tyrosine phosphatases during oligodendrocyte differentiation. *Mol. Cell. Neuro.*, **7**, 404.

42. Shock, L. P., Bare, D. J., Klinz, S. G., and Maness, P. F. (1995). Protein tyrosine phosphatases expressed in developing brain and retinal Müller glia. *Mol. Brain Res.*, **28**, 110.

43. Schaapveld, R. Q. J., Schepens, J. T. G., Bachner, D., Attema, J., Wieringa, B., Jap, P. H. K., *et al.* (1998). Developmental expression of the cell adhesion molecule-like protein tyrosine phosphatases LAR, RPTPδ and RPTPσ in the mouse. *Mech. Dev.*, **77**, 59.

44. Johnson, K. G. and Holt, C. E. (2000). Expression of CRYP-alpha, LAR, PTP-delta, and PTP-rho in the developing *Xenopus* visual system. *Mech. Dev.*, **92**, 291.

45. Sahin, M. and Hockfield, S. (1993). Protein tyrosine phosphatases expressed in the developing rat brain. *J. Neurosci.*, **13**, 4968.

46. Walton, K., Martell, K. J., Kwak, S. P., Dixon, J. E., and Largent, B. L. (1993). A novel receptor-type protein tyrosine phosphatase is expressed during neurogenesis in the olfactory neuroepithelium. *Neuron*, **11**, 387.

47. Stoker, A. W., Gehrig, B., Haj, F., and Bay, B. H. (1995). Axonal localisation of the CAM-like tyrosine phosphatase CRYPα: a signaling molecule of embryonic growth cones. *Development*, **121**, 1833.

48. Ledig, M. M., McKinnell, I. W., Mrsic-Flogel, T., Wang, J., Alvares, C., Mason, I., *et al.* (1999). Expression of receptor tyrosine phosphatases during development of the retinotectal projection of the chick. *J. Neurobiol.*, **39**, 81.

49. Haworth, K., Shu, K. K., Stokes, A., Morris, R., and Stoker, A. (1998). The expression of receptor tyrosine phosphatases is responsive to sciatic nerve crush. *Mol. Cell. Neurosci.*, **12**, 93.

50. Ahmad, F. and Goldstein, B. J. (1997). Functional association between the insulin receptor and the transmembrane protein-tyrosine phosphatase LAR in intact cells. *J. Biol. Chem.*, **272**, 448.

51. Kulas, D. T., Goldstein, B. J., and Mooney, R. A. (1996). The transmembrane protein-tyrosine phosphatase LAR modulates signaling by multiple receptor tyrosine kinases. *J. Biol. Chem.*, **271**, 748.

52. O'Grady, P., Krueger, N. X., Streuli, M., and Saito, H. (1994). Genomic organization of the human LAR protein tyrosine phosphatase gene and alternative splicing in the extracellular fibronectin type-III domains. *J. Biol. Chem.*, **269**, 25193.

53. Zhang, J. S. and Longo, F. M. (1995). LAR tyrosine phosphatase receptor: alternative splicing is preferential to the nervous system, coordinated with cell growth and generates novel isoforms containing extensive CAG repeats. *J. Cell Biol.*, **128**, 415.

54. Tabiti, K., Cui, L., Chhatwal, V., Moochhala, S., Ngoi, S. S., and Pallen, C. J. (1996). Novel alternative splicing predicts a secreted extracellular isoform of the human receptor-like protein tyrosine phosphatase LAR. *Gene*, **175**, 7.

55. Honkaniemi, J., Zhang, J. S., Yang, T., Zhang, C., Tisi, M. A., and Longo, F. M. (1998). LAR tyrosine phosphatase receptor: proximal membrane alternative splicing is co-ordinated with regional expression and intraneuronal localization. *Mol. Brain Res.*, **60**, 1.

56. O'Grady, P., Thai, T. C., and Saito, H. (1998). The laminin-nidogen complex is a ligand for a specific splice isoform of the transmembrane protein tyrosine phosphatase LAR. *J. Cell Biol.*, **141**, 1675.

57. Kamiguchi, H. and Lemmon, V. (1997). Neural cell adhesion molecule L1: signaling pathways and growth cone motility. *J. Neurosci. Res.*, **49**, 1.

58. Schaapveld, R. Q. J., Schepens, J. T. G., Robinson, G. W., Attema, J., Oerleman, F. T. J. J., Fransen, J. A. M., *et al.* (1997). Impaired mammary gland development and function in mice lacking LAR receptor-like tyrosine phosphatase activity. *Dev. Biol.*, **188**, 134.

59. Yeo, T. T., Yang, T., Massa, S. M., Zhang, J. S., Honkaniemi, J., Butcher, L. L., *et al.* (1997). Deficient LAR expression decreases basal forebrain cholinergic neuronal size and hippocampal cholinergic innervation. *J. Neurosci. Res.*, **47**, 348.

60. Zinn, K. (1993). *Drosophila* protein tyrosine phosphatases. *Semin. Cell Biol.*, **4**, 397.

61. Tian, S.-S., Tsoulfas, P., and Zinn, K. (1991). Three receptor-linked protein-tyrosine phosphatases are selectively expressed on central nervous system axons in the *Drosophila* embryo. *Cell*, **67**, 675.

62. Desai, C. J., Popova, E., and Zinn, K. (1994). A *Drosophila* receptor tyrosine phosphatase expressed in the embryonic CNS and larval optic lobes is a member of the set of proteins bearing the 'HRP' carbohydrate epitope. *J. Neurosci.*, **14**, 7272.

63. Gebbink, M., van Etten, I., Hateboer, G., Suijkerbuijk, R., Beijersbergen, R., van Kessel, A., *et al.* (1991). Cloning, expression and chromosomal localization of a new putative receptor-like protein tyrosine phosphatase. *FEBS Lett.*, **290**, 123.

64. Brady-Kalnay, S. M., Rimm, D. L., and Tonks, N. K. (1995). The receptor protein tyrosine phosphatase PTPμ associates with cadherins and catenins *in vivo*. *J. Cell Biol.*, **130**, 977.

65. Brady-Kalnay, S. M., Mourton, T., Nixon, J. P., Kinch, M., Chen, H., Brackenbury, R., *et al.* (1998). Dynamic interaction of PTPμ with multiple cadherins *in vivo*. *J. Cell Biol.*, **141**, 287.

66. Campman, M., Yoshizumi, M., Seidah, N. G., Lee, M. E., Bianchi, C., and Haber, E. (1996). Increased proteolytic processing of protein tyrosine phosphatase μ in confluent vascular endothelial cells: the role of PC5, a member of the subtilisin family. *Biochemistry*, **35**, 3797.

67. Sommer, L., Rao, M., and Anderson, D. J. (1997). RPTPδ and the novel protein tyrosine phosphatase RPTPΨ are expressed in restricted regions of the developing nervous system. *Dev. Dyn.*, **208**, 48.

68. Gebbink, M., Zondag, G., Koningstein, G., Feiken, E., Wubbolts, R., and Moolenaar, W. (1995). Cell surface expression of receptor protein tyrosine phosphatase RPTPμ is regulated by cell-cell contact. *J. Cell Biol.*, **131**, 251.

69. Bianchi, C., Sellke, F. W., and Neel, B. G. (1999). Receptor-type protein-tyrosine phosphatase mu is expressed in specific vascular endothelial beds *in vivo*. *Exp. Cell Res.*, **248**, 329.

70. Jiang, Y., Wang, H., D'Eustachio, P., Musacchio, J., Schlessinger, J., and Sap, J. (1993). Cloning and characterization of R-PTP-κ, a new member of the receptor protein tyrosine phosphatase family with a proteolytically cleaved cellular adhesion molecule-like extracellular region. *Mol. Cell. Biol.*, **13**, 2942.

71. Yang, Y., Gil, M. C., Choi, E. Y., Park, S. H., Pyun, K. H., and Ha, H. (1997). Molecular cloning and chromosomal localization of a human gene homologous to the murine R-PTP-κ, a receptor-type protein tyrosine phosphatase. *Gene*, **77**, 77.

72. Lu, J., Li, Q., Donadel, G., Notkins, A. L., and Lan, M. S. (1998). Profile and differential expression of protein tyrosine phosphatases in mouse pancreatic islet tumor cell lines. *Pancreas*, **16**, 515.

73. McAndrew, P. E., Frostholm, A., Evans, J. E., Zdilar, D., Goldowitz, D., Chiu, I. M., *et al.* (1998). Novel receptor protein tyrosine phosphatase (RPTPρ) and acidic fibroblast growth factor (FGF-1) transcripts delineate a rostrocaudal boundary in the granule cell layer of the murine cerebellar cortex. *J. Comp. Neurol.*, **391**, 444.

74. Burden-Gulley, S. M. and Brady-Kalnay, S. M. (1999). PTPμ regulates N-cadherin-dependent neurite outgrowth. *J. Cell Biol.*, **144**, 1323.

75. Li, J., Tullai, J. W., Yu, W. H., and Salton, S. R. (1998). Regulated expression during development and following sciatic nerve injury of mRNAs encoding the receptor tyrosine phosphatase HPTPzeta/RPTPbeta. *Brain Res. Mol. Brain Res.*, **60**, 77.

76. Levy, J. B., Cannoll, P. D., Silbennoinen, O., Barnea, G., Morse, B., Honegger, A. M., *et al.* (1993). The cloning of a receptor-type protein tyrosine phosphatase expressed in the central nervous system. *J. Biol. Chem.*, **268**, 10573.

77. Canoll, P. D., Barnea, G., Levy, J., Sap, U., Ehrlich, M., Silvennoinen, O., *et al.* (1993). The expression of a novel receptor-type tyrosine phosphatase suggests a role in morphogenesis and plasticity of the nervous system. *Dev. Brain Res.*, **75**, 293.

78. Maeda, N. and Noda, M. (1996). 6B4 proteoglycan/phosphacan is a repulsive substratum but promotes morphological differentiation of cortical neurons. *Development*, **122**, 647.

79. Sakurai, T., Friedlander, D. R., and Grumet, M. (1996). Expression of polypeptide variants of receptor-type protein tyrosine phosphatase β: the secreted form, phosphacan, increases dramatically during embryonic development and modulates glial cell behavior *in vitro*. *J. Neuro. Res.*, **43**, 694.

80. Sakurai, T., Lustig, M., Nativ, M., Hemperly, J. J., Schlessinger, J., Peles, E., *et al.* (1997). Induction of neurite outgrowth through contactin and Nr-CAM by extracellular regions of glial receptor tyrosine phosphatase β. *J. Cell Biol.*, **136**, 907.

81. Shintani, T., Watanabe, E., Maeda, N., and Noda, M. (1998). Neurons as well as astrocytes express proteoglycan-type protein tyrosine phosphatase zeta/RPTPbeta: analysis of mice in which the PTPzeta/RPTPbeta gene was replaced with the LacZ gene. *Neurosci. Lett.*, **247**, 135.

82. Grumet, M., Milev, P., Sakurai, T., Karthikeyan, L., Bourdon, M., Margolis, R. K., *et al.* (1994). Interactions with tenascin and differential effects on cell adhesion of neurocan and phosphacan, two major chondroitin sulfate proteoglycans of nervous tissue. *J. Biol. Chem.*, **269**, 12142.

83. Milev, P., Friedlander, D., Sakurai, T., Karthikeyan, L., Flad, M., Margolis, R. K., *et al.* (1994). Interactions of the chondroitin sulfate proteoglycan phosphacan, the extracellular domain of a receptor-type protein tyrosine phosphatase, with neurons, glia and neural cell adhesion molecules. *J. Cell Biol.*, **127**, 1703.

84. Shitara, K., Yamada, H., Watanabe, K., Shimonaka, M., and Yamaguchi, Y. (1994). Brain-specific receptor-type protein-tyrosine phosphatase RPTPβ is a chondroitin sulfate proteoglycan *in vivo*. *J. Biol. Chem.*, **269**, 20189.

85. Canoll, P. D., Petanceska, S., Schlessinger, J., and Musacchio, J. M. (1996). Three forms of RPTP-beta are differentially expressed during gliogenesis in the developing rat brain and during glial cell differentiation in culture. *J. Neurosci. Res.*, **44**, 199.

86. Shintani, T., Maeda, N., Nishiwaki, T., and Noda, M. (1997). Characterization of rat receptor-like protein tyrosine phophastase gamma isoforms. *Biochem. Biophys. Res. Commun.*, **230**, 419.

87. LaForgia, S., Morse, B., Levy, J., Barnea, G., Cannizzaro, L. A., Li, F., *et al.* (1991). Receptor protein-tyrosine phosphatase γ is a candidate tumor suppressor gene at human chromosome region 3p21. *Proc. Natl. Acad. Sci. USA*, **88**, 5036.

88. Kaplan, R., Morse, B., Huebner, K., Croce, C., Howk, R., Ravera, M., *et al.* (1990). Cloning of three human tyrsoine phsophatases reveals a multigene family of receptor-linked protein-tyrosine-phosphatases expressed in brain. *Proc. Natl. Acad. Sci. USA*, **87**, 7000.

89. Matthews, R. J., Cahir, E. D., and Thomas, M. L. (1990). Identification of an additional member of the protein-tyrosine-phosphatase family: evidence for alternative splicing in the tyrosine phosphatase domain. *Proc. Natl. Acad. Sci. USA*, **87**, 4444.

90. Sap, J., D'Eustachio, P. D., Givol, D., and Schlessinger, J. (1990). Cloning and expression of a widely expressed receptor tyrosine phosphatase. *Proc. Natl. Acad. Sci. USA*, **87**, 6112.

91. Yang, C. Q. and Friesel, R. (1998). Identification of a receptor-like protein tyrosine phosphatase expressed during *Xenopus* development. *Dev. Dyn.*, **212**, 403.

92. Justement, L. B. (1997). The role of CD45 in signal transduction. *Adv. Immunol.*, **66**, 1.

93. Brady-Kalnay, S., Flint, A. J., and Tonks, N. K. (1993). Homophilic binding of PTPμ, a receptor-type protein tyrosine phosphatase, can mediate cell–cell aggregation. *J. Cell Biol.*, **122**, 961.

94. Gebbink, M. F. B. G., Zondag, G. C. M., Wubbolts, R. W., Beijersbergen, R. L., van Etten, I., and Moolenaar, W. H. (1993). Cell–cell adhesion mediated by a receptor-like protein tyrosine phosphatase. *J. Biol. Chem.*, **268**, 16101.

95. Sap, J., Jiang, Y. P., Friedlander, D., Grumet, M., and Schlessinger, J. (1994). Receptor tyrosine phosphatase R-PTP-κ mediates homophilic binding. *Mol. Cell. Biol.*, **14**, 1.

96. Brady-Kalnay, S. and Tonks, N. K. (1994). Identification of the homophilic binding site of the receptor protein tyrosine phosphatase PTPμ. *J. Biol. Chem.*, **269**, 28472.

97. Wang, J. and Bixby, J. L. (1999). Receptor tyrosine phosphatase-delta is a homophilic, neurite-promoting cell adhesion molecule for CNS neurons. *Mol. Cell. Neurosci.*, **14**, 370.

98. Zondag, G., Koningstein, G., Jiang, Y. P., Sap, J., Moolenaar, W. H., and Gebbink, M. (1995). Homophilic interactions mediated by receptor tyrosine phosphatases μ and κ. *J. Biol. Chem.*, **270**, 14247.

99. He, Z. and Tessier-Lavigne, M. (1997). Neuropilin is a receptor for the axonal chemorepellent semaphorin III. *Cell*, **90**, 739.

100. Kolodkin, A. L., Levengood, D. V., Rowe, E. G., Tai, Y. T., Giger, R. J., and Ginty, D. D. (1997). Neuropilin is a semaphorin III receptor. *Cell*, **90**, 753.

101. Fujisawa, H. and Kitsukawa, T. (1998). Receptors for collapsin/semaphorins. *Curr. Opin. Neurobiol.*, **8**, 587.

102. Gennarini, G., Cibelli, G., Rougon, G., Mattei, M. G., and Goridos, C. (1989). The mouse neuronal cell surface protein F3: a phosphatidylinositol-anchored member of the immunoglobulin superfamily related to chicken contactin. *J. Cell Biol.*, **109**, 755.

103. Reid, R. A., Bronson, D. D., Young, K. M., and Hemperly, J. J. (1994). Identification and characterization of the human cell adhesion molecule contactin. *Brain Res. Mol. Brain Res.*, **21**, 1.

104. Zeng, L., D'Alessandri, L., Kalousek, M. B., Vaughan, L., and Pallen, C. J. (1999). Protein tyrosine phosphatase alpha (PTPalpha) and contactin form a novel neuronal receptor complex linked to the intracellular tyrosine kinase fyn. *J. Cell Biol.*, **147**, 707.

105. Barnea, G., Grumet, M., Milev, P., Silvennoinen, O., Levy, J., Sap, J., *et al.* (1994). Receptor tyrosine phosphatase β is expressed in the form of proteoglycan and binds to the extracellular matrix protein tenascin. *J. Biol. Chem.*, **269**, 14349.

106. Milev, P., Chiba, A., Haring, M., Rauvala, H., Schachner, M., Ranscht, B., *et al.* (1998). High affinity binding and overlapping localization of neurocan and phosphacan/protein-tyrosine phosphatase-ζ/β with tenascin-R, amphoterin, and the heparin-binding growth-associated molecule. *J. Biol. Chem.*, **273**, 6998.

107. Milev, P., Maurel, P., Haring, M., Margolis, R. K., and Margolis, R. U. (1996). TAG-1/Axonin-1 is a high-affinity ligand of neurocan, phosphacan/protein-tyrosine phosphatase-ζ/β, and N-CAM. *J. Biol. Chem.*, **271**, 15716.

108. Maeda, N., Nishiwaki, T., Shintani, T., Hamanaka, H., and Noda, M. (1996). 6B4 proteoglycan/phosphacan, an extracellular variant of receptor-like protein-tyrosine phosphatase ζ/RPTPβ, binds pleiotrophin/heparin-binding growth-associated molecule (HB-GAM). *J. Biol. Chem.*, **271**, 21446.

109. Lorenz, H. M., Harrer, T., Lagoo, A. S., Baur, A., Eger, G., and Kalden, J. R. (1993). CD45 mAb induces cell adhesion in peripheral blood mononuclear cells via lymphocyte function-associated antigen-1 (LFA-1) and intercellular cell adhesion molecule 1 (ICAM-1). *Cell. Immunol.*, **147**, 110.

110. Bernard, G., Zoccola, D., Ticchioni, M., Breittmayer, J., Aussel, C., and Bernard, A. (1994). Engagement of the CD45 molecule induces homotypic adhesion of human thymocytes through a LFA-1/ICAM-3-dependent pathway. *J. Immunol.*, **151**, 5162.

111. Spertini, F., Wang, A. V. T., Chatila, T., and Geha, R. S. (1994). Engagement of the common leukocyte antigen CD45 induces homotypic adhesion of activated human T cells. *J. Immunol.*, **153**, 1593.

112. Zapata, J. M., Campanero, M. R., Marazuela, M., Sanchez-Madrid, F., and de Landazuri, M. O. (1995). B-cell homotypic adhesion through exon-A restricted epitopes of CD45 involves LFA-1/ICAM-1, ICAM-3 interactions, and induces coclustering of CD45 and LFA-1. *Blood*, **86**, 1861.

113. Wagner, N., Engel, P., and Tedder, T. F. (1993). Regulation of the tyrosine kinase-dependent adhesion pathway in human lymphocytes through CD45. *J. Immunol.*, **150**, 4887.

114. Arroyo, A. G., Campanero, M. R., Sanchez-Mateos, P., Zapata, J. M., Angeles Ursa, M. A., del Pozo, M. A., *et al.* (1994). Induction of tyrosine phosphorylation during ICAM-3 and LFA-1-mediated intercellular adhesion, and its regulation by the CD45 tyrosine phosphatase. *J. Cell Biol.*, **126**, 1277.

115. Iida, N., Lokeshwar, V. B., and Bourguignon, L. Y. W. (1994). Mapping the fodrin binding domain in CD45, a leukocyte membrane-associated tyrosine phosphatase. *J. Biol. Chem.*, **269**, 28576.

116. Lorenz, H. M., Lagoo, A. S., Lagoo-Deenadalayan, S. A., Barber, W. H., Kalden, J. R., and Hardy, K. J. (1998). Epitope-specific signaling through CD45 on T lymphocytes leads to cAMP synthesis in monocytes after ICAM-1-dependent cellular interaction. *Eur. J. Immunol.*, **28**, 2300.

117. Roach, T., Slater, S., Koval, M., White, L., McFarland, E. C., Okumura, M., *et al.* (1997). CD45 regulates Src family member kinase activity associated with macrophage integrin-mediated adhesion. *Curr. Biol.*, **7**, 408.

118. Sgroi, D., Koretzky, G. A., and Stamenkovic, I. (1995). Regulation of CD45 engagement by the B-cell receptor CD22. *Proc. Natl. Acad. Sci. USA*, **92**, 4026.

119. Tedder, T. F., Steeber, D. A., Chen, A., and Engel, P. (1995). The selectins: vascular adhesion molecules. *FASEB J.*, **9**, 866.

120. Varki, A. (1997). Selectin ligands: will the real ones please stand up? *J. Clin. Invest.*, **99**, 158.

121. Rosen, S. D. and Bertozzi, C. R. (1994). The selectins and their ligands. *Curr. Opin. Cell Biol.*, **6**, 663.

122. Coombe, D. R., Watt, S. M., and Parish, C. R. (1994). Mac-1 (CD11b/CD18) and CD45 mediate the adhesion of hematopoietic progenitor cells to stromal cell elements via recognition of stromal heparan sulfate. *Blood*, **3**, 739.

123. Juneja, H. S., Schmalstieg, F. C., Lee, S., and Chen, J. (1998). CD45 partially mediates heterotypic adhesion between murine leukemia/lymphoma cell line L5178Y and marrow stromal cells. *Leuk. Res.*, **22**, 805.

124. Walzel, H., Schulz, U., Neels, P., and Brock, J. (1999). Galectin-1, a natural ligand for the receptor-type protein tyrosine phosphatase CD45. *Immunol. Lett.*, **67**, 193.

125. Herold, C., Elhabazi, A., Bismuth, G., Bensussan, A., and Boumsell, L. (1996). CD100 is associated with CD45 at the surface of human T lymphocytes. *J. Immunol.*, **157**, 5262.

126. Kolodkin, A. L. (1998). Semaphorin-mediated neuronal growth cone guidance. *Prog. Brain Res.*, **117**, 115.

127. Giger, R. J., Pasterkamp, R. J., Holtmaat, A. J., and Verhaagen, J. (1998). Semaphorin III: role in neuronal development and structural plasticity. *Prog. Brain Res.*, **117**, 133.

128. Yu, H. H. and Kolodkin, A. L. (1999). Semaphorin signaling: a little less per-plexin. *Neuron*, **22**, 11.

129. Delaire, S., Elhabazi, A., Bensussan, A., and Boumsell, L. (1998). CD100 is a leukocyte semaphorin. *Cell. Mol. Life Sci.*, **54**, 1265.

130. Petrone, A. and Sap, J. (2000). Emerging issues in receptor protein tyrosine phosphatase function: lifting fog or simply shifting? *J. Cell Sci.*, **113**, 2345.

131. Bilwes, A. M., den Hertog, J., Hunter, T., and Noel, J. P. (1996). Structural basis for inhibition of receptor protein-tyrosine phosphatase-α by dimerization. *Nature*, **382**, 555.

132. Jiang, G., den Hertog, J., and Hunter, T. (2000). Receptor-like protein tyrosine phosphatase alpha homodimerizes on the cell surface. *Mol. Cell. Biol.*, **20**, 5917.

133. Majeti, R., Bilwes, A. M., Noel, J. P., Hunter, T., and Weiss, A. (1998). Dimerization-induced inhibition of receptor protein tyrosine phosphatase function through an inhibitory wedge. *Science*, **279**, 88.

134. Majeti, R., Xu, Z., Parslow, T. G., Olson, J. L., Daikh, D. I., Killeen, N., *et al.* (2000). An inactivating point mutation in the inhibitory wedge of CD45 causes lymphoproliferation and autoimmunity. *Cell*, **103**, 1059.

135. Nam, H. J., Poy, F., Krueger, N. X., Saito, H., and Frederick, C. A. (1999). Crystal structure of the tandem phosphatase domains of RPTP LAR. *Cell*, **97**, 449.

136. Hoffmann, K. M. V., Tonks, N. K., and Barford, D. (1997). The crystal structure of domain 1 of receptor protein-tyrosine phosphatase μ. *J. Biol. Chem.*, **272**, 27505.

137. Feiken, E., van Etten, I., Gebbink, M. F., Moolenaar, W. H., and Zondag, G. C. (2000). Intramolecular interactions between the juxtamembrane domain and phosphatase domains of receptor protein-tyrosine phosphatase RPTPmu. Regulation of catalytic activity. *J. Biol. Chem.*, **275**, 15350.

138. Tracy, S., van der Geer, P., and Hunter, T. (1995). The receptor-like protein-tyrosine phosphatase, RPTP alpha, is phosphorylated by protein kinase C on two serines close to the inner face of the plasma membrane. *J. Biol. Chem.*, **270**, 10587.

139. den Hertog, J., Sap, J., Pals, C. E., Schlessinger, J., and Kruijer, W. (1995). Stimulation of receptor protein-tyrosine phosphatase alpha activity and phosphorylation by phorbol ester. *Cell Growth Differ.*, **6**, 303.

140. Lim, K. L., Kolatkar, P. R., Ng, K. P., Ng, C. H., and Pallen, C. J. (1998). Interconversion of the kinetic identities of the tandem catalytic domains of receptor-like protein-tyrosine phophatase PTPα by two point mutations is synergistic and substrate-dependent. *J. Biol. Chem.*, **273**, 28986.

141. Buist, A., Zhang, Y. L., Keng, Y. F., Wu, L., Zhang, Z. Y., and den Hertog, J. (1999). Restoration of potent protein-tyrosine phosphatase activity into the membrane-distal domain of receptor protein-tyrosine phosphatase α. *Biochemistry*, **38**, 914.

142. Liang, L., Lim, K. L., Seow, K. T., Ng, C. H., and Pallen, C. J. (2000). Calmodulin binds to and inhibits the activity of the membrane distal catalytic domain of receptor protein-tyrosine phosphatase alpha. *J. Biol. Chem.*, **275**, 30075.

143. Wallace, M. J., Fladd, C., Batt, J., and Rotin, D. (1998). The second catalytic domain of protein tyrosine phosphatase δ (PTPδ) binds to and inhibits the first catalytic domain of PTPσ. *Mol. Cell. Biol.*, **18**, 2608.

144. Brady-Kalnay, S. and Tonks, N. K. (1994). Receptor protein tyrosine phosphatases, cell adhesion and signal transduction. *Adv. Protein Phosphatases*, **8**, 241.

145. Streuli, M., Krueger, N., Ariniello, P., Tang, M., Munro, J., Blattler, W., *et al.* (1992). Expression of the receptor-linked protein tyrosine phosphatase LAR: proteolytic cleavage and shedding of the CAM-like extracellular region. *EMBO J.*, **11**, 897.

146. Yu, Q., Lenardo, T., and Weinberg, R. A. (1992). The N-terminal and C-terminal domains of a receptor tyrosine phosphatase are associated by non-covalent linkage. *Oncogene*, **7**, 1051.

147. Serra-Pages, C., Saito, H., and Streuli, M. (1994). Mutational analysis of proprotein processing, subunit association, and shedding of the LAR transmembrane protein tyrosine phosphatase. *J. Biol. Chem.*, **269**, 23632.

148. Aicher, B., Lerch, M. M., Müller, T., Schilling, J., and Ulrich, A. (1997). Cellular redistribution of protein tyrosine phosphatases LAR and PTPs by inducible proteolytic processing. *J. Cell Biol.*, **138**, 681.

149. Brady-Kalnay, S. and Tonks, N. K. (1993). Purification and characterization of the human protein tyrosine phosphatase, PTPμ, from a baculovirus expression system. *Mol. Cell. Biochem.*, **127/128**, 131.

150. Ostman, A., Yang, Q., and Tonks, N. K. (1994). Expression of DEP-1, a receptor-like protein-tyrosine-phosphates, is enhanced with increasing cell density. *Proc. Natl. Acad. Sci. USA*, **91**, 9680.

151. Schaller, M. D. and Parsons, J. T. (1993). Focal adhesion kinase: an integrin-linked protein tyrosine kinase. *Trends Cell Biol.*, **3**, 258.

152. Dunlevy, J. R. and Couchman, J. R. (1993). Controlled induction of focal adhesion disassembly and migration in primary fibroblasts. *J. Cell Sci.*, **105**, 489.

153. Volberg, T., Zick, Y., Dror, R., Sabanay, I., Gilon, C., Levitzki, A., *et al.* (1992). The effect of tyrosine-specific protein phosphorylation on the assembly of adherens-type junctions. *EMBO J.*, **11**, 1733.

154. Bennett, P., Dixon, R., and Kellie, S. (1993). The phosphotyrosine phosphatase inhibitor vanadyl hydroperoxide induces morphological alterations, cytoskeletal rearrangements and increased adhesiveness in rat neutrophil leucocytes. *J. Cell Sci.*, **106**, 891.

155. Serra-Pages, C., Kedersha, N. L., Fazikas, L., Medley, Q., Debant, A., and Streuli, M. (1995). The LAR transmembrane protein tyrosine phosphatase and a coiled-coil LAR-interacting protein colocalize at focal adhesions. *EMBO J.*, **14**, 2827.

156. Serra-Pages, C., Medley, Q. G., Tang, M., Hart, A., and Streuli, M. (1998). Liprins, a family of LAR transmembrane protein-tyrosine phosphatase-interacting proteins. *J. Biol. Chem.*, **273**, 15611.

157. Debant, A., Serra-Pages, C., Seipel, K., O'Brien, S., Tang, M., Park, S. H., *et al.* (1996). The multidomain protein Trio binds the LAR transmembrane tyrosine phosphatase, contains a protein kinase domain, and has separate rac-specific and rho-specific guanine nucleotide exchange factor domains. *Proc. Natl. Acad. Sci. USA*, **93**, 5466.

158. Lin, M. Z. and Greenberg, M. E. (2000). Orchestral maneuvers in the axon: trio and the control of axon guidance. *Cell*, **101**, 239.

159. Anderson, P. (1997). Kinase cascades regulating entry into apoptosis. *Micro. Mol. Biol. Rev.*, **61**, 33.

160. Weng, L. P., Yuan, J., and Yu, Q. (1998). Overexpression of the transmembrane tyrosine phosphatase LAR activates the caspase pathway and induces apoptosis. *Curr. Biol.*, **8**, 247.

161. Huang, T. S., Shu, C. H., Shih, Y. L., Huang, H. C., Su, Y. C., Chao, Y., *et al.* (1996). Protein tyrosine phosphatase activities are involved in apoptotic cancer cell death induced by GL331, a new homolog of etoposide. *Cancer Lett.*, **110**, 77.

162. Zheng, X., Wang, Y., and Pallen, C. (1992). Cell transformation and activation of pp60 *c-src* by overexpression of a protein tyrosine phosphatase. *Nature*, **359**, 336.

163. den Hertog, J., Pals, C. E. G. M., Peppelenbosch, M. P., Tertoolen, L. G. J., de Laat, S. W., and Kruijer, W. (1993). Receptor protein tyrosine phosphatase α activates pp60^{c-src} and is involved in neuronal differentiation. *EMBO J.*, **12**, 3789.

164. den Hertog, J., Blanchetot, C., Buist, A., Overvoorde, J., van der Sar, A., and Tertoolen, L. G. (1999). Receptor protein-tyrosine phosphatase signalling in development. *Int. J. Dev. Biol.*, **43**, 723.

165. Bhandari, V., Lim, K. L., and Pallen, C. J. (1998). Physical and functional interactions between receptor-like protein-tyrosine phosphatase α and p59^{fyn}. *J. Biol. Chem.*, **273**, 8691.

166. Harder, K. W., Moller, N. P. H., Peacock, J. W., and Jirik, F. R. (1998). Protein tyrosine phosphatase α regulates Src family kinases and alters cell-substratum adhesion. *J. Biol. Chem.*, **273**, 31890.

167. Helmke, S., Lohse, K., Mikule, K., Wood, M. R., and Pfenninger, K. H. (1998). SRC binding to the cytoskeleton, triggered by growth cone attachment to laminin, is protein tyrosine phosphatase-dependent. *J. Cell Sci.*, **111**, 2465.

168. Irie-Sasaki, J., Sasaki, T., Matsumoto, W., Opavsky, A., Cheng, M., Welstead, G., *et al.* (2001). CD45 is a JAK phosphatase and negatively regulates cytokine receptor signalling. *Nature*, **409**, 349.

169. Kashio, N., Matsumoto, W., Parker, S., and Rothstein, D. M. (1998). The second domain of the CD45 protein tyrosine phosphatase is critical for interleukin-2 secretion and substrate recruitment of TCR-zeta *in vivo*. *J. Biol. Chem.*, **273**, 33856.

170. Kishihara, K., Penninger, J., Wallace, V. A., Kundig, T. M., Kawai, K., Wakeham, A., *et al.* (1993). Normal B lymphocyte development but impaired T cell maturation in CD45-exon6 protein tyrosine phosphatase-deficient mice. *Cell*, **74**, 143.

171. Cyster, J. G., Healy, J. I., Kishihara, K., Mak, T. W., Thomas, M. L., and Goodnow, C. C. (1996). Regulation of B-lymphocyte negative and positive selection by tyrosine phosphatase CD45. *Nature*, **381**, 325.

172. Knudsen, K. and Wheelock, M. J. (1992). Plakoglobin, or an 83-kD homologue distinct from β catenin, interacts with E-cadherin and N-cadherin. *J. Cell Biol.*, **118**, 671.

173. Pipenhagen, P. A. and Nelson, W. J. (1993). Defining E-cadherin-associated protein complexes in epithelial cells: plakoglobin, β and γ catenin are distinct components. *J. Cell Sci.*, **104**, 751.

174. Kemler, R. (1993). From cadherins to catenins: cytoplasmic protein interactions and regulation of cell adhesion. *Trends Genet.*, **9**, 317.

175. Shibamoto, S., Hayakawa, M., Takeuchi, K., Hori, T., Oku, N., Miyazawa, K., *et al.* (1994). Tyrosine phosphorylation of β catenin and plakoglobin enhanced by hepatocyte growth factor and epidermal growth factor in human carcinoma cells. *Cell Adhes. Commun.*, **1**, 295.

176. Takeda, H., Nagafuchi, A., Yonemura, S., Tsukita, S., Behrens, J., Birchmeier, W., *et al.* (1995). V-src kinase shifts the cadherin-based cell adhesion from the strong to the weak state and β catenin is not required for the shift. *J. Cell Biol.*, **131**, 1839.

177. Takeichi, M. (1993). Cadherins in cancer: implications for invasion and metastasis. *Curr. Opin. Cell Biol.*, **5**, 806.

178. Kypta, R., Su, H., and Reichardt, L. (1996). Association between a transmembrane protein tyrosine phosphatase and the cadherin–catenin complex. *J. Cell Biol.*, **134**, 1519.

179. Müller, T., Choidas, A., and Ullrich, A. (1999). Phosphorylation and free pool of beta-catenin are regulated by tyrosine kinases and tyrosine phosphatases during epithelial cell migration. *J. Biol. Chem.*, **274**, 10173.

180. Meng, K., Rodriguez-Pena, A., Dimitrov, T., Chen, W., Yamin, M., Noda, M., *et al.* (2000). Pleiotrophin signals increased tyrosine phosphorylation of beta-catenin through inactivation of the intrinsic catalytic activity of the receptor-type protein tyrosine phosphatase beta/zeta. *Proc. Natl. Acad. Sci. USA*, **97**, 2603.

181. Balsamo, J., Leung, T. C., Ernst, H., Zanin, M. K. B., Hoffman, S., and Lilien, J. (1996). Regulated binding of a PTP1B-like phosphatase to N-cadherin: control of cadherin-mediated adhesion by dephosphorylation of β catenin. *J. Cell Biol.*, **134**, 801.

182. Balsamo, J., Arregui, C., Leung, T. C., and Lilien, J. (1998). The nonreceptor protein tyrosine phosphatase PTP1B binds to the cytoplasmic domain of N-cadherin and regulates the cadherin–actin linkage. *J. Cell Biol.*, **143**, 523.

183. Ozawa, M., Baribault, H., and Kemler, R. (1989). The cytoplasmic domain of the cell adhesion molecule uvomorulin associates with three independent proteins structurally related in different species. *EMBO J.*, **8**, 1711.

184. Provost, E. and Rimm, D. L. (1999). Controversies at the cytoplasmic face of the cadherin-based adhesion complex. *Curr. Opin. Cell Biol.*, **11**, 567.

185. Fuchs, M., Müller, T., Lerch, M. M., and Ulrich, A. (1996). Association of human protein-tyrosine phosphatase κ with members of the armadillo family. *J. Biol. Chem.*, **271**, 16712.

186. Zondag, G. C., Reynolds, A. B., and Moolenaar, W. H. (2000). Receptor protein-tyrosine phosphatase RPTPmu binds to and dephosphorylates the catenin p120(ctn). *J. Biol. Chem.*, **275**, 11264.

187. Hiscox, S. and Jiang, W. G. (1998). Association of PTPμ with catenins in cancer cells: a possible role for E-cadherin. *Int. J. Oncol.*, **13**, 1077.

188. Hiscox, S. and Jiang, W. G. (1999). Hepatocyte growth factor/scatter factor disrupts epithelial tumour cell–cell adhesion: involvement of beta-catenin. *Anticancer Res.*, **19**, 509.

189. Finnemann, S., Mitrik, I., Hess, M., Otto, G., and Wedlich, D. (1997). Uncoupling of XB/U-Cadherin–catenin complex formation from its function in cell–cell adhesion. *J. Biol. Chem.*, **272**, 11856.

190. Stappert, J. and Kemler, R. (1994). A short core region of E-cadherin is essential for catenin binding and is highly phosphorylated. *Cell Adhes. Commun.*, **2**, 319.

191. Jou, T. S., Stewart, D. B., Stappert, J., Nelson, W. J., and Marrs, J. A. (1995). Genetic and biochemical dissection of protein linkages in the cadherin–catenin complex. *Proc. Natl. Acad. Sci. USA*, **92**, 5067.

192. Barth, A. I. M., Nathke, I. S., and Nelson, W. J. (1997). Cadherins, catenins and APC protein: interplay between cytoskeletal complexes and signaling pathways. *Curr. Opin. Cell Biol.*, **9**, 683.

193. Daniel, J. M. and Reynolds, A. B. (1997). Tyrosine phosphorylation and cadherin/catenin function. *Bioessays*, **19**, 883.

194. Xu, Y., Guo, D. F., Davidson, M., Inagami, T., and Carpenter, G. (1997). Interaction of the adaptor protein shc and the adhesion molecule cadherin. *J. Biol. Chem.*, **272**, 13463.

195. Rimm, D. L. and Morrow, J. S. (1994). Molecular cloning of human E-cadherin suggests a novel subdivision of the cadherin superfamily. *Biochem. Biophys. Res. Commun.*, **200**, 1754.

196. Mourton, T., Hellberg, C., Burden-Gulley, S., Hinman, J., Rhee, A., and Brady-Kalnay, S. M. (2001). The PTPµ protein tyrosine phosphatase binds and recruits the scaffolding protein RACK1 to cell–cell contacts. *J. Biol. Chem.*, **276**, 14896.

197. Garcia-Higuera, I., Fenoglio, J., Li, Y., Lewis, C., Panchenko, M. P., Reiner, O., *et al.* (1996). Folding of proteins with WD-repeats: comparison of six members of the WD-repeat superfamily to the G protein beta subunit. *Biochemistry*, **35**, 13985.

198. Disatnik, M. H., Hernandez-Sotomayor, S. M. T., Jones, G., Carpenter, G., and Mochly-Rosen, D. (1994). Phospholipase C-γ1 binding to intracellular receptors for ativated protein kinase C. *Proc. Natl. Acad. Sci. USA*, **91**, 559.

199. Chang, B. Y., Conroy, K. B., Machleder, E. M., and Cartwright, C. A. (1998). RACK1, a receptor for activated C kinase and a homolog of the β subunit of G proteins, inhibits activity of Src tyrosine kinases and growth of NIH 3T3 cells. *Mol. Cell. Biol.*, **18**, 3245.

200. Yarwood, S. J., Steele, M. R., Scotland, G., Houslay, M. D., and Bolger, G. B. (1999). The RACK1 signaling scaffold protein selectively interacts with the cAMP-specific phosphodiesterase PDE4D5 isoform. *J. Biol. Chem.*, **274**, 14909.

201. Liliental, J. and Chang, D. D. (1998). Rack1, a receptor for activated protein kinase C, interacts with integrin beta subunit. *J. Biol. Chem.*, **273**, 2379.

202. Geijsen, N., Spaargaren, M., Raaijmakers, J. A., Lammers, J. W., Koenderman, L., and Coffer, P. J. (1999). Association of RACK1 and PKCbeta with the common beta-chain of the IL- 5/IL-3/GM-CSF receptor. *Oncogene*, **18**, 5126.

203. Rodriguez, M. M., Ron, D., Touhara, K., Chen, C. H., and Mochly-Rosen, D. (1999). RACK1, a protein kinase C anchoring protein, coordinates the binding of activated protein kinase C and select pleckstrin homology domains *in vitro*. *Biochemistry*, **38**, 13787.

204. Van Vactor, D. (1998). Protein tyrosine phosphatases in the developing nervous system. *Curr. Opin. Cell Biol.*, **10**, 174.

205. Stoker, A. and Dutta, R. (1998). Protein tyrosine phosphatases and neural development. *Bioessays*, **20**, 463.

206. Bixby, J. L. (2000). Receptor tyrosine phosphatases in axon growth and guidance. *Neuroreport*, **11**, R5.

207. Drosopoulos, N. E., Walsh, F. S., and Doherty, P. (1999). A soluble version of the receptor-like protein tyrosine phosphatase kappa stimulates neurite outgrowth via a Grb2/MEK1-dependent signaling cascade. *Mol. Cell Neurosci.*, **13**, 441.

208. Hirata, T., Takagi, S., and Fujisawa, H. (1993). The membrane protein A5, a putative neuronal recognition molecule, promotes neurite outgrowth. *Neurosci. Res.*, **17**, 159.

209. Desai, C. J., Sun, Q., and Zinn, K. (1997). Tyrosine phosphorylation and axon guidance: of mice and flies. *Curr. Opin. Neuro.*, **7**, 70.

210. Desai, C., Gindhart, J. G., Goldstein, L. S. B., and Zinn, K. (1996). Receptor tyrosine phosphatases are required for motor axon guidance in the *Drosophila* embryo. *Cell*, **84**, 599.

211. Krueger, N. X., Van Vactor, D., Wan, H. I., Gelbart, W. M., Goodman, C. S., and Saito, H. (1996). The transmembrane tyrosine phosphatase DLAR controls motor axon guidance in *Drosophila*. *Cell*, **84**, 611.

212. Garrity, P. A., Lee, C. H., Salecker, I., Robertson, H. C., Desai, C. J., Zinn, K., *et al.* (1999). Retinal axon target selection in *Drosophila* is regulated by a receptor protein tyrosine phosphatase. *Neuron*, **22**, 707.

213. Wills, Z., Bateman, J., Korey, C. A., Comer, A., and Van Vactor, D. (1999). The tyrosine kinase Abl and its substrate enabled collaborate with the receptor phosphatase Dlar to control motor axon guidance. *Neuron*, **22**, 301.

214. Tian, S. and Zinn, K. (1994). An adhesion molecule-like protein that interacts with an is a substrate for a *Drosophila* receptor-linked protein tyrosine phosphatase. *J. Biol. Chem.*, **269**, 28478.
215. Gershon, T. R., Baker, M. W., Nitabach, M., and Macagno, E. R. (1998). The leech receptor protein tyrosine phosphatase HmLAR2 is concentrated in growth cones and is involved in process outgrowth. *Development*, **125**, 1183.
216. Baker, M. W. and Macagno, E. R. (2000). The role of a LAR-like receptor tyrosine phosphatase in growth cone collapse and mutual-avoidance by sibling processes. *J. Neurobiol.*, **44**, 194.
217. Sun, Q. L., Wang, J., Bookman, R. J., and Bixby, J. L. (2000). Growth cone steering by receptor tyrosine phosphatase delta defines a distinct class of guidance cue. *Mol. Cell Neurosci.*, **16**, 686.
218. Ledig, M. M., Haj, F., Bixby, J. L., Stoker, A. W., and Mueller, B. K. (1999). The receptor tyrosine phosphatase CRYPalpha promotes intraretinal axon growth. *J. Cell Biol.*, **147**, 375.
219. Haj, F., McKinnell, I., and Stoker, A. (1999). Retinotectal ligands for the receptor tyrosine phosphatase CRYPalpha. *Mol. Cell Neurosci.*, **14**, 225.
220. Mueller, B. K., Ledig, M. M., and Wahl, S. (2000). The receptor tyrosine phosphatase CRYPalpha affects growth cone morphology. *J. Neurobiol.*, **44**, 204.
221. Maeda, N. and Noda, M. (1998). Involvement of receptor-like protein tyrosine phosphatase ζ/RPTPβ and its ligand pleiotrophin/heparin-binding growth-associated molecule (HB-GAM) in neuronal migration. *J. Cell Biol.*, **142**, 203.
222. Zinn, K. (1998). Receptor tyrosine phosphatases: the worm clears the picture. *Curr. Biol.*, **8**, R725.
223. Kokel, M., Borland, C. Z., DeLong, L., Horvitz, H. R., and Stern, M. J. (1998). *clr-1* encodes a receptor tyrosine phosphatase that negatively regulates an FGF receptor signaling pathway in *C. elegans*. *Genes Dev.*, **12**, 1425.
224. Norman, S. A., Golfinos, J. G., and Scheck, A. C. (1998). Expression of a receptor protein tyrosine phosphatase in human glial tumors. *J. Neuro-Oncol.*, **36**, 209.
225. Yamakawa, T., Kurosawa, N., Kadomatsu, K., Matsui, T., Itoh, K., Maeda, N., *et al.* (1999). Levels of expression of pleiotrophin and protein tyrosine phosphatase ζ are decrease in human colorectal cancers. *Cancer Lett.*, **135**, 91.
226. Urushibara, N., Karasaki, H., Nakamura, K., Mizuno, Y., Ogawa, K., and Kikuchi, K. (1998). The selective reduction in PTPdelta expression in hepatomas. *Int. J. Oncol.*, **12**, 603.
227. Zhai, Y., Wirth, J., Kang, S., Welsch, C. W., and Esselman, W. J. (1995). LAR-PTPase cDNA transfection suppression of tumor growth of neu oncogene-transformed human breast carcinoma cells. *Mol. Carcinog.*, **14**, 103.
228. Notkins, A. L., Zhang, B., Matsumoto, Y., and Lan, M. S. (1997). Comparison of IA-2 with IA-2β and with six other members of the protein tyrosine phosphatase family: recognition of antigenic determinants by IDDM sera. *J. Autoimmun.*, **10**, 245.
229. Wishart, M. J. and Dixon, J. E. (1998). Gathering STYX: phosphatase-like form predicts functions for unique protein-interaction domains. *Trends Biochem. Sci.*, 8301.
230. Brückner, K. and Klein, R. (1998). Signaling by Eph receptors and their ephrin ligands. *Curr. Opin. Neurobiol.*, **8**, 375.

8 | The adherens junction

CARA J. GOTTARDI,* CARIEN M. NIESSEN,* and
BARRY M. GUMBINER

*Contributed equally to this chapter

1. Introduction

The adherens junction (AJ) is a specialized type of membrane cytoskeletal-based intercellular contact required for adhesion, tissue organization, and cell–cell communication. Most often discussed in the context of polarized epithelial cells (where a form of it is known as the zonula adherens), this junction has been characterized in numerous cell and tissue types where it serves distinct functions (Fig. 1). In Schwann cells, adherens junctions anchor membranes at the paranodal loops, contributing to the successive wrapping of myelin around an axon (1), while at neuronal synapses, these junctions are thought to contribute to the specificity of neural connectivity (2, 3). By coupling cell–cell adhesion to the cytoskeleton, adherens junctions can create a transcellular network that may be fundamental for co-ordinating the behaviour of a population of cells. For example, in polarized epithelial cells, a 'purse-string'-like constriction of the zonula adherens in a single cell could drive the morphogenetic movements of adjoining cells. In heart tissue, adherens junctions are part of the intercalated disks (where they are known as fascia adherens) which connect individual cardiac myocytes so that these cells can contract synchronously (4). The variety and near ubiquity of adherens junctions underscores their importance to the organization of cells into tissues. In this chapter, we cover the most fundamental aspects of adherens junction structure and function.

Adherens junctions were originally defined by morphological criteria distinguished at the electron microscopic level. More recently, our understanding of adherens junctions has been advanced by the identification and characterization of its molecular components. We will primarily focus on how the ultrastructural and molecular organization of adherens junctions contribute to adhesive function. How cell–cell adhesion might influence cell behaviour and tissue morphogenesis is less well understood. Towards this end, we will discuss the role that a particular junction-associated component, β-catenin, plays in a signal transduction pathway required for tissue patterning. This might provide a framework for understanding the relationship between adherens junctions and the signalling of its components.

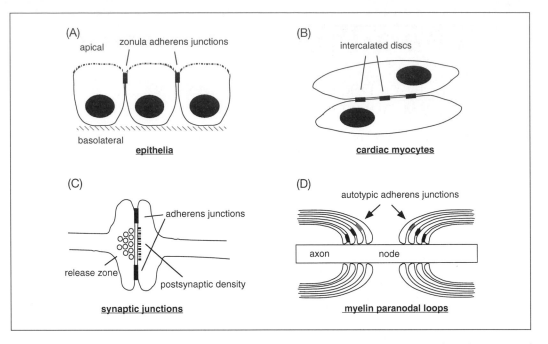

Fig. 1 Types of adherens junctions. Several types of adherens junctions as they appear in various types of tissues. (A) Zonula adherens junctions of epithelial cells. This junction forms a circumferential belt around the cell near the boundary between the apical and the basolateral cell surfaces. (B) Intercalated disks of cardiac myocytes. These sites consist of fascia adherens and desmosomes, thus providing strong intercellular attachment and anchorage of both intermediate filaments and contractile filaments to the plasma membrane. (C) Synaptic junctions in the central nervous system. These adherens junctions flank the 'active zone' where neurotransmitter is released and hold the synapse together. (D) Autotypic adherens junction of the Schwann cell paranodal loops. Adherens junctions form between different membrane regions of the same Schwann cell, lining up between the successive layers of the myelin sheath.

2. Ultrastructure of the adherens junction

By transmission electron microscopic analysis (TEM), the adherens junction is a region of close cell–cell contact where the membranes appear perfectly parallel over an intercellular space of $\sim$ 200 Å and extending anywhere between 0.2–0.5 μm in length (5). The intercellular space is occupied by a weakly electron dense material that often appears amorphous, but in some preparations looks filamentous (6, 7). The cytoplasmic side of the adherens contact is characterized by a plate-like densification or 'plaque' into which numerous microfilaments (4–7 nm in diameter) feed.

Although adherens junctions are observed in various cell contexts, the most well known example is the zonular adherens junction, or *zonula adherens* of polarized epithelial cells (Fig. 2). The zonula adherens was described by Farquhar and Palade (5) as the intermediate component of a tripartite junctional complex commonly found between the apices of adjacent cells from most epithelia. The two other components of this complex, the more lumenally localized zonula occludens (or tight

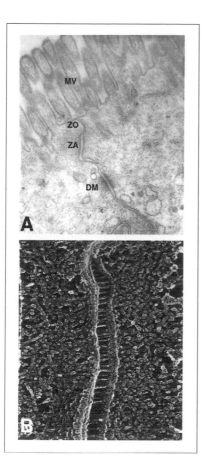

Fig. 2 Adherens junction ultrastructure. (A) Transmission electron microscopic analysis of the zonula adherens (ZA) described by Farquhar and Palade (1963) as part of a tripartite junctional complex. In this complex, the ZA is located between the zonula occludens (ZO) and the desmosome (DM) just beneath the microvillar (MV) domain. (B) Quick-freeze, deep-etch image of the adherens junction between intestinal epithelial cells. Note the presence of 'rod-like' bridging structures extending between adjacent cells at the attachment zone. Micrographs were kindly provided by Dr M. G. Farquhar (A) and Dr N. Hirokawa (B) and reprinted with permission from *J. Cell Biol.* (refs 5 and 39).

junction) and the basally localized desmosome are compositionally and functionally distinct and will be discussed elsewhere in this volume (see Chapters 11 and 10, respectively). At the zonula adherens, actin microfilaments are particularly prominent, and appear to be continuous with the bundle of actin filaments that hugs the cytoplasmic surface of the region of cell–cell contact and encircles the cell. While the zonula adherens is continuous or belt-like in most epithelia, outside of the tripartite junctional complex, adherens junctions are often discontinuous, or 'spot-like'. These latter types of adherens contacts can be observed scattered along the entire lateral surface of epithelial cells (6) or in non-epithelial cell types such as the synaptic junctions of neurons (2, 3). From the standpoint of mediating intercellular adhesion, we have no reason to believe that there are fundamental differences between a 'zonular' versus 'spot-like' adherens junction, and have therefore not distinguished between them in this chapter. However, in line with the notion that subtle changes in structure are often functionally driven, it is intriguing to consider that these ultrastructurally distinct adherens junctions might reflect different states of mechanical strength or different states of cytoskeletal organization.

3. Description of junctional components

Electron microscopic examination of the adherens junction predicted that three fundamental classes of proteins would be required to organize such a structure:

(a) Adhesive components spanning the intercellular space of the junction and linking adjacent cells.

(b) A cytoskeletal network anchoring the adhesive components.

(c) Plaque-associated proteins linking the adhesive component to the cytoskeleton and/or mediating junctional regulation.

This prediction has held up with the identification of a number of junction-localizing proteins ranging from structural components, such as cadherin-type adhesion molecules and their associated catenins and actin-organizing proteins, to potential regulatory components of the junctions, such as growth factor receptors and kinases. In this section, we will primarily describe the components that appear to contribute fundamentally to adherens junction structure and function. The localization of putative signalling-type molecules to adherens junctions will be briefly discussed.

3.1 Cadherins and catenins as a 'core complex' for adhesion

Classical cadherins were the first family of adhesion molecules found in the adherens junction. These molecules are type I, single-pass transmembrane glycoproteins that mediate Ca^{2+}-dependent adhesion (for an extensive discussion of the cadherin superfamily, see Chapter 3). Specific adhesive binding is conferred by the ecto-domain, which for the classical cadherins contains five 110 amino acid repeats. The cadherin ectodomain projects into the intercellular cleft of the adherens junction (8) where it is thought to engage an identical molecule on the adjacent cell's surface. The cytoplasmic domain likely connects with the cytoskeleton through its association with proteins known as catenins. The catenins were originally identified as the three major constituents associated with the cadherin by immunoprecipitation analysis (9). Importantly, transfection of a cadherin into a cadherin/catenin negative fibroblastic cell line resulted in the dramatic up-regulation of the catenins, their association with the cadherin, and acquisition of cell–cell adhesion (10, 11). Thus, together these molecules are thought to constitute a multiprotein complex for adhesion (Fig. 3).

β-Catenin interacts directly with the cadherin, and belongs to the armadillo family of proteins. Members of this family contain a central region consisting of a repeating amino acid motif, termed armadillo repeats. These repeats were originally identified in the *Drosophila* segment polarity gene product, Armadillo (12), and have since been found in other cadherin-associated proteins, such as plakoglobin (also known as γ-catenin) and p120[ctn], as well as in proteins with seemingly different cellular functions (13). X-ray crystallographic analysis of the central repeat region of β-catenin revealed a superhelix of stacked alpha helices that forms a long positively charged groove (14). It is in this groove where the cadherin cytoplasmic tail has been proposed to

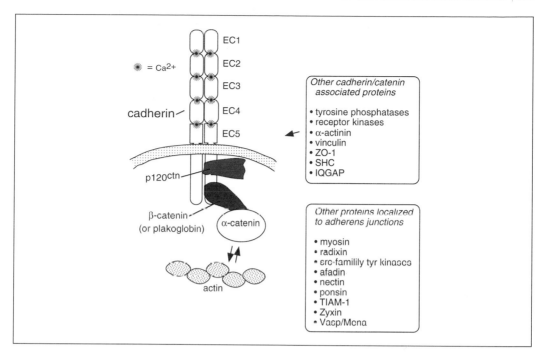

Fig. 3 The cadherin–catenin complex and associated molecules. Classical cadherins are transmembrane molecules that consist of five extracellular repeats (EC1–5). Ca^{2+} likely bridges each of these repeats, contributing to the rigid elongated shape of the molecule. The C-terminus of the cytoplasmic tail binds tightly to β-catenin, which in turn binds α-catenin, forming a stable complex. α-Catenin probably mediates the interaction with the cytoskeleton. The juxtamembrane region binds to p120[ctn]. Additional proteins have been found to be associated with the cadherin core complex, such as tyrosine phosphatases, and receptor kinases. Other proteins localize to adherens junctions, but their relationship to the core complex is unknown. Some of these proteins, such as nectin, and L-afadin, are found to be more specific for the zonula adherens than the cadherin complex.

interact with β-catenin. Like β-catenin, plakoglobin has been shown to interact with the carboxy terminus of classical cadherins, but interestingly, these two proteins form mutually exclusive cadherin complexes (15). While plakoglobin appears to substitute for β-catenin's adhesive role in some tissues of mice lacking β-catenin, the functional implications for two distinct cadherin–catenin complexes are presently unclear (16). The association of plakoglobin with classical cadherins seems to be of greatest significance in endothelial cells where plakoglobin is a major component of adherens junctions. In most other cell types, plakoglobin is predominantly localized to desmosomes where it is associated with non-classical cadherins (see Chapter 10).

α-Catenin is associated with the complex through its direct binding with the amino terminus of β-catenin (or plakoglobin). α-Catenin is structurally unrelated to β-catenin and most likely provides the cadherin complex with a link to the actin cytoskeleton, since α-catenin has been shown to directly bind and bundle F-actin *in vitro* (17) as well as interact with the actin-associated proteins, vinculin, α-actinin, and ZO-1 (18–20).

More recently, another catenin has been shown to interact with the cytoplasmic domain of cadherins, p120[ctn] (21). In contrast with β-catenin and plakoglobin, p120[ctn] binds to a region of the cytoplasmic domain that is more proximal to the transmembrane region of the cadherin and does not interact with α-catenin (22, 23). Moreover, unlike α/β-catenin, the levels of p120[ctn] are not dependent on cadherin expression. While the stoichiometry of an α-catenin: β-catenin: cadherin complex has been estimated at a ~ 1:1:1 ratio, the recovery of p120[ctn] with the complex is lower, suggesting either that not every cadherin is occupied by p120[ctn], or that the affinity of its interaction with the cadherin is low.

p120[ctn] belongs to a subfamily of armadillo proteins and is characterized by the existence of multiple, alternatively spliced mRNA isoforms (24). These isoforms show no obvious differences in cadherin binding. Conspicuously, it is the N- and C-termini of p120[ctn] that are subject to extensive alternative splicing suggesting that these domains may be important for modulating p120[ctn] and/or cadherin function (see Section 4). While it remains to be determined whether or not p120[ctn] is a structural or regulatory component of the cadherin complex, it is consistently found to be associated with the cadherin. We consider, therefore, that a cadherin trans-membrane glycoprotein associated with β-catenin (or less often, plakoglobin), α-catenin, and p120[ctn] constitute a 'core complex' for adhesion (Fig. 3).

3.2 Actin-associated proteins

It is clear from the electron micrograph in Fig. 2 that actin and a dense, submem-branous material feature prominently at the adherens junction. Despite its visual prominence, the precise molecular composition and arrangement of this material in relation to the site of cell contact is poorly understood. A number of actin-associated proteins have been localized to this region by both immunofluorescence and biochemical techniques. Examples include vinculin, α-actinin, ZO-1 (at the fascia adherens in heart), and members of the ezrin–radixin–moesin family of membrane–cytoskeletal linker proteins. We have chosen to highlight those proteins which may link α-catenin to the actin cytoskeleton.

Vinculin is an actin binding protein localized to both the adherens junction and focal contacts (see Chapter 9). α-Actinin is also found in both adhesion structures and has actin-crosslinking activity. Interestingly, both vinculin and α-actinin share an overlapping binding site on α-catenin (18, 19), although it is not yet clear if these proteins bind in a mutually exclusive fashion. ZO-1 has also been shown to interact *in vitro* with both α-catenin and actin. ZO-1 was originally identified as a peripheral component of tight junctions in polarized epithelial cells, and was also found to localize to the intercalated disk of cardiac myocytes (25). ZO-1 belongs to a family of MAGUK proteins (membrane-associated guanylate kinase homologues) which share a conserved modular organization of PDZ-type protein–protein interaction motifs and a region of guanylate kinase homology (see Chapter 11). Therefore, in light of the aforementioned *in vitro* binding interactions, there may be distinct ways to link α-catenin (as a part of the cadherin core complex) to the actin cytoskeleton. The consequences of most of these molecular interactions for adherens junction structure and function remain to be elucidated.

3.3 Signalling molecules

The numerous cell–cell rearrangements which seem to accompany the development of a tissue have long suggested that adhesion must be regulated (see Section 4). In addition, cell contact in and of itself has emerged as an important way to regulate cellular behaviour, for example, by bringing juxtacrine ligand–receptor signalling molecules in direct apposition (reviewed in ref. 26). In the context of such bidirectional junctional regulation (i.e. inside-out and outside-in), it is perhaps not surprising that a number of signalling molecules have been localized to adherens junctions: these include growth factor receptor tyrosine kinases such as EGFR (27), Src family kinases (28), tyrosine phosphatases (see Chapter 7); adaptor molecules such as Shc (29); a nucleotide exchange factor for Rac, Tiam-1 (30), and an effector of both CDC42 and Rac, IQGAP1 (31). Both the EGFR and several protein tyrosine phosphatases (PTPs) have been shown to interact directly with the cadherin core complex via either β-catenin or the cadherin cytoplasmic domain. Evidence exists that a number of these proteins can affect junction formation, adhesion, and perhaps contact-dependent signalling.

4. Role of junctional components in cell adhesion

With an ultrastructural framework and the basic molecular components of adherens junctions at hand, experimental dissection of the cadherin 'core complex' has told us much about how this complex mediates adhesion.

4.1 The extracellular domain of cadherins and adhesion

Cadherins are homophilic adhesion molecules, meaning that one cadherin recognizes an identical cadherin on an adjacent cell. There is evidence from a number of studies that the extracellular segment of cadherins forms lateral dimers on the same cell and that this dimer is essential for its adhesive function. Crystal structures of both the first cadherin domain of N-cadherin (EC1) and the EC1–2 domain of E-cadherin revealed the presence of lateral dimers (32, 33). Biochemical evidence for dimers was also obtained after chemical crosslinking and gel filtration using the entire extracellular domain (EC1–5) of *Xenopus* C-cadherin (34). More recently, *in vivo* evidence for cadherin multimers has been obtained (35, 36). The lateral dimer is likely to be the basic unit necessary for adhesion because dimerization of the extracellular domain was required for effective bead aggregation (34). In addition, a pentameric E-cadherin extracellular fusion molecule, which was observed to form dimers at its tips by electron microscopy, could mediate cell attachment when coated as a substrate (37).

How does dimerization contribute to adhesion? Dimerization may be necessary to form a homophilic binding site which would allow the formation of a higher order structure. In the EC1 crystal structure of N-cadherin, dimer subunits packed in an antiparallel orientation to form an extended ribbon (32). This observation lead to the 'linear zipper' model for cell–cell adhesion, which proposes that lateral dimers from adjacent cells may interdigitate, thus 'zippering-up' the two membranes. Alternatively, dimerization may be required to form discrete higher order structures. In the deep-etch electron microscopic image shown in Fig. 2B, the intercellular space at the adherens junction is connected by numerous cylinder-like projections with no apparent linear interdigitations (38), consistent with the latter alternative. Clearly, we will need to learn more about cadherin structure in order to understand how dimers contribute to a higher order junctional structure.

4.2 Function of cadherin cytoplasmic domain and its interaction with catenins

As mentioned above, the extracellular domain of cadherins contains the minimal information necessary for homotypic recognition and binding. While this binding activity was experimentally quantifiable, mutant cadherins encoding only the extracellular and transmembrane domain are not sufficient to support wild-type adhesion, in either assays of cell–cell aggregation or assays of cell attachment to immobilized recombinant cadherin ectodomain (10, 11, 34). Therefore, the cytoplasmic domain is somehow required to convert intrinsic binding activity into robust cell–cell adhesion. Mutational analysis mapped this activity to the catenin binding domain, strongly implicating the catenins in adhesion (10, 11). Genetic loss-of-function studies from both cancer cells and *Drosophila* have shown that both α-catenin and β-catenin are indeed essential for strong cell–cell adhesion (39, 40).

Several experiments have demonstrated that the catenin binding domain may not act alone in contributing to adhesion. Overexpression of the juxtamembrane domain of N-cadherin, which does not bind α- and β-catenin, also disrupted adhesion presumably due to dominant inhibitory activity (41). Moreover, motility was perturbed when a similar construct was overexpressed in both neurons and cell lines (42, 43). Furthermore, a cadherin mutant encoding only the juxtamembrane region of the cytoplasmic domain showed adhesive activity (23, 44). Cells expressing this mutant exhibited ligand-dependent clustering, meaning that the mutant cadherin accumulated in cell surface clusters when plated on a cadherin ectodomain substrate (23). Since artificially-induced clustering of cadherins has been shown to strengthen adhesion (45), the juxtamembrane region may increase adhesion by driving the clustering of cadherin molecules. p120^{ctn} appears to be the most prominent protein associated with this juxtamembrane region, and therefore is considered a likely candidate for mediating clustering (23). Indeed, for cells expressing an E-cadherin mutant in which p120 binding was abolished by site-directed mutagenesis, the strengthening of adhesion appeared to be diminished (46). However, a number of recent studies have provided a different view of how the juxtamembrane region and/or p120^{ctn} contribute to cadherin-based adhesion. In some mutagenesis studies, deletion of the juxtamembrane/p120^{ctn} binding domain of the cadherin actually increased adhesion, suggesting that this domain might inhibit adhesion (47, 48). Moreover, overexpression of an N-terminally deleted p120^{ctn} molecule was able to restore adhesion in an otherwise non-adherent colon carcinoma cell line (48). Since this cell line expresses all the core components of the cadherin complex, the interpretation is that p120^{ctn} plays a role in the negative regulation of adhesion.

Presently, it is difficult to encompass all of these findings within one simple model. However, it is clear that both the juxtamembrane region and the catenin binding domain can contribute independently to adhesion. How these two domains collaborate with the extracellular domain of the cadherin to mediate physiological adhesion remains to be elucidated.

4.3   Role for actin binding proteins

The requirement for polymerized actin in cadherin-mediated adhesion has been inferred from the observation that the actin filament disrupting drug, cytochalasin D, interfered with cadherin function. However, it is difficult to glean molecular meaning from this observation since an intact cytoskeleton is likely critical for general plasma membrane support. As discussed in Section 3 of this chapter, a number of actin and α-catenin binding proteins have been localized to the adherens junction, and some studies suggest they may regulate adhesion and junction formation. The interaction of vinculin or α-actinin with α-catenin does not seem to be essential for cadherin-mediated adhesion in non-epithelial cells, since deletion of the vinculin/α-actinin interacting domain of α-catenin did not abrogate aggregation. Although this mutant moderately affected cell–cell adhesion in epithelial cells (20), vinculin may be more important for the positioning of junctions (see Section 5). The actin binding MAGUK protein, ZO-1, has been implicated in cadherin function, since ZO-1 has been shown to interact with α-catenin directly *in vitro*, and overexpression of an α-catenin mutant lacking the ZO-1 binding domain in L cell fibroblasts reduced the strengthening of adhesion (20). However relevant this interaction may be in L cells, it is important to point out that ZO-1 is not likely to have a general role in cadherin function. In most epithelial cell types, ZO-1 accumulates at the zonula occludens, or tight junction, and little if any of the protein is localized to adherens junctions (49). In this regard, deletion of the ZO-1 binding domain of α-catenin had little effect on adhesion in epithelial cells (20).

4.4   Regulatory mechanisms in adhesion

Cadherin-mediated cell–cell adhesion is not a static condition that is either on or off, but rather is dynamic, manifesting varying states of adhesiveness in response to internal or external cues. One classic example of adhesive regulation is the process of compaction in the early pre-implantation stage of the mouse embryo. During this process, blastomeres flatten against each other and this 'compaction' of cells correlates with increased E-cadherin adhesive activity (50). Cadherin-based adhesive regulation is also required for morphogenetic movements in developing tissues such as convergence–extension intercellular rearrangements in *Xenopus*. When animal cap tissue explants are treated with the TGF-β-like growth factor, activin, the tissue will elongate as a result of cells converging towards the midline resulting in the extension of the tissue. This intercellular rearrangement correlated with a decrease in C-cadherin adhesive activity (51). Importantly, incubation of the activin-treated tissue explants with an antibody that activates C-cadherin-mediated adhesion inhibited the elongation of this tissue, showing that the decrease in C-cadherin activity is actually required to permit intercellular movements (52). Other factors shown to negatively regulate cadherin function are hepatocyte growth factor (HGF), EGF, and v-src (53–55). Neurite outgrowth of retinal ganglion cells on N-cadherin could be strongly inhibited by overexpressing either antisense RPTPμ (receptor protein tyrosine phosphatase), or a phosphatase dead RPTPμ, suggesting that tyrosine phosphatases might regulate cell–cell adhesion (56) (also see Chapter 7).

While there are a number of examples where cadherin function is regulated, the molecular mechanisms underlying regulation are not well understood. In many cases where cadherin function is altered (e.g. during compaction, activin-induced convergence–extension movements, Src overexpression, or EGF stimulation), no obvious change in cadherin surface expression or association with catenins has been detected (50, 54). In some studies, however, differences in the association of either β-catenin, α-catenin, or p120[ctn] have been correlated with changes in cell behaviour (e.g. during gastrulation movements in sea urchin embryos) (57). Taken together, these findings imply that the molecular mechanisms for regulating adhesion do not necessarily require stoichiometric changes in the core complex. Perhaps post-translational modification, or the manner in which the core complex may be organized into higher order cytoskeletal structures may underlie adhesive changes.

4.4.1 Phosphorylation and adhesion

As mentioned previously, several growth factor receptors, kinases, and tyrosine phosphatases have been localized to the adherens junction and shown to modulate cadherin-mediated adhesion. The effect on adhesiveness has often correlated with phosphorylation of core complex components, in particular β-catenin and p120[ctn] (reviewed in ref. 58). Such observations have long suggested that adhesive function might be regulated by phosphorylation. To date, however, no study has established whether or not the observed phosphorylations constitute key steps in the alteration of cadherin function, or are merely non-causal epiphenomena. For example, cells transfected with v-src were shown to exhibit a decrease in cell–cell adhesion in association with tyrosine phosphorylation of β-catenin suggesting that this modification negatively regulated adhesion (53). However, a chimeric molecule in which the extracellular and juxtamembrane regions of the cadherin were fused directly with α-catenin was still susceptible to v-src regulation, even though it entirely bypassed the requirement for β-catenin (59). Thus, identifying the functionally relevant substrates of kinases that modulate cell–cell adhesion will be essential to our understanding of cadherin regulation.

4.4.2 Small GTPases and cell–cell adhesion

The Rho subfamily of small GTPases has been implicated in several processes that involve actin assembly and disassembly. Because of actin's proximity to the adherens junction, it came as no surprise that members of this family, such as CDC42, Rac, and Rho, might influence cadherin-mediated cell–cell adhesion. Overexpression of constitutively active Rac resulted in more intense staining of E-cadherin, β-catenin, and actin at the lateral membranes of epithelial cells, while dominant-negative Rac had the opposite effect (60, 61). Similarly, overexpression of a dominant-negative Rac blocked actin accumulation at the adherens junction in *Drosophila* cells (62). Tiam-1, a nucleotide-exchange factor for Rac, has been localized to the adherens junctions of MDCK cells. Overexpression of Tiam-1 or activated Rac increased E-cadherin-mediated adhesion, as measured by cell aggregation assays (30). Rho and CDC42 have also been found to influence cell–cell adhesion, but their roles have been less consistent.

The underlying mechanism of how the Rho subfamily of proteins regulates cell–cell adhesion is not yet clear. Their effects on actin polymerization could certainly regulate adhesion indirectly. However, one study provided evidence that CDC42 may directly affect the cadherin complex. IQGAP, an effector of both CDC42 and Rac, can bind to β-catenin *in vitro* and compete for its binding to α-catenin. Overexpression of IQGAP rendered the cells non-adhesive and resulted in less α-catenin associated with the cadherin complex. Co-expression with dominant active CDC42 was found to rescue adhesion (31). This suggests that the small GTPases can regulate adherens junction formation at the level of α-catenin's association with the complex.

5. Assembly of junctions

From embryonic development throughout adulthood, there are numerous occasions where cells form or leave a tissue. Examples include neural tube formation, gastrulation, wound healing, and transmigration of leukocytes across an endothelium. This implies that adherens junctions are not fixed entities, but rather must be disrupted and re-established during the course of such cell movements. Therefore, elucidating the mechanisms underlying junctional assembly and disassembly will be important for understanding many developmental and physiological processes. Observations from a number of studies suggest that cadherin-mediated junction assembly can be viewed as a multistep process.

5.1 Cadherin clustering upon ligand binding

Since a cadherin dimer has been shown to be required for adhesion, the initial step for adhesion, and concomitant junction formation, likely involves the recognition of apposed lateral dimers. After this initial adhesion step, more cadherins might be recruited into the adhesive complex, serving as a nucleation site for adherens

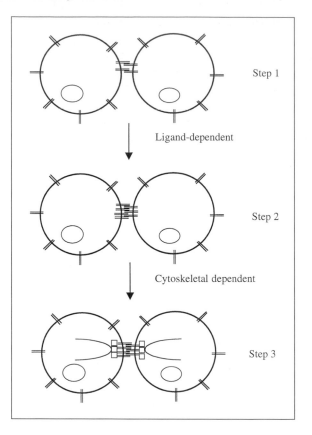

Fig. 4 Model for stepwise assembly of adherens junctions. Lateral cadherin dimers on one cell recognize a similar cadherin dimer on another cell (step 1). This results in ligand-dependent clustering in the plane of the membrane, which is probably mediated by the juxtamembrane region (step 2). Subsequently, actin (represented by a half circle) and actin binding proteins (represented by the squares) associate with the cadherin complex, leading to increased clustering and thus mature adherens junctions (step 3).

junction formation. Evidence for such a multistep assembly process is supported by the observation that C-cadherin is induced to form clusters at the cell surface only when plated on the C-cadherin extracellular domain. The juxtamembrane region of the cadherin was sufficient for initial ligand-dependent clustering, whereas the catenin binding domain did not have this activity (23). Since the catenin binding domain recruits actin and associated proteins to the complex via the catenins, we suggest that junctional assembly might entail a sequence of steps: ectodomain engagement, juxtamembrane-mediated clustering, and α-, β-catenin-mediated tethering to the cytoskeleton (Fig. 4).

5.2   Macro-assembly of adherens junctions

Observations from both mammalian cell culture and *Drosophila* cellularization studies suggest a stepwise junctional assembly process that starts with discrete regions of cadherin-based cell contact (sometimes referred to as spot-like adherens junctions) where local actin polymerization can occur and which ultimately coalesce into a more continuous, belt-like structure (7, 63, 64). The cadherin, α- and β-catenin all appear to localize at the cell contact concurrently, suggesting that these proteins arrive at the membrane as a complex (65). Curiously, the manner in which actin is localized to the adherens junction appears to change over the course of adherens junction maturation. More specifically, actin filaments approach initial cell contacts 'end-to-membrane' (perpendicularly) while mature contacts are characterized by a 'side-to-membrane' (parallel) relationship (7, 66). Whether these two distinct ways of organizing actin at adherens junctions are functionally important is not yet clear.

In polarized epithelial cells, cadherin-mediated adhesion is not only critical for the formation of adherens junctions, but is also required for the establishment and maintenance of other types of lateral junctions, such as tight junctions, gap junctions, and desmosomes. For example, when E-cadherin function was perturbed by the addition of blocking antibodies, the cells were not able to form any of the three types of junctions (67). Similarly, in an epithelial-derived cell line that is devoid of α-catenin, tight junctions and desmosomes were restored only after cadherin function was rescued by α-catenin transfection (68). Recently, an α-catenin–vinculin linkage has been implicated in both formation of an epithelial zonular adherens junction and the co-ordinate formation of the zonular tight junction (19). In this study, an α-catenin negative cell line failed to form both adherens junctions and tight junctions. Transfection of α-catenin mutants containing a vinculin binding domain or direct fusion of vinculin to α-catenin induced zonula adherens formation and tight junction alignment. Interestingly, vinculin binding to α-catenin is not essential for pure cadherin-based adhesion since L cell fibroblasts did not require this interaction to aggregate (20). Thus, recruitment of vinculin to the adherens junction might be important for zonula adherens assembly, which would in turn co-ordinate tight junction alignment with adherens junction formation.

Genetic studies from *Drosophila* have identified several other genes required for proper zonula adherens assembly: stardust, bazooka, crumbs. Stardust has yet to be cloned. Bazooka encodes a PDZ-containing protein that is related to Par-3, a protein involved in asymmetric cell divisions (69). Crumbs encodes a single-pass trans-membrane protein that does not itself localize to the zonula adherens, but is localized more apically (70). Interestingly, mutations in any one of the aforementioned genes do not affect the ability of the cadherin core complex to be detected at the cell surface, but rather prevent the conversion of these spot adherens junctions into a mature zonula adherens (reviewed in ref. 71). These results emphasize that zonula adherens formation and cadherin-based adherens formation may be somewhat independent processes. Intriguingly, these mutants also show a dramatic loss of cell polarity, suggesting that there may be a relationship between the zonula adherens and the establishment of a polarized epithelium. Something other than basic cadherin function, perhaps a higher order cadherin-containing structure organized at the zonula adherens may be important for generating and maintaining cell polarity.

While the bulk of this chapter is focused on the role of the cadherin/catenin complex in adherens junctions and the zonula adherens, it is perhaps important to point out that a new member of the IgG superfamily of calcium-independent adhesion molecules (see Chapter 1), nectin, and a molecule that interacts with its cytoplasmic domain, AF6/afadin, were shown to localize to the zonula adherens

(72). Indeed, they are even more highly concentrated at the zonula adherens than the E-cadherin complex, which is also found distributed over the entire lateral membrane. AF6/afadin is a modular protein that has been also found to interact with Ras and the junction-associated proteins ZO-1 and ponsin. A possible role for AF6/afadin comes from studies with mice carrying a null mutation in the AF-6/afadin locus. In these mice, the classically defined tripartite junctional complex and overall epithelial polarity are disrupted without significantly altering the subcellular localization of E-cadherin (73, 74). Thus as with the *Drosophila* mutants described above, it appears that the zonula adherens and cadherin-based adherens junctions may serve distinct functions, and in particular, there is a relationship between zonula adherens formation and polarity. How or whether AF6/afadin and nectin collaborate with the cadherin complex to ensure proper formation of a zonula adherens is not clear.

6. Communication between the junctional complex and the nucleus via catenins

Up until now we have discussed how a cell might regulate the cadherin complex and adhesion. However, it is equally important to consider how the state of cell–cell adhesion might be communicated from the outside of a cell to a cell's interior. Might the cadherin complex transduce a signal that ultimately affects changes in gene expression? And if so, what cellular components might transduce such a signal?

The notion that cadherins might communicate signals to a cell's interior was given some credence when it was discovered that β-catenin was found to be highly homologous to Armadillo, a *Drosophila* protein required for proper segmentation of the ventral epidermis in the developing fly embryo (75). Armadillo is one of a number of proteins which co-operate to transduce a signal initiated by a secreted factor known as wingless (Wg). Wg (and its vertebrate homologues, wnts) are used many times throughout the early development of an organism to activate the expression of target genes that will instruct cells to adopt particular cell fates. Certain components of this signal transduction pathway (namely, the adenomatous polyposis coli tumour suppressor gene product, APC and β-catenin) have been the focus of much recent attention as their dysregulation has been strongly implicated in the formation and/or progression of various tumour types (76).

The molecular components of Wg/wnt signal transduction are highly conserved from flies to vertebrates, and a convergence of genetic epistasis, biochemical, and human cancer studies have provided us with a linear representation of the essential players in this pathway (see Fig. 5) (reviewed in ref. 77). A secreted Wg/wnt acts through cell surface receptors of the Frizzled family. These receptors are seven-pass transmembrane proteins that topologically resemble G protein-coupled receptors. The ultimate consequence of Frizzled activation by Wg/wnt appears to be the activation of a particular family of high mobility group (HMG)-type transcription factors known as LEF/TCF (lymphocyte enhancer factor/T cell factor). Between

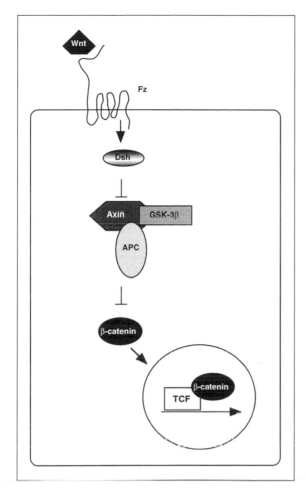

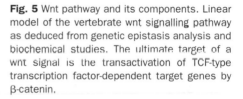

Fig. 5 Wnt pathway and its components. Linear model of the vertebrate wnt signalling pathway as deduced from genetic epistasis analysis and biochemical studies. The ultimate target of a wnt signal is the transactivation of TCF-type transcription factor-dependent target genes by β-catenin.

receptor activation at the plasma membrane and gene activation in the nucleus lie a number of cytoplasmic effectors: a novel protein, dishevelled; a serine-threonine kinase, glycogen synthase kinase-3β (GSK-3β); the tumour suppressor APC; another novel protein, axin and β-catenin. How these proteins come together to transduce a Wg/wnt signal into TCF-dependent gene activation has been the subject of intense investigation. What has become increasingly clear is that β-catenin is a key player in this pathway, and that the collective function of GSK-3β, APC, and axin is to regulate β-catenin's role in mediating Wg/wnt signalling.

The discovery that a cadherin-associated molecule also served an essential role in Wg/wnt signal transduction led to one early hypothesis that signalling occurred via β-catenin at the adherens junction. However, observations from both *Drosophila* and *Xenopus* systems have since indicated that an non-cadherin associated pool of β-catenin is important for signalling. For example, in the absence of a Wg signal, most of the armadillo was associated with cadherins at zonula adherens junctions, while cells receiving a Wg signal accumulated high levels of soluble Armadillo in the cytosol (78). Mutant forms of β-catenin/armadillo that interact poorly or not at all with cadherins or adherens junctions were found to exhibit signalling activity. In addition, immunolocalization studies revealed that exogenously overexpressed β-catenin could be localized to a cell's nucleus (79), and endogenous β-catenin was found localized to nuclei from cells receiving a wnt signal (80). Consistent with its nuclear localization, β-catenin was ultimately determined to interact directly with LEF/TCF-type transcription factors and mediate transcriptional transactivation (81, 82). Thus, there are at least two functionally distinct pools of β-catenin in cells: one that is cadherin-associated at the membrane; another that is cytoplasmic/nuclear. The latter appears most critical for Wg/wnt signal transduction. While there is no evidence to date that β-catenin signalling can be mediated through its association with cadherins, it is important to emphasize that β-catenin signalling is not entirely independent of cadherins (see Section 6.3).

6.1 β-Catenin phosphorylation as a key regulatory step

Understanding how a separate, cytosolic/nuclear pool of β-catenin is generated and targeted to the nucleus has been the focus of much recent investigation. Phosphorylation appears to be one important aspect of this regulation. Genetic epistasis analysis in *Drosophila* (83) and comparable studies in vertebrates (84) places armadillo/β-catenin downstream of a serine/threonine protein kinase, zeste-white-3/shaggy/ GSK-3β. The observation that armadillo could be phosphorylated on serine/threonine residues, and that less phosphorylation was detected in ZW3 mutants (85) led to the early suggestion that armadillo might be regulated by phosphorylation. Since ZW3 mutants displayed a loss-of-function phenotype opposite to that of armadillo and wg mutants, a model was suggested whereby the kinase negatively regulates armadillo activity (78). Thus, a Wg/wnt signal would serve to activate armadillo/β-catenin by antagonizing the constitutive, inhibitory activity of ZW3/ GSK-3β. There is some evidence that β-catenin may be a direct substrate of GSK-3β (86) and this is consistent with the observation that certain serine/threonine residues in the amino terminal region of β-catenin fit a minimally restrictive consensus motif for phosphorylation by GSK-3β. Other groups, however, have found β-catenin to be a poor substrate for GSK-3β *in vitro* (87–89), suggesting that GSK-3β may work indirectly through other components during this important regulatory step.

Independent of whether β-catenin is the direct target of GSK-3β, it became clear that certain serine/threonine residues in the amino terminus of β-catenin regulate its signalling function. Loss or missense mutation of these serine/threonine residues was found associated with certain types of cancer (76), and these mutations correlated with increased signalling of β-catenin using LEF/TCF-dependent reporter gene assays, or axis duplication activity in the frog (86, 90). Together, these observations provided rather compelling evidence that phosphorylation of certain residues in the amino terminus of β-catenin negatively regulates its signalling activity.

6.2 Relationship between β-catenin phosphorylation state and protein levels: roles for β-TrCP, GSK-3β, axin, and APC

How does the phosphorylation state of β-catenin alter its signalling activity? A number of studies noted a positive correlation between the steady state levels of cytoplasmic β-catenin and signalling. For example, *Drosophila* epidermal cells receiving a Wg signal appeared to stabilize a soluble, cytoplasmic form of armadillo/ β-catenin (78, 91). Serine/threonine point mutants in the amino terminal region of β-catenin that resulted in reduced steady state phosphorylation of the protein exhibited a longer than normal half-life in pulse chase experiments (90). Lastly, cell lines mutant for the adenomatous polyposis coli, tumour suppressor gene product (APC), accumulated higher than normal levels of cytoplasmic and nuclear pools of β-catenin (92). This strong correlation between signalling and the stability of β-catenin suggested that the actual mechanism by which phosphorylation inhibited β-catenin signalling might be through a tightly controlled degradation mechanism.

A number of cellular processes from cell cycle regulation to cytokine-activated immune responses (e.g. the NFκB signalling pathway) rely on regulated degradation as a way to control protein activities (reviewed in ref. 93). A commonly used pathway for regulating the degradation of proteins involves the ubiquitin–proteosome pathway (reviewed in ref. 94). The key aspects of this degradation pathway link serine phosphorylation of the target molecule to the covalent attachment of a ubiquitin polypeptide chain. Ubiquitin-conjugated substrates are ultimately degraded by the 26S proteosome. Consistent with this mechanism of turnover, β-catenin is indeed ubiquitinated, and its turnover rate can be slowed upon addition of specific proteosomal inhibitors (95). Phosphorylation of the putative GSK-3β sites within the amino terminus was considered important for ubiquitin-mediated degradation, since mutants lacking these serine/threonine residues were not found to be ubiquitin modified (95). The critical link that N-terminal phosphorylation specifically targets β-catenin for rapid degradation came from studies of a mutation in the *Drosophila* gene, slimb, which resulted in excess Wg signalling (96). Slimb encodes an F-box, WD-40-repeat protein related to the budding yeast protein, cdc4p, which has been shown to target cell cycle regulators for degradation by the ubiquitin proteosome pathway (reviewed in ref. 97). Recent studies with the mammalian homologue of slimb, β-TrCP, demonstrated that this protein directly recognizes the amino terminally phosphorylated β-catenin, and brings this substrate to the ubiquitin conjugation machinery (98), thus providing compelling evidence that one important consequence of β-catenin phosphorylation is ubiquitin-mediated degradation (Fig. 6).

If it is clear that phosphorylation negatively regulates β-catenin signalling through degradation, how is this process precisely controlled? What (if not GSK-3β alone) controls β-catenin phosphorylation? Analysis of two distinct β-catenin binding proteins, the adenomatous polyposis coli, tumour suppressor gene product (APC) and a novel gene product, axin, has shed recent light on this question.

The APC gene was discovered as a mutated locus in families known to develop multiple benign intestinal polyps (99). Mutations in APC were also found in most

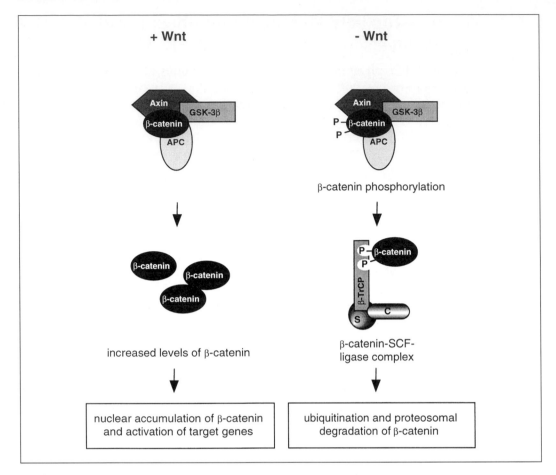

Fig. 6 Turnover model for β-catenin signalling. In the absence of wnt (*left*), GSK-3β is constitutively active. β-Catenin becomes phosphorylated at its amino terminus. Phosphorylated β-catenin is specifically recognized by the β-TrCP F-box protein as part of an SCF–ubiquitin ligase complex. Ubiquitination targets β-catenin for proteolysis by the proteosome. In the presence of wnt (*right*), GSK-3β-induced phosphorylation and degradation is inhibited, resulting in increased levels of β-catenin protein available for signalling.

sporadic colorectal cancers suggesting a critical role in the initiation of this type of cancer (100). Biochemical analysis of APC revealed that it is a large, multidomain protein, that can bind β-catenin directly (101) and promote its degradation upon transfection into cancer cell lines harbouring APC mutations (92). Axin was discovered as a naturally occurring mouse mutation (formerly known as *fused*) that resulted in axial duplications of embryos (102). Importantly, this gene product binds β-catenin, APC, and GSK-3β and can antagonize a wnt signal by promoting the turnover of β-catenin (103). The fact that axin can promote β-catenin phosphorylation by GSK-3β (104) and that APC, GSK-3β, and β-catenin can form a multi-molecular complex has lead to the suggestion that axin may serve as a 'scaffold' for β-catenin regulatory molecules, and together this complex may regulate the critical serine/threonine phosphorylation required for β-TrCP recognition and degradation. While this model is consistent with much of the available data, APC and an APC-related protein were found to *facilitate* wnt signalling in *Xenopus* and *C. elegans*, respectively (105, 106). This suggests that the relationship between APC and β-catenin signalling may be complex, and that APC may not function exclusively as a negative regulator of β-catenin.

6.3 Relationship between the cadherin complex and β-catenin signalling

The existence of two compositionally distinct pools of β-catenin in the cell, one associated with the plasma membrane as an integral part of the cadherin core complex, and the other a cytoplasmic/nuclear pool, which serves to transduce a membrane to nuclear signal, raises intriguing questions as to whether or not there is a dynamic relationship between these two pools. For instance, might the changes in adhesion which accompany morphogenetic movements during development result in increased β-catenin nuclear signalling? While there is no direct evidence for such signalling, experimental manipulations have revealed that cadherin expression and β-catenin signalling are at least potentially interrelated. For example, overexpression of the cadherin can antagonize β-catenin signalling activity in *Xenopus* embryos, and this appears to work through binding and sequestering β-catenin from the cytoplasmic signalling pool (107, 108). Under normal physiological circumstances, it is possible that the cadherin pool might serve as a 'sink' for β-catenin, so that β-catenin levels would have to rise beyond a threshold of cadherin expression in order to signal. Alternatively, a cell could increase its rate of cadherin synthesis in order to dampen β-catenin signalling. Since TCF binding sites have been identified in the E-cadherin promoter (109), it is intriguing to consider a negative feedback mechanism where β-catenin nuclear signalling could activate E-cadherin transcription, which would in turn inhibit β-catenin signalling through sequestration. In this regard, transgenic overexpression of a constitutively active form of β-catenin was associated with the up-regulation of E-cadherin (110).

Teasing out the relationship between cadherin expression, adhesion, and β-catenin signalling may be facilitated by an understanding of how these factors contribute to tumour progression. Many studies have reported loss of, or mutation within the E-cadherin/catenin complex in numerous epithelial cancers (reviewed by ref. 111). In many cases, E-cadherin loss correlated with the invasive component of a given tumour, suggesting that E-cadherin loss-of-function might promote tumour progression, local invasion, and metastasis (112). More recently, germline mutations in E-cadherin have been found associated with familial gastric cancer, underscoring its importance as a true tumour suppressor gene (113). Importantly, re-introduction of E-cadherin into either invasive or tumorigenic E-cadherin negative cell lines significantly reduced the transformed properties of these cells (114, 115), strongly implicating E-cadherin in both tumour and invasion suppression. Furthermore, using a mouse model system, a causal role for E-cadherin in tumour progression has been firmly established (see Chapter 3 for additional discussion of the role of cadherins in cancer) (116). While it is now clear that E-cadherin plays a critical role in tumour progression, the molecular means through which growth and/or invasion suppression are achieved are not clearly understood. It has previously been reasoned that the tumour/invasion suppressor activities of E-cadherin would be mediated through maintaining cell–cell adhesion. However, the recent finding that β-catenin is an oncogene, and that constitutive signalling is associated with both colon cancers and melanomas (90, 117) has led to the suggestion that an equally important role for E-cadherin in tumour suppression might be through antagonizing the nuclear signalling activity of β-catenin. Distinguishing between such possibilities will enable us to dissect how E-cadherin-based adhesion, junction formation, and effects on β-catenin signalling individually alter cell behaviour.

7. Emerging concepts

7.1 APC–axin–GSK-3β as a complex regulator of β-catenin signalling

It has become increasingly clear that β-catenin-mediated wnt signalling can be negatively regulated by a complex of proteins, APC–axin–GSK-3β. However, APC has been found to *promote* β-catenin signalling in some developmental systems (105, 106). In addition, some studies have shown that regulating the levels of β-catenin is not the sole mechanism for controlling its signalling activity (118, 119; Guger and Gumbiner, in preparation). It is therefore tempting to speculate that an APC-containing complex might also have the capacity to positively regulate β-catenin signalling. Precisely how such a complex might activate β-catenin is presently unknown but could conceivably regulate steps that are required for β-catenin signalling, such as nuclear import/export or transcriptional activation. In this context, it is perhaps worth mentioning that GSK-3β has been found to influence the nuclear distributions of a number of signalling proteins (120, 121).

Recently a number of other proteins have been shown to interact with members of the APC complex, such as dishevelled, protein phosphatase 2A (PP2A), GSK-3β binding protein (GBP), and casein kinase Iε (122–125). This raises the possibility that all of these proteins may coexist in a large, multiprotein complex. Alternatively, these proteins might form related, but compositionally distinct complexes. The existence of such a complex(es) suggests that β-catenin may be subject to signalling pathways other than those initiated by wnts. For example, TCF/β-catenin nuclear signalling can be inhibited by a MAP kinase-related pathway (126, 127). Also, there is preliminary evidence that the Alzheimer's associated presenilins and the integrin-linked kinase (ILK) may influence β-catenin nuclear signalling (128, 129). Whether or not these proteins transduce their signals through an APC–axin–GSK-3β-like complex is not known.

7.2 Nuclear function for armadillo repeat proteins

In addition to β-catenin, a number of armadillo repeat proteins have been found to exhibit both nuclear and junctional localizations, including p120^{ctn}, plakoglobin, and plakophilin (130, 131). Recently, p120^{ctn} has been found to interact with a putative transcription factor using yeast two-hybrid analysis (132). While potential signalling pathways and nuclear targets have not yet been identified for these putative transducers, it is possible that, like β-catenin, many of these armadillo repeat proteins may serve both nuclear and adhesive functions. Another group of proteins with armadillo repeats and the related 'HEAT' repeats is the family of nuclear import receptors, or importins. In this regard, it is notable that β-catenin is imported into the nucleus independent of a nuclear localization signal or the known importins, and interacts directly with the nuclear pore (133). Thus, one attractive hypothesis is that all armadillo repeat proteins are evolutionarily and perhaps functionally related to nuclear import receptors and can interact directly with the nuclear import machinery.

7.3 Modulation of cadherin function

It has become increasingly clear that cadherin-mediated cell–cell adhesion is a dynamic process that is subject to regulation by multiple factors. In a few cases, differences in the association of β-catenin, α-catenin, or p120^{ctn} with the cadherin complex were found, whereas in most cases, no such changes could be observed. These findings alone suggest that cadherin function is not modulated by one specific mechanism. The recruitment of actin to the cadherin complex may be another possible way to regulate adhesion. The observation that dominant-negative Rac inhibited actin localization to the adherens junctions in *Drosophila* epithelial cells without changing the composition of the cadherin complex is consistent with this possibility (62). Adhesion could also be regulated through a conformational change in the cadherin ectodomain that could either decrease or increase the affinity of cadherin-specific binding. Interesting in this regard is an antibody to *Xenopus* C-cadherin which activates C-cadherin-mediated adhesion (52). Thus, it appears that the cell employs multiple mechanisms to regulate cadherin function.

7.4 Emerging ideas: transducing signals from the adherens junction

Many signalling molecules appear to be enriched at the adherens junction or are associated with cadherins. In this chapter, we have largely discussed how these molecules may alter cadherin-dependent adhesion. One should keep in mind, however, that cadherins might allow these signalling molecules to transduce possible cell contact-dependent signals. By simply bringing two cells together, cadherins may enable receptor–membrane anchored ligand pairs to be brought into close proximity to permit juxtacrine signalling (26). Alternatively, cadherin function might be analogous to that of integrins which serve both as adhesion and signalling molecules. Clustering of integrins upon ligand binding results in the recruitment and subsequent activation of cytosolic signalling molecules, such as FAK, Shc, or Src (see Chapter 9). Whether signalling proteins are recruited as a consequence of cadherin clustering, and the type of signals these proteins might transduce (e.g. contact inhibition of growth, or cell differentiation) remain to be elucidated.

Acknowledgements

We thank past and present members of the Gumbiner laboratory for critically reading this chapter. We also thank M. Farquhar (UCSD) and N. Hirokawa (Univ. of Tokyo) for kindly providing us with the micrographs shown in Fig. 2.

References

1. Fannon, A. M., Sherman, D. L., Ilyina-Gragerova, G., Brophy, P. J., Friedrich, V. L. J., and Colman, D. R. (1995). Novel E-cadherin mediated adhesion in peripheral nerve: Schwann cell architecture is stabilized by autotypic adherens junctions. *J. Cell Biol.*, **129**, 189–202.
2. Fannon, A. M. and Colman, D. R. (1996). A model for central synaptic junctional complex formation based on the differential adhesive specificities of the cadherins. *Neuron*, **17**, 423–34.
3. Uchida, N., Honjo, Y., Johnson, K. R., Wheelock, M. J., and Takeichi, M. (1996). The catenin/cadherin adhesion system is localized in synaptic junctions bordering transmitter release zones. *J. Cell Biol.*, **135**, 767–79.
4. Volk, T. and Geiger, B. (1986). A-CAM: a 135-kD receptor of intercellular adherens junctions. I. Immunoelectron microscopic localization and biochemical studies. *J. Cell Biol.*, **103**, 1441–50.
5. Farquhar, M. G. and Palade, G. E. (1963). Junctional complexes in various epithelia. *J. Cell Biol.*, **17**, 375–412.
6. Drenckhahn, D. and Franz, H. (1986). Identification of actin-, alpha-actinin-, and vinculin-containing plaques at the lateral membrane of epithelial cells. *J. Cell Biol.*, **102**, 1843–52.
7. Yonemura, S., Itoh, M., Nagafuchi, A., and Tsukita, S. (1995). Cell-to-cell adherens junction formation and actin filament organization: similarities and differences between non-polarized fibroblasts and polarized epithelial cells. *J. Cell Sci.*, **108**, 127–42.
8. Boller, K., Vestweber, D., and Kemler, R. (1985). Cell-adhesion molecule uvomorulin is localized in the intermediate junctions of adult intestinal epithelial cells. *J. Cell Biol.*, **100**, 327–32.
9. Ozawa, M., Baribault, H., and Kemler, R. (1989). The cytoplasmic domain of the cell adhesion molecule uvomorulin associates with three independent proteins structurally related in different species. *EMBO J.*, **8**, 1711–17.
10. Nagafuchi, A. and Takeichi, M. (1988). Cell binding function of E-cadherin is regulated by the cytoplasmic domain. *EMBO J.*, **7**, 3679–84.
11. Ozawa, M., Ringwald, M., and Kemler, R. (1990). Uvomorulin-catenin complex formation is regulated by a specific domain in the cytoplasmic region of the cell adhesion molecule. *Proc. Natl. Acad. Sci. USA*, **87**, 4246–50.

12. Riggleman, B., Wieschaus, E., and Schedl, P. (1989). Molecular analysis of the armadillo locus: uniformly distributed transcripts and a protein with novel internal repeats are associated with a *Drosophila* segment polarity gene. *Genes Dev.*, **3**, 96–113.

13. Peifer, M., Berg, S., and Reynolds, A. (1994). A repeating amino acid motif shared by proteins with diverse cellular roles [letter]. *Cell*, **76**, 789–91.

14. Huber, A. H., Nelson, W. J., and Weis, W. I. (1997). Three-dimensional structure of the armadillo repeat region of beta-catenin. *Cell*, **90**, 871–82.

15. Hinck, L., Nathke, I. S., Papkoff, J., and Nelson, W. J. (1994). Dynamics of cadherin/catenin complex formation: novel protein interactions and pathways of complex assembly. *J. Cell Biol.*, **125**, 1327–40.

16. Haegel, H., Larue, L., Ohsugi, M., Fedorov, L., Herrenknecht, K., and Kemler, R. (1995). Lack of beta-catenin affects mouse development at gastrulation. *Development*, **121**, 3529–37.

17. Rimm, D. L., Koslov, E. R., Kebriaei, P., Cianci, C. D., and Morrow, J. S. (1995). α(E)-catein is an actin-binding and -bundling protein mediating the attachment of F-actin to the membrane adhesion complex. *Proc. Natl. Acad. Sci. USA*, **92**, 8813–17.

18. Nieset, J. E., Redfield, A. R., Jin, F., Knudsen, K. A., Johnson, K. R., and Wheelock, M. J. (1997). Characterization of the interactions of alpha-catenin with alpha- actinin and beta-catenin/plakoglobin. *J. Cell Sci.*, **110**, 1013–22.

19. Watabe-Uchida, M., Uchida, N., Imamura, Y., Nagafuchi, A., Fujimoto, K., Uemura, T., *et al.* (1998). alpha-Catenin-vinculin interaction functions to organize the apical junctional complex in epithelial cells. *J. Cell Biol.*, **142**, 847–57.

20. Imamura, Y., Itoh, M., Maeno, Y., Tsukita, S., and Nagafuchi, A. (1999). Functional domains of alpha-catenin required for the strong state of cadherin-based cell adhesion. *J. Cell Biol.*, **144**, 1311–22.

21. Reynolds, A. B., Daniel, J., McCrea, P. D., Wheelock, M. J., Wu, J., and Zhang, Z. (1994). Identification of a new catenin: the tyrosine kinase substrate p120cas associates with E-cadherin complexes. *Mol. Cell. Biol.*, **14**, 8333–42.

22. Daniel, J. M. and Reynolds, A. B. (1995). The tyrosine kinase substrate p120cas binds directly to E-cadherin but not to the adenomatosis polyposis coli protein or α-catenin. *Mol. Cell. Biol.*, **15**, 4819–24.

23. Yap, A. S., Niessen, C. M., and Gumbiner, B. M. (1998). The juxtamembrane region of the cadherin cytoplasmic tail supports lateral clustering, adhesive strengthening, and interaction with p120ctn. *J. Cell Biol.*, **141**, 779–89.

24. Keirsebilck, A., Bonne, S., Staes, K., van Hengel, J., Nollet, F., Reynolds, A., *et al.* (1998). Molecular cloning of the human p120ctn catenin gene (CTNND1): expression of multiple alternatively spliced isoforms. *Genomics*, **50**, 129–46.

25. Itoh, M., Yonemura, S., Nagafuchi, A., and Tsukita, S. (1991). A 220-kD undercoat-constitutive protein: its specific localization at cadherin-based cell–cell adhesion sites. *J. Cell Biol.*, **115**, 1449–62.

26. Fagotto, F. and Gumbiner, B. M. (1996). Cell contact-dependent signaling. *Dev. Biol.*, **180**, 445–54.

27. Hoschuetzky, H., Aberle, H., and Kemler, R. (1994). Beta-catenin mediates the interaction of the cadherin-catenin complex with epidermal growth factor receptor. *J. Cell Biol.*, **127**, 1375–80.

28. Tsukita, S., Oishi, K., Akiyama, T., Yamanashi, Y., and Yamamoto, T. (1991). Specific proto-oncogenic tyrosine kinases of src family are enriched in cell-to-cell adherens junctions where the level of tyrosine phosphorylation is elevated. *J. Cell Biol.*, **113**, 867–79.

29. Xu, Y., Guo, D. F., Davidson, M., Inagami, T., and Carpenter, G. (1997). Interaction of the adaptor protein Shc and the adhesion molecule cadherin. *J. Biol. Chem.*, **272**, 13463–6.

30. Hordijk, P. L., ten Klooster, J. P., van der Kammen, R. A., Michiels, F., Oomen, L. C., and Collard, J. G. (1997). Inhibition of invasion of epithelial cells by Tiam1-Rac signaling. *Science*, **278**, 1464–6.

31. Kuroda, S., Fukata, M., Nakagawa, M., Fujii, K., Nakamura, T., Ookubo, T., *et al.* (1998). Role of IQGAP1, a target of the small GTPases Cdc42 and Rac1, in regulation of E-cadherin- mediated cell–cell adhesion. *Science*, **281**, 832–5.

32. Shapiro, L., Fannon, A. M., Kwong, P. D., Thompson, A., Lehmann, M. S., Grubel, G., *et al.* (1995). Structural basis of cell–cell adhesion by cadherins. *Nature*, **374**, 327–37.

33. Nagar, B., Overduin, M., Ikura, M., and Rini, J. M. (1996). Structural basis of calcium-induced E-cadherin rigidification and dimerization. *Nature*, **380**, 360–4.

34. Brieher, W. M., Yap, A. S., and Gumbiner, B. M. (1996). Lateral dimerization is required for the homophilic binding activity of C-cadherin. *J. Cell Biol.*, **135**, 487–96.

35. Norvell, S. M. and Green, K. J. (1998). Contributions of extracellular and intracellular domains of full length and chimeric cadherin molecules to junction assembly in epithelial cells. *J. Cell Sci.*, **111**, 1305–18.

36. Chitaev, N. A. and Troyanovsky, S. M. (1998). Adhesive but not lateral E-cadherin complexes require calcium and catenins for their formation. *J. Cell Biol.*, **142**, 837–46.

37. Tomschy, A., Fauser, C., Landwehr, R., and Engel, J. (1996). Homophilic adhesion of E-cadherin occurs by a co-operative two-step interaction of N-terminal domains. *EMBO J.*, **15**, 3507–14.

38. Hirokawa, N. and Heuser, J. E. (1981). Quick-freeze, deep-etch visualization of the cytoskeleton beneath surface differentiations of intestinal epithelial cells. *J. Cell Biol.*, **91**, 399–409.

39. Cox, R. T., Kirkpatrick, C., and Peifer, M. (1996). Armadillo is required for adherens junction assembly, cell polarity, and morphogenesis during *Drosophila* embryogenesis. *J. Cell Biol.*, **134**, 133–48.

40. Nagafuchi, A., Ishihara, S., and Tsukita, S. (1994). The roles of catenins in the cadherin-mediated cell adhesion: functional analysis of E-cadherin-alpha catenin fusion molecules. *J. Cell Biol.*, **127**, 235–45.

41. Kintner, C. (1992). Regulation of embryonic cell adhesion by the cadherin cytoplasmic domain. Cell, **69**, 225–36.

42. Riehl, R., Johnson, K., Bradley, R., Grunwald, G. B., Cornel, E., Liliebaum, A., *et al.* (1996). Cadherin function is required for axon outgrowth in retinal ganglion cells *in vivo*. *Neuron*, **17**, 837–48.

43. Chen, H., Paradies, N. E., Fedor-chaiken, M., and Brackenbury, R. (1997). E-cadherin mediates adhesion and suppresses cell motility via distinct mechanisms. *J. Cell Sci.*, **110**, 345–56.

44. Navarro, P., Caveda, L., Breviario, F., Mandoteanu, I., Lampugnani, M. G., and Dejana, E. (1995). Catenin-dependent and -independent functions of vascular endothelial cadherin. *J. Biol. Chem.*, **270**, 30965–72.

45. Yap, A. S., Brieher, W. M., Pruschy, M., and Gumbiner, B. M. (1997). Lateral clustering of the adhesive ectodomain: a fundamental determinant of cadherin function. *Curr. Biol.*, **7**, 308–15.

46. Thoreson, M. A., Anastasiadis, P. Z., Daniel, J. M., Ireton, R. C., Wheelock, M. J., Johnson, K. R., *et al.* (2000). Selective uncoupling of p120(ctn) from E-cadherin disrupts strong adhesion [In Process Citation]. *J. Cell Biol.*, **148**, 189–202.

47. Ozawa, M. and Kemler, R. (1998). The membrane-proximal region of the E-cadherin cytoplasmic domain prevents dimerization and negatively regulates adhesion activity. *J. Cell Biol.*, **142**, 1605–13.

48. Aono, S., Nakagawa, S., Reynolds, A. B., and Takeichi, M. (1999). p120(ctn) Acts as an inhibitory regulator of cadherin function in colon carcinoma cells. *J. Cell Biol.*, **145**, 551–62.

49. Stevenson, B. R., Siliciano, J. D., Mooseker, M. S., and Goodenough, D. A. (1986). Identification of ZO-1: a high molecular weight polypeptide associated with the tight junction (zonula occludens) in a variety of epithelia. *J. Cell Biol.*, **103**, 755–66.

50. Vestweber, D., Gossler, A., Boller, K., and Kemler, R. (1987). Expression and distribution of cell adhesion molecule uvomorulin in mouse preimplantation embryos. *Dev. Biol.*, **124**, 451–6.

51. Brieher, W. M. and Gumbiner, B. M. (1994). Regulation of C-cadherin function during activin induced morphogenesis of *Xenopus* animal caps. *J. Cell Biol.*, **126**, 519–27.

52. Zhong, Y., Brieher, W. M., and Gumbiner, B. M. (1999). Analysis of C-cadherin regulation during tissue morphogenesis with an activating antibody. *J. Cell Biol.*, **144**, 351–9.

53. Matsuyoshi, N., Hamaguchi, M., Taniguchi, S., Nagafuchi, A., Tsukita, S., and Takeichi, M. (1992). Cadherin-mediated cell–cell adhesion is perturbed by v-src tyrosine phosphorylation in metastatic fibroblasts. *J. Cell Biol.*, **118**, 703–14.

54. Shibamoto, S., Hayakawa, M., Takeuchi, K., Hori, T., Oku, N., Miyazawa, K., *et al.* (1994). Tyrosine phosphorylation of beta-catenin and plakoglobin enhanced by hepatocyte growth factor and epidermal growth factor in human carcinoma cells. *Cell Adhes. Commun.*, **1**, 295–305.

55. Fujii, K., Furukawa, F., and Matsuyoshi, N. (1996). Ligand activation of overexpressed epidermal growth factor receptor results in colony dissociation and disturbed E-cadherin function in HSC-1 human cutaneous squamous carcinoma cells. *Exp. Cell Res.*, **223**, 50–62.

56. Burden-Gulley, S. M. and Brady-Kalnay, S. M. (1999). PTPμ regulates N-cadherin-dependent neurite outgrowth. *J. Cell Biol.*, **144**, 1323–36.

57. Miller, J. R. and McClay, D. R. (1997). Changes in the pattern of adherens junction-associated beta-catenin accompany morphogenesis in the sea urchin embryo. *Dev. Biol.*, **192**, 310–22.

58. Daniel, J. M. and Reynolds, A. B. (1997). Tyrosine phosphorylation and cadherin/catenin function. *Bioessays*, **19**, 883–91.

59. Takeda, H., Nagafuchi, A., Yonemura, S., Tsukita, S., Behrens, J., and Birchmeier, W. (1995). V-src kinase shifts the cadherin-based cell adhesion from the strong to the weak state and beta catenin is not required for the shift. *J. Cell Biol.*, **131**, 1839–47.

60. Braga, V. M., Machesky, L. M., Hall, A., and Hotchin, N. A. (1997). The small GTPases Rho and Rac are required for the establishment of cadherin-dependent cell–cell contacts. *J. Cell Biol.*, **137**, 1421–31.

61. Takaishi, K., Sasaki, T., Kotani, H., Nishioka, H., and Takai, Y. (1997). Regulation of cell–cell adhesion by rac and rho small G proteins in MDCK cells. *J. Cell Biol.*, **139**, 1047–59.

62. Eaton, S., Auvinen, P., Luo, L., Jan, Y. N., and Simons, K. (1995). CDC42 and Rac1 control different actin-dependent processes in the *Drosophila* wing disc epithelium. *J. Cell Biol.*, **131**, 151–64.

63. Tepass, U. and Hartenstein, V. (1994). The development of cellular junctions in the *Drosophila* embryo. *Dev. Biol.*, **161**, 563–96.

64. Vasioukhin, V., Bauer, C., Yin, M., and Fuchs, E. (2000). Directed actin polymerization is the driving force for epithelial cell–cell adhesion. *Cell*, **100**, 209–19.

65. Adams, C. L., Nelson, W. J., and Smith, S. J. (1996). Quantitative analysis of cadherin-catenin-actin reorganization during development of cell-cell adhesion. *J. Cell Biol.*, **135**, 1899–911.

66. Adams, C. L., Chen, Y. T., Smith, S. J., and Nelson, W. J. (1998). Mechanisms of epithelial cell–cell adhesion and cell compaction revealed by high-resolution tracking of E-cadherin-green fluorescent protein. *J. Cell Biol.*, **142**, 1105–19.

67. Gumbiner, B., Stevenson, B., and Grimaldi, A. (1988). The role of the cell adhesion molecule uvomorulin in the formation and maintenance of the epithelial junctional complex. *J. Cell Biol.*, **107**, 1575–87.

68. Watabe, M., Nagafuchi, A., Tsukita, S., and Takeichi, M. (1994). Induction of polarized cell–cell association and retardation of growth by activation of the E-cadherin-catenin adhesion system in a dispersed carcinoma line. *J. Cell Biol.*, **127**, 247–56.

69. Kuchinke, U., Grawe, F., and Knust, E. (1998). Control of spindle orientation in *Drosophila* by the Par-3-related PDZ- domain protein Bazooka. *Curr. Biol.*, **8**, 1357–65.

70. Wodarz, A., Grawe, F., and Knust, E. (1993). CRUMBS is involved in the control of apical protein targeting during *Drosophila* epithelial development. *Mech. Dev.*, **44**, 175–87.

71. Tepass, U. (1997). Epithelial differentiation in *Drosophila*. *Bioessays*, **19**, 673–82.

72. Takahashi, K., Nakanishi, H., Miyahara, M., Mandai, K., Satoh, K., Satoh, A., *et al.* (1999). Nectin/PRR: an immunoglobulin-like cell adhesion molecule recruited to cadherin-based adherens junctions through interaction with Afadin, a PDZ domain-containing protein. *J. Cell Biol.*, **145**, 539–49.

73. Ikeda, W., Nakanishi, H., Miyoshi, J., Mandai, K., Ishizaki, H., Tanaka, M., *et al.* (1999). Afadin: a key molecule essential for structural organization of cell–cell junctions of polarized epithelia during embryogenesis. *J. Cell Biol.*, **146**, 1117–32.

74. Zhadanov, A. B., Provance, D. W., Jr., Speer, C. A., Coffin, J. D., Goss, D., Blixt, J. A., *et al.* (1999). Absence of the tight junctional protein AF-6 disrupts epithelial cell–cell junctions and cell polarity during mouse development. *Curr. Biol.*, **9**, 880–8.

75. McCrea, P., Turck, C., and Gumbiner, B. (1991). A homolog of the armadillo protein in *Drosophila* (plakoglobin) associated with E-cadherin. *Science*, **254**, 1359–61.

76. Polakis, P. (1999). The oncogenic activation of beta-catenin. *Curr. Opin. Genet. Dev.*, **9**, 15–21.

77. Wodarz, A. and Nusse, R. (1998). Mechanisms of Wnt signaling in development. *Annu. Rev. Cell Dev. Biol.*, **14**, 59–88.

78. Peifer, M., Sweeton, D., Casey, M., and Wieschaus, E. (1994). wingless signal and Zeste-white 3 kinase trigger opposing changes in the intracellular distribution of Armadillo. *Development*, **120**, 369–80.

79. Funayama, N., Fagotto, F., McCrea, P., and Gumbiner, B. (1995). Embryonic axis induction by the armadillo repeat domain of beta-catenin: evidence for intracellular signaling. *J. Cell Biol.*, **128**, 959–68.

80. Schneider, S., Steinbeisser, H., Warga, R., and Hausen, P. (1996). Beta-catenin translocation into nuclei demarcates the dorsalizing centers in frog and fish embryos. *Mech. Dev.*, **57**, 191–8.

81. Molenaar, M., van de Wetering, M., Oosterwegel, M., Peterson-Maduro, J., Godsave, S., Korinek, V., *et al.* (1996). XTcf-3 transcription factor mediates beta-catenin-induced axis formation in *Xenopus* embryos. *Cell*, **86**, 391–9.

82. Behrens, J., von Kries, J., Kuhl, M., Bruhn, L., Wedlich, D., Grosschedl, R., *et al.* (1996). Functional interaction of beta-catenin with the transcription factor LEF-1. *Nature*, **382**, 638–42.

83. Siegfried, E., Wilder, E. L., and Perrimon, N. (1994). Components of wingless signalling in *Drosophila*. *Nature*, **367**, 76–80.

84. He, X., Saint-Jeannet, J., Woodgett, J., Varmus, H., and Dawid, I. (1995). Glycogen synthase kinase-3 and dorsoventral patterning in *Xenopus* embryos [published erratum appears in *Nature* 1995, May 18; 375(6528), 253]. *Nature*, **374**, 617–22.

85. Peifer, M., Pai, L., and Casey, M. (1994). Phosphorylation of the *Drosophila* adherens junction protein Armadillo: roles for wingless signal and zeste-white 3 kinase. *Dev. Biol.*, **166**, 543–56.

86. Yost, C., Torres, M., Miller, J., Huang, E., Kimelman, D., and Moon, R. (1996). The axis-inducing activity, stability, and subcellular distribution of beta-catenin is regulated in *Xenopus* embryos by glycogen synthase kinase 3. *Genes Dev.*, **10**, 1443–54.

87. Stambolic, V., Ruel, L., and Woodgett, J. (1996). Lithium inhibits glycogen synthase kinase-3 activity and mimics wingless signalling in intact cells. *Curr. Biol.*, **6**, 1664–8.

88. Rubinfeld, B., Albert, I., Porfiri, E., Fiol, C., Munemitsu, S., and Polakis, P. (1996). Binding of GSK3beta to the APC-beta-catenin complex and regulation of complex assembly [see comments]. *Science*, **272**, 1023–6.

89. Pai, L., Orsulic, S., Bejsovec, A., and Peifer, M. (1997). Negative regulation of Armadillo, a Wingless effector in *Drosophila*. *Development*, **124**, 2255–66.

90. Morin, P., Sparks, A., Korinek, V., Barker, N., Clevers, H., Vogelstein, B., *et al.* (1997). Activation of beta-catenin-Tcf signaling in colon cancer by mutations in beta-catenin or APC [see comments]. *Science*, **275**, 1787–90.

91. van Leeuwen, F., Samos, C., and Nusse, R. (1994). Biological activity of soluble wingless protein in cultured *Drosophila* imaginal disc cells. *Nature*, **368**, 342–4.

92. Munemitsu, S., Albert, I., Souza, B., Rubinfeld, B., and Polakis, P. (1995). Regulation of intracellular beta-catenin levels by the adenomatous polyposis coli (APC) tumor-suppressor protein. *Proc. Natl. Acad. Sci. USA*, **92**, 3046–50.

93. Maniatis, T. (1999). A ubiquitin ligase complex essential for the NF-kappaB, Wnt/Wingless, and Hedgehog signaling pathways. *Genes Dev.*, **13**, 505–10.

94. Hershko, A. and Ciechanover, A. (1998). The ubiquitin system. *Annu. Rev. Biochem.*, **67**, 425–79.

95. Aberle, H., Bauer, A., Stappert, J., Kispert, A., and Kemler, R. (1997). beta-catenin is a target for the ubiquitin-proteasome pathway. *EMBO J.*, **16**, 3797–804.

96. Jiang, J. and Struhl, G. (1998). Regulation of the Hedgehog and Wingless signalling pathways by the F-box/WD40-repeat protein Slimb. *Nature*, **391**, 493–6.

97. Patton, E. E., Willems, A. R., and Tyers, M. (1998). Combinatorial control in ubiquitin-dependent proteolysis: don't Skp the F-box hypothesis. *Trends Genet.*, **14**, 236–43.

98. Hart, M., Concordet, J. P., Lassot, I., Albert, I., del los Santos, R., Durand, H., *et al.* (1999). The F-box protein beta-TrCP associates with phosphorylated beta-catenin and regulates its activity in the cell. *Curr. Biol.*, **9**, 207–10.

99. Kinzler, K. W., Nilbert, M. C., Vogelstein, B., Bryan, T. M., Levy, D. B., Smith, K. J., *et al.* (1991). Identification of a gene located at chromosome 5q21 that is mutated in colorectal cancers [see comments]. *Science*, **251**, 1366–70.

100. Kinzler, K. W. and Vogelstein, B. (1997). Cancer-susceptibility genes. Gatekeepers and caretakers [news; comment]. *Nature*, **386**, 761, 763.

101. Rubinfeld, B., Souza, B., Albert, I., Muller, O., Chamberlain, S. H., Masiarz, F. R., *et al.* (1993). Association of the APC gene product with beta-catenin. *Science*, **262**, 1731–4.

102. Zeng, L., Fagotto, F., Zhang, T., Hsu, W., Vasicek, T. J., Perry W. L. III, *et al.* (1997). The mouse Fused locus encodes Axin, an inhibitor of the Wnt signaling pathway that regulates embryonic axis formation. *Cell*, **90**, 181–92.

103. Hart, M. J., de los Santos, R., Albert, I. N., Rubinfeld, B., and Polakis, P. (1998). Down-regulation of beta-catenin by human Axin and its association with the APC tumor suppressor, beta-catenin and GSK3 beta. *Curr. Biol.*, **8**, 573–81.

104. Ikeda, S., Kishida, S., Yamamoto, H., Murai, H., Koyama, S., and Kikuchi, A. (1998). Axin, a negative regulator of the Wnt signaling pathway, forms a complex with GSK-3beta and beta-catenin and promotes GSK-3beta-dependent phosphorylation of beta-catenin. *EMBO J.*, **17**, 1371–84.

105. Vleminckx, K., Wong, E., Guger, K., Rubinfeld, B., Polakis, P., and Gumbiner, B. (1997). Adenomatous polyposis coli tumor suppressor protein has signaling activity in *Xenopus laevis* embryos resulting in the induction of an ectopic dorsoanterior axis. *J. Cell Biol.*, **136**, 411–20.

106. Rocheleau, C. E., Downs, W. D., Lin, R., Wittmann, C., Bei, Y., Cha, Y. H., *et al.* (1997). Wnt signaling and an APC-related gene specify endoderm in early *C. elegans* embryos [see comments]. *Cell*, **90**, 707–16.

107. Heasman, J., Crawford, A., Goldstone, K., Garner-Hamrick, P., Gumbiner, B., McCrea, P., *et al.* (1994). Overexpression of cadherins and underexpression of beta-catenin inhibit dorsal mesoderm induction in early *Xenopus* embryos. *Cell*, **79**, 791–803.

108. Fagotto, F., Funayama, N., Gluck, U., and Gumbiner, B. (1996). Binding to cadherins antagonizes the signaling activity of beta-catenin during axis formation in *Xenopus*. *J. Cell Biol.*, **132**, 1105–14.

109. Huber, O., Korn, R., McLaughlin, J., Ohsugi, M., Herrmann, B. G., and Kemler, R. (1996). Nuclear localization of beta-catenin by interaction with transcription factor LEF-1. *Mech. Dev.*, **59**, 3–10.

110. Wong, M. H., Rubinfeld, B., and Gordon, J. I. (1998). Effects of forced expression of an NH2-terminal truncated beta-Catenin on mouse intestinal epithelial homeostasis. *J. Cell Biol.*, **141**, 765–77.

111. Christofori, G. and Semb, H. (1999). The role of the cell-adhesion molecule E-cadherin as a tumour-suppressor gene. *Trends Biochem. Sci.*, **24**, 73–6.

112. Birchmeier, C., Birchmeier, W., and Brand-Saberi, B. (1996). Epithelial-mesenchymal transitions in cancer progression. *Acta Anat.*, **156**, 217–26.

113. Berx, G., Becker, K. F., Hofler, H., and van Roy, F. (1998). Mutations of the human E-cadherin (CDH1) gene. *Hum. Mutat.*, **12**, 226–37.

114. Navarro, P., Gomez, M., Pizarro, A., Gamallo, C., Quintanilla, M., and Cano, A. (1991). A role for the E-cadherin cell–cell adhesion molecule during tumor progression of mouse epidermal carcinogenesis. *J. Cell Biol.*, **115**, 517–33.

115. Vleminckx, K., Vakaet, L., Jr., Mareel, M., Fiers, W., and van Roy, F. (1991). Genetic manipulation of E-cadherin expression by epithelial tumor cells reveals an invasion suppressor role. *Cell*, **66**, 107–19.

116. Perl, A. K., Wilgenbus, P., Dahl, U., Semb, H., and Christofori, G. (1998). A causal role for E-cadherin in the transition from adenoma to carcinoma. *Nature*, **392**, 190–3.

117. Rubinfeld, B., Robbins, P., El-Gamil, M., Albert, I., Porfiri, E., and Polakis, P. (1997). Stabilization of beta-catenin by genetic defects in melanoma cell lines [see comments]. *Science*, **275**, 1790–2.

118. Young, C. S., Kitamura, M., Hardy, S., and Kitajewski, J. (1998). Wnt-1 induces growth, cytosolic beta-catenin, and Tcf/Lef transcriptional activation in Rat-1 fibroblasts. *Mol. Cell. Biol.*, **18**, 2474–85.

119. Nelson, R. W. and Gumbiner, B. M. (1999). A cell-free assay system for beta-catenin signaling that recapitulates direct inductive events in the early *Xenopus laevis* embryo. *J. Cell Biol.*, **147**, 367–74.

120. Beals, C. R., Sheridan, C. M., Turck, C. W., Gardner, P., and Crabtree, G. R. (1997). Nuclear export of NF-ATc enhanced by glycogen synthase kinase-3. *Science*, **275**, 1930–4.

121. Diehl, J. A., Cheng, M., Roussel, M. F., and Sherr, C. J. (1998). Glycogen synthase kinase-3beta regulates cyclin D1 proteolysis and subcellular localization. *Genes Dev.*, **12**, 3499–511.

122. Yost, C., Farr, G. H., III, Pierce, S. B., Ferkey, D. M., Chen, M. M., and Kimelman, D. (1998). GBP, an inhibitor of GSK-3, is implicated in *Xenopus* development and oncogenesis. *Cell*, **93**, 1031–41.

123. Hsu, W., Zeng, L., and Costantini, F. (1999). Identification of a domain of Axin that binds to the serine / threonine protein phosphatase 2A and a self-binding domain. *J. Biol. Chem.*, **274**, 3439–45.

124. Kishida, S., Yamamoto, H., Hino, S., Ikeda, S., Kishida, M., and Kikuchi, A. (1999). DIX domains of Dvl and axin are necessary for protein interactions and their ability to regulate beta-catenin stability. *Mol. Cell. Biol.*, **19**, 4414–22.

125. Peters, J. M., McKay, R. M., McKay, J. P., and Graff, J. M. (1999). Casein kinase I transduces Wnt signals. *Nature*, **401**, 345–50.

126. Rocheleau, C. E., Yasuda, J., Shin, T. H., Lin, R., Sawa, H., Okano, H., *et al.* (1999). WRM-1 activates the LIT-1 protein kinase to transduce anterior / posterior polarity signals in *C. elegans*. *Cell*, **97**, 717–26.

127. Ishitani, T., Ninomiya-Tsuji, J., Nagai, S., Nishita, M., Meneghini, M., Barker, N., *et al.* (1999). The TAK1-NLK-MAPK-related pathway antagonizes signalling between beta-catenin and transcription factor TCF. *Nature*, **399**, 798–802.

128. Nishimura, M., Yu, G., Levesque, G., Zhang, D. M., Ruel, L., Chen, F., *et al.* (1999). Presenilin mutations associated with Alzheimer disease cause defective intracellular trafficking of beta-catenin, a component of the presenilin protein complex. *Nature Med.*, **5**, 164–9.

129. Novak, A., Hsu, S. C., Leung-Hagesteijn, C., Radeva, G., Papkoff, J., Montesano, R., *et al.* (1998). Cell adhesion and the integrin-linked kinase regulate the LEF-1 and beta-catenin signaling pathways. *Proc. Natl. Acad. Sci. USA*, **95**, 4374–9.

130. Karnovsky, A. and Klymkowsky, M. W. (1995). Anterior axis duplication in *Xenopus* induced by the over-expression of the cadherin-binding protein plakoglobin. *Proc. Natl. Acad. Sci. USA*, **92**, 4522–6.

131. Mertens, C., Kuhn, C., and Franke, W. W. (1996). Plakophilins 2a and 2b: constitutive proteins of dual location in the karyoplasm and the desmosomal plaque. *J. Cell Biol.*, **135**, 1009–25.

132. Daniel, J. M. and Reynolds, A. B. (1999). The catenin p120(ctn) interacts with kaiso, a novel BTB / POZ domain zinc finger transcription factor. *Mol. Cell. Biol.*, **19**, 3614–23.

133. Fagotto, F., Gluck, U., and Gumbiner, B. (1998). Nuclear localization signal-independent and importin / karyopherin-independent nuclear import of beta-catenin. *Curr. Biol.*, **8**, 181–90.

9 | Focal adhesions and focal complexes

LYNDA PETERSON and KEITH BURRIDGE

1. Introduction

Focal adhesions were first identified in electron micrographs of the leading lamellae of motile chick-heart fibroblasts (1). These images showed electron dense plaques containing longitudinal oblique filaments at regions of close approach between the ventral cell membrane and the substratum. Later studies revealed these areas of close approach as discrete darkened regions in interference reflection microscopy (IRM) images of adherent cells in tissue culture (2–4). Several alternative names for these structures have been used, such as adhesion plaque (1) and focal contact (3), however, focal adhesion has become the most commonly used name. More recently, the term focal complex has been introduced to describe related structures that form at the leading edge of cells in response to the activity of the low molecular weight G proteins, Rac1 and Cdc42 (5). Focal complexes often mature into focal adhesions. In this review, we will use the terms focal adhesion and focal complex to differentiate the two types of structure that are under distinct modes of regulation (5). In addition, we will briefly discuss 'fibrillar adhesions', a type of adhesion that also relates to focal adhesions and that are made between the cell and fibronectin fibrils (6–8).

IRM measurements revealed that focal adhesions are regions where the cell membrane comes to within 10–15 nm of the underlying substratum (3). Observations from electron microscopy and IRM revealed that focal adhesions are sites where large bundles of actin filaments (stress fibres) terminate (4). The first proteins found to be concentrated within focal adhesions were α-actinin (9, 10) and vinculin (11, 12). α-Actinin localizes both to the ends of stress fibres and periodically along their length, whereas vinculin is confined to the focal adhesion. EM micrographs of thin sectioned cells that have been immunolabelled to detect vinculin and α-actinin revealed that vinculin localizes closest to the cell membrane while α-actinin is concentrated more distally within the focal adhesion (6). Because of its prominence in these structures, vinculin has been used frequently as an immunofluorescent marker for focal adhesions. However, vinculin is also prominent in cadherin-based

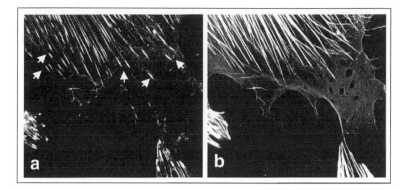

Fig. 1 Immunofluorescent images of focal adhesions and stress fibres. (a) Focal adhesions (*arrows*) visualized with an antibody against talin. (b) The same cell stained with fluorescent phalloidin to visualize actin filaments organized in stress fibres.

cell–cell junctions. Other focal adhesion proteins, such as talin or paxillin that are not found in cell–cell junctions, provide more definitive markers of focal adhesions (Fig. 1). A key to understanding the structure and function of focal adhesions was the identification of integrins as the major transmembrane components that are concentrated within these structures (13–16). Integrins are a family of heterodimeric proteins that function as receptors for many different extracellular matrix (ECM) proteins (see Chapter 4 for detailed discussion). The specific integrins that are clustered within focal adhesions are determined by the nature of the ECM adsorbed onto the surface on which cells are growing (17–20). The integrin found within focal adhesions of cells grown in the presence of serum is the major vitronectin receptor, $\alpha v\beta 3$, because vitronectin is the protein adsorbed to tissue culture surfaces under these conditions. In many experimental situations, cells are plated on fibronectin-coated surfaces, and earlier results indicated that the integrin in these focal adhesions is $\alpha 5\beta 1$, the high affinity fibronectin receptor (20).

For many years focal adhesions drew interest because of their obvious structural function anchoring bundles of actin filaments to the plasma membrane at sites where cells adhere to the ECM. Many of the first focal adhesion proteins to be isolated, such as vinculin and talin, appeared to have a structural role. For the past decade, however, it has become increasingly clear that focal adhesions and focal complexes are also major sites of signal transduction. Several signalling pathways become activated in response to integrin-mediated adhesion, and the components of these pathways are concentrated within focal adhesions. In this review, we will discuss the organization and assembly of focal adhesions, how these structures are regulated by the Rho family of low molecular weight GTPases, and how multiple signalling pathways are initiated from focal adhesions. For more detailed information, there are longer reviews that deal with the structure of focal adhesions (22, 23), the regulation of cytoskeletal organization by Rho family GTPases (24–26), and integrin-mediated signalling (23, 27–29).

2. *In situ* focal adhesion homologues

Junctions between cells and the ECM *in situ* reveal structural and functional similarities with focal adhesions. Both are sites where tension generated within bundles of actin filaments is transmitted across the membrane to the ECM. For example, myotendinous junctions of skeletal muscle and the dense plaques of smooth muscle not only anchor actin filaments to the plasma membrane and across it to the surrounding ECM, but they also contain many of the same proteins as focal adhesions (30). These proteins are also enriched at costameres, the lateral attachments of myofibrils to the plasma membranes of skeletal and cardiac muscles at Z discs (31–33). Although it is not known whether tension is transmitted across the postsynaptic regions of neuromuscular junctions, many focal adhesion proteins have been localized within these sites as well (34–36). Certainly, tension is transmitted to the ECM during contraction of blood clots by platelets and it is notable that platelets are a rich source of focal adhesion proteins.

The type of cultured cell that most commonly displays prominent stress fibres and focal adhesions is the fibroblast. It is interesting that fibroblasts rarely possess these structures *in situ*, although stress fibres are seen in granulation tissue that develops during wound healing (37, 38). Similarly, stress fibres and focal adhesions are prominent in endothelial cells growing in culture, but are absent from most endothelial cells lining blood vessels *in situ*. An exception to this is seen when endothelial cells are examined from blood vessels experiencing very high blood pressure conditions (39–41).

3. Assembly and maturation of focal complexes and focal adhesions

3.1 Low molecular weight G proteins control focal adhesion assembly

Studies of the mechanisms involved in focal adhesion formation have utilized two model systems: stimulation of serum-starved cells that have lost focal adhesions and stress fibres (despite remaining fully spread), and spreading or migrating cells. The former system has revealed the importance of Rho family GTPases in regulating the assembly of focal adhesions and focal complexes (24). In their GTP-bound states, RhoA stimulates the assembly of stress fibres and focal adhesions (42), whereas Rac1 induces formation of lamellipodia and ruffles (43), and Cdc42 causes the assembly of filopodia (Fig. 2) (5, 44). Focal complexes develop during the assembly of lamellipodia and filopodia (5) and contain many of the same components as focal adhesions. They differ in that focal complexes are usually smaller, more transient, and under the regulation of Rac1 and Cdc42, as opposed to RhoA. Many focal complexes develop at or just behind the leading edge of cells. A few focal complexes mature into focal adhesions, while most disassemble before this occurs (45, 46). The

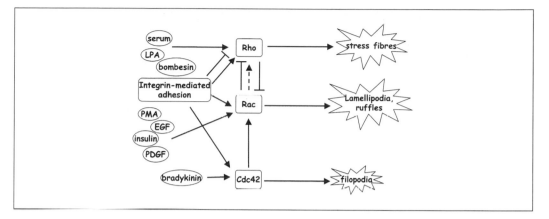

Fig. 2 Rho family GTPases regulate the assembly of stress fibres, ruffles, and filopodia. A variety of soluble factors, as well as integrin-mediated adhesion, activate RhoA, Rac1, and Cdc42. A complex interaction between Rac and Rho is indicated. In some situations, Rac1 has been shown to be upstream of RhoA activation (indicated by the *dotted arrow*), but in many situations the effects of Rac1 and RhoA are antagonistic to each other. (See abbreviations list for LPA, PMA, EGF, and PDGF.)

interplay between Rho family members is complex. Originally, evidence was presented that a cascade occurred in which activation of Cdc42 led to the activation of Rac1, that, in turn, led to the activation of RhoA (5). This pathway corresponds to the temporal sequence in which cells plated on an ECM typically organize their actin cytoskeletons: filopodia are extended first, followed by ruffles (lamellipodia), and with time the cells develop stress fibres and focal adhesions. However, in many systems the Rho phenotype is one of contraction, whereas the Rac1 and Cdc42 phenotypes involve cell extension. These phenotypes are opposite and antagonistic. This is most clearly seen in cultured nerve cells, where Cdc42 and Rac1 promote neurite outgrowth, but RhoA induces neurite retraction (47, 48). Antagonistic interactions between Rac1 and RhoA have been demonstrated in fibroblasts as well, with Rac1 suppressing RhoA activity (46, 49).

Rho family proteins are active in the GTP-bound form and possess intrinsic GTPase activity that hydrolyses the GTP to GDP, thereby inactivating these signalling proteins. Rho proteins thus function as 'molecular switches' that cycle between the active GTP-bound and the inactive GDP-bound states. Activation of Rho proteins is mediated by guanine nucleotide exchange factors (GEFs) and inactivation is promoted by GTPase activating proteins (GAPs). A large number of GEFs and GAPs have been identified. Some have broad specificity, acting on RhoA, Rac, and Cdc42, while others act on single members of the family. Much interest is currently focused on identifying the signalling pathways that lead to GEF activation. Many of the soluble factors that stimulate RhoA activity, such as LPA, thrombin, and sphingosine-1-phosphate act on G protein-coupled receptors. Several RhoA-specific GEFs have been identified that interact with, and are stimulated by, heterotrimeric G_α subunits (50). In several situations, activation of RhoA is regulated by tyrosine phosphorylation (51–55). Calpeptin, an inhibitor of the calcium-activated protease, calpain, was found to be a potent inducer of stress fibres and focal adhesions. Surprisingly this effect was shown to be due not to its action on calpain, but rather to its inhibition of PTPases (56). Subsequent work demonstrated that calpeptin elevated the level of active RhoA and that the target for calpeptin was the PTPase, Shp-2 (57). Interestingly, earlier studies found that cells with dysfunctional Shp-2 exhibited decreased cell spreading and migration, as well as increased focal adhesions (58). These findings are consistent with Shp-2 negatively regulating RhoA activity. Some GEFs, such as members of the Vav family (59, 221), are regulated by tyrosine phosphorylation, and it has been speculated that Shp-2 may be regulating a tyrosine phosphorylated GEF (57). However, it is also possible that Shp-2 may be affecting the activity of a tyrosine phosphorylated GAP or some other upstream component.

3.2 Assembly of focal adhesions

Using the quiescent, serum-starved cell model, it was shown that agents that activate RhoA, or microinjection of activated RhoA, rapidly induce focal adhesion and stress fibre assembly (42). Earlier it had been argued that stress fibres are a manifestation of isometric tension (60), which arises from actomyosin-induced contractility generating tension between sites of tight adhesion (focal adhesions) to a rigid substratum (60). To explore whether the effects of RhoA were due to increased contractility, the action of agents that inhibit actin–myosin interaction, and hence contraction, were examined. It was found that inhibiting contractility blocked RhoA-induced formation of stress fibres and focal adhesions (61). In addition, it was shown that agents that activate RhoA stimulate increased tension in fibroblasts and other cells, and this is accompanied by increased phosphorylation of the myosin light chain (61–63). Myosin light chain phosphorylation is a biochemical correlate of myosin activation. These observations have led to a model of focal adhesion assembly in which the clustering of integrins is driven by RhoA-activated contractility (61, 64).

How might RhoA stimulate the actomyosin system? Many downstream effectors of RhoA have been identified (26), but key to the stimulation of contractility are members of the Rho kinase family (Rho kinase, p160ROCK, ROKα, and ROKβ) (65–70). Kaibuchi's laboratory demonstrated that a target for Rho kinase is the myosin phosphatase, a PP1-type phosphatase that removes phosphates from Ser 19 and Thr 18 on the myosin light chain in both smooth muscle and non-muscle cells (Fig. 3) (67). Phosphorylation of the myosin phosphatase by Rho kinase inhibits the activity of the phosphatase, thereby elevating levels of myosin light chain phosphorylation. Subsequently, Rho kinase was shown to directly phosphorylate the myosin light chain on Ser 19 (71), potentially replacing the action of the myosin light chain kinase (MLCK), which is normally responsible for this phosphorylation. Work from the Matsumura group has shown that both Rho kinase and MLCK play regulatory roles during the formation of stress fibres. In the more peripheral regions of the cell, myosin activity is governed by Ca^{2+}-calmodulin regulated MLCK, whereas in central regions myosin activity is regulated by RhoA activation of Rho

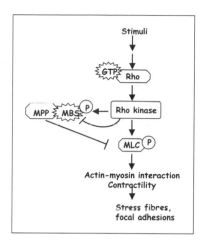

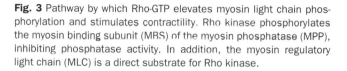

Fig. 3 Pathway by which Rho-GTP elevates myosin light chain phosphorylation and stimulates contractility. Rho kinase phosphorylates the myosin binding subunit (MBS) of the myosin phosphatase (MPP), inhibiting phosphatase activity. In addition, the myosin regulatory light chain (MLC) is a direct substrate for Rho kinase.

kinase (72). Phosphorylation of Ser 19 and Thr 18 are associated with a conformational change in the myosin molecule that promotes its assembly into filaments (73). In addition, light chain phosphorylation stimulates the actin-activated myosin ATPase, thereby increasing the force generated by the myosin heads on the actin filaments (73). Myosin filaments are very effective at crosslinking and bundling actin filaments. The clustering of integrins to form the core of focal adhesions probably results both from the tension transmitted to the integrins via their attached actin filaments and from the bundling of the actin filaments by the myosin as well. This model is illustrated in Fig. 4. It is relevant that the size of a focal adhesion correlates with the size of the attached stress fibre. We envisage that integrins anchored to the actin cytoskeleton are pulled together as their associated actin filaments are bundled into a stress fibre (61, 64). Expression of constitutively active forms of Rho kinase stimulates the assembly of stress fibres and focal adhesions in quiescent cells (65, 74, 75). However, the organization of these structures in these cells is abnormal, suggesting that additional targets downstream from RhoA are required for the normal organization and assembly of stress fibres and focal adhesions.

One possible pathway involves Rho-induced activation of PIP 5-kinase and elevated PIP_2 levels. PIP_2 influences the behaviour of many proteins, including the focal adhesion protein, vinculin (discussed later). Another RhoA effector that induces cytoskeletal changes is mDia1, the mammalian homologue of the *Drosophila* protein Diaphanous (76, 77). Watanabe and colleagues found that mDia1 co-operates with Rho kinase (ROCK) to induce formation of stress fibres and focal adhesions (76). Expression of dominant-negative forms of either Rho kinase or mDia1 inhibits the formation of stress fibres and focal adhesions, indicating that activities of both Rho kinase and mDia1 are necessary for the assembly of these structures (77). However, when either of the two RhoA effectors is overexpressed as an active form and is expressed alone, a distinctive F-actin phenotype results. In the presence of the RhoA inhibitor C3, overexpression of active Rho kinase induces thick, relatively sparse stress fibres that are often organized in a stellate configuration. In contrast, co-

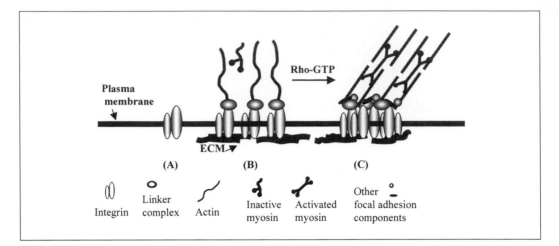

Fig. 4 Model for focal adhesion assembly. Integrin heterodimers are portrayed in three states. (A) The integrin is not bound to extracellular ligands and therefore is not associated with the cytoskeleton. (B) The integrins engage ECM on the outside inducing coupling to actin filaments on the inside. However, the actin filaments are not bundled or under tension, and so the integrins are dispersed in the plane of the membrane. (C) RhoA becomes activated (Rho-GTP), inducing assembly of myosin filaments and their interaction with integrin-bound actin filaments. The bundling of actin filaments by myosin, as well as myosin-induced tension, clusters the integrins, thereby generating a focal adhesion. Other focal adhesion proteins are recruited to the clustered, ligand-bound integrins.

expression of C3 and active mDia1 induces numerous, thin stress fibres that are diffusely distributed. Rho kinase and mDia1 thus function as antagonists towards each other in the thickness and density of the stress fibres they induce. By varying the ratio of Rho kinase to mDia1 expression, stress fibre phenotype can be pushed more towards either distinctive phenotype. At a ratio of 1:1 the induced stress fibres most resemble those induced by RhoA (76). However, further work has shown that co-expression of active Rho kinase and active mDia1 still does not result in a stress fibre phenotype that is identical to that induced by active RhoA (77). Rho kinase has been shown to induce stress fibre formation by regulating actin–myosin contractility (see above). However, the role of mDia1 in stress fibre formation is not so clear. One possibility is that mDia1 could be involved in mediating actin polymerization by its interactions with actin binding proteins such as profilin (78). However, since there appears to be little actin polymerization occurring in focal adhesions, the importance of this functional role for mDia1 in focal adhesion assembly is less apparent.

MDia2 is a homologue of mDia1 that has been identified as a RhoA effector and as a Src binding partner (79). Both mDia1 and mDia2 are necessary for RhoA-dependent formation of stress fibres in HeLa as well as NIH 3T3 cells (79). When active forms of mDia2 and Rho kinase were co-expressed with an inactive Src mutant, stress fibres assembled in a stellate pattern reminiscent of when active Rho kinase is expressed alone (79). In addition, expression of an mDia2 mutant lacking the Src binding domain also disrupted normal stress fibre formation and mimicked the active Rho kinase phenotype (79). This suggests that mDia2 (and homologues such as mDia1) but not Rho kinase, requires Src activity for its co-operative function in stress fibre formation and, by implication, focal adhesion assembly. Thus, members of the Diaphanous family of proteins might provide a link between RhoA signalling and signalling pathways involving Src family kinases.

3.3 Assembly of focal complexes

Much less is known about the assembly of focal complexes. It has been demonstrated, however, that inhibitors of contractility block Rac-induced assembly of focal complexes (46). Interestingly, inhibitors of Rho kinase do not block focal complex formation. This suggests that although, myosin-induced clustering of integrins is required for focal complex formation, this assembly is controlled by a distinct pathway (46). Rac1 and Cdc42 interact with a number of serine/threonine kinases, as well as with other effectors. One downstream kinase for both Rac1 and Cdc42 that has been implicated in myosin light chain phosphorylation is the p21-activated kinase (PAK), but here the data are controversial, with conflicting results and conclusions drawn by different groups. Direct phosphorylation of the myosin light chain has been described *in vitro* (80), and increased light chain phosphorylation has been observed in cells expressing activated PAK (81). In contrast, another laboratory observed decreased light chain phosphorylation in PAK expressing cells and PAK was shown to phosphorylate and inhibit the MLCK (82). The reason for these opposite results is not clear. However, Kiosses *et al.* (83) found that expression of dominant-negative (DN) PAK or constitutively active PAK, both inhibited cell migration, but had opposite effects on contractility and the formation of stress fibres and focal adhesions. DN PAK expression stabilized stress fibres and focal adhesions while active PAK disrupted these structures. Furthermore, active PAK stimulated increased MLC phosphorylation, but DN PAK had no effect on levels of light chain phosphorylation. This suggests that the effects on stress fibres and focal adhesions observed with DN PAK expression, involved a regulatory pathway other than through phosphorylation of MLC. Multiple PAK substrates and pathways could in part explain different effects on light chain phosphorylation concurrent with similar effects on cell migration/adhesion. Additionally, Kiosses and colleagues found that the effects of introducing active PAK into cells were cell type specific (83).

Activated PAK1 has been localized to the tips of filopodia, the leading edge of lamellipodia and to focal adhesions in fibroblasts that have been stimulated with PDGF or activated by wounding of the fibroblast monolayer (84). This seems to suggest that PAK activation is commensurate with activation of the low molecular weight G proteins that regulate formation of all three of these structures, Cdc42, Rac, and RhoA respectively. Filopodia and lamellipodia are dynamic features of the cortical actin network where active actin polymerization is occurring in spreading or migrating cells. Focal complexes typically form at such sites (46). In contrast, focal adhesions have very little actin polymerization and as discussed above, RhoA activation is often antagonistic to Cdc42 and Rac phenotypes. Sells and colleagues also found that inhibitors of PI3 kinase and Src family kinases blocked PAK activation stimulated by wounding (84).

Early work on the development of focal complexes correlated immunofluorescent images with those from IRM or differential interference contrast (45, 85, 86). This work revealed an F-actin-rich, rib-like focal adhesion precursor or focal complex that formed in lamellipodia of motile and spreading cells. Developing focal complexes were observed to divide into two portions. A distal part continued to move with the leading edge, and a proximal part remained stationary (45, 85), presumably because it became engaged with the underlying ECM. In a free leading edge that is unattached to the ECM, the integrins are not engaged with their ECM ligands and consequently, should not be attached at their cytoplasmic face to the actin cytoskeleton (see below). When a leading edge makes contact with the ECM, integrins will bind their ligands and be stimulated to interact with cytoskeletal components. Where the cytoskeletal components are already bundled into rib-like precursors, coupling to ligand-bound integrins should rapidly lead to a small cluster of integrins tethered to the tip of the bundled filaments. The resulting structure is a focal complex. Similarly, a focal complex should also arise when the tip of a bundle of actin filaments making up the core of a filopodium engages with integrins that are ligand-bound.

One major difference between focal complexes and focal adhesions may be the amount of actin polymerization that occurs at these sites. Originally, it was thought that focal adhesions are major regions of actin polymerization since focal adhesions are associated with large bundles of actin filaments. However, although some polymerization may occur in focal adhesions, this is not an important event in their assembly (87). In contrast, the leading edges of lamellipodia and filopodia are major sites of polymerization. The Arp2/3 complex has been implicated in nucleating actin polymerization at these sites (88, 89), but the predicted concentration of the Arp2/3 complex in focal complexes themselves has yet to be demonstrated. If this is the case, the Arp2/3 complex would be a useful marker for distinguishing focal complexes from focal adhesions. Nevertheless, proteins associated with actin polymerization, such as zyxin, VASP, and mena, are found in both focal complexes and focal adhesions (90–92).

VASP and its relations in the Ena/VASP family are recruited to the tail of *Listeria monocytogenes* (92, 93), where rapid actin polymerization is harnessed by this bacterium to propel it through the cytoplasm of host cells (94, 95). Much is being learned about the regulation of actin polymerization from studies using *Listeria*. A key component regulating actin polymerization in this system is the bacterial protein, ActA, which is concentrated at one end of the bacterium. The N-terminal functional domain of ActA has been shown to interact with PIP_2, *in vitro*, suggesting PIP_2 may induce an activating conformational change in a manner similar to its effect on vinculin function (96). ActA has two domains that affect actin polymerization. One domain recruits the Arp2/3 complex and is involved in nucleating polymerization (95). The second domain, a proline-rich FP_4 motif, accelerates polymerization and binds VASP and related proteins, such as mena, through the Ena/VASP homology (EVH1) domain of these proteins (95). In turn, VASP binds the actin monomer binding protein, profilin. Deletion of the VASP binding motifs in ActA inhibits F-actin accumulation at the base of the *Listeria* and decreases bacterial motility (97). Additionally, microinjection of peptides that block VASP and mena binding to their FP_4-containing ligands causes VASP and mena to be displaced from focal adhesions and induces retraction of membrane protrusions (97). VASP contains a second EVH domain (EVH2) that has been shown to bind and bundle F-actin *in vitro* (98). Thus, VASP, and its relatives could provide a link between F-actin and the actin polymerizing complex.

The region within ActA that interacts with VASP is a proline-rich repeat sequence, which is found almost identically in zyxin (99). Indeed, antibodies against ActA cross-react with zyxin (99). Experiments in which the sequences of zyxin related to ActA were expressed and targeted to the cytoplasmic face of the plasma membrane induced actin-rich surface protrusions, further supporting a role for zyxin in actin polymerization (99). Zyxin and the VASP family of proteins are likely to have an important function in regulating actin polymerization in focal complexes. Rottner and colleagues (100) found evidence to support this idea by using GFP-VASP expressed in B16 melanoma, Swiss 3T3, and goldfish fibroblasts to study the dynamics during lamellipodia protrusion (100). Their analysis of the GFP-VASP localization showed a sharp peak of fluorescence along the immediate front of the leading edge of the lamellipodia as well as at the tip of filopodia. Immunoelectron microscopy further identified these regions as the anterior boundary of the actin meshwork in the lamellipodia and the electron dense material at the tip of the filopodia. A second, smaller peak of fluorescence was seen at the posterior base of the lamellipodia. This was identified as GFP-VASP incorporated into focal complexes that developed at the base of the lamellipodia as it moved onwards (100). These results suggest that VASP and its associated proteins such as zyxin and vinculin may provide a stabilizing and docking complex for the rapidly growing ends of the actin filaments at the leading edge and in the developing focal complexes. The above evidence, particularly from the *Listeria* model system, has indicated a role for VASP and Mena in promoting actin polymerization, and this has led to the prediction that these proteins contribute to cell migration by promoting cell protrusion. However, studies with VASP/Mena null cells have given unexpected results. Rather than showing decreased migration, the cells lacking these proteins migrate more rapidly (101). Conversely, overexpression of these proteins retarded the rate of migration. The function of these proteins, particularly at sites such as focal adhesions remains to be determined.

3.4   Maturation of focal complexes into focal adhesions

Observations of live cells reveal that many focal complexes form at the leading lamellae of migrating cells but that only a few of these will typically mature into the larger and more stable focal adhesions (45, 102). In some cases, focal complexes mature by elongating centripetally towards the nuclear region, but in others there is a fusion of focal complexes giving rise to larger focal adhesions. Increasing contractility can stimulate growth of a focal complex into a focal adhesion (46, 103). Focal complexes induced by activated Rac1 in quiescent cells can be further stimulated by activated RhoA to undergo centripetal growth, converting them into focal adhesions (46). Additional work has also shown that as new focal adhesions are formed near the edge of stationary cells, some established focal adhesions move linearly towards the centre of the cell (104). This focal adhesion motility is a result of contraction of the associated actin stress fibres (104).

3.5   Focal adhesions and fibrillar adhesions

A frequently asked question about focal adhesions concerns how heterogeneous they are. For a long time it has been known that different integrins occur in focal adhesions reflecting the nature of the ECM components to which the cells are adhering. However, the assumption has often been that most of the cytoskeletal and signalling components at the cytoplasmic face of focal adhesions are the same, regardless of the integrin that is present. That view is now changing with the application of ratio imaging to study the distribution of particular components within focal adhesions (7). In addition, a number of types of adhesion are now being distinguished which formerly were all grouped together as focal adhesions. These include focal complexes, discussed above, as well as 'fibrillar adhesions'. These latter are adhesions made to fibrillar ECM, particularly fibronectin fibrils that may assemble on either the ventral or dorsal surface of cells (7, 8, 105). In fact, this type of adhesion has been observed previously. Fibronectin fibrils on the surface of cells were noted to align with cytoplasmic bundles of actin filaments (106). Upon microinjection into cells, fluorescent vinculin was seen to assemble into a fibrillar pattern that correlated closely with the pattern of fibronectin fibrils on the upper or lower surfaces of the injected cells and that was distinct from focal adhesions (12). Studying the organization of adhesions by electron microscopy, Chen and Singer (6) distinguished several types of adhesion that differed in their separation from the substratum as well as in their content of α-actinin, vinculin, and fibronectin. They identified two types of adhesion to ECM fibrils. Both of these contained α-actinin, but one of the adhesions had relatively low amounts of vinculin. This latter type of organization probably corresponds to the fibrillar adhesions that have been analysed more recently, which again tend to show low vinculin content, although vinculin can be recruited to these structures following its introduction by microinjection.

In general, fibrillar adhesions differ from focal adhesions in having low amounts of vinculin, paxillin, FAK, and phosphotyrosine (7). In contrast, focal adhesions are prominent sites of tyrosine phosphorylation and are enriched for FAK, paxillin, and vinculin. Fibrillar adhesions are relatively rich in tensin, whereas this is typically lower in focal adhesions. Whether this relative distribution of components in the different adhesions is universal remains to be determined. Additionally, the integrins involved in these adhesions usually differ. Focal adhesions in well spread cells typically have αvβ3 as the major integrin in the core of the adhesion, reflecting adhesion to vitronectin. Only slight amounts of α5β1, the fibronectin receptor, are localized at the periphery of focal adhesions in what has been termed the 'needle eye' configuration (13, 21). Using electron microscopy to study the distribution of β1 integrins in wet cleaved preparations, Meijne and co-workers observed that β1 was localized in discrete clusters at the periphery of focal adhesions, as well as in clusters associated with fibronectin fibrils (21). Not surprisingly, α5β1 is the integrin found in fibrillar adhesions where it binds to fibronectin fibrils. When cells are plated on fibronectin-coated surfaces, the fibronectin is cleared from the substratum if serum is present and replaced by vitronectin (108). Interestingly, if fibronectin is covalently coupled to the substratum such that it cannot be cleared, then α5β1 is prominent in the focal adhesions that form on this rigid substrate and these focal adhesions do not accumulate αvβ3 (8). Under these circumstances, formation of fibrillar adhesions is inhibited.

The development of fibrillar adhesions has been studied (105, 109). Fibronectin fibrils are drawn out from the proximal region of focal adhesions in a manner that depends on actin–myosin contractility moving the fibril centripetally away from focal adhesions. This requires engagement of the α5β1 integrin with fibronectin. Pankov and co-workers have shown that the assembly of these adhesions is blocked by a dominant-negative tensin construct (105). The role of rigidity of the ECM in development of the different types of adhesion is intriguing. It appears that if the matrix is pliable, the matrix component, e.g. fibronectin, will be drawn out and assembled into fibrils. However, if the matrix is rigid, such as occurs with vitronectin adsorbed to glass or with crosslinked fibronectin, then isometric tension develops (60) and the resulting adhesion is a focal adhesion. The response to substrates of different plasticity may explain why in tissue culture focal adhesions are very common but are rarely seen in tissues in the body. In tissue culture, cells adhere typically to vitronectin that adsorbs very tightly to the underlying plastic or glass surface, thereby generating a rigid matrix. In contrast, within tissues most ECM is relatively pliable with the result that the adhesions that develop will tend to be of the fibrillar adhesion type.

4. Integrin–cytoskeletal links

As noted earlier, key components of focal adhesions are integrins. The integrins found in focal adhesions are α/β heterodimers with each subunit spanning the membrane once and possessing a short cytoplasmic domain (17). The functions of the cytoplasmic domains have been studied using chimeric constructs and mutational analysis. Deletion of the β subunit cytoplasmic domain blocks integrin localization to focal adhesions (110–112), whereas deletion of α cytoplasmic domains drives the integrin into pre-existing focal adhesions (113, 114). Expression of integrin β cytoplasmic domains as chimeras with irrelevant transmembrane and extracellular domains targets these chimeras to focal adhesions (115, 116). These results indicate that the cytoplasmic domain of β integrins mediates localization of integrins to focal adhesions, which is opposed by the α cytoplasmic domain. This antagonism is relieved when integrins bind their extracellular ligands (115). This has led to the idea that ligand binding induces a conformational change in the integrin dimer, such that the β cytoplasmic domain is unrestrained by the α cytoplasmic domain and thus able to interact with cytoskeletal components.

The interaction of integrin β subunit cytoplasmic domains with cytoskeletal proteins has been studied using several techniques. Potential interactions, as well as binding sites within β cytoplasmic sequences, have been identified by various binding assays. The relatively low affinity of the interactions, however, has posed a problem for identifying which associations occur *in vivo*. Techniques such as co-immunoprecipitation, which have been used to identify associations between other membrane receptors and cytoskeletal proteins, have generally failed when used with integrins. Two major cytoskeletal proteins, talin and α-actinin, bind to β1 integrins *in vitro* (117, 118). Both proteins interact with other focal adhesion proteins and each also binds actin, suggesting that both could provide direct links between integrins and actin filaments (Figs 5 and 6). To examine cytoplasmic domain interactions, Ginsberg's laboratory created chimeric β cytoplasmic domain constructs that exist as dimers and used these to precipitate binding proteins from cell lysates (119). Interestingly, this approach confirmed an interaction with talin, but failed to identify binding of α-actinin to the integrin cytoplasmic domains. With respect to talin binding, it was found that an alternatively spliced variant of the β1 cytoplasmic domain, β1D, demonstrated stronger binding. This was consistent with previous work which used both *in vitro* binding as well as immunoprecipitation to demonstrate that this muscle-specific form of β1 integrin bound talin with higher affinity than the more widespread β1A isoform (120). The lack of α-actinin binding to the dimeric cytoplasmic domain constructs was surprising (119). In leukocytes, co-immunoprecipitation of α-actinin with β2 integrins has been observed, but it depends on activation of the leukocytes (121). In resting leukocytes, the β2 integrin is preferentially associated with talin, but upon activation, talin becomes cleaved by calpain and α-actinin binding was detected (122). However, the role of α-actinin as a link between actin filaments and β1 integrins has been questioned by experiments of Lewis and Schwartz (123). These investigators expressed chicken β1 integrins in mouse fibroblasts and then examined the co-distribution of cytoskeletal proteins with the clustered chicken integrins. They found that deletion of the 13 most C-terminal residues from the β1 cytoplasmic domain prevented the co-localization of talin and actin with the clustered integrin but did not affect the co-clustering of α-actinin. These data support the idea that talin provides a link between integrins and actin filaments. However, the data also suggest α-actinin alone is not sufficient to recruit actin to the clustered integrins (123). Deletion of an additional 15 residues prevented α-actinin from co-localizing with the clustered integrins, indicating an interaction between α-actinin and integrin cytoplasmic domains (123).

The dimeric β1 cytoplasmic constructs that were observed to pull down talin from cell lysates also precipitated the cytoskeletal protein filamin (119). Previous work identified an interaction between filamin and the β2 integrin cytoplasmic domain (124). Filamin (actin binding protein) is a major cytoskeletal protein that is found

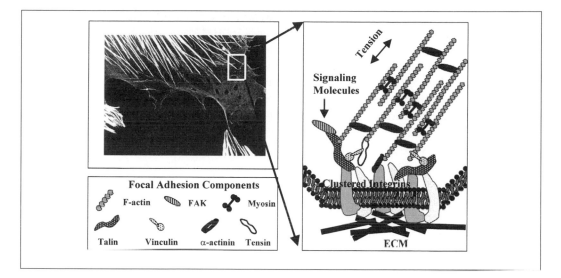

Fig. 5 Diagram of some of the potential links between actin filaments and integrins in focal adhesions. Only a few focal adhesion proteins are represented in this diagram. Emphasis is placed on those that have been implicated in binding to integrin cytoplasmic domains.

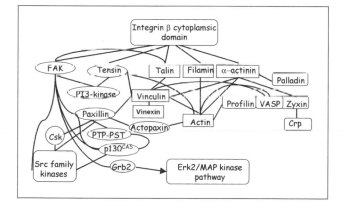

Fig. 6 Diagram illustrating interactions between focal adhesion components. Lines between proteins indicate interactions that have been identified *in vitro* or *in vivo*. (VASP, vasodilator-stimulated phosphoprotein; PTP-PEST, protein tyrosine phosphatase with PEST repeats; FAK, focal adhesion kinase; Csk, c-src kinase; Grb2, adaptor protein; PI3 kinase, phosphoinositide 3-kinase; MAP kinase, mitogen-associated protein kinase; Crp, cysteine-rich protein.)

along stress fibres, but it does not appear to be enriched within focal adhesions. Filamin is prominent in membrane ruffles and lamellipodia, raising the possibility that it may participate in linking integrins to the actin cytoskeleton within these structures and possibly in focal complexes rather than in focal adhesions. It is noteworthy that in platelets, filamin has been shown to mediate a link between actin filaments and the non-integrin receptor glycoprotein Ib (125). Tensin is another protein that has been suggested to interact directly with integrin β cytoplasmic domains, based on tensin co-clustering with aggregated integrins (126). Tensin is an actin binding protein of focal adhesions and fibrillar adhesions, and the presence of tensin has been shown to increase the rate of actin polymerization *in vitro* (127–129).

Most work has focused on the interactions of the β subunit cytoplasmic domain with cytoskeletal proteins. However, Liu and co-workers have found that the cytoplasmic tail of the integrin α4 subunit binds paxillin. Interestingly, α4 is rarely, if ever, found in focal adhesions. They expressed a chimeric protein composed of the αIIb extracellular and transmembrane domains and the α4 cytoplasmic tail, which paired with the β3 subunit. Expression of this construct bound paxillin, increased the phosphorylation of FAK, and increased cell migration when these cells were plated on fibrinogen. Conversely, cell spreading and the formation of stress fibres and focal adhesions decreased (130, 131). A point mutation in the α4 tail disrupted paxillin binding and reversed these effects on cytoskeletal organization and cell adhesion/ migration (130).

Much remains to be learned about how cytoskeletal proteins link to integrin cytoplasmic domains. However, we now know a bit about the roles played by some of these proteins in integrin–cytoskeletal linkages, and subsequent regulation of cell behaviour. Talin is a prominent component of focal adhesions and has been shown to bind to integral membrane proteins such as integrins (117, 119, 120, 132, 133) and layilin (134), as well as to actin filaments and other focal adhesion proteins such as vinculin (135, 136) and FAK (134). Thus, talin has long been thought to play a role linking membrane proteins with the cytoskeleton and providing a platform for signalling molecules. The role of talin as a link between integrins and actin has been investigated, using embryonic stem (ES) cells deficient in intact talin (137). In their undifferentiated form, talin (–/–) ES cells display extensive membrane blebbing, significantly reduced β1 integrin expression and a reduced capacity for adhesion to gelatin or laminin. These talin null cells were unable to assemble focal adhesions or stress fibres upon attachment to FN, supporting the notion that talin is a critical component for focal adhesion assembly. However, the inhibition of β1 expression could also account for the inability to develop focal adhesions. An unexpected result was that upon differentiation these cells were able to form small focal adhesions, indicating that in some situations talin is not needed for focal adhesion assembly. The role of talin in integrin expression is not clear. Evidence has been presented that talin may promote stabilization of newly synthesized β1 integrin at the level of the ER (138). One study suggests that talin regulates exit of α5β1 integrin from early compartments in the secretory pathway (139).

Talin is a large, elongated protein that can form homodimers (140), and can be cleaved into a small head domain plus large tail domain by the calcium-dependent protease, calpain (141). Binding sites for actin, as well as vinculin and integrin, have been localized to the tail domain (117, 135, 136, 141, 142). Until recently, only FAK and PIP$_2$ were known to interact with the N-terminal talin head domain (134). Calderwood and colleagues (133) have now identified an integrin binding site spanning the head domain as well as integrin binding sites within the tail region. The binding of the talin head to integrin was shown to be specific since a point mutation within the NPXY motif of the integrin cytoplasmic domain abrogated talin head binding. The same point mutation blocked integrin localization to focal adhesions as well. Overexpression of a talin construct that included the head domain and part of the tail domain was able to activate ligand binding affinity of αIIbβ3 integrin exogenously expressed in CHO cells (133). However, a similar construct that lacked the head domain was unable to activate these integrins when co-expressed in the same cells (133). These results indicate that the binding of talin to the β3 cytoplasmic domain may play a role in regulating the affinity of αIIbβ3 for its extracellular ligand. It has been hypothesized that talin may be involved with regulating platelet adhesion (143). When platelets are activated, talin is redistributed to the plasma membrane from the cytoplasm (143). However, talin redistribution is integrin-independent because the same pattern of localization occurs in Glanzman thrombasthenic platelets that lack αIIbβ3 expression (143). Clearly talin can interact with other membrane components that affect its cellular localization. In addition to binding integrin cytoplasmic domains, talin has recently been shown to interact with another transmembrane protein, layilin, which is prominent in membrane ruffles (134).

One study using leukocytes determined that β2 integrin activation is mediated in part by interactions with cytoskeletal proteins (122). In resting leukocytes the β2 integrins were constitutively associated with actin filaments by talin-mediated linkage. Activation of the leukocytes with PMA induced proteolytic cleavage of talin and its dissociation from the integrins. However, the integrins re-established links with actin through α-actinin (122). Mutational analysis suggested that β2 affinity for α-actinin is regulated by a conformational change in the integrin cytoplasmic tail that unmasked a cryptic α-actinin binding site (122). Another situation in which cleavage of talin has been observed is in smooth muscle cells responding to addition of collagen fragments (144). In the presence of collagen fragments, focal adhesions are disassembled and this correlates with calpain cleavage of talin, as well as FAK and paxillin. Although proteolysis of talin may be central to the disassembly of focal adhesions in this instance, the disassembly could be the result of other signalling pathways activated by the collagen fragments. Clearly integrins are able to form multiple modes of linkage with the actin cytoskeleton. However, the particular links and the context in which some might regulate integrin activation are not yet fully understood.

Vinculin is one of the major components of focal adhesions and has many binding partners including talin, paxillin, α-actinin, tensin, VASP, vinexin, ponsin (220), F-actin, PIP$_2$, and itself (reviewed in ref. 23). Folding of vinculin, mediated through the binding of its N-terminal globular head domain to its C-terminal tail, negatively regulates the function of this protein by masking protein binding sites (145, 146). Interaction with lipids such as PIP$_2$ (147, 148) releases vinculin from its repressed conformation, allowing access to potential binding partners. RhoA activates PIP 5-kinase resulting in elevation of PIP$_2$ levels (149). The resulting enhancement of vinculin's interactions with its binding partners may contribute to the stimulation of focal adhesion formation by RhoA. A recently identified protein, vinexin, binds via an SH3 domain to the proline-rich, hinge region of vinculin (150). Vinexin binding to vinculin may also induce a conformational change in vinculin, enhancing its interactions with other proteins. Overexpression of vinexin promotes cell spreading and increases the size of focal adhesions (150). Contrary to early expectations, vinculin is not essential for formation of focal adhesions, as cells lacking this protein retain the ability to assemble these adhesions, although vinculin null cells display enhanced migration (151–153).

The role of vinculin in cell locomotion has been investigated by Xu and colleagues (154). Whereas reintroduction of intact vinculin into vinculin null F9 cells restored wild-type characteristics, expression of the head domain exaggerated the mutant phenotype, doubling the locomotion rate of the mutant cells. Expression of the vinculin tail, however, reduced cell migration (154). These findings complement work performed by Laine and colleagues (155) who identified vinculin as a protein recruited to the bacterial surface of the intracellular pathogen *Shigella flexneri*, and used by the bacterium for actin-based movement. *Shigella* infection results in proteolytic cleavage of vinculin to generate the 90 kDa head domain. It is only this fragment of vinculin that promotes bacterial movement. Links between vinculin and actin polymerization are suggested by vinculin's interaction with VASP (156), which, as mentioned earlier, is implicated in actin polymerization through its interaction with profilin (157).

Ezrin, radixin, and moesin (ERM) are three closely related proteins that function in membrane–cytoskeletal interactions, particularly at sites of membrane protrusion (reviewed in ref. 158). This family of proteins has been implicated in the assembly of focal adhesions. Moesin was identified as a critical factor required for reconstitution of RhoA-induced focal adhesion and stress fibre assembly in a permeabilized cell system (159). This result was unexpected because, although ERM proteins are found in many sites of actin–membrane interaction, they are not prominent in focal adhesions. How ERM proteins contribute to the assembly of focal adhesions and stress fibres remains to be determined. The function of ERM proteins is regulated by an intramolecular head-to-tail interaction (160–162), reminiscent of vinculin regulation. Release of this interaction unmasks potential binding sites for both actin and membrane proteins, such as CD44 (163). The conformational change in ERM proteins is promoted both by Rho kinase phosphorylation (164) and by PIP$_2$ (165, 166). In some cells, RhoA activation stimulates the assembly of apical membrane protrusions and ERM proteins appear to be directly involved in this RhoA response (167).

New proteins that target to focal adhesions continue to be identified including two recently isolated focal adhesion components, actopaxin (168) and palladin (169). Actopaxin binds both paxillin and actin, and co-localizes with paxillin in focal adhesions and focal complexes at the leading edge of migrating cells. Ectopic expression of truncated actopaxin that does not bind paxillin, interferes with cell adhesion and spreading, indicating that actopaxin plays a role in these cell processes (168). Palladin is concentrated in focal adhesions but is also found periodically distributed along stress fibres. It has been detected in cell–cell junctions and embryonic Z-lines as well (169). Attenuation of palladin expression with antisense causes cytoskeletal disruption and cell rounding (169), suggesting that palladin either performs a key structural function or is signalling to the cytoskeleton possibly via the Rho family.

5. Hierarchical structure of focal adhesions

Focal adhesions are sites where integrins are both clustered and engaged with their extracellular ligands. Miyamoto and colleagues investigated the relative roles of clustering versus ligand occupancy (126, 170). Their approach was to aggregate integrins using beads coated with antibodies directed against integrin extracellular epitopes or integrin ligands. Two types of antibodies were used that were similar in their abilities to cluster integrins but differed in their capacities to mimic soluble ligand-activation of integrins. Inhibitory antibodies are those that block cell adhesion by interacting with and blocking ligand binding sites, whereas non-inhibitory antibodies do not block ligand binding. Recruitment of cytoskeletal or signalling components to antibody-coated beads was assessed by immunofluorescence microscopy. It was observed that clustering of integrins with non-inhibitory antibodies recruited tensin and FAK to the beads. However, clustering integrins with inhibitory antibodies, or with non-inhibitory antibodies together with soluble ligands, resulted in the additional recruitment of actin, α-actinin, talin, and vinculin to the beads (126). These results suggested that FAK and tensin are associated with integrins in the absence of ligand binding, although whether this association is constitutive or induced by clustering has not been determined. On the other hand, ligand occupancy appears to be critical for the association of actin, α-actinin, talin, and vinculin with integrins. An interaction between the N-terminal domain of FAK and integrin cytoplasmic domains was independently identified using synthetic peptides and *in vitro* binding assays (171). This latter work found a potential binding site on the integrin β1 cytoplasmic domain close to the transmembrane sequence. However, in a separate set of experiments, deleting the most C-terminal 13 amino acids from the cytoplasmic domain prevented the co-clustering of FAK with aggregated integrins (123). Interestingly, the recruitment of FAK to focal adhesions does not involve the N-terminal domain that binds integrins *in vitro*, but is mediated by the C-terminal domain of FAK (172).

Following ligand binding and clustering of integrins, many signalling pathways are triggered. A prominent event is the tyrosine phosphorylation of multiple proteins concentrated within focal adhesions. Applying beads coated with antibodies against integrins to cells, it was reported that the recruitment of a large number of signalling components to the region of clustered integrins depended on tyrosine phosphorylation. Recruitment of these signalling molecules is inhibited by tyrosine kinase inhibitors (170). Several of the signalling proteins had not previously been detected in focal adhesions or focal complexes, which may be due to their transient association with these structures or to some aspect of the assay. Nevertheless, these results support the idea that one function of tyrosine phosphorylation within focal adhesions is to recruit additional components that bind to specific tyrosine phosphorylation sites on other proteins such as FAK, paxillin, or p130cas (173–175).

The above results suggest a hierarchy of assembly, in which integrin occupancy, clustering, and the tyrosine phosphorylation of various 'scaffold' proteins together contribute to the formation of focal complexes and focal adhesions (170). An important caveat to this hierarchy, however, is that although integrins can be experimentally clustered from the outside, clustering is normally dependent on Rho family GTPases acting from within the cell (176). It has often been thought that clustering is driven by the ECM that acts like a large multivalent ligand. However, cells can adhere to ECM without clustering their integrins (61, 176). Our interpretation is that ECM is critical in providing the extracellular ligand for integrins, because ligand occupancy is necessary for coupling integrins to the cytoskeleton. Intracellular cytoskeletal forces drive aggregation of integrins into focal complexes and focal adhesions (46, 129). Upon binding to ECM and becoming linked to the cytoskeleton, whether a particular integrin becomes clustered or not depends on the state of the cytoskeleton, which is regulated, in turn, by RhoA, Rac, Cdc42, and other factors. Integrin-mediated adhesion to ECM, itself affects the activation state of Rho family GTPases (see below), with the potential for differing degrees of positive or negative feedback on the clustering of integrins.

6. Integrin-mediated signalling and focal adhesions

For the past decade, research on focal adhesions has been dominated by investigations of their role in signal transduction and by studies of signalling pathways initiated in response to integrin-mediated adhesion. This area is too extensive to cover adequately here, but we will discuss briefly a few central topics. Other reviews of integrin-mediated signalling should be referred to for greater detail (23, 27–29).

A major consequence of integrin-mediated adhesion is a series of tyrosine phosphorylation events. The major tyrosine kinases activated are FAK and members of the Src family (177), but other kinases such as Abl (178) and Syk (179) have also been found to be activated depending on the cell type. For many cells, adhesion to ECM stimulates autophosphorylation of FAK. This provides a binding site for SH2 domains of Src family kinases, thereby recruiting these kinases into a complex (174). In turn, Src family kinases phosphorylate additional sites within FAK that act to recruit other SH2 domain-containing signalling proteins (180). The FAK/Src complex also phosphorylates other focal adhesion proteins such as paxillin and p130cas, which similarly recruit other binding partners (173). What are the consequences of the tyrosine phosphorylation that occurs within focal adhesions? Initially, it was argued that FAK and its kinase activity contributed to the assembly of the structural components of focal adhesions and focal complexes. This idea has been largely dispelled, not only because FAK null cells assemble particularly prominent focal adhesions (181), but also because displacement of FAK from focal adhesions does not disrupt these structures or prevent their formation (182). Clues to the function of integrin-mediated tyrosine phosphorylation come from several directions. The FAK null cells reveal decreased motility (181), as do cells in which FAK has been displaced from focal adhesions (182). Conversely, when FAK is overexpressed in cells or when activated forms of FAK are expressed, there is an increase in cell migration (183). These observations have led to the idea that FAK tyrosine phosphorylation within focal adhesions and focal complexes may contribute to their disassembly, an event necessary for cell migration. FAK null fibroblasts fail to transiently inhibit RhoA activity when these cells are plated on fibronectin.

A large body of data indicates that FAK and its downstream targets provide signals that synergize with growth factor signalling (184, 185). There is an extensive literature concerning anchorage-dependent growth control, the phenomenon in which normal cells fail to respond to growth factors when they are not adhering to an ECM. Integrin-mediated adhesion is associated with activation of the MAP kinase pathway (180, 186, 187). Evidence for multiple pathways leading to MAP kinase activation has been presented (184, 185) but some data indicate that FAK tyrosine phosphorylation generates a binding site (tyrosine 925) for the adapter Grb2 (180), which activates the Ras/MAPK pathway. Displacement of FAK from focal adhesions decreases the response of cells to growth factors, as judged by DNA synthesis (182). Expression of a constitutively active form of FAK in epithelial cells converted them to an anchorage-independent cell line (188). These observations argue for a role for FAK in the synergy between growth factor and integrin signalling. Not only is anchorage (i.e. integrin engagement) required for normal cells to respond to growth factors, but also sustained loss of anchorage is frequently a signal for apoptosis (189, 190). Epithelial cells that enter apoptosis upon suspension can be rescued by expression of activated FAK (188), whereas strategies to decrease FAK levels or to perturb FAK function promote apoptosis (191, 192). These findings indicate that FAK is important both in the normal anchorage-dependent response of cells to growth factors and as a survival factor, antagonizing apoptosis.

Other work has revealed that integrin-mediated adhesion activates Rho family GTPases (Fig. 7). It has been known for a long time that cell adhesion to ECM promotes the extension of filopodia and membrane ruffles (193), processes now recognized as being controlled by Cdc42 and Rac1 respectively (24). This led to the prediction that adhesion to ECM would activate Cdc42 and Rac1, and this has been experimentally confirmed (194, 195). The pathway from integrins to activation of Rac1 and Cdc42 has not been resolved, but in the case of Rac1 activation may involve the FAK/p130cas/Crk interaction mentioned above (196). One protein that binds to Crk in response to integrin-mediated adhesion is DOCK180 (197, 198), a protein that interacts with Rac1 and increases Rac-GTP levels (Fig. 7) (199).

Whether integrin engagement activates RhoA has been more controversial. It was observed that adhesion of quiescent cells to fibronectin failed to induce the assembly of stress fibres and focal adhesions, unless RhoA was activated by addition of soluble agents (176). However, other investigators noted that prolonged adhesion of similarly quiescent cells induced focal adhesions and stress fibres, indicating eventual RhoA activation (200). Development of techniques to measure RhoA-GTP levels made it possible to analyse RhoA activity directly in response to adhesion to fibronectin (201). Interestingly, this work demonstrated that upon plating on fibronectin, there is an initial dip in RhoA activity, followed by an increase at longer time points. This biphasic response in RhoA activity is consistent with the two earlier studies that had seemed in conflict. Thus, short times of adhesion to fibronectin do not activate

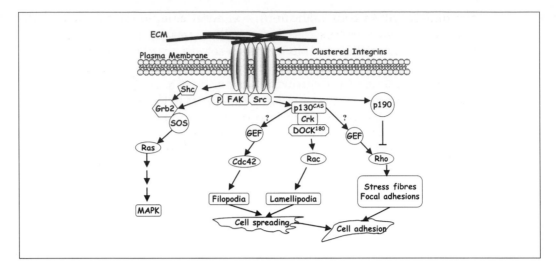

Fig. 7 Diagram illustrating some of the signalling pathways that emanate from focal adhesions. Integrin-mediated signalling synergizes with growth factor signalling probably via FAK (focal adhesion kinase) and Shc (Src homology domain carboxy terminal), and their interactions with Grb2 (adaptor protein), which in turn links to SOS (son of sevenless exchange factor) and to activation of the Ras/MAP kinase pathway. The pathways by which Cdc42, Rac1, and RhoA become activated have not been established, but we presume that in each case a guanine nucleotide exchange factor (GEF) must be involved immediately upstream of the GTPase. (Crk, Src homology domain-containing adaptor protein; DOCK180, 180 kDa protein downstream of Crk, p190/p190RhoGAP.)

RhoA (176), whereas longer times of adhesion to fibronectin increase RhoA activity (202). The increase in RhoA activity in response to prolonged adhesion to fibronectin is weak compared to the activation induced by soluble factors such as lyso-phosphatidic acid or thrombin (201).

Pursuing the pathway by which integrin engagement signals to RhoA, Arthur and co-workers found that suspended cells respond to soluble integrin ligands with a transient decrease in RhoA activity, similar to that seen in response to adhesion to fibronectin (203). This response to integrin ligands was blocked by inhibitors of Src family tyrosine kinases and was absent in cells from which Src, Yes, and Fyn, the three widespread Src family kinases, had been deleted. Re-expression of Src, however, restored the response. Downstream from Src activation, p190RhoGAP was identified as becoming tyrosine phosphorylated and activated in response to integrin engagement in a Src-dependent manner. A dominant-negative form of p190RhoGAP blocked the integrin-induced decrease in RhoA activity (203). In parallel work, Ren and co-workers noted that the dip in RhoA activity failed to occur in cells lacking FAK. Notably these cells have pronounced focal adhesions, consistent with elevated RhoA activity (204). In other work, p190RhoGAP has been detected binding to FAK (205), raising the possibility that FAK acts as a scaffold for both Src and p190 RhoGAP, bringing these two components together downstream from integrin engagement. It has been speculated that the transient decrease in RhoA activity in response to integrin-mediated adhesion may have a physiological function during cell migration. As cells migrate, the leading lamellipodium engages the ECM via integrins, but this protrusion is initially unstable. High contractile force generated in response to active RhoA would tend to retract this protrusion and thus antagonize migration. In order to sustain productive protrusions, it would be advantageous for the cell to suppress RhoA activity transiently following integrin engagement.

In the context of signalling from focal adhesions, most attention has been directed toward the activation of kinases and their downstream cascades. However, other modes of signal transduction may also be initiated from focal adhesions. For example, zyxin contains a nuclear export signal and this protein has been shown to shuttle between the nucleus and focal adhesions (206). The significance of this shuttling has not been established. Zyxin contains LIM domains that are also prominent in several transcription factors (207). The precedent of proteins functioning within the nucleus to regulate transcription, as well as at regions of cell adhesion has been established with β-catenin (208). It remains to be determined whether zyxin regulates transcription, or functions to transfer information from focal adhesions into the nucleus, or from the nucleus to focal adhesions.

7. Syndecan-4-mediated signalling in focal adhesion assembly

In addition to integrins, another type of transmembrane cell surface receptor, syndecan-4, has been identified in focal adhesions and appears to have a role in focal adhesion dynamics. Syndecan-4 is a heparan sulfate proteoglycan (HSPG) that interacts with a variety of ECM components and co-localizes with integrins in focal adhesions (209). In serum-starved cells, syndecan-4 is recruited to focal adhesions by the activity of PKCα (210). There may be a feedback mechanism involved in this process since the cytoplasmic domain of syndecan-4 can bind to PKCα and potentiate its kinase activity as well (211, 212). In conjunction with integrins, syndecan-4 participates in the control of cell adhesion and cytoskeletal organization. Overexpression of full-length syndecan-4 core protein in CHO cells increases cell spreading, and potentiates stress fibre and focal adhesion formation (213, 214). These cells show decreased motility and a more flattened morphology. However, in contrast, overexpression of the syndecan-4 core protein with partial or full deletion of the cytoplasmic domain decreases adhesion and the presence of stress fibres and focal adhesions (213, 214). These effects were independent of the composition of the ECM and the specific integrins engaged (213, 214). The syndecan-4 cytoplasmic domain is thus essential for focal adhesion formation in these cells. Syndecan-4 cytoplasmic domain also directly interacts with PIP_2 that, in turn, activates PKCα (215).

Activation of PKCα potentiates cell spreading and, in some circumstances, PKC activation has been shown to synergize with RhoA in the development of focal adhesions, as for example when cells are plated on anti-integrin antibodies (216). The activation of PKC by syndecan-4 suggests that it may signal co-operatively with integrins with respect to the organization of actin during cell adhesion. As discussed above syndecan-4 has been shown to localize to focal adhesions in a PKC-dependent manner and in turn to activate PKC activity (211, 212). This suggests that integrins and syndecan-4 function as co-receptors for ECM components and may signal co-operatively in actin reorganization during cell adhesion. It has long been known that integrins and heparan sulfate proteoglycans bind two different portions of the fibronectin (FN) molecule (217). Integrins bind to the RGD sequence within the large 120 kDa proteolytic fragment of FN, whereas syndecan-4 binds the FN heparin binding segment. These two fibronectin domains have been used to examine the relative importance of integrin and syndecan-4 in adhesion-mediated signalling and the formation of stress fibres and focal adhesions. Saoncella and colleagues (218) used fibronectin null mouse fibroblasts (FN$-/-$) that they plated on the RGD-containing 120 kDa FN domain or on antibodies directed against the β1 integrin extracellular domain. FN ($-/-$) cells plated on the 120 kDa fragment failed to spread, remained rounded, and did not form stress fibres and focal adhesions. When these same cells were then supplemented with the 40 kDa heparin binding fragment in solution, the adherent cells then spread and formed stress fibres and focal adhesions. Similar results were obtained when an antibody directed against the extracellular domain of syndecan-4 was used in solution to supplement FN ($-/-$) cells plated on the 120 kDa fragment. Furthermore, when the cells were plated on anti-β1 integrin antibodies and supplemented with soluble anti-syndecan-4 antibodies, these cells spread fully, and formed stress fibres and focal adhesions as well. Thus, occupancy of both integrins and syndecan-4 is required for full cell spreading and the assembly of stress fibres and focal adhesions on fibronectin. Cells plated on the 120 kDa fragment and supplemented with LPA were able to spread and organize actin stress fibres and focal adhesions as well. This suggests that syndecan-4 occupancy stimulates RhoA activation. Strikingly, when these experiments were repeated in the presence of the RhoA inhibitor C3 exotransferase, supplementing the adherent cells with either the heparin binding FN fragment or the anti-syndecan-4 antibody failed to induce cell spreading on the 120 kDa substrate and blocked the assembly of stress fibres and focal adhesions. These results further indicate that RhoA activation occurs downstream from syndecan-4 engagement (218).

Finally, results from another study indicate that the role of syndecan-4 in stress fibre and focal adhesion assembly is more complex than the above results suggest. Using fibroblasts from syndecan-4-deficient mice, Ishiguro and associates (219) found that these cells were able to assemble normal stress fibres and focal adhesions when plated on fibronectin, or on a mixture of the 120 kDa and heparin binding fragments. This suggests that some other molecule(s) is able to compensate in the absence of syndecan-4. However, a further complication is indicated since stress fibre and focal adhesion assembly was inhibited when the syndecan-4 null fibroblasts were plated on the 120 kDa FN fragment and supplemented with soluble heparin binding fragments. Thus, syndecan-4 may be important when the signal from the FN heparin binding domain is delivered in a soluble form as opposed to a substrate-bound form. Clearly, further elucidation of these signalling dynamics is needed.

8. Concluding remarks

The prominence of stress fibres and focal adhesions is striking in cells grown in tissue culture compared with cells in tissues. As mentioned earlier, stress fibres are also seen in the granulation tissue of healing wounds. Our interpretation is that stress fibres and focal adhesions are prominent in cultured cells because tissue culture mimics the wound environment. Very frequently, cells are grown in the presence of serum as a source of growth factors. However, serum contains many products of platelet secretion that would normally function in a wound response. For example, PDGF is a growth factor that stimulates proliferation of many cells, but it also activates Rac1. This promotes cell migration and the formation of focal complexes. Also in serum are factors that potently activate RhoA, such as LPA and thrombin. These promote contractility, thereby contributing to wound closure. In tissue culture, however, the enhanced contractility, combined with strong adhesion to a rigid substrate, generates isometric tension and the formation of stress fibres and focal adhesions. In a tissue, the factors released by platelets should contribute to the closure of wounds by stimulating migration (mediated by Rac1) and the contraction of the edges of a wound (mediated by RhoA). Focal adhesions in cultured cells may thus reflect the 'wound' environment of tissue culture, together with artefactual adhesion to a rigid substratum of plastic or glass. Nevertheless, these structures provide models for studying both the cytoskeletal links to integrins and the signalling complexes that assemble in response to integrin engagement and clustering. Many unanswered questions remain concerning the organization of these complex structures and the signalling pathways that emanate from them. Rather than slowing down, the pace of research in this area has quickened. As new structural components continue to be found and novel signalling pathways discovered, the next few years promise to be exciting for research on focal adhesions and focal complexes.

Acknowledgements

We gratefully acknowledge Sarita Sastry and Becky Worthylake for their comments and advice on this manuscript. The authors have been supported by NIH grants GM29860 and HL45100.

References

1. Abercrombie, M., Heaysman, J., and Pegrum, S. M. (1971). The locomotion of fibroblasts in culture. *Exp. Cell Res.*, **67**, 359.
2. Abercrombie, M. and Dunn, G. A. (1975). Adhesions of fibroblasts to substratum during contact inhibition observed by interference reflection microscopy. *Exp. Cell Res.*, **92**, 57.
3. Izzard, C. S. and Lochner, L. R. (1976). Cell-to-substrate contacts in living fibroblasts: an interference reflexion study with an evaluation of the technique. *J. Cell Sci.*, **21**, 129.

4. Heath, J. P. and Dunn, G. A. (1978). Cell to substratum contacts of chick fibroblasts and their relation to the microfilament system. A correlated interference-reflexion and high-voltage electron-microscope study. *J. Cell Sci.*, **29**, 197.

5. Nobes, C. D. and Hall, A. (1995). Rho, rac, and cdc42 GTPases regulate the assembly of multimolecular focal complexes associated with actin stress fibers, lamellipodia, and filopodia. *Cell*, **81**, 53.

6. Chen, W. T. and Singer, S. I. (1982). Immunoelectron microscopic studies of the sites of cell–substratum and cell–cell contacts in cultured fibroblasts. *J. Cell Biol.*, **95**, 205.

7. Zamir, E., Katz, B. Z., Aota, S., Yamada, K. M., Geiger, B., and Kam, Z. (1999). Molecular diversity of cell–matrix adhesions. *J. Cell Sci.*, **112**, 1655.

8. Katz, B. Z., Zamir, E., Bershadsky, A., Kam, Z., Yamada, K. M., and Geiger, B. (2000). Physical state of the extracellular matrix regulates the structure and molecular composition of cell-matrix adhesions. *Mol. Biol. Cell*, **11**, 1047.

9. Lazarides, E. and Burridge, K. (1975). Alpha-actinin: immunofluorescent localization of a muscle structural protein in nonmuscle cells. *Cell*, **6**, 289.

10. Wehland, J., Osborn, M., and Weber, K. (1979). Cell-to-substratum contacts in living cells: a direct correlation between interference-reflexion and indirect-immunofluoescence microscopy using antibodies against actin and alpha-actinin. *J. Cell Sci.*, **37**, 257.

11. Geiger, B. (1979). A 130K protein from chicken gizzard: its localization at the termini of microfilament bundles in cultured chicken cells. *Cell*, **18**, 193.

12. Burridge, K. and Feramisco, J. R. (1980). Microinjection and localization of a 130K protein in living fibroblasts: a relationship to actin and fibronectin. *Cell*, **19**, 587.

13. Damsky, C. H., Knudsen, K. A., Bradley, D., Buck, C. A., and Horwitz, A. F. (1985). Distribution of the cell substratum attachment (CSAT) antigen on myogenic and fibroblastic cells in culture. *J. Cell Biol.*, **100**, 1528.

14. Chen, W. T., Hasegawa, E., Hasegawa, T., Weinstock, C., and Yamada, K. M. (1985). Development of cell surface linkage complexes in cultured fibroblasts. *J. Cell Biol.*, **100**, 1103.

15. Giancotti, F. G., Comoglio, P. M., and Tarone, G. (1986). A 135,000 molecular weight plasma membrane glycoprotein involved in fibronectin-mediated cell adhesion. Immunofluorescence localization in normal and RSV-transformed fibroblasts. *Exp. Cell Res.*, **163**, 47.

16. Kelly, T., Molony, L., and Burridge, K. (1987). Purification of two smooth muscle glycoproteins related to integrin. Distribution in cultured chicken embryo fibroblasts. *J. Biol. Chem.*, **262**, 17189.

17. Hynes, R. O. (1992). Integrins: versatility, modulation, and signaling in cell adhesion. *Cell*, **69**, 11.

18. Singer, I. I., Scott, S., Kawka, D. W., Kazazis, D. M., Gailit, J., and Ruoslahti, E. (1988). Cell surface distribution of fibronectin and vitronectin receptors depends on substrate composition and extracellular matrix accumulation. *J. Cell Biol.*, **106**, 2171.

19. Dejana, E., Colella, S., Conforti, G., Abbadini, M., Gaboli, M., and Marchisio, P. C. (1988). Fibronectin and vitronectin regulate the organization of their respective Arg-Gly-Asp adhesion receptors in cultured human endothelial cells. *J. Cell Biol.*, **107**, 1215.

20. Fath, K. R., Edgell, C. J., and Burridge, K. (1989). The distribution of distinct integrins in focal contacts is determined by the substratum composition. *J. Cell Sci.*, **92**, 67.

21. Meijne, A. M., Casey, D. M., Feltkamp, C. A., and Roos, E. (1994). Immuno-EM localization of the beta 1 integrin subunit in wet-cleaved fibronectin-adherent fibroblasts. *J. Cell Sci.*, **107**, 1229.

22. Jockusch, B. M., Bubeck, P., Giehl, K., Kroemker, M., Moschner, J., Rothkegel, M., *et al.* (1995). The molecular architecture of focal adhesions. *Annu. Rev. Cell Biol.*, **11**, 379.

23. Burridge, K. and Chrzanowska-Wodnicka, M. (1996). Focal adhesions, contractility and signaling. *Annu. Rev. Cell Dev. Biol.*, **12**, 463.

24. Hall, A. (1998). Rho GTPases and the actin cytoskeleton. *Science*, **279**, 509.

25. Mackay, D. J. G. and Hall, A. (1998). Rho GTPases. *J. Biol. Chem.*, **273**, 20685.

26. Van Aelst, L. and D'Souza-Schorey, C. (1997). Rho GTPases and signaling networks. *Genes Dev.*, **11**, 2295.

27. Clark, E. A. and Brugge, J. S. (1995). Integrins and signal transduction pathways: the road taken. *Science*, **268**, 233.

28. Schwartz, M. A., Schaller, M. D., and Ginsberg, M. H. (1995). Integrins: Emerging paradigms of signal transduction. *Annu. Rev. Cell Biol.*, **11**, 549.

29. Schoenwaelder, S. M. and Burridge, K. (1999). Bidirectional signaling between the cytoskeleton and integrins. *Curr. Opin. Cell Biol.*, **11**, 274.

30. Geiger, B., Volk, T., and Volberg, T. (1985). Molecular heterogeneity of adherens junctions. *J. Cell Biol.*, **101**, 1523.

31. Pardo, J. V., Siliciano, J. D., and Craig, S. W. (1983). Vinculin is a component of an extensive network of myofibril-sarcolemma attachment regions in cardiac muscle fibers. *J. Cell Biol.*, **97**, 1081.

32. Pardo, J. V., Siliciano, J. D., and Craig, S. W. (1983). A vinculin-containing cortical lattice in skeletal muscle. transverse lattice elements ('costameres') mark sites of attachment between myofibrils and sarcolemma. *Proc. Natl. Acad. Sci. USA*, **80**, 1008.

33. Tidball, J. G., O'Halloran, T., and Burridge, K. (1986). Talin at myotendinous junctions. *J. Cell Biol.*, **103**, 1465.

34. Bloch, R. J. and Hall, Z. W. (1983). Cytoskeletal components of the vertebrate neuromuscular junction: vinculin, alpha-actinin, and filamin. *J. Cell Biol.*, **97**, 217.

35. Sealock, R., Paschal, B., Beckerle, M. C., and Burridge, K. (1986). Talin is a post-synaptic component of the rat neuromuscular junction. *Exp. Cell Res.*, **163**, 143.

36. Rochlin, M. W., Chen, Q. M., Tobler, M., Turner, C. E., Burridge, K., and Peng, H. B. (1989). The relationship between talin and acetylcholine receptor clusters in *Xenopus* muscle cells. *J. Cell Sci.*, **92**, 461.

37. Gabbiani, G., Majno, G., and Ryan, G. B. (1973). The fibroblast as a contractile cell: the myofibroblast. In *Biology of fibroblast* (ed. E. Kulonen and J. Pikkarainen), p. 139. Academic Press, New York.

38. Welch, M. P., Odland, G. F., and Clark, R. A. F. (1990). Temporal relationships of F-actin bundle formation, collagen and fibronectin matrix assembly, and fibronectin receptor expression to wound contraction. *J. Cell Biol.*, **110**, 133.

39. Gabbiani, G., Badonell, M. C., and Rona, G. (1975). Cytoplasmic contractile apparatus in aortic endothelial cells of hypertensive rats. *Lab. Invest.*, **32**, 227.

40. White, G. E., Gimbrone, M. A., and Fujiwara, K. (1983). Factors influencing the expression of stress fibers in vascular endothelial cells *in situ*. *J. Cell Biol.*, **97**, 416.

41. Wong, A. J., Pollard, T. D., and Herman, I. M. (1983). Actin filament stress fibers in vascular endothelial cells *in vivo*. *Science*, **219**, 867.

42. Ridley, A. J. and Hall, A. (1992). The small GTP-binding protein rho regulates the assembly of focal adhesions and actin stress fibers in response to growth factors. *Cell*, **70**, 389.

43. Ridley, A. J., Paterson, H. F., Johnston, C. L., Diekmann, D., and Hall, A. (1992). The small GTP-binding protein rac regulates growth factor-induced membrane ruffling. *Cell*, **70**, 401.

44. Kozma, R., Ahmed, S., Best, A., and Lim, L. (1995). The Ras-related protein Cdc42Hs and bradykinin promote formation of peripheral actin microspikes and filopodia in Swiss 3T3 fibroblasts. *Mol. Cell. Biol.*, **15**, 1942.

45. DePasquale, J. A. and Izzard, C. S. (1987). Evidence for an actin-containing cytoplasmic precursor of the focal contact and the timing of incorporation of vinculin at the focal contact. *J. Cell Biol.*, **105**, 2803.

46. Rottner, K., Hall, A., and Small, J. V. (1999). Interplay between Rac and Rho in the control of substrate contact dynamics. *Curr. Biol.*, **9**, 640.

47. Kozma, R., Sarner, S., Ahmed, S., and Lim, L. (1997). Rho family GTPases and neuronal growth cone remodelling—relationship between increased complexity induced by Cdc42hs, Rac1, and acetylcholine and collapse induced by RhoA and lysophosphatidic acid. *Mol. Cell. Biol.*, **17**, 1201.

48. Jalink, K., van Corven, E. J., Hengeveld, T., Morii, N., Narumiya, S., and Moolenaar, W. H. (1994). Inhibition of lysophosphatidate- and thrombin-induced neurite retraction and neuronal cell rounding by ADP ribosylation of the small GTP-binding protein Rho. *J. Cell Biol.*, **126**, 801.

49. Sander, E. E., ten Klooster, J. P., van Delft, S., van der Kammen, R. A., and Collard, J. G. (1999). Rac downregulates Rho activity: reciprocal balance between both GTPases determines cellular morphology and migratory behavior. *J. Cell Biol.*, **147**, 1009.

50. Hart, M. J., Jiang, X., Kozasa, T., Roscoe, W., Singer, W. D., Gilman, A. G., *et al.* (1998). Direct stimulation of the guanine nucleotide exchange activity of p115 RhoGEF by Galpha13. *Science*, **280**, 2112.

51. Ridley, A. J. and Hall, A. (1994). Signal transduction pathways regulating Rho-mediated stress fibre formation: requirement for a tyrosine kinase. *EMBO J.*, **13**, 2600.

52. Barry, S. T. and Critchley, D. R. (1994). The RhoA-dependent assembly of focal adhesions in Swiss 3T3 cells is associated with increased tyrosine phosphorylation and the recruitment of both pp125FAK and protein kinase C-delta to focal adhesions. *J. Cell Sci.*, **107**, 2033.

53. Chrzanowska-Wodnicka, M. and Burridge, K. (1994). Tyrosine phosphorylation is involved in reorganization of the actin cytoskeleton in response to serum or LPA stimulation. *J. Cell Sci.*, **107**, 3643.

54. Nobes, C. D., Hawkins, P., Stephens, L., and Hall, A. (1995). Activation of the small GTP-binding proteins rho and rac by growth factor receptors. *J. Cell Sci.*, **108**, 225.

55. Schneider, G. B., Gilmore, A. P., Lohse, D. L., Romer, L. H., and Burridge, K. (1998). Microinjection of protein tyrosine phosphatases into fibroblasts disrupts focal adhesions and stress fibers. *Cell Adhes. Commun.*, **5**, 207.

56. Schoenwaelder, S. M. and Burridge, K. (1999). Evidence for a calpeptin-sensitive protein-tyrosine phosphatase upstream of the small GTPase Rho. *J. Biol. Chem.*, **274**, 14359.

57. Schoenwaelder, S. M., Petch, L. A., Williamson, D., Shen, R., Feng, G. S., and Burridge, K. (2000). The protein tyrosine phosphatase Shp-2 regulates RhoA activity. *Curr. Biol.*, **10**, 1523.

58. Yu, D. H., Qu, C. K., Henegariu, O., Lu, X., and Feng, G. S. (1998). Protein-tyrosine phosphatase Shp-2 regulates cell spreading, migration, and focal adhesion. *J. Biol. Chem.*, **273**, 21125.

59. Schoenwaelder, S. M. and Burridge, K. (1999). Bidirectional signaling between the cytoskeleton and integrins. *Curr. Opin. Cell Biol.*, **11**, 274.

60. Burridge, K. (1981). Are stress fibres contractile? *Nature*, **294**, 691.

61. Chrzanowska-Wodnicka, M. and Burridge, K. (1996). Rho-stimulated contractility drives the formation of stress fibers and focal adhesions. *J. Cell Biol.*, **133**, 1403.

62. Kolodney, M. S. and Elson, E. L. (1993). Correlation of myosin light chain phosphorylation with isometric contraction of fibroblasts. *J. Biol. Chem.*, **268**, 23850.

63. Goeckeler, Z. M. and Wysolmerski, R. B. (1995). Myosin light chain kinase-regulated endothelial cell contraction: the relationship between isometric tension, actin polymerization, and myosin phosphorylation. *J. Cell Biol.*, **130**, 613.

64. Burridge, K., Chrzanowska-Wodnicka, M., and Zhong, C. (1997). Focal adhesion assembly. *Trends Cell Biol.*, **7**, 342.

65. Leung, T., Chen, X. Q., Manser, E., and Lim, L. (1996). The p160 RhoA-binding kinase ROKalpha is a member of a kinase family and is involved in the reorganization of the cytoskeleton. *Mol. Cell. Biol.*, **16**, 5313.

66. Leung, T., Manser, E., Tan, L., and Lim, L. (1995). A novel serine/threonine kinase binding the ras-related RhoA GTPase which translocates the kinase to peripheral membranes. *J. Biol. Chem.*, **270**, 29051.

67. Kimura, K., Ito, M., Amano, M., Chihara, K., Fukata, Y., Nakafuku, M., *et al.* (1996). Regulation of myosin phosphatase by Rho and Rho-associated kinase (Rho-kinse). *Science*, **273**, 245.

68. Matsui, T., Amano, M., Yamamoto, T., Chihara, K., Nakafuku, M., Ito, M., *et al.* (1996). Rho-associated kinase, a novel serine/threonine kinase, as a putative target for the small GTP binding protein Rho. *EMBO J.*, **15**, 2208.

69. Ishizaki, T., Maekawa, M., Fujisawa, K., Okawa, K., Iwamatsu, A., Fujita, A., *et al.* (1996). The small GTP-binding protein Rho binds to and activates a 160 kDa Ser/Thr protein kinase homologous to myotonic dystrophy kinase. *EMBO J.*, **15**, 1885.

70. Nakagawa, O., Fujisawa, K., Ishizaki, T., Saito, Y., Nakao, K., and Narumiya, S. (1996). Rock-I and Rock-II, two isoforms of Rho-associated coiled-coil forming protein serine/threonine kinase in mice. *FEBS Lett.*, **392**, 189.

71. Amano, M., Ito, M., Kimura, K., Fukata, Y., Chihara, K., Nakano, T., *et al.* (1996). Phosphorylation and activation of myosin by Rho-associated kinase (Rho-kinase). *J. Biol. Chem.*, **271**, 20246.

72. Totsukawa, G., Yamakita, Y., Yamashiro, S., Hartshorne, D. J., Sasaki, Y., and Matsumura, F. (2000). Distinct roles of ROCK (Rho-kinase) and MLCK in spatial regulation of MLC phosphorylation for assembly of stress fibers and focal adhesions in 3T3 fibroblasts. *J. Cell Biol.*, **150**, 797.

73. Tan, J. L., Ravid, S., and Spudich, J. A. (1992). Control of nonmuscle myosins by phosphorylation. *Annu. Rev. Biochem.*, **61**, 721.

74. Amano, M., Chihara, K., Kimura, K., Fukata, Y., Nakamura, N., Matsuura, Y., *et al.* (1997). Formation of actin stress fibers and focal adhesions enhanced by Rho-kinase. *Science*, **275**, 1308.

75. Ishizaki, T., Naito, M., Fujisawa, K., Maekawa, M., Watanabe, N., Saito, Y., *et al.* (1997). p160ROCK, a Rho-associated coiled-coil forming protein kinase, works downstream of Rho and induces focal adhesions. *FEBS Lett.*, **404**, 118.

76. Watanabe, N., Kato, T., Fujita, A., Ishizaki, T., and Narumiya, S. (1999). Cooperation between mDia1 and ROCK in Rho-induced actin reorganization. *Nature Cell Biol.*, **1**, 136.

77. Nakano, K., Takaishi, K., Kodama, A., Mammoto, A., Shiozaki, H., Monden, M., T., *et al.* (1999). Distinct actions and cooperative roles of ROCK and mDia in Rho small G protein-induced reorganization of the actin cytoskeleton in Madin-Darby canine kidney cells. *Mol. Biol. Cell*, **10**, 2481.

78. Watanabe, N., Madaule, P., Reid, T., Ishizaki, T., Watanabe, G., Kakizuka, A., *et al*. (1997). p140mDia, a mammalian homolog of *Drosophila* diaphanous, is a target protein for Rho small GTPase and is a ligand for profilin. *EMBO J.*, **16**, 3044.

79. Tominaga, T., Sahai, E., Chardin, P., McCormick, F., Courtneidge, S., and Alberts, A. S. (2000). Diaphanous-related formins bridge Rho GTPase and Src tyrosin kinase signaling. *Mol. Cell*, **5**, 13.

80. Chew, T. L., Masaracchia, R. A., Goeckeler, Z. M., and Wysolmerski, R. B. (1998). Phosphorylation of non-muscle myosin II regulatory light chain by p21-activated kinase (gamma-PAK). *J. Muscle Res. Cell Motil.*, **19**, 839.

81. Sells, M. A., Boyd, J. T., and Chernoff, J. (1999). p21-activated kinase 1 (Pak1) regulates cell motility in mammalian fibroblasts. *J. Cell Biol.*, **145**, 837.

82. Sanders, L. C., Matsumura, F., Bokoch, G. M., and de Lanerolle, P. (1999). Inhibition of myosin light chain kinase by p21-activated kinase. *Science*, **283**, 2083.

83. Kiosses, W. B., Daniels, R. H., Otey, C., Bokoch, G. M., and Schwartz, M. A. (1999). A role for p21-activated kinase in endothelial cell migration. *J. Cell Biol.*, **147**, 831.

84. Sells, M. A., Pfaff, A., and Chernoff, J. (2000). Temporal and spatial distribution of activated Pak1 in fibroblasts. *J. Cell Biol.*, **151**, 1449.

85. Izzard, C. S. (1988). A precursor of the focal contact in cultured fibroblasts. *Cell Motil. Cytoskel.*, **10**, 137.

86. Izzard, C. S. and Lochner, L. R. (1980). Formation of cell-to-substrate contacts during fibroblast motility: an interference-reflexion study. *J. Cell Sci.*, **42**, 81.

87. Machesky, L. M. and Hall, A. (1997). Role of actin polymerization and adhesion to extracellular matrix in Rac- and Rho-induced cytoskeletal reorganization. *J. Cell Biol.*, **138**, 913.

88. Welch, M. D., DePace, A. H., Verma, S., Iwamatsu, A., and Mitchison, T. J. (1997). The human Arp2/3 complex is composed of evolutionarily conserved subunits and is localized to cellular regions of dynamic actin filament assembly. *J. Cell Biol.*, **138**, 375.

89. Mullins, R. D., Heuser, J. E., and Pollard, T. D. (1998). The interaction of Arp2/3 complex with actin: nucleation, high affinity pointed end capping, and formation of brachning networks of filaments. *Proc. Natl. Acad. Sci. USA*, **95**, 6181.

90. Beckerle, M. C. (1986). Identification of a new protein localized at sites of cell-substrate adhesion. *J. Cell Biol.*, **103**, 1679.

91. Reinhard, M., Halbrugge, M., Scheer, U., Wiegand, C., Jockusch, B. M., and Walter, U. (1992). The 46/50 kDa phosphoprotein VASP purified from human platelets is a novel protein associated with actin filaments and focal contacts. *EMBO J.*, **11**, 2063.

92. Gertler, F. B., Niebuhr, K., Reinhard, M., Wehland, J., and Soriano, P. (1996). Mena, a relative of VASP and *Drosophila* Enabled is implicated in the control of microfilament dynamics. *Cell*, **87**, 227.

93. Smith, G. A., Theriot, J. A., and Portnoy, D. A. (1996). The tandem repeat domain in the *Listeria monocytogenes* ActA protein controls the rate of actin-based motility, the percentage of moving bacteria, and the localization of vasodilator-stimulated phosphoprotein and profilin. *J. Cell Biol.*, **135**, 647.

94. Theriot, J. A., Mitchison, T. J., Tilney, L. G., and Portnoy, D. A. (1992). The rate of actin-based motility of intracellular *Listeria monocytogenes* equals the rate of actin polymerization. *Nature*, **357**, 257.

95. Beckerle, M. C. (1998). Spatial control of actin filament assembly: lessons from *Listeria*. *Cell*, **95**, 741.

96. Steffen, P., Schafer, D. A., David, V., Goum, E., Cooper, J. A., and Cossart, P. (2000). *Listeria monocytogenes* ActA protein interacts with phosphatidylinositol 4,5-bisphosphate *in vitro. Cell Motil. Cytoskel.*, **45**, 58.

97. Niebuhr, K., Ebel, F., Frank, R., Reinhard, M., Domann, E., Carl, U. D., *et al.* (1997). A novel proline-rich motif present in ActA of *Listeria monocytogenes* and cytoskeletal proteins is the ligand for the EVH1 domain, a protein module present in the Ens/VASP family. *EMBO J.*, **16**, 5433.

98. Bachmann, C., Fischer, L., Walter, U., and Reinhard, M. (1999). The EVH2 domain of the vasodilator-stimulated phosphoprotein mediates tetramerization, F-actin binding, and actin bundle formation. *J. Biol. Chem.*, **274**, 23549.

99. Golsteyn, R. M., Beckerle, M. C., Koay, T., and Friederich, E. (1997). Structural and functional similarities between the human cytoskeletal protein zyxin and the ActA protein of *Listeria monocytogenes. J. Cell Sci.*, **110**, 1893.

100. Rottner, K., Behrendt, B., Small, J. V., and Wehland, J. (1999). VASP dynamics during lamelliodia protrusion. *Nature Cell Biol.*, **1**, 321.

101. Bear, J. E., Loureiro, J. J., Libova, I., Fassler, R., Wehland, J., and Gertler, F. B. (2000). Negative regulation of fibroblast motility by Ena/VASP proteins. *Cell*, **101**, 717.

102. DePasquale, J. A. and Izzard, C. S. (1991). Accumulation of talin in nodes at the edge of the lamellipodium and separate incorporation into adhesion plaques at focal contacts in fibroblasts. *J. Cell Biol.*, **113**, 1351.

103. Pletjushkina, O., Belkin, A. M., Ivanova, O. J., Oliver, T., Jacobson, K., and Vasiliev, J. M. (1998). Maturation of cell-substratum focal adhesions induced by depolymerization of microtubules is induced by increased cortical tension. *Cell Adhes. Commun.*, **5**, 121.

104. Smilenov, L. B., Mikhailov, A., Pelham Jr., R. J., Marcantonio, E. E., and Gundersen, G. G. (1999). Focal adhesions motility revealed in stationary fibroblasts. *Science*, **286**, 1172.

105. Pankov, R., Cukierman, E., Katz, B. Z., Matsumoto, K., Lin, D. C., Lin, S., *et al.* (2000). Integrin dynamics and matrix assembly: Tensin-dependent translocation of alpha(5) beta(1) integrins promotes early fibronectin fibrillogenesis. *J. Cell Biol.*, **148**, 1075.

106. Hynes, R. O. and Destree, A. T. (1978). Relationships between fibronectin (LETS protein) and actin. *Cell*, **15**, 875.

107. Feramisco, J. R. and Burridge, K. (1980). A rapid purification of alpha-actinin, filamin, and a 130,000-dalton protein from smooth muscle. *J. Biol. Chem.*, **255**, 1194.

108. Avnur, Z. and Geiger, B. (1981). The removal of extracellular fibronectin from areas of cell-substrate contact. *Cell*, **25**, 121.

109. Zamir, E., Katz, M., Posen, Y., Erez, N., Yamada, K. M., Katz, B. Z., *et al.* (2000). Dynamics and segregation of cell-matrix adhesions in cultured fibroblasts. *Nature Cell Biol.*, **2**, 191.

110. Solowska, J., Guan, J. L., Marcantonio, E. E., Trevithick, J. E., Buck, C. A., and Hynes, R. O. (1989). Expression of normal and mutant avian integrin subunits in rodent cells [published erratum appears in *J. Cell Biol.* 1989, Oct, 109(4 Pt 1), 1187]. *J. Cell Biol.*, **109**, 853.

111. Marcantonio, E. E., Guan, J. L., Trevithick, J. E., and Hynes, R. O. (1990). Mapping of the functional determinants of the integrin beta 1 cytoplasmic domain by site-directed mutagenesis. *Cell Regul.*, **1**, 597.

112. Hayashi, Y., Haimovich, B., Reszka, A., Boettiger, D., and Horwitz, A. (1990). Expression and function of chicken integrin beta 1 subunit and its cytoplasmic domain mutants in mouse NIH 3T3 cells. *J. Cell Biol.*, **110**, 175.

113. Briesewitz, R., Kern, A., and Marcantonio, E. E. (1993). Ligand-dependent and -independent integrin focal contact localization: the role of the alpha chain cytoplasmic domain. *Mol. Biol. Cell*, **4**, 593.

114. Ylanne, J., Chen, Y., O'Toole, T. E., Loftus, J. C., Takada, Y., and Ginsberg, M. H. (1993). Distinct functions of integrin alpha and beta subunit cytoplasmic domains in cell spreading and formation of focal adhesions. *J. Cell Biol.*, **122**, 223.

115. LaFlamme, S. E., Akiyama, S. K., and Yamada, K. M. (1992). Regulation of fibronectin receptor distribution. *J. Cell Biol.*, **117**, 437.

116. Geiger, B., Salomon, D., Takeichi, M., and Hynes, R. O. (1992). A chimeric N-cadherin/beta 1-integrin receptor which localizes to both cell–cell and cell–matrix adhesions. *J. Cell Sci.*, **103**, 943.

117. Horwitz, A., Duggan, K., Buck, C., Beckerle, M. C., and Burridge, K. (1986). Interaction of plasma membrane fibronectin receptor with talin—a transmembrane linkage. *Nature*, **320**, 531.

118. Otey, C. A., Pavalko, F. M., and Burridge, K. (1990). An interaction between alpha-actinin and the beta 1 integrin subunit *in vitro*. *J. Cell Biol.*, **111**, 721.

119. Pfaff, M., Liu, S. C., Erle, D. J., and Ginsberg, M. H. (1998). Integrin beta cytoplasmic domains differentially cytoskeletal proteins. *J. Biol. Chem.*, **273**, 6104.

120. Belkin, A. M., Retta, S. F., Pletjushkina, O. Y., Balzac, F., Silengo, L., Fassler, R., *et al.* (1997). Muscle beta1d integrin reinforces the cytoskeleton-matrix link—modulation of integrin adhesive function by alternative splicing. *J. Cell Biol.*, **139**, 1583.

121. Pavalko, F. M. and LaRoche, S. M. (1993). Activation of human neutrophils induces an interaction between the integrin beta 2-subunit (CD18) and the actin binding protein alpha-actinin. *J. Immunol.*, **151**, 3795.

122. Sampath, R., Gallagher, P. J., and Pavalko, F. M. (1998). Cytoskeletal interactions with the leukocyte integrin beta2 cytoplasmic tail: activation-dependent regulation of associations with talin and alpha-actinin. *J. Biol. Chem.*, **273**, 33588.

123. Lewis, J. M. and Schwartz, M. A. (1995). Mapping *in vivo* associations of cytoplasmic proteins with integrin beta 1 cytoplasmic domain mutants. *Mol. Biol. Cell*, **6**, 151.

124. Sharma, C. P., Ezzell, R. M., and Arnaout, M. A. (1995). Direct interaction of filamin (ABP-280) with the beta 2-integrin subunit CD18. *J. Immunol.*, **154**, 3461.

125. Andrews, R. K. and Fox, J. E. (1991). Interaction of purified actin-binding protein with the platelet membrane glycoprotein Ib-IX complex. *J. Biol. Chem.*, **266**, 7144.

126. Miyamoto, S., Akiyama, S. K., and Yamada, K. M. (1995). Synergistic roles for receptor occupancy and aggregation in integrin transmembrane function. *Science*, **267**, 883.

127. Chuang, J. Z., Lin, D. C., and Lin, S. (1995). Molecular cloning, expression, and mapping of the high affinity actin-capping domain of chicken cardiac tensin. *J. Cell Biol.*, **128**, 1095.

128. Wilkins, J. A., Risinger, M. A., and Lin, S. (1986). Studies on proteins that co-purify with smooth muscle vinculin: identification of immunologically related species in focal adhesions of nonmuscle and Z-lines of muscle cells. *J. Cell Biol.*, **103**, 1483.

129. Lo, S. H., Janmey, P. A., Hartwig, J. H., and Chen, L. B. (1994). Interactions of tensin with actin and identification of its three distinct actin-binding domains. *J. Cell Biol.*, **125**, 1067.

130. Liu, S., Thomas, S. M., Woodside, D. G., Rose, D. M., Kiosses, W. B., Pfaff, M., *et al.* (1999). Binding of paxillin to alpha4 integrins modifies integrin-dependent biological responses. *Nature*, **402**, 676.

131. Liu, S. and Ginsberg, M. H. (2000). Paxillin binding to a conserved sequence motif in the alpha 4 integrin cytoplasmic domain. *J. Biol. Chem.*, **275**, 22736.

132. Knezevic, I., Leisner, T. M., and Lam, S. C.-T. (1996). Direct binding of the platelet integrin alphaIIbbeta3 (GPIIB-IIIA) to talin. *J. Biol. Chem.*, **271**, 16416.

133. Calderwood, D. A., Zent, R., Grant, R., Jasper, D., Rees, G., Hynes, R. O., *et al.* (1999). The talin head domain binds to integrin beta subunit cytoplasmic tails and regulates integrin activation. *J. Biol. Chem.*, **274**, 28071.

134. Borowsky, M. L. and Hynes, R. O. (1998). Layilin, a novel talin-binding transmembrane protein homologous with C-type lectins, is localized in membrane ruffles. *J. Cell Biol.*, **143**, 429.

135. Gilmore, A. P., Jackson, P., Waites, G. T., and Crtichley, D. R. (1992). Further characterization of the talin binding site in the cytoskeletal protein vinculin. *J. Cell Sci.*, **103**, 719.

136. Gilmore, A. P., Wood, C., Ohanian, V., Jackson, P., Patel, B., Rees, D. J. G., *et al.* (1993). The cytoskeletal protein talin contains at least two distinct vinculin binding domains. *J. Cell Biol.*, **122**, 337.

137. Priddle, H., Hemmings, L., Monkley, S., Woods, A., Patel, B., Sutton, S., *et al.* (1998). Disruption of the talin gene compromises focal adhesion assembly in undifferentiated but not differentiated embryonic stem cells. *J. Cell Biol.*, **142**, 1121.

138. Albiges-Rizo, C., Frachet, P., and Block, M. R. (1995). Down regulation of talin alters cell adhesion and the processing of the $\alpha5\beta1$ integrin. *J. Cell Sci.*, **108**, 3317.

139. Martel, V., Vignoud, L., Dupe, S., Frachet, P., Block, M. R., and Albiges-Rizo, C. (2000). Talin controls the exit of integrin alpha5 beta1 from an early compartment of the secretory pathway. *J. Cell Sci.*, **113**, 1951.

140. Muguruma, M., Nishimuta, S., Tomisaka, Y., Ito, T., and Matsumura, S. (1995). Organization of the functional domains in membrane cytoskeletal protein talin. *J. Biochem.*, **117**, 1036.

141. Rees, D. J. G., Ades, S. E., Singer, S. J., and Hynes, R. O. (1990). Sequence and domain structure of talin. *Nature*, **347**, 685.

142. Hemmings, L., Rees, D. J. G., Ohanian, V., Bolton, S. J., Gilmore, A. P., Patel, B., *et al.* (1996). Talin contains three actin-binding sites each of which is adjacent to a vinculin-binding site. *J. Cell Sci.*, **109**, 2715.

143. Beckerle, M., Miller, D. E., Bertagnolli, M. E., and Locke, S. J. (1989). Activation-dependent redistribution of the adhesion plaque protein, talin, in intact human platelets. *J. Cell Biol.*, **109**, 3333.

144. Carragher, N. O., Levkau, B., Ross, R., and Raines, E. W. (1999). Degraded collagen fragments promote rapid disassembly of smooth muscle focal adhesions that correlate with cleavage of pp125[FAK], paxillin, and talin. *J. Cell Biol.*, **147**, 619.

145. Johnson, R. P. and Craig, S. W. (1995). F-actin binding site masked by the intramolecular association of vinculin head and tail domains. *Nature*, **373**, 261.

146. Johnson, R. P. and Craig, S. W. (1994). An intramolecular association between the head and tail domains of vinculin modulates talin binding. *J. Biol. Chem.*, **269**, 12611.

147. Gilmore, A. P. and Burridge, K. (1996). Regulation of vinculin binding to talin and actin by phosphatidylinositol-4-5-bisphosphate. *Nature*, **381**, 531.

148. Weekes, J., Barry, S. T., and Critchley, D. R. (1996). Acidic phospholipids inhibit the intramolecular association between the N- and C-terminal regions of vinculin, exposing actin-binding and protein kinase C phosphorylation sites. *Biochem. J.*, **314**, 827.

149. Chong, L. D., Traynor-Kaplan, A., Bokoch, G. M., and Schwartz, M. A. (1994). The small GTP-binding protein Rho regulates a phosphatidylinositol 4-phosphate 5-kinase in mammalian cells. *Cell*, **79**, 507.

150. Kioka, N., Sakata, S., Kawauchi, T., Amachi, T., Akiyama, S. K., Okazaki, K., *et al.* (1999). Vinexin: a novel vinculin-binding protein with multiple SH3 domains enhances actin cytoskeletal organization. *J. Cell Biol.*, **144**, 59.

151. Volberg, T., Geiger, B., Kam, Z., Pankov, R., Simcha, I., Sabanay, H., *et al.* (1995). Focal adhesion formation by F9 embryonal carcinoma cells after vinculin gene disruption. *J. Cell Sci.*, **108**, 2253.

152. Coll, J. L., Ben-Ze'ev, A., Ezzell, R. M., Rodriguez Fernandez, J. L., Baribault, H., Oshima, R. G., *et al.* (1995). Targeted disruption of vinculin genes in F9 and embryonic stem cells changes cell morphology, adhesion, and locomotion. *Proc. Natl. Acad. Sci. USA*, **92**, 9161.

153. Samuels, M., Ezzell, R. M., Cardozo, T. J., Critchley, D. R., Coll, J. L., and Adamson, E. D. (1993). Expression of chicken vinculin complements the adhesion-defective phenotype of a mutant mouse F9 embryonal carcinoma cell. *J. Cell Biol.*, **121**, 909.

154. Xu, W., Coll, J. L., and Adamson, E. D. (1998). Rescue of the mutant phenotype by reexpression of full-length vinculin in null F9 cells; effects on cell locomotion by domain deleted vinculin. *J. Cell Sci.*, **111**, 1535.

155. Laine, R. O., Zeile, W., Kang, F., Purich, D. L., and Southwick, F. S. (1997). Vinculin proteolysis unmasks an ActA homolog for actin-based *Shigella* motility. *J. Cell Biol.*, **138**, 1255.

156. Brindle, N. P. J., Holt, M. R., Davies, J. E., Price, C. J., and Critchley, D. R. (1996). The focal-adhesion vasodilator-stimulated phosphoprotein (VASP) binds to the proline-rich domain of vinculin. *Biochem. J.*, **318**, 753.

157. Reinhard, M., Giehl, K., Abel, K., Haffner, C., Jarchau, T., Hoppe, V., *et al.* (1995). The proline-rich focal adhesion and microfilament protein VASP is a ligand for profilins. *EMBO J.*, **14**, 1583.

158. Tsukita, S. and Yonemura, S. (1997). ERM (ezrin/radixin/moesin) family: from cytoskeleton to signal transduction. [Review] [42 refs]. *Curr. Opin. Cell Biol.*, **9**, 70.

159. Mackay, D. J., Esch, F., Furthmayr, H., and Hall, A. (1997). Rho- and rac-dependent assembly of focal adhesion complexes and actin filaments in permeabilized fibroblasts: an essential role for ezrin/radixin/moesin proteins. *J. Cell Biol.*, **138**, 927.

160. Gary, R. and Bretscher, A. (1995). Ezrin self-association involves binding of an N-terminal domain to a normally masked C-terminal domain that includes the F-actin binding site. *Mol. Biol. Cell*, **6**, 1061.

161. Magendantz, M., Henry, M. D., Lander, A., and Solomon, F. (1995). Interdomain interactions of radixin *in vitro*. *J. Biol. Chem.*, **270**, 25324.

162. Andreoli, C., Martin, M., Le Borgne, R., Reggio, H., and Mangeat, P. (1994). Ezrin has properties to self-associate at the plasma membrane. *J. Cell Sci.*, **107**, 2509.

163. Tsukita, S., Oishi, K., Sato, N., Sagara, J., and Kawai, A. (1994). ERM family members as molecular linkers between the cell surface glycoprotein CD44 and actin-based cytoskeletons. *J. Cell Biol.*, **126**, 391.

164. Matsui, T., Maeda, M., Doi, Y., Yonemura, S., Amano, M., Kaibuchi, K., *et al.* (1998). Rho-kinase phosphorylates COOH-terminal threonines of ezrin/radixin/moesin (ERM) proteins and regulates their head-to-tail association. *J. Cell Biol.*, **140**, 647.

165. Hirao, M., Sato, N., Kondo, T., Yonemura, S., Monden, M., Sasaki, T., *et al.* (1996). Regulation mechanism of ERM (ezrin/radixin/moesin) protein/plasma membrane association: possible involvement of phosphatidylinositol turnover and Rho-dependent signaling pathway. *J. Cell Biol.*, **135**, 37.

166. Niggli, V., Andreoli, C., Roy, C., and Mangeat, P. (1995). Identification of a phosphatidylinositol-4,5-bisphosphate-binding domain in the N-terminal region of ezrin. *FEBS Lett.*, **376**, 172.

167. Shaw, R. J., Henry, M., Solomon, F., and Jacks, T. (1998). RhoA-dependent phosphorylation and relocalization of ERM proteins into apical membrane/actin protrusions in fibroblasts. *Mol. Biol. Cell*, **9**, 403.

168. Nikolopoulos, S. N. and Turner, C. E. (2000). Actopaxin, a new focal adhesion protein that binds paxillin LD motifs and actin and regulates cell adhesion. *J. Cell Biol.*, **151**, 1435.

169. Parast, M. M. and Otey, C. (2000). Characterization of palladin, a novel protein localized to stress fibers and cell adhesions. *J. Cell Biol.*, **150**, 643.

170. Miyamoto, S., Teramoto, H., Coso, O. A., Gutkind, J. S., Burbelo, P. D., Akiyama, S. K., *et al.* (1995). Integrin function—molecular hierarchies of cytoskeletal and signaling molecules. *J. Cell Biol.*, **131**, 791.

171. Schaller, M. D., Otey, C. A., Hildebrand, J. D., and Parsons, J. T. (1995). Focal adhesion kinase and paxillin bind to peptides mimicking beta integrin cytoplasmic domains. *J. Cell Biol.*, **130**, 1181.

172. Hildebrand, J. D., Schaller, M. D., and Parsons, J. T. (1993). Identification of sequences required for the efficient localization of the focal adhesion kinase, pp125FAK, to cellular focal adhesions. *J. Cell Biol.*, **123**, 993.

173. Schaller, M. D. and Parsons, J. T. (1995). pp125FAK-dependent tyrosine phosphorylation of paxillin creates a high-affinity binding site for Crk. *Mol. Cell. Biol.*, **15**, 2635.

174. Schaller, M. D., Hildebrand, J. D., Shannon, J. D., Fox, J. W., Vines, R. R., and Parsons, J. T. (1994). Autophosphorylation of the focal adhesion kinase, pp125FAK, directs SH2-dependent binding of pp60src. *Mol. Cell. Biol.*, **14**, 1680.

175. Richardson, A. and Parsons, J. T. (1995). Signal transduction through integrins: a central role for focal adhesion kinase? *Bioessays*, **17**, 229.

176. Hotchin, N. A. and Hall, A. (1995). The assembly of integrin adhesion complexes requires both extracellular matrix and intracellular Rho/Rac GTPases. *J. Cell Biol.*, **131**, 1857.

177. Parsons, J. T. (1996). Integrin-mediated signalling—Regulation by protein tyrosine kinases and small GTP-binding proteins. *Curr. Opin. Cell Biol.*, **8**, 146.

178. Lewis, J. M., Baskaran, R., Taagepera, S., Schwartz, M. A., and Wang, J. Y. J. (1996). Integrin regulation of c-Abl tyrosine kinase activity and cytoplasmic-nuclear transport. *Proc. Natl. Acad. Sci. USA*, **93**, 15174.

179. Gao, J., Zoller, K. E., Ginsberg, M. H., Brugge, J. S., and Shattil, S. J. (1997). Regulation of the pp72syk protein tyrosine kinase by platelet integrin alpha IIb beta 3. *EMBO J.*, **16**, 6414.

180. Schlaepfer, D. D., Hanks, S. K., Hunter, T., and van der Geer, P. (1994). Integrin-mediated signal transduction linked to Ras pathway by GRB2 binding to focal adhesion kinase. *Nature*, **372**, 786.

181. Ilic, D., Furuta, Y., Kanazawa, S., Takeda, N., Sobue, K., Nakatsuji, N., *et al.* (1995). Reduced cell motility and enhanced focal contact formation in cells from FAK-deficient mice. *Nature*, **377**, 539.

182. Gilmore, A. P. and Romer, L. H. (1996). Inhibition of FAK signalling in focal adhesions decreases cell motility and proliferation. *Mol. Biol. Cell*, **7**, 1209.

183. Cary, L., Chang, J., and Guan, J. L. (1996). Stimulation of cell migration by over-expression of focal adhesion kinase and its association with Src and Fyn. *J. Cell Sci.*, **109**, 1787.

184. Howe, A., Aplin, A. E., Alahari, S. K., and Juliano, R. L. (1998). Integrin signaling and cell growth control. *Curr. Opin. Cell Biol.*, **10**, 220.

185. Schwartz, M. A. and Baron, V. (1999). Interactions between mitogenic stimuli, or, a thousand and one connections. *Curr. Opin. Cell Biol.*, **11**, 197.

186. Chen, Q., Kinch, M. S., Lin, T. H., Burridge, K., and Juliano, R. L. (1994). Integrin-mediated cell adhesion activates mitogen-activated protein kinases. *J. Biol. Chem.*, **269**, 26602.

187. Zhu, X. and Assoian, R. K. (1995). Integrin-dependent activation of MAP kinase: a link to shape-dependent cell proliferation. *Mol. Biol. Cell*, **6**, 273.

188. Frisch, S. M., Vuori, K., Ruoslahti, E., and Chan-Hui, P. Y. (1996). Control of adhesion-dependent cell survival by focal adhesion kinase. *J. Cell Biol.*, **134**, 793.

189. Meredith, J. E., Jr., Fazeli, B., and Schwartz, M. A. (1993). The extracellular matrix as a cell survival factor. *Mol. Biol. Cell*, **4**, 953.

190. Frisch, S. M. and Francis, H. (1994). Disruption of epithelial cell-matrix interactions induces apoptosis. *J. Cell Biol.*, **124**, 619.

191. Xu, L.-H., Owens, L. V., Sturge, G. C., Yang, X., Liu, E. T., Craven, R. J., *et al.* (1996). Attenuation of the expression of the focal adhesion kinase induces apoptosis in tumor cells. *Cell Growth Differ.*, **7**, 413.

192. Hungerford, J. E., Compton, M. T., Matter, M. L., Hoffstrom, B. G., and Otey, C. A. (1996). Inhibition of pp125(FAK) in cultured fibroblasts results in apoptosis. *J. Cell Biol.*, **135**, 1383.

193. Albrecht-Buehler, G. (1976). Filopodia of spreading 3T3 cells: Do they have a substrate-exploring function? *J. Cell Biol.*, **69**, 275.

194. Clark, E. A., King, W. G., Brugge, J. S., Symons, M., and Hynes, R. O. (1998). Integrin-mediated signals regulated by members of the Rho family of GTPases. *J. Cell Biol.*, **142**, 573.

195. Price, L. S., Leng, J., Schwartz, M. A., and Bokoch, G. M. (1998). Activation of Rac and Cdc42 by integrins mediates cell spreading. *Mol. Biol. Cell*, **9**, 1863.

196. Klemke, R., Leng, J., Molander, R., Brooks, P. C., Vuori, K., and Cheresh, D. A. (1998). CAS/Crk coupling serves as a 'molecular switch' for induction of cell migration. *J. Cell Biol.*, **140**, 961.

197. Kiyokawa, E., Hashimoto, Y., Kurata, T., Sugimura, H., and Matsuda, M. (1998). Evidence that DOCK180 up-regulates signals from the CrkII-p130^{cas} complex. *J. Biol. Chem.*, **273**, 24479.

198. Dolfi, F., Garcia-Guzman, M., Ojaniemi, M., Nakarmura, H., Matsuda, M., and Vuori, K. (1998). The adaptor protein Crk connects muliple cellular stimuli to the JNK signaling pathway. *Proc. Natl. Acad. Sci. USA*, **95**, 15394.

199. Kiyokawa, E., Hashimoto, Y., Kobayashi, S., Sugimura, H., Kurata, T., and Matsuda, M. (1998). Activation of Rac1 by a Crk SH3-binding protein, DOCK180. *Genes Dev.*, **12**, 3331.

200. Barry, S. T., Flinn, H. M., Humphries, M. J., Critchley, D. R., and Ridley, A. J. (1997). Requirement for Rho in integrin signalling. *Cell Adhes. Commun.*, **4**, 387.

201. Ren, X., Kiosses, W. B., and Schwartz, M. A. (1999). Regulation of the small GTPase binding protein Rho by cell adhesion and the cytoskeleton. *EMBO J.*, **18**, 578.

202. Barry, S. T., Flinn, H. M., Humphries, M. J., Critchley, D. R., and Ridley, A. J. (1997). Requirement for Rho in integrin signaling. *Cell Adhes. Commun.*, **4**, 387.

203. Arthur, W. T., Petch, L. A., and Burridge, K. (2000). Integrin engagement suppresses RhoA activity via a c-Src-dependent mechanism. *Curr. Biol.*, **19**, 719.

204. Ren, X., Kiosses, W. B., Sieg, D. J., Otey, C., Schlaefer, D. D., and Schwartz, A. M. (2000). Focal adhesion kinase suppresses Rho activity to promote focal adhesion turnover. *J. Cell Sci.*, **113**, 3673.

205. Masiero, L., Lapidos, K., Ambudkas, I., and Kohn, E. C. (1999). Regulation of the RhoA pathway in human endothelial cell spreading on type IV collagen: role of calcium influx. *J. Cell Sci.*, **112**, 3205.

206. Nix, D. A. and Beckerle, M. C. (1997). Nuclear-cytoplasmic shuttling of the focal contact protein, zyxin: a potential mechanism for communication between sites of cell adhesion and the nucleus. *J. Cell Biol.*, **138**, 1139.

207. Beckerle, M. C. (1997). Zyxin: zinc fingers at sites of cell adhesion. *Bioessays*, **19**, 949.

208. Gumbiner, B. M. (1995). Signal transduction of beta-catenin. *Curr. Opin. Cell Biol.*, **7**, 634.

209. Woods, A. and Couchman, J. R. (1994). Syndecan 4 heparan sulfate proteoglycan is a selectively enriched and widespread focal adhesion component. *Mol. Biol. Cell*, **5**, 183.

210. Baciu, P. C. and Goetinck, P. F. (1995). Protein kinase C regulates the recruitment of syndecan-4 into focal contacts. *Mol. Biol. Cell*, **6**, 1503.

211. Oh, E. S., Woods, A., and Couchman, J. R. (1997). Multimerization of the cytoplasmic domain of syndecan-4 is required for its ability to activate protein kinase C. *J. Biol. Chem.*, **272**, 11805.

212. Oh, E. S., Woods, A., and Couchman, J. R. (1997). Syndecan-4 proteoglycan regulates the distribution and activity of protein kinase C. *J. Biol. Chem.*, **272**, 8133.

213. Longley, R. L., Woods, A., Fleetwood, A., Cowling, G. J., Gallagher, J. T., and Couchman, J. R. (1999). Control of morphology, cytoskeleton and migration by syndecan-4. *J. Cell Sci.*, **112**, 3421.

214. Echtermeyer, F., Baciu, P. C., Saoncella, S., Ge, Y., and Goetinck, P. F. (1999). Syndecan-4 core protein is sufficient for the assembly of focal adhesions and actin stress fibers. *J. Cell Sci.*, **112**, 3433.

215. Oh, E. S., Woods, A., Lim, S. T., Theibert, A. W., and Couchman, J. R. (1998). Syndecan-4 proteoglycan cytoplasmic domain and phosphatidylinositol 4,5-bisphosphate coordinately regulate protein kinase C activity. *J. Biol. Chem.*, **273**, 10624.

216. Defilippi, P., Venturino, M., Gulino, D., Duperray, A., Boquet, P., Fiorentini, C., *et al.* (1997). Dissection of pathways implicated in integrin-mediated actin cytoskeleton assembly, involvement of protein kinase C, Rho GTPase, and tyrosine phosphorylation. *J. Biol. Chem.*, **272**, 21726.

217. Woods, A., Couchman, J. R., Johansson, S., and Hook, M. (1986). Adhesion and cytoskeletal organization of fibroblasts in response to fibronectin fragments. *EMBO J.*, **5**, 665.

218. Saoncella, S., Echtermeyer, F., Denhez, F., Nowlen, J. K., Mosher, D. F., Robinson, S. D., *et al.* (1999). Syndecan-4 signals cooperatively with integrins in a Rho-dependent manner in the assembly of focal adhesions and stress fibers. *Proc. Natl. Acad. Sci. USA*, **96**, 2805.

219. Ishiguro, K., Kadomatsu, K., Kojima, T., Mursmatsu, H., Tsuzuki, S., Nakamura, E., *et al.* (2000). Syndecan-4 deficiency impairs focal adhesion formation only under restricted conditions. *J. Biol. Chem.*, **275**, 5249.

220. Mandai, K., Nakanishi, H., Satoh, A., Takahashi, K., Satoh, K., Mizoguchi, A., *et al.* (1999). Ponsin/SH3P12: an 1-afadin and vinculin-binding protein localized at cell-cell and cell-matrix adherens junctions. *J. Cell Biol.*, **144**, 1001.

221. Bustelo, X. R. (2000). Regulatory and signaling properties of the Vav family. *Mol. Cell. Biol.*, **20**, 1461.

10 | Desmosomes and hemidesmosomes

LESLIE J. BANNON, LAWRENCE E. GOLDFINGER,
JONATHAN C. R. JONES, and KATHLEEN J. GREEN

1. Introduction

Desmosomes and hemidesmosomes are specialized structures requisite for the establishment and maintenance of adhesion in tissues of epithelial origin. Desmosomes (from the Greek *desmo* = bound) are sites of intercellular adhesion, whereas hemidesmosomes mediate cell–substratum adhesion at the basal cell surface. These junctions both serve as prominent attachment sites for intermediate filaments to the plasma membrane; however, their molecular composition and properties vastly differ (1–3). Hemidesmosomes are largely restricted to epithelia, including the epidermis, bladder epithelium, lung epithelium, and various glandular epithelia (2, 4). In contrast, the desmosome has been identified not only in epithelia, but also in cardiac muscle, in the arachnoid and pia meninges, and in lymphoid follicular dendritic cells (5).

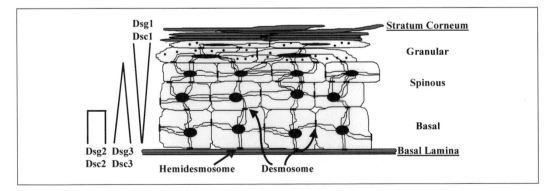

Fig. 1 Structure of the skin: the epidermis is a stratified tissue comprising multiple epithelial layers. The expression profile of the desmosomal cadherins is indicated on the left. Desmosomes and hemidesmosomes contribute to the supracellular intermediate filament architecture, imparting mechanical strength to the tissue as a whole. (Dsg = desmoglein, Dsc = desmocollin.)

In epithelial sheets of one cell thickness, desmosomes are found at the lateral surface of adjacent cells. In multilayered epithelia such as epidermis (Fig. 1), these junctions are found both at the lateral and apical surfaces of basal cells, and in suprabasal cells they are present on all aspects. By associating at sites of tight intercellular adhesion, the intermediate filament scaffolding forms a supracellular network necessary for the maintenance of tissue integrity and resistance to mechanical stress (6).

Epithelial cells are separated from connective tissue by a specialized extracellular matrix called the basement membrane (basal lamina), which is composed of a variety of matrix molecules including proteoglycans, collagen, and laminin isoforms (7). The multiprotein hemidesmosomal complex provides a source of mechanical integrity by connecting the cytoskeletal elements of the cell and the extracellular matrix proteins that constitute the underlying substrata, thereby mediating firm attachment of basal epithelial cells and providing sites of integration for epithelia and connective tissues.

2. The desmosome

Ultrastructurally, the desmosome (Fig. 2) appears as two parallel tripartite plaques, at the centre of which is a 30 nm extracellular space bisected by an electron dense median line (6, 8, 9). The centre core of the desmosome comprises the mirror image plasma membranes and associated extracellular portions of the desmosomal transmembrane components (10). At the intracellular aspect of the desmosome, next to the plasma membrane, lies a 15–20 nm thick outer dense plaque. This outer plaque is separated from a second, inner dense plaque by an 8–10 nm electron lucent region (11). The plaque is decorated with intermediate filaments which do not appear to terminate at the plasma membrane but instead appear to loop through the plaque at a distance of 20–40 nm from the membrane and radiate back out into the cytoplasm

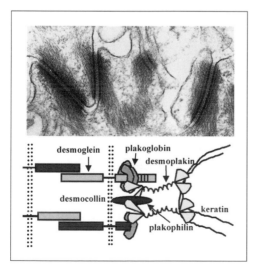

Fig. 2 Desmosome protein composition, structure, and ultrastructural properties: an electron micrograph of bovine tongue epidermis is shown on the top. On the bottom panel, opposing cell membranes are indicated by parallel dotted lines, and a desmosomal plaque (including intermediate filament linkage) from one cell is depicted. The placement of desmosomal proteins is approximate; however, currently documented protein–protein interactions are indicated.

(6). *En face*, the desmosome appears as a circular, or elliptical, spot weld which serves to rivet adjacent cells to one another.

2.1 Molecular components

2.1.1 Desmosomal cadherins

The transmembrane components of the desmosome, the desmogleins (Dsg) and desmocollins (Dsc), belong to the cadherin superfamily of cell–cell adhesion mediators. These type I membrane glycoproteins each exist as three distinct isoforms, termed Dsg1–3 and Dsc1–3, which are the products of different genes and are expressed in a cell type and differentiation-specific manner (reviewed in ref. 12). In the case of the Dscs, variation in the C-terminus is generated by alternative splicing, resulting in a longer 'a' form, and a shorter 'b' form (13–15).

Like the classical cadherins (see Chapter 3 of this volume), the desmosomal cadherins contain in their extracellular domain four repeats of ± 110 amino acids, termed EI–EIV (16)(Fig. 3). These repeats are thought to be involved in binding to Ca^{2+}, which, in the case of classical cadherins is necessary for adhesive function and may be essential for the maintenance of tertiary structure of the extracellular domain (reviewed in ref. 17). To date, little is known about the specific structural features of the desmosomal cadherins.

Intracellularly and proximal to the membrane, lies the intracellular anchor region (IA) which has been implicated in contributing to cadherin-based adhesion (reviewed in ref. 17). The last segment common to classical cadherins, Dsgs, and Dscs is the ICS, or intracellular cadherin segment (also referred to as the catenin binding

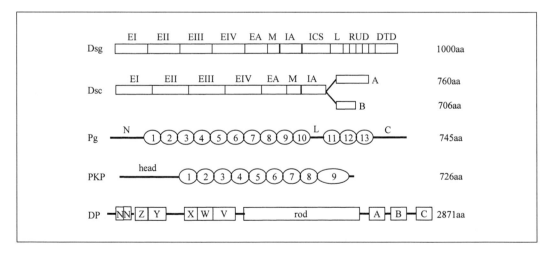

Fig. 3 Protein components of the desmosome: approximate domain structures of the five major desmosomal constituents are depicted here. N-termini are oriented at the left. Where multiple isoforms exist, the structure is representative of the '1' isoform. Pg domain structure follows that reported in (42), and PKP domain structure follows (80). Residue numbers for desmosomal cadherins correspond to the mature, processed polypeptides. (Dsg = desmoglein 1, Dsc = desmocollin 1, Pg = plakoglobin, PKP = plakophilin 1, DP = desmoplakin 1.)

domain; CBD). It is at this region that members of the catenin family have been shown to bind (18). The desmogleins are unique in the cadherin superfamily in that they contain three additional domains at their cytoplasmic tail, a proline-rich linker (L), a repeating unit domain (RUD), and a desmoglein terminal domain (DTD). The RUD comprises a segment of 29 ± 1 residue repeats. Dsg1 contains five such repeats, Dsg2 contains six, and Dsg3 contains two (19). Predictions based on amino acid analysis (20) suggest that these repeat domains each form two β-strands. In addition, homoassociation of recombinant RUD has been observed by rotary shadowing, implying that this region may mediate homodimerization of the desmoglein tail (21). The true functional significance of the L, RUD, and DTD domains is not known, and its elucidation will likely be important to understanding desmoglein's contribution to the regulation of desmosomal plaque assembly.

The expression pattern of the individual Dsg and Dsc isoforms is complex, and is presumed to be related to control of the differentiation program in stratified epithelia (22, 23). In tissues in which two or more isoforms of a Dsg or Dsc is expressed, there generally exists considerable overlap. Additionally, individual desmosomes have been found to contain more than one isoform (24). In general, Dsg2 and Dsc2 mRNA is ubiquitously expressed in tissues which make desmosomes, and this pair is found both in simple epithelia and the basal layer of the skin (Fig. 1) as well as in the myocardium of the heart (12, 25–28). In adult skin, Dsg3 and Dsc3 are found predominantly in the basal and suprabasal layers of stratified epithelia (Fig. 1) (26, 29), whereas Dsg3 is detected in all layers of neonatal skin (30). In contrast, expression of the Dsg1/Dsc1 pair appears to be predominantly detected in the upper, more differentiated layers (Fig. 1) (25, 26, 31). Several studies have indicated that the requirements for desmosomal assembly of individual isoforms are specific. In A431 cells, which express endogenous Dsg2, ectopic expression of Dsg1 results in dissolution of desmosomes (32, 33). Additionally, the presence of this inappropriate isoform results in a reduction in endogenous cadherin levels (33).

As previously mentioned, the desmosomal cadherins, like the classical cadherins, contain a catenin binding domain. The catenin known to directly associate with Dsgs and the 'a' form of Dscs during normal desmosome assembly is the armadillo family protein, plakoglobin (γ-catenin) (Pg) (34–36). Deletion analysis has identified a 19 amino acid stretch in the Dsg ICS domain that is required for mediating this association (36, 37).

The importance of Dsg's role in contributing to epithelial tissue integrity was recently emphasized by the identification of a variant of the heterogeneous group of genodermatoses known as striate palmoplantar keratodermas (SPPK). In this variant of SPPK, hyperkeratosis on the hands and feet has been shown to be the result of loss of part of the N-terminal domain of desmoglein 1 (38). Specifically, a point mutation results in aberrant splicing of the transcript, and consequent loss of much of the first cadherin repeat, including a putative calcium binding site. This represents the first genetic disease known to be caused by a mutation in a desmosomal cadherin gene, and underscores the importance of Dsg1 in contributing to desmosomal function. Dsg's involvement in epidermal blistering diseases is discussed later.

2.1.2 Plakoglobin

Plakoglobin belongs to the well-known armadillo family of structurally related proteins which includes β-catenin, the APC tumour suppressor protein, p120, and plakophilin (for review see refs 39, 40). Originally identified in bovine muzzle epidermis as 'band 5' of the desmosomal components, plakoglobin was shown early on to be a constituent both of desmosomes, and adherens junctions (41). Initial comparisons revealed that Pg closely resembles β-catenin; both were reported to contain 13 central repeats termed ARM repeats (42) with repeats 10 and 11 separated by a flexible linker domain (Fig. 3). High resolution structural analysis of β-catenin's ARM domain redefined the region to contain 12 repeats which together form a superhelix of helices exhibiting a positively charged shallow groove through which protein–protein interactions are facilitated (43). Although the plakoglobin structure has not yet been solved, the ARM domain of Pg has also been shown to be important for the mediation of protein–protein interactions (44). The Pg ARM domain is flanked by divergent N- and C-terminal domains which do not appear to be essential for Pg's assembly into the desmosome (45). Deletion of the N-terminal domain results in a subcellular redistribution of Pg, possibly in part reflective of removal of sequences important for regulation of its stability (45). As assessed by deletion analysis, function of the C-terminus (in which the greatest divergence is noted between armadillo family members) appears to be related to regulation of the length of desmosomes by limiting lateral assembly (45).

Because it is the only major structural component present in both desmosomes and adherens junctions (see Chapter 8 of this volume), plakoglobin is likely to be a mediator both for regulation of assembly of these complexes and segregation of their junctional components. In order to perform these functions, Pg would need to be both promiscuous and selective, able to bind to a variety of molecules but with some mechanism of discrimination. Binding to the classical cadherins has been localized to the central Pg repeat domain (46–48). Pg's co-ordinate interaction with α-catenin in the adherens junction is governed by the Pg N-terminus, specifically that portion encoded by exon 3 (48, 49). Binding to the desmosomal cadherins, however, has been localized to the first three repeats (50, 51), with contributions from the C-terminus required for stable interactions (51). Furthermore, when bound to a desmosomal cadherin, plakoglobin appears unable to interact with α-catenin, providing a possible means by which binding selectivity is achieved between the two groups of junctional components (52, 53). Finer analysis of this region of plakoglobin indicates that of nine hydrophobic amino acids required for binding to Dsg and Dsc, eight are also involved in binding to α-catenin, suggesting that binding selectivity between junctions is mediated by mutual exclusion (54).

More recently, Pg has been shown to directly interact with desmoplakin (53, 55), a member of the plakin family of cytolinkers (for review see ref. 56). This mediation of a link between the cadherins and a cytoskeletal filament binding protein appears to be a key structural role for plakoglobin, both in the adherens junction and the desmosome (Fig. 2). Finally, the presence of Pg has also been documented in non-desmosome containing cells, suggesting the possibility of yet undetermined functions (41).

2.1.3   Plakins

Members of the emerging plakin family link cytoskeletal elements to one another and to sites of intercellular and cell–substrate contact. These proteins share similarity in their structure and function, and several are thought to help anchor IF to desmosomes and hemidesmosomes (56). Included in this family are periplakin and envoplakin, which have been localized to the desmosomal periphery (57, 58); analysis of transglutaminase-mediated crosslinks suggests that these proteins may be involved in forming the cornified cell envelope (59). The family member plectin, which is similar and may be identical to IFAP300 (60), has also been reported in desmosomes (61, 62), but its function there has not yet been well-characterized. Although several plakin family members may be necessary for assembling the IF–desmosomal plaque scaffold, the best characterized plakin family member is desmoplakin.

Desmoplakin exists in two distinct forms, DPI and DPII, which are splice variants of the same gene (63, 64). DPI, the longer of the two and the form more abundant in the plaque, is an obligate component of desmosomes, whereas DPII's expression pattern is more restricted (3, 65). DPs are a large proteins, with predicted molecular weights of 332 kDa and 259 kDa, for DPI and II respectively (66). The amino acid sequence of DP suggests that it likely comprises globular head and tail domains flanking a central rod domain (Fig. 3). It is proposed to dimerize through this central α-helical coiled-coil rod, and these parallel dimers may then form higher ordered filaments mediated by the rod (67–69).

Early evidence suggesting that DP might be capable of interacting with the intermediate filament cytoskeleton came from studies in which segments of DP were ectopically expressed and analysed for their ability to co-align with keratin and vimentin networks (69). Deletion analysis revealed regions in the carboxyl terminus which were required for co-alignment with either keratin or vimentin filament networks, and suggested that the amino terminus of DP was required for localization to the plaque but not for filament alignment (66). Subsequent overlay experiments showed evidence of a direct interaction between the DP C-terminus and the amino terminal head of type II epidermal keratins (70). More recent yeast two-hybrid analysis showed that in addition to interacting with type II epidermal keratins, DP also interacts with the simple epithelial keratins, K8 and K18, and this interaction requires the presence of both keratin subunits (71). To test the hypothesis that DP was, in fact, responsible for mediating a link between desmosomal plaque components and the intermediate filament cytoskeleton, a dominant-negative construct comprising the N-terminal desmoplakin polypeptide was expressed in A431 epithelioid carcinoma cells and resulted in the near ablation of tonofilament connection to intercellular junctions (72). These molecular genetic data have more recently been expanded upon by gene targeting experiments which confirm that not only is desmoplakin necessary for IF linkage to the plasma membrane, but that it is also required for desmosome assembly (Fig. 4) (73).

The critical role played by desmoplakin in the maintenance of epithelial tissue integrity was recently underscored by the identification of two other variants of

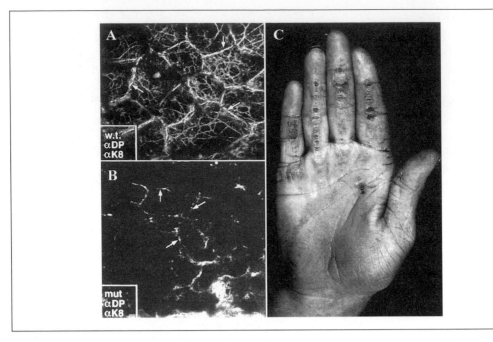

Fig. 4 Desmoplakin is required for intermediate filament organization, desmosome assembly, and epidermal integrity. (A) A (confocal) surface view of a normal mouse embryo, where staining for desmoplakin (*arrows*) and keratin (K8) can be seen in visceral endoderm. (B) Endoderm of embryos homozygous null for desmoplakin exhibits loss of intracellular keratin filament bundle organization. Keratin appears to be aggregated near the cell periphery in mutant embryos. (Please refer to the journal for colour micrographs.) Reproduced from ref. 73, by copyright permission from The Rockfeller University Press. (C) Palmar evidence of striate palmoplantar keratoderma on a human exhibiting haploinsufficiency in desmoplakin expression. Reproduced from ref. 74, by copyright permission of Oxford University Press.

striate palmoplantar keratoderma: a large kindred of manual labourers exhibiting striations on the palms of the hands and soles of the feet have been identified, and these symptoms have been determined to be the result of a haploinsuffiency of desmoplakin (Fig. 4) (74). Analysis of the epidermal desmosomes of this family has reflected a loss of the outer plaque as well as aberrant intermediate filament connections. In a second kindred, a single mutation resulting in premature stop and elimination of the C-domain of the desmoplakin tail produces a recessive disorder marked by perturbations in intermediate filament linkage, keratoderma, woolly hair, and cardiomyopathy (75).

2.1.4 Plakophilins

Other members of the *armadillo* gene family have more recently emerged as key components of the desmosome. Originally referred to as 'band 6' in bovine muzzle desmosome preparations, plakophilin 1 was first shown to be an accessory to the desmosomal plaque (76) and later determined to share similarity with other armadillo proteins (77, 78). Subsequent searches have identified PKP2 (79), PKP3 (80, 81).

Structurally, the PKPs share a general organization in their content of nine ARM repeats preceded by a head domain (39, 80). PKP1 and PKP2 are each known to exist as two variants (PKP1a, PKP1b and PKP2a, PKP2b) which are the products of alternative splicing (79, 82). All three PKPs have been localized by immuno-cytochemistry to the desmosomal plaque (80, 83, 84), however their tissue distri-bution appears to be somewhat specific. PKP2 appears to be the most widespread, as it is found in the desmosomes of simple epithelia, non-epithelial tissues, and cultured cells (although its absence in the suprabasal layers of stratified epithelia is noted), and it is also present in every cell type tested to date (79). In contrast, PKP1 is notably absent from desmosomes of simple epithelia and basal cell epithelia but is detected in suprabasal cells, appearing to be differentiation specific (83). Perhaps the most surprising and intriguing result of the PKP localization studies is that all three PKPs appear to be present in the nucleus (79, 81, 82). The search for nuclear binding partners of the plakophilins is currently underway.

PKPs 1 and 2 have been shown to bind to intermediate filaments *in vitro* (55, 76–78, 85), and to the N-terminus of DP through its non-ARM head region (55, 86). Further-more, transfection studies have indicated that PKP1 enhances recruitment of DP to the cell membrane (Fig. 5), and its presence appears to facilitate association of 10 nm intermediate filaments with the plaque, as evidenced at the ultrastructural level (87). These observations together provide evidence for a role for PKP1 in lateral clustering of DP, and therefore enhancement of the number of IF binding sites (86). Conversely, PKP's ability to associate with the desmosomal cadherins, remains controversial. Although overlay assays have shown binding of PKP1 to some desmosomal cad-herins (37, 55), detection of interactions by yeast two-hybrid suggest that they are less robust than those with desmoplakin, and of all of the desmosomal cadherins, only interactions between Dsg1 and PKP1 are detectable in this way (86, 88). Although PKP1 appears to interfere with the binding of Pg to DP, both PKP1 and Pg appear to be required for organizing Dsg1 and DP into plaque-like structures (87). Most PKP1 interactions characterized to date have been mediated through the non-ARM head domain, however a recent study suggests that overexpression of its central ARM region results in PKP/actin filament co-localization and a phenotype characterized by filopodia formation (88).

The importance of PKP in the assembly and maintenance of desmosomes was highlighted by the discovery of a human lacking PKP1 (89). At birth, this child exhibited a type of ectodermal dysplasia/skin fragility syndrome characterized by peeling of the skin, erosions on the feet, and dysplasia affecting the skin, hair, and nails (89). Microscopic analysis of the skin revealed widening of the keratinocyte intercellular spaces, and a reduction in the size and number of desmosomes. Subsequent immunohistochemistry using antibodies against desmosome-specific proteins showed a complete absence of staining for plakophilin 1, and concomitant loss of desmoplakin at cell borders. This phenotype suggests that PKP1 is important to the function of the desmosomal plaque, may play an important role during morphogenesis of the ectoderm, and appears to be involved in the organization of desmoplakin at intercellular junctions (89).

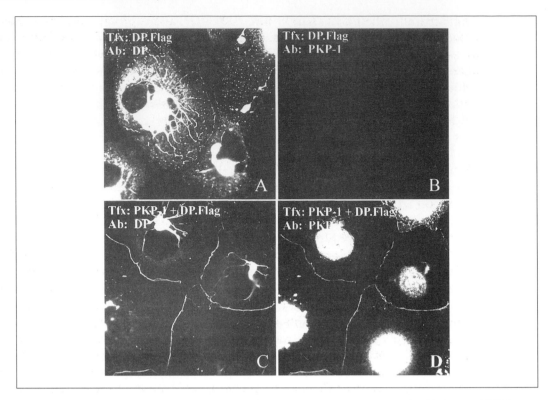

Fig. 5 Plakophilin 1 promotes recruitment of desmoplakin to cell borders: full-length DP.Flag and PKP1 were transiently expressed (Tfx) in COS cells. Double label immunofluorescence (Ab) was performed to detect DP or PKP1. (A) DP.Flag expressed by itself was found in cytoplasmic aggregates, or co-localized with intermediate filaments in cables or perinuclear aggregates. (B) No endogenous PKP1 was detected. When co-expressed with PKP1 (which by itself localizes to cell–cell borders; not shown), DP.Flag was efficiently redistributed to borders (C), where it co-localized with PKP1 (D), suggesting an interaction between the two proteins. Reproduced from ref. 87, by copyright permission of The Company of Biologists, Ltd.

A number of studies have identified other proteins which may play potentially critical structural or regulatory roles in the desmosome, such as desmocalmin/ keratocalmin and pinin (90–92). Although space limitations do not permit discussion of these studies here, we refer the interested reader to an earlier review of this literature (17).

2.1.5 A molecular map

Assembly of all of the identified interactions between desmosomal components results in the emergence of a hypothetical blueprint of desmosomal architecture. In order to verify the accuracy of this proposed layout of the desmosomal plaque, North *et al.* have established a map of the desmosome via quantitative analysis of immunogold electron micrographs using antibodies specific for terminal domains of the desmosomal components (93). The authors propose a model in which the desmoplakin N-terminus lies within the outer dense plaque, and the C-terminus near the presumed intermediate filament attachment region, desmoglein 3 tail spans the outer plaque, desmocollin 'a' form is near the region in which desmoplakin and plakoglobin are detected, and plakoglobin overlaps with desmoglein and desmo-collin C-terminal domains. Unexpectedly, in light of the biochemical evidence for interactions between intermediate filaments and plakophilin, plakophilin 1 appeared to localize deep within the outer plaque, near the plasma membrane (93).

2.2 Adhesive function of the desmosome

It has long been assumed that a primary function of the desmosome is in the mediation of intercellular adhesion. For instance, by virtue of their structural similarities to the classical cadherins, the desmosomal cadherins were hypothesized to mediate calcium-dependent homophilic adhesion. As the following points will indicate, proving this seemingly logical hypothesis has been unexpectedly difficult. Early studies directed toward proving that classical cadherins were mediators of cell–cell adhesion utilized a system in which cDNA encoding E-cadherin was introduced into non-adhesive L cell fibroblasts (94). These experiments elegantly demonstrated that in these cells, ectopic E-cadherin was sufficient for the mediation of strong homophilic intercellular adhesion (coincident with the consequent up-regulation of catenin binding partners). Similar experiments were subsequently performed for the desmosomal cadherins. First, a chimeric molecule comprising the extracellular domain of Dsg3 and the cytoplasmic domain of E-cadherin was shown to mediate only weak aggregation of L cells (95), markedly different from the strong intercellular adhesion and compaction seen in the E-cadherin expressing cells. In similar experiments, neither Dsc2 nor a Dsc2.Ecad chimera was capable of conferring adhesive properties (96). One interpretation of these data was that coincident expression of other transmembrane components of the desmosome (presumably other desmosomal cadherin isoforms) was required for formation of adhesive structures. To test whether the presence of both a Dsg and a Dsc would be sufficient to mediate adhesion, Dsg1, Dsc2a, and plakoglobin were co-expressed in L cells. Again, these cells failed to aggregate (97). However, when cells expressing the same components were generated using a dexamethasone-inducible Dsc2a construct, adhesion was observed (98), suggesting that perhaps the correct stoichiometry of desmosomal components was critical for assembling a functional adhesive interface. Most recently, the combination of Dsg1, Dsc1a and Dsc1b, and Pg was expressed in non-adhesive L929 cells, and conferred not only measurable adhesion but also appeared to have an invasion-suppressive effect *in vitro* (99). These details provide important clues about the nature of the adhesive interface of the desmosome, however the exact intermolecular interactions involved in assembling an adhesive desmosomal structure are far from understood. Recent work suggesting that heterophilic (Dsg-Dsc) interactions, both on a single cell and between neighbouring cells, may be important (100, 101), lends a further level of complexity to the problem. Furthermore, such evidence implies that the functional nature of the desmosomal cadherins and classical cadherins may well be quite different, and that caution should be invoked when making assumptions about desmosomal cadherin structure and function based on what has been learned about the classical cadherins.

Further support for the desmosome's function in adhesion comes from auto-immune diseases, in which targeting of specific desmosomal cadherins results in the loss of intercellular attachment (102–104). The clinical manifestation of this loss of contact is the formation of blisters or erosions (the result of separation between epithelial layers), also known as acantholysis. These autoimmune diseases are collectively known as pemphigus (105). Pemphigus vulgaris (PV) is the most common type of pemphigus, and is marked by involvement of pathogenic antibodies against Dsg3 (106, 107). Histologically, PV is characterized by deep epidermal acantholysis (Fig. 6). The pathogenic antibodies of PV involving both mucous membranes and skin are known to be directed toward Dsg3 and Dsg1, although those of PV involving only mucous membranes appear to be exclusively against Dsg3 (reviewed in ref. 105). Pemphigus foliaceus (PF) results from pathogenic antibodies directed against Dsg1 (reviewed in ref. 105). These antibodies elicit acantholysis in the uppermost layers of the epidermis where Dsg1 is most abundantly expressed (Fig. 6), and unlike PV, PF is almost never manifested in oral tissues. This histological characterization correlates well with the expression profile of the desmosomal cadherins (Fig. 1). Interesting recent work has demonstrated that in tissues (such as suprabasal epidermis) which express both Dsg1 and Dsg3, antibodies against both of these isoforms are required for blistering (108). Similarly, neonates of pregnant PF patients appear to be protected from epidermal blistering by virtue of the fact that Dsg3 is distributed throughout all layers of neonatal epidermis (30). These data not only evidence the adhesive function of the desmosomal cadherins, but also support the idea that these isoforms can support adhesion individually (108). Another type of pemphigus, known as paraneoplastic pemphigus (PNP), is a specific syndrome seen primarily in patients with malignant lymphoproliferative disease; although PNP is marked by a cocktail of autoantibodies directed against both desmoglein and plakin family

Fig. 6 Histological characterization of acantholysis in pemphigus vulgaris and pemphigus foliaceus. (A) Pemphigus vulgaris is characterized by autoantibodies directed against desmosomal cadherins in basal and suprabasal epidermal keratinocytes (Dsg3 and Dsg1), resulting in loss of adhesion between these layers. (B) Pemphigus foliaceus autoantibodies directed against Dsg1 result in loss of adhesion between the uppermost granular layers. Reproduced with permission from Dr J. Stanley.

members (109), it appears that anti-Dsg antibodies are the pathogenic agents (110). Finally, recent work has shown Dsg1 to be the target of another blistering disease, staphylococcal scalded skin syndrome (SSSS). The observation that SSSS blistering resembles that of PF led to a study in which the causative bacterial toxin, exfoliative toxin A (ETA), was shown to proteolytically cleave Dsg1 (but not Dsg3) in its extracellular domain (111).

2.3 Role of the junctional elements as revealed by gene targeting

Targeted disruption of specific elements of the desmosome has lent further insight not only into determining how each of the junctional components contributes to junctional integrity, but has also underscored the critical role these junctions play in maintenance of cell shape, tissue architecture, and apparent adhesive strength of the junctions. A common theme emergent in these studies is the involvement of desmosomal elements in linkage of intermediate filaments to the membrane, and in the resistance to mechanical insult.

2.3.1 Desmoplakin

Although the previously described biochemical studies have demonstrated that this protein is critical for intermediate filament linkage to the plaque, its absolute necessity during embryonic development was shown through ablation of the DP gene (73). Mouse homozygous null for DP do not survive beyond embryonic day 6. In the mutant embryos, keratin intermediate filaments appeared disorganized and were found mainly around the cytoplasmic periphery (Fig. 4). In addition, the mutant embryos also displayed very few desmosomes, indicating that DP has a critical role in mediating the formation and/or stabilization of desmosomal structure (73).

2.3.2 Plakoglobin

Animals null for plakoglobin expression also exhibit embryonic lethality, however gross morphological development appears to be normal prior to death. Around embryonic day 12, Pg (−/−) animals begin to die of apparent defects in cardiac development and functional integrity (112, 113). Closer examination of cardiac tissues reveals a paucity of typical desmosomes, and instead, overlapping adherens junctions containing desmosomal components are present. Surprisingly, epithelial desmosomes in epidermis and gut appear to assemble normally, and in earlier embryos, the integrity of these tissues does not appear to be compromised in the absence of Pg. In an alternative genetic background, mice lived longer and appeared to have some epithelial dysfunction including skin fragility and subcorneal acantholysis (112). It is logical to speculate that a compensatory mechanism for Pg's role in normal desmosomes must exist, and indeed, further characterization of Pg null mice has shown that β-catenin appears to associate with desmoglein in the absence of Pg (114).

to Troyanovsky *et al.*, the cytoplasmic domain of desmocollin appears to be important for recruitment of plaque components (32, 131). A more recent study showed that when co-expressed in L cell fibroblasts, ectopic desmoplakin N-terminus can effectively cluster ectopic desmosomal cadherin/plakoglobin complexes through direct binding of plakoglobin (53). Taken together, these data begin to confirm that some protein–protein interactions precede others, and suggest that regulated junction assembly is the result of a cascade of events facilitated by the binding and subsequent recruitment of specific molecules to specific spaces at specific times.

Another potential mechanism for the direct regulation of desmosome assembly, probably in part through direct control of protein–protein interactions, is by post-translational modification of desmosomal proteins. One report suggested that phosphorylation of a specific C-terminal serine residue in desmoplakin inhibits its association with keratin filament network (132). It is interesting therefore that the phosphatase inhibitor, okadaic acid has been shown to inhibit the late stages of plaque assembly, suggesting that removal of phosphate from proteins such as DP may be required for stable association and oligomerization with intermediate filaments (133). Another study has suggested that activation of PKC signalling can trigger desmosome formation, even in the absence of intact adherens junctions (134). Additionally, phosphorylation of plakoglobin may also be an important mechanism involved in regulating desmosome assembly: Pg phosphorylation has been documented in response to stimulation by growth factors (135, 136). Documentation of the association of plakoglobin with protein tyrosine phosphatases such as PTPκ (137) lends further evidence that modification of plakoglobin may be an important mechanism of junctional regulation.

2.5 Signalling through the desmosome

In addition to their presumed adhesive function, cadherin-based junctions and their constituent proteins are now hypothesized to be a portal for the transduction of signals from the extracellular environment to the intercellular space. There are two major ways in which signalling might occur: first, via the initiation of a cascade originating at the plasma membrane that is dependent on the cadherins and/or associated junction proteins, and secondly, via the involvement of junction accessory proteins in non-junctional cascades. Evidence for the first type of signalling comes from observations that PV antibodies directed against Dsg3 elicit changes in intracellular calcium and phospholipid metabolism in keratinocytes (138, and references therein).

Of the non-transmembrane components of the desmosome, one that has emerged as a favourite candidate for the transduction of intracellular signals is plakoglobin. Though originally identified as a structural protein (41), plakoglobin is now known to share sequence identity with the adherens junction protein β-catenin, and its invertebrate homologue, *armadillo* (139). Armadillo has been firmly established in the wingless pathway in *Drosophila*, where it plays an indispensible role in the establishment of correct segmental patterning. β-Catenin has been similarly placed in the orthologous Wnt pathway in vertebrates (139). Plakoglobin, β-catenin, and armadillo are 60–70% identical, and share similar structures (140). Plakoglobin's ability to mimic β-catenin's signalling potential in certain systems has implied that it, too, may play an important role in this regard (141); however, the means by which these signals may be elicited, is highly controversial (142–144). One clue that Pg does in fact have some role in the regulation of cellular signalling is that like β-catenin, plakoglobin's metabolic stability appears to be tightly regulated (35) (145, and references therein). It is not at all clear, however, whether Pg's ability to elicit cellular signals is partly indirect, mediated via its ability to replace β-catenin in its various binding capacities: its association at junctions, ability to activate transcription, or its degradation (for review see ref. 146). Since, like β-catenin, Pg is known to be able to mediate transcription via association with LEF/TCF-1 factors, it has long been hypothesized that Pg may in fact be a potent signaller on its own, activating targets in a manner distinct from that of β-catenin (147). Indeed, in one recent study, high levels of cellular plakoglobin produced unregulated growth, and coincident up-regulation of the anti-apoptotic protein, BCL-2 in SCC-9 cells (148). In contrast, β-catenin is known to up-regulate proteins which promote proliferation, such as c-Myc and cyclin D1 (149–151). Secondly, Pg's activation of c-Myc via a pathway independent of β-catenin has been reported (152). Finally, a somewhat surprising effect was observed when Pg was altered to increase its stability in a transgenic mouse model: epithelial cells expressing the transgene appeared to grow more slowly than did normal cells (116). Clearly, the exact nature of plakoglobin's potential in eliciting cellular signalling events remains to be determined.

3. The hemidesmosome

Under the electron microscope, a hemidesmosome can be seen as having a triangular-shaped electron dense cytoplasmic plaque adjacent to the basal cell surface, a sub-basal dense plate just below and external to the plasma membrane, and thin extracellular filaments, called anchoring filaments, which extend from the plate into a specialized extracellular matrix that separates epithelium from underlying connective tissue, known as the basement membrane (1, 2)(Fig. 7). In contrast to focal adhesions, which connect the extracellular matrix to the actin microfilament network, hemidesmosomes link the matrix to keratin intermediate filaments (153, 154). The electron dense plaque of the hemidesmosome comprises an inner cytoplasmic-most plaque through which intermediate filaments appear to loop, and a perimembrane plaque consisting of the cytoplasmic tails of trans-membrane components of the hemidesmosome (153).

Each hemidesmosome connects the keratin intermediate filament cytoskeleton of basal epithelial cells to a particular protein in the basement membrane, an isoform of laminin called laminin-5. Laminin-5–hemidesmosome interaction is believed to promote stable epithelial attachment to basement membrane, maintaining the integrity of the epithelial tissue (153). The components of hemidesmosomes will be discussed next.

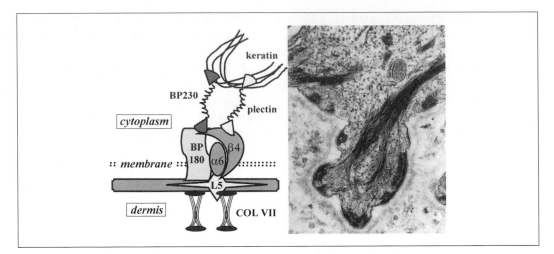

Fig. 7 Hemidesmosome protein composition, structure, and ultrastructural properties. A schematic diagram of a hemidesmosome is shown on the left, and an electron micrograph displaying hemidesmosome ultrastructure is shown on the right. Placement of molecules in the schematic diagram are approximate and are intended to indicate the most current model of intermolecular interactions. The basal epithelial cell plasma membrane is indicated by a dotted line.

3.1 Molecular components

3.1.1 α6β4 Integrin

The principal agent for integrating the cytoskeleton and extracellular matrix at the site of the hemidesmosome is the α6β4 integrin (2, 4, 155) (Fig. 8). The integrin α6β4 serves as a transmembrane receptor for the extracellular matrix component, laminin-5, and together, α6β4 integrin and laminin-5 not only provide the core of hemidesmosome structure but are believed to be the conduit for hemidesmosome-mediated cell signalling (156–160) (for integrin structure, see Chapter 4 of this volume).

The α6 integrin subunit is widely expressed during development and in adult tissues, but its principal expression is in developing and mature human skin and cornea, the central and peripheral nervous system, and renal epithelia (153). The mRNA encoding the α6 integrin subunit is subject to alternative splicing in many tissues, certainly in epithelia, resulting in splice variants α6A and α6B which differ only in their cytoplasmic domains (161, 162). The α6A isoform is more prevalent, and is the only isoform expressed in many epithelial cells, including 804G rat bladder carcinoma cells which assemble hemidesmosomes *in vitro* (161, 163, 164). Some earlier studies in mice suggested that these splice variants are necessary for the proper development of distinct tissues (165). Recent gene targeting studies have demonstrated that the presence of an intact α6 integrin is required for the development of skin and lymphocytes, as well as for long-term survival (166, 167). Although the α6 integrin is able to dimerize with both the β1 and β4 subunits, it has been shown that in cells which express all three subunits, α6 prefers to bind to the β4 subunit over the β1 subunit. Thus, in cells containing α6, β1, and β4 integrin subunits, the α6β4 integrin pair is in abundance, whereas little α6β1 is present (168). α6β4 integrin is expressed primarily in epithelial tissues, although it is also found in Schwann cells and some endothelial cells (see, for example refs 169, 170). The β4 integrin subunit also has the unique feature of a large cytoplasmic tail domain, of about 1000 amino acids. This feature appears to allow the α6β4 integrin to mediate several specialized functions.

α6β4 integrin is a receptor for laminin-1, -2, and -4, but shows highest affinity for laminin-5 (156, 171). In intact epithelia, α6β4 integrin is typically found in the basal cell layer, and in particular is polarized at the basal cell surface (172, 173). This integrin is also a central component of hemidesmosomes and is essential for their formation, as well as for the maintenance of attachment of epithelial cells to basement membrane (173–175).

Antibodies to the α6β4 extracellular domain block hemidesmosome assembly in cultured cells (173). In addition, a recombinant β4 integrin which lacks a cytoplasmic tail disrupts existing hemidesmosomes and prevents the formation of new ones, but does not block α6β4-dependent cell adhesion to laminin-5 (176). Together, these results imply that the β4 cytoplasmic tail and the α6β4 extracellular domains are essential for hemidesmosome formation, but the extracellular domains of the α6β4 integrin are sufficient to interact with laminin-5 and mediate cell adhesion. Thus, hemidesmosome formation is dependent not only upon integrin–matrix ligation, but also on integrin-mediated intracellular events. Another laminin-5 receptor, α3β1,

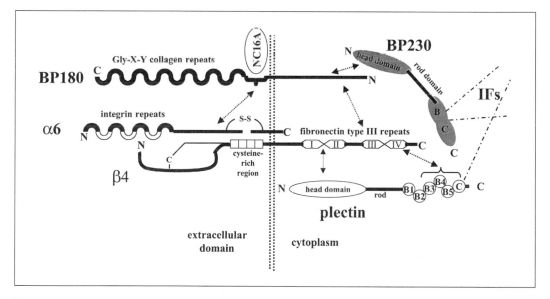

Fig. 8 Protein components of the hemidesmosome. Protein–protein interactions are shown by *arrows*, indicating approximate binding regions. The juxtaposition of the various proteins represents the current model for hemidesmosome organization. BP180 is depicted in monomeric form for clarity, but is likely to exist as a homotrimer *in vivo* through the extracellular collagenous domains. A dicysteine bridge in the extracellular domain of β4 integrin is shown. (IF = intermediate filament.)

may also play a role in hemidesmosome regulation, although it is not itself located at the hemidesmosome.

3.1.2 BP180

There is a second transmembrane member of the hemidesmosomal complex. Bullous pemphigoid antigen II (BP180, BPAG2, or type XVII collagen) is a 180 kDa type II transmembrane protein recognized by autoantibodies in the sera of bullous pemphigoid patients (BP) (177–181). In its C-terminal extracellular domain lie a series of collagenous Gly–X–Y repeats, hence its designation in the collagen family (178, 180, 181). BP180 assembles as a trimer through triple helical coiled-coil formation by the extracellular collagenous domain, and is believed to contribute to the structure of anchoring filaments (181–183). Experimental evidence suggests that BP180 may associate with β4 integrin in the cytoplasm, and that this interaction appears to involve sequences toward the extreme N-terminus of BP180 (184, 185) (Fig. 8). Recently, molecular genetics were used to demonstrate that BP180 interacts with α6 integrin (186). Subsequent studies have demonstrated a direct interaction between a region (NC-16A) in the non-collagenous ectodomain of BP180, which is part of one of the epitopes recognized by pathogenic autoantibodies in BP, and α6 (187). This BP180/α6 integrin interaction can be inhibited by peptides encoding sequences in the NC-16A domain, and application of these peptides to cultured epithelial cells appears to prevent hemidesmosome formation. Thus, BP180 can be considered a novel transmembrane ligand of α6β4 integrin, and evidence suggests that this BP180/α6 interaction is necessary for the formation and/or stabilization of hemidesmosomes in culture (187).

Several human diseases have been found to result from loss of function of BP180. Bullous pemphigoid is a class of diseases resulting from the targeting of autoantibodies to components (such as BP180) of the hemidesmosomal adhesion complex. The consequence of this antibody targeting is subepidermal blistering, or separation of the skin between the basal cell layer and the underlying matrix (188). In addition, a 120 kDa collagenous target of autoantibodies in the sera of linear IgA dermatosis patients (LAD), is now thought to be a direct result of protease cleavage of the BP180 extracellular domain (189).

3.1.3 BP230

Linkage of the hemidesmosome to the keratin intermediate filament network involves a second bullous pemphigoid antigen, BP230 (BPAG1). Whereas BP180 is a transmembrane protein, BP230 is localized at the inner cytoplasmic plaque of the hemidesmosome, as judged by immunoelectron microscopy (177). By virtue of its structure, BP230 is now classified as a member of the growing family of cytolinker proteins called plakins, which are thought to be involved in organization of the cytoskeleton (56). Like other plakin family members, BP230 is predicted to contain a central coiled-coil region flanked by globular head and tail domains. The C-terminus of BP230 is capable of associating with intermediate filaments (190), a function consistent with the fact that BP230-deficient mice have poorly formed hemidesmosomes to which few keratin filaments attach (191). Very recent work using both a yeast two-hybrid approach and recombinant proteins has established that BP180 and BP230 directly interact through their N-terminal domains (192) (Fig. 8). As BP180 and keratin filaments do not seem to interact, these data suggest a model in which BP230 acts as a connecting segment between intermediate filaments (through binding to the BP230 C-terminus), and BP180 and the hemidesmosomal plaque (by binding of the BP230 N-terminus to the BP180 N-terminus).

3.1.4 Plectin

Plectin is a 500 kDa protein, which is widely expressed and shows structural similarity to members of the plakin family of cytolinkers (56, 193). Like desmoplakin, plectin's predicted structure is that of an α-helical coiled-coil rod flanked by globular N-and C-terminal regions, giving plectin the overall appearance of a dumbbell (193). Immunoelectron microscopy has established that plectin, and the related IFAP300 molecule (60) are localized both to the desmosome and hemidesmosome (194, 195). The necessity of plectin function for the maintenance of epidermal integrity is exemplified by the hereditary epidermolysis bullosa simplex disease, EBS-MD, in which plectin defects result in severe skin blistering combined with muscular dystrophy (196–200). Other variants of epidermolysis bullosa simplex result from mutations in the keratin 5 and keratin 14 genes, and are marked by epidermal blistering resulting from cytolysis in the basal layer (201).

By virtue of its association both with itself and with a large number of other molecules, plectin has been proposed to be an important integrator of cytoskeletal networks (reviewed in refs 193, 202). Its interaction with intermediate filament proteins (203, 204), taken together with the establishment of interactions between plectin and multiple domains of the β4 integrin tail (202, 205)(Fig. 8) suggest that it is likely a key mechanism by which the intermediate filament network is linked to the hemidesmosomal plaque. More recent work has also pointed to a potential role for plectin in clustering α6β4 integrins at the basal cell edge, an event which may be critical for the establishment of hemidesmosomes (206). The combined observations that plectin connects intermediate filaments to the β4 cytodomain, and that BP230 performs the same function for BP180 (and thus indirectly for the α6 integrin subunit), suggest that these two plakin proteins are required for maintaining multiple keratin filament connections to a single hemidesmosome. These findings further indicate the necessity of both linker proteins for hemidesmosome stability.

3.1.5 CD151

It has also been proposed that CD151, a member of the tetraspan family, is a component of the hemidesmosome (207). Immunoelectron microscopic localization of CD151 shows it to be concentrated at hemidesmosomes. This localization appears to be dependent on association between α6 and β4, and may involve binding of CD151 to this integrin pair, although the exact role of tetraspans in the assembly of hemidesmosomal structures remains unclear.

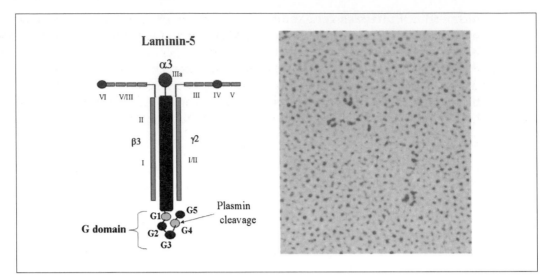

Fig. 9 The laminin-5 heterotrimer. (A) Schematic diagram of laminin-5 depicting the cross-shaped structure composed of $\alpha 3$, $\beta 3$, and $\gamma 2$ subunits. The C-terminal portion of the $\alpha 3$ subunit forms a compact globular domain, called the G domain, which is divided into subdomains G1 through G5. The putative plasmin cleavage site in the G4 subdomain is indicated by an *arrow*. (B) Rotary-shadowed electron micrograph of purified rat laminin-5.

3.1.6 Laminin-5

The extracellular matrix creates a framework that is essential for the maintenance of tissue integrity. Emerging evidence implies that matrix proteins play more than just structural roles and suggests their importance in events involved in the regulation of adhesion, migration, proliferation, differentiation, and gene expression in adjacent cells (208).

The laminins are a family of proteins comprising a major component of extracellular matrices. Laminins are large (400–600 kDa) heterotrimeric glycoproteins composed of a α, β, and γ subunit (Fig. 9). The basic shape of most laminin molecules is of a cruciform or cross, formed by covalent and non-covalent interactions between each subunit. The N-termini of the three subunits comprise the short arms of the cross. The long arm of laminin, sometimes called the rod domain, is created by the association of the central region of the α chain and the central and C-terminal portions of the β and γ. Particular laminin isoforms are dictated by the specific α, β, and γ subunits which compose each molecule (209).

Laminin-5 (epiligrin, nicein, kalinin) is a heterotrimer made of the $\alpha 3$, $\beta 3$, and $\gamma 2$ subunits (Fig. 9). The laminin $\alpha 3$ chain is a component of laminin-5 ($\alpha 3 \beta 3 \gamma 2$), laminin-6 ($\alpha 3 \beta 1 \gamma 1$), and laminin-7 ($\alpha 3 \beta 2 \gamma 1$). Although the N-terminus of the $\alpha 3$ chain is truncated, a small region known as the IIIa domain is present which presumably extends above the long arm domain, although such an extension has not been detected by electron microscopy (209). Two distinct transcripts for the $\alpha 3$ chain, α3A and α3B, have also been identified, which apparently encode gene products with different N-terminal sequences. Miner *et al.* (210) have described a novel α3 isoform of 300 kDa encoded by the mouse Lama3 gene and containing a large N-terminal region; however this α3 subunit has yet to be fully characterized. In either case, adjacent to the IIIa region is a α-helical region composed of the domains II/I; this region is incorporated into the triple helical coiled-coil of the laminin long arm (211, 212). The remainder of the molecule, containing amino acids 793–1713, makes up a large globular domain known as the G domain (211). Although this domain comprises 54% of the residues of the α3 chain, the G domain displays a compact ball-shaped structure when viewed in the electron microscope, suggesting that it is tightly folded (212, 213). The α3 chain G domain can be divided into five sub-domains, related to those described for the α1 and α2 chains (209, 211). However, the sequences of the α3 chain G subdomains are quite divergent from those in the α1 and α2 subunits, suggesting that the G domain of the α3 chain may provide this laminin chain with unique functions. The skin blistering diseases junctional epidermolysis bullosa and cicatricial pemphigoid, incurred from mutations in or autoantibodies against laminin-5, respectively, are both characterized by a complete separation of the skin epidermis from the underlying dermis (214–219, 220) and provide clear evidence that laminin-5 is necessary for firm attachment of basal epithelial cells to the subepithelial basement membrane.

3.2 Hemidesmosome interactions with the basement membrane and connective tissue

All the protein components described above are important contributors to hemidesmosome structure, but they are not sufficient to mediate connections between hemidesmosomes and the subjacent tissue. As described, the transmembrane integrin receptor α6β4 can interact with BP180, as well as with laminin-5 in the basement membrane. But how does integrin binding to laminin-5 connect the hemidesmosome with the underlying tissue? Besides interacting with laminin-6 and -7 and some other basement membrane proteins, laminin-5 and perhaps BP180 also help to anchor the hemidesmosome into the connective tissue. Anchoring filaments, which are composed of either laminin-5 and/or BP180, extend from the sub-basal dense plate of the hemidesmosome through the lamina lucida layer of the basement membrane (153). An interaction between BP180 and laminin-5 is also possible. These filaments are thus thought to link the sub-basal dense plate with the lamina densa. Anchoring fibrils, which are thick banded fibres (by electron microscopy) composed of anti-parallel dimers of collagen type VII molecules, extend from the lamina densa downward into the connective tissue, which in the case of skin, is called the dermis (Fig. 7) (221–223). It is possible that laminin-5 in anchoring filaments forms a bridge between the epithelial (epidermal in skin) hemidesmosome and the stromal (dermal) anchoring fibrils. Indeed, laminin-5 has been shown to interact directly with the NC-1 domain at the N-terminus of type VII collagen (223, 224). Since the NC-1 domains point outwards from the ends of collagen VII anchoring fibrils, a role for laminin-5/collagen VII interaction in maintaining epidermal–dermal connections is likely to be valid (225, 226). Consistent with this hypothesis, the skin diseases dystrophic epidermolysis bullosa and epidermolysis bullosa acquisita are characterized by mutation in, and autoantibodies against, collagen VII molecules, respectively. Consequent disruption of anchoring fibril stability results in separation of the epidermis from the dermis, leading to skin blisters (222, 227, 228).

3.3 Hemidesmosome functions

It has long been assumed that hemidesmosomes play an important role in establishing and maintaining adhesion of epithelial tissue to the basement membrane. This assumption is based partly on the fact that hemidesmosomes appear as spot welds via electron microscopy (9). This hypothetical mediation of adhesion has gained support through the discovery that perturbation in hemidesmosomal integrity is a feature of a number of skin diseases characterized by epithelial cell/basement membrane dysadhesion and blister formation. Additionally, hemidesmosomes are absent both from those epithelial cells which are repopulating wounds, as well as invasive epithelial tumour cells. Thus, it has been suggested that hemidesmosomes are required for stable epithelial cell anchorage and that their loss is a prerequisite for cell motility (see for example ref. 229).

However, clearly hemidesmosomes are more than just inert spot welds. There are emerging data that they are involved in both 'inside-out' and 'outside-in' signalling phenomena and may therefore affect such diverse cellular activities as gene expression, cell proliferation, and differentiation (160).

3.3.1 Hemidesmosomes in a physiological context

In addition to adhesion and cell signalling, hemidesmosomes apparently play an important role in morphogenetic events including those involved in breast epithelial tubule formation (230). Several lines of evidence indicate that hemidesmosomes are not completely static *in vivo* but are actually more dynamic cellular structures. Experimental depolymerization of actin microfilaments leads to a redistribution of hemidesmosomes to the cell periphery in 804G rat bladder carcinoma cells, but these hemidesmosomes remain connected to intermediate filament bundles. Thus it seems that the hemidesmosomal plaques are dynamic and can move laterally within the plasma membrane (231).

3.4 Transmembrane signalling via hemidesmosomes

Transduction of signals at the basal cell surface can be mediated via the hemidesmosomal $\alpha6\beta4$ integrin pair. In particular, the long $\beta4$ cytoplasmic tail appears to be the point of initiation of several hemidesmosome-based signalling cascades. Expression of $\alpha6\beta4$ is restricted to proliferating keratinocytes (169), and detachment and subsequent migration to the upper layers of the skin results in cell cycle exit

(232). Thus, it has long been suspected that ligation of basal keratinocytes to matrix components such as laminin-5 is linked to the control of cell cycle progression. Early evidence for α6β4's involvement in control of the cell cycle came from work showing that binding of laminin-5 to α6β4 leads to the recruitment of Shc/Grb2 (early effectors of a *ras* signalling pathway) to the integrin, and subsequent phosphorylation of the β4 cytoplasmic domain (158). The recruitment of Shc has since been mapped to interaction between two phosphotyrosine residues with the Shc SH2 domain, and two phosphotyrosine residues with the Shc PTP domain (233). This β4 phosphorylation is thought to be involved in the regulation of hemidesmosome formation (158). One study in which the β4 tyrosine-based activation motif (TAM) was mutated, reported junctional assembly (234). More recently it was shown that these phosphorylation events appear to antagonize the formation of hemi-desmosomes (233). It has also been determined that ligation of α6β4 results in subsequent activation of the Ras-MAP kinase pathway and consequent transit through G1 (235). Additionally, there are contradicting reports of the consequences of loss of expression of the β4 tail, implying the existence of multiple signalling pathways. In one study, post-mitotic enterocytes of mice expressing a tail-less β4 subunit have been shown to have increased levels of the cyclin-dependent kinase inhibitor, p27^{Kip} (236), whereas another study suggested that reexpression of β4 in cells deficient in its expression induced up-regulation of the G1 cyclin-dependent kinase inhibitor p21 (237).

The α6β4 integrin pair has also been implicated in signalling events related to cell survival (238, 239). Previous lessons from integrin signalling have demonstrated that integrins can promote cell survival by inhibiting apoptotic signals (240). However, a recent study suggests that in certain carcinoma cells, ectopic expression of α6β4 can stimulate p53 function and consequent cell death (241). This contradiction might be explained by the fact that α6β4's ability to stimulate apoptosis correlates with wild-type but not mutant p53 status (241). A subsequent study suggests that α6β4 integrin survival signals are abrogated by wild-type p53 through caspase-dependent cleavage of AKT/PKB kinase (242). Other work has indicated that stimulation of epithelioid carcinoma cells with epidermal growth factor initiates a protein kinase C-dependent cascade of events resulting in the redistribution of α6β4 prior to chemotactic cell migration (243). However, recent knock-out studies in which both α6β4 and α3β1 were eliminated suggest that the contribution made by α6β4 in skin development may be simply structural. The phenotype of these mice suggested that prior to loss of basal cell attachment to the basement membrane at E15, processes previously ascribed to integrin ligation proceeded normally (244). In an oppositely oriented event, a membrane-associated protein of 80 kDa becomes dephosphorylated following dissociation of α6β4 integrin from laminin-5, implying that an 'inside-out' signalling mechanism may be important for α6β4/laminin-5 ligation (245). Taken together the above data provide further evidence that the α6β4 integrin function is important not only for hemidesmosome formation and epithelial cell attachment, but also for the propagation of various signalling pathways which are involved in morphogenesis, development, and tumour progression (160).

3.5   Regulation of hemidesmosome assembly

The regulation of hemidesmosome assembly appears to be dependent on the spatially and temporally controlled initiation and co-ordination of interactions among junctional constituents. Although many of these interactions have been described in previous sections, several key concepts have recently evolved in the literature and will be addressed here.

One of the early clues on regulation came from experiments using 804G rat bladder epithelial cells (163). These cells readily assemble hemidesmosomes in culture, and have therefore provided a valuable tool. When plated along with antibodies to α6β4 integrin, 804G cells lose their ability to assemble hemidesmosomes, providing the first direct evidence that α6β4 is required for junctional assembly (173). Subsequent work suggested that a likely early event in the initiation of hemidesmosome assembly is the clustering of integrins (176, 246). More recent evidence suggests that portions of the β4 tail may be required for recruitment of BP180 and BP230 (247). Indeed, expression of β4 in keratinocytes lacking endogenous β4 induced clustering of α6β4 integrin pairs along with plectin, BP180, and BP230, suggesting again that α6β4 is important for the initiation of hemidesmosome assembly (247). The interaction of β4 with plectin has also been suggested to be an important event in the intiation of hemidemosome assembly, as cells expressing mutants which could not recruit plectin, could neither recruit BP180 to a potential nucleation site (247).

The mechanism for strong adherence of basal keratinocytes is believed to be based on the interaction of laminin-5 with hemidesmosomes (153, 213). The binding of laminin-5 to α6β4 integrin receptors initiates formation of the core of hemidesmosomes (155, 174, 175). In fact, laminin-5 is capable of, and necessary for, inducing the assembly of hemidesmosomal structures (229, 246, 248). More specifically, an antibody (CM6) binding at the G-domain of the α3 chain of laminin-5 prevents hemidesmosome formation, suggesting that laminin-5's ability to initiate nucleation of hemidesmosomes involves this specific domain (212). Previous work has shown that the laminin-5 secreted by 804G cells was capable of inducing the formation of hemidesmosomes in cells which secreted laminin-5 but did not normally assemble hemidesmosomes (246, 248). This suggested that there must exist some difference in the laminin-5 capable of inducing junction assembly. Indeed, it was demonstrated that the α3 chain of laminin-5 secreted by 804G cells has been proteolytically processed to a shorter fragment, and that this processing is required to initiate hemidesmosome assembly and discourage motility (249). This proteolysis was determined to be mediated through a plasmin-dependent mechanism which involves activation of plasminogen (Pg) by tissue-type plasminogen activator (tPA) (249). More recently, high affinity binding of Pg and tPA to laminin-5 has been mapped to the G1 subdomain of the N-terminus of laminin's α3 subunit (250). This binding results in 32-fold enhancement of the catalytic efficiency of Pg, and consequent increase in plasmin activity.

Laminin-5 also mediates attachment of epithelial cells by interaction with a non-hemidesmosomal integrin receptor, α3β1 (159, 251). Antibodies to this integrin have been shown to inhibit the adhesion of various cell lines to laminin-5 substrates (213, 251, 252), and it has been suggested that α3β1 might be an important initiator of hemidemosome assembly. Evidence for α3β1's potential involvement in the formation of cell–substratum attachment came from a targeted inactivation of the α3 integrin subunit in which the skin was marked by a loss of adherence to the basement membrane (253). However, these cells assemble fairly typical hemidesmosomes. More recently, conditional ablation of β1 integrin in skin produced a strikingly different phenotype marked by epidermal blistering, hair defects, disorganization of basement membrane, and impairment in cell proliferation (254). Hemidesmosomes in these animals were significantly reduced in number. These data suggest that minor αβ1 integrin pairs may be important for proper basement membrane assembly which necessarily precedes the formation of stable hemidesmosomes. Finally, it has also been suggested that interaction between α3β1 integrin and unproteolysed laminin-5 may be involved in driving the forward migration of epithelial cells at the leading edge of wounds; in these cells, α6β4 integrin is laterally polarized and there is an absence of hemidesmosomes at the basal surface (Fig. 10) (255).

Involvement of the BP antigens has also been implicated in the regulation of hemidesmosome assembly. The fact that some tumour cells have been shown to express a truncated version of the BP230 molecule, has led to the speculation that expression of the truncated isoform, rather than the intact BP230 molecule, may allow a cell to regulate hemidesmosome assembly (256). BP230 also appears to be an important early initiator of binding to other hemidesmosomal molecules: expression of a GFP-tagged N-terminal fragment of BP230 can actually disrupt the localization

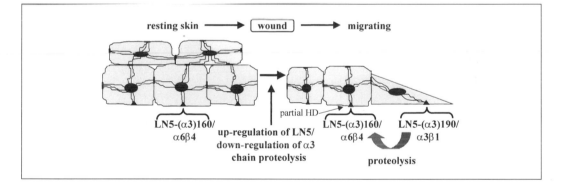

Fig. 10 Model for hemidesmosomes in epithelial wound closure: in static, resting skin, basal epithelial cells are anchored to the basement membrane through the binding of basal α6β4 integrin in hemidesmosomes to laminin-5 in the extracellular matrix. Following wounding, production of laminin-5 is up-regulated, and/or a down-regulation of proteolysis of the laminin-5 α3 chain occurs, resulting in an increased concentration of unproteolysed laminin-5 at the wound edge. This unprocessed laminin-5 interacts with α3β1 integrin along the leading edge of the wound to drive forward migration of epithelial cells. Hemidesmosomes disassemble in these leading edge cells and α6β4 integrin redistributes to lateral cell surfaces. In cells distal from the wound edge, some α6β4 integrin remains basally located where it can interact with laminin-5 in rudimentary hemidesmosome structures, which may assist in maintaining the stability of the migrating epithelial sheet. Adapted from ref. 255 by copyright permission of The Company of Biologists, Ltd.

pattern of endogenous BP180, but not that of endogenous BP230 or α6β4, suggesting that BP180 and BP230 might normally interact in the cytoplasm prior to their incorporation into the hemidesmosome (192).

3.6 Lessons from gene targeting

3.6.1 BP230

Much of what is currently known about the nature of BP230's contribution to hemidesmosomal structure and assembly stemmed from the generation of the BP230 knock-out mouse. Targeted disruption of the BP230 gene produced hemidesmosomes lacking both an inner plaque structure and intermediate filament connections (191). These mice displayed tissue fragility in epidermis and oral epithelia, as well as unexpected symptoms of neural degeneration. This neuronal phenotype was later found to be due to concomitant ablation of neural BP230 (dystonin), and resultant lack of cytoskeletal organization in neuronal filament networks (257).

3.6.2 Plectin

Ablation of plectin in an animal model has clearly confirmed its role as an integral mediator of epithelial tissue integrity. Animals lacking plectin died two to three days after birth, due in large part to severe skin blistering (258). Histological studies revealed structural abnormalities in both skin and skeletal muscle tissue. Subsequent cellular degeneration, marked by a loss of IF cytoskeletal organization is probably due to inefficient bridging of cytoskeletal filaments both between one another and to membrane-associated molecules. Interestingly, the ultrastructural appearance of desmosomes and hemidesmosomes is normal in the plectin null animals, suggesting that formation of these junctions may not be strictly dependent on plectin function (258).

3.6.3 α6β4 Integrin

Targeted disruption of the β4 gene results in severe blistering due to separation at the dermo-epidermal junction (259, 260). The epidermal defect is likely due to trauma during birth, thus underscoring the importance of hemidesmosomal integrity to the resistance of mechnical insult. Death of these mice occurred within hours of birth, and examination of their tissues revealed a complete loss of hemidesmosomal structures. The loss of β4 also produced a reduction in α6 protein levels, and loss of typical BP230 staining, suggesting that the stability and/or subcellular organization of both of these proteins may be dependent on β4. In contrast, BP180 appeared to be present, suggesting that at least at birth, its stability may be independent of α6β4 interaction. Taken together, these data emphasize that α6β4 is clearly a requirement for hemidesmosomal assembly, and that hemidesmosomes are indispensible for the proper function of the skin (259, 260).

Targeted deletion of the cytoplasmic domain of β4 integrin indicated retention of ability to bind to laminin-5, but loss of cytoskeletal linkage (236). Histological examination revealed severe epidermal detachment, and this detachment is likely to be the cause of death within several hours of birth. Additionally, these mice displayed a defect in cellular proliferation consistent with a role for the β4 tail in conducting cellular signals (see above).

Ablation of the α6 gene also results in severe epidermal blistering and subsequent death of the mice (166). Most notably, histological evaluation reveals that similar to the β4 null animals, hemidesmosomes are absent in mice lacking α6. This confirms other already reviewed evidence for the dependence of hemidesmosome assembly on an intact α6β4 subunit. Interestingly, targeted inactivation of sequences allowing for expression of the 'A' splice variant of the α6 integrin subunit appeared to have little to no effect on development of epithelial tissues, epidermal differentiation, or keratinocyte migration suggesting that either this variant is not involved in these processes or that the 'B' variant can effectively compensate (167).

Finally, recent double knock-out experiments in which both α6β4 and α3β1 pairs were eliminated suggest that although blisters formed at stage E15.5, the program of epidermal differentiation and skin morphogenesis proceeded normally prior to this stage, suggesting that at early stages of development, these integrin pairs appear to play a chiefly structural role (244).

3.6.4   Laminin α3 chain

Deletion of the LAMA3 gene resulted in neonatal lethality presumed to result from severe abnormalities in epithelial tissue, including epidermal blistering (261). Null animals reveal that the laminin-5 α3 chain appears to be necessary both for epidermal integrity and for the proper formation of hemidesmosomes. Additionally, potential defects in integrin-mediated signalling were implied by a decreased survival rate of keratinocytes derived from these animals; these defects could be rescued with exogenous laminin-5, or antibody ligation of α6β4. However, double knock-out of both laminin-5 receptors, α6β4, and α3β1, revealed that in areas of detached epidermis basal keratinocytes were apoptotic, however in attached epidermis, no apoptosis was detected. These data suggest that adhesion-dependent survival does not require ligation of α6β4 or α3β1, but instead requires another basal receptor/ligand interaction (244).

4.   Conclusion

In summary, despite nearly complete dissimilarity in the structural composition of these junctions, the emergent themes in desmosome and hemidesmosome biology are surprisingly parallel. First, in both cases, analysis of human disease and gene targeting experiments suggest that these junctions play crucial mechanical roles. However, since direct physical demonstration and quantification of mechanical strength conferred by these cell junctions has never been reported, the biomechanical characterization of these junctions as well as their individual components will be an important contribution to the field. Secondly, both junctions have been strongly implicated in cellular signalling, although the impact of these signalling events on cellular processes is still not well understood. It appears likely that a complicated picture will emerge in which junctional components play both structural and transductional roles, and that these functions will overlap. Additionally, much work is still needed to understand the events involved in regulation of assembly and disassembly of the junctions. This regulation seems paramount to many epithelial cell processes including growth and morphogenesis, wound repair and tissue remodelling, maintenance of barrier function and homeostasis, and the transition toward a motile phenotype in metastatic disease.

Acknowledgements

We thank those who contributed work prior to its publication and apologize for work not included due to space limitations. We thank Xinyu Chen, and Drs Elayne Bornslaeger, Claire Gaudry, Ken Ishii, and Andrew Kowalczyk for critical reading of the manuscript and insightful discussion; we also thank Connie Corcoran for assistance with preparation of the figures and Ian Gallicano and John Stanley for contributing images. This work was supported by grants from the National Institutes of Health (R01 AR43380, AR41836 to K. J. G.) (R01 GM38470 to J. C. R. J.) (P01 DE12328 to K. J. G. & J. C. R. J.). L. J. B. and L. E. G. were supported by a training grant from the National Cancer Institute (T32 CA09560).

References

1. Garrod, D. R. (1993). Desmosomes and hemidesmosomes. *Curr. Opin. Cell Biol.*, **5**, 30.
2. Green, K. J. and Jones, J. C. R. (1996). Desmosomes and hemidesmosomes: structure and function of molecular components. *FASEB J.*, **10**, 871.
3. Schwarz, M. A., Owaribe, K., Kartenbeck, J., and Franke, W. W. (1990). Desmosomes and hemidesmosomes: constitutive molecular components. *Annu. Rev. Cell Biol.*, **6**, 461.
4. Borradori, L. and Sonnenberg, A. (1999). Structure and function of hemidesmosomes: more than simple adhesion complexes. *J. Invest. Dermatol.*, **112**, 411.
5. Schmidt, A., Heid, H. W., Schafer, S., Nuber, U. A., Zimbelmann, R., and Franke, W. W. (1994). Desmosomes and cytoskeletal architecture in epithelial differentiation: cell type-specific plaque components and intermediate filament anchorage. *Eur. J. Cell Biol.*, **65**, 229.
6. Kelly, D. E. (1966). Fine structure of desmosomes, hemidesmosomes and an adepidermal globular layer in developing newt epidermis. *J. Cell Biol.*, **28**, 51.
7. Ashkenas, J., Muschler, J., and Bissell, M. J. (1996). The extracellular matrix in epithelial biology: shared molecules and common themes in distant phyla. *Dev. Biol.*, **180**, 433.
8. Farquhar, M. G. and Palade, G. E. (1963). Junctional complexes in various epithelia. *J. Cell Biol.*, **17**, 375.
9. Staehelin, L. A. (1974). Structure and function of intercellular junctions. *Int. Rev. Cytol.*, **39**, 191.
10. Gorbsky, G. and Steinberg, M. S. (1981). Isolation of the intercellular glycoproteins of desmosomes. *J. Cell Biol.*, **90**, 243.
11. Steinberg, M. S., Shida, H., Giudice, G. J., Shida, M., Patel, N. H., and Blaschuk, O. W. (1987). On the molecular organization, diversity and functions of desmosomal proteins. *Ciba Found. Symp.*, **125**, 3.

12. Koch, P. J., Goldschmidt, M. D., Zimbelmann, R., Troyanovsky, R., and Franke, W. W. (1992). Complexity and expression patterns of the desmosomal cadherins. *Proc. Natl. Acad. Sci. USA*, **89**, 353.

13. Collins, J. E., Legan, P. K., Kenny, T. P., MacGarvie, J., Holton, J. L., and Garrod, D. R. (1991). Cloning and sequence analysis of desmosomal glycoproteins 2 and 3 (desmocollins): cadherin-like desmosomal adhesion molecules with heterogeneous cytoplasmic domains. *J. Cell Biol.*, **113**, 381.

14. Mechanic, S., Raynor, K., Hill, J. E., and Cowin, P. (1991). Desmocollins form a distinct subset of the cadherin family of cell adhesion molecules. *Proc. Natl. Acad. Sci. USA*, **88**, 4476.

15. Parker, A. E., Wheeler, G. N., Arnemann, J., Pidsley, S., Ataliotis, P., Thomas, C. L., *et al.* (1991). Desmosomal glycoproteins II and III: Cadherin-like junctional molecules generated by alternative splicing. *J. Biol. Chem.*, **266**, 10438.

16. Koch, P. J., Walsh, M. J., Schmelz, M., Goldschmidt, M. D., Zimbelmann, R., and Franke, W. W. (1990). Identification of desmoglein, a constitutive desmosomal glycoprotein, as a member of the cadherin family of cell adhesion molecules. *Eur. J. Cell Biol.*, **53**, 1.

17. Kowalczyk, A. P., Bornslaeger, E. A., Norvell, S. M., Palka, H. L., and Green, K. J. (1999). Desmosomes: intercellular adhesive junctions specialized for attachment of intermediate filaments. *Int. Rev. Cytol.*, **185**, 237.

18. Ozawa, M., Ringwald, M., and Kemler, R. (1990). Uvomorulin-catenin complex formation is regulated by a specific domain in the cytoplasmic region of the cell adhesion molecule. *Proc. Natl. Acad. Sci. USA*, **87**, 4246.

19. Koch, P. J. and Franke, W. W. (1994). Desmosomal cadherins: another growing multigene family of adhesion molecules. *Curr. Opin. Cell Biol.*, **6**, 682.

20. Nilles, L. A., Parry, D. A. D., Powers, E. E., Angst, B. D., Wagner, R. M., and Green, K. J. (1991). Structural analysis and expression of human desmoglein: a cadherin-like component of the desmosome. *J. Cell Sci.*, **99**, 809.

21. Rutman, A. J., Buxton, R. S., and Burdett, I. D. J. (1994). Visualisation by electron microscopy of the unique part of the cytoplasmic domain of a desmoglein, a cadherin-like protein of the desmosome type of cell junction. *FEBS Lett.*, **353**, 194.

22. Garrod, D. R. and Collins, J. E. (1994). Desmosomes in differentiation and development. In *Molecular biology of desmosomes and hemidesmosomes* (ed. J. E. Collins and D. R. Garrod), p. 1. R. G. Landes Co., Austin.

23. Denning, M. F., Guy, S. G., Ellerbroek, S. M., Norvell, S. M., Kowalczyk, A. P., and Green, K. J. (1998). The expression of desmoglein isoforms in cultured human keratinocytes is regulated by calcium, serum, and protein kinase C. *Exp. Cell Res.*, **239**, 50.

24. North, A. J., Chidgey, M. A. J., Clarke, J. P., Bardsley, W. G., and Garrod, D. R. (1996). Distinct desmocollin isoforms occur in the same desmosomes and show reciprocally graded distributions in bovine nasal epidermis. *Proc. Natl. Acad. Sci. USA*, **93**, 7701.

25. Schafer, S., Koch, P. J., and Franke, W. W. (1994). Identification of the ubiquitous human desmoglein, Dsg2, and the expression catalogue of the desmoglein subfamily of desmosomal cadherins. *Exp. Cell Res.*, **211**, 391.

26. Arnemann, J., Sullivan, K. H., Magee, A. I., King, I. A., and Buxton, R. S. (1993). Stratification-related expression of isoforms of the desmosomal cadherins in human epidermis. *J. Cell Sci.*, **104**, 741.

27. Theis, D. G., Koch, P. J., and Franke, W. W. (1993). Differential synthesis of type 1 and type 2 desmocollin mRNAs in human stratified epithelia. *Int. J. Dev. Biol.*, **37**, 101.

28. Schafer, S., Stumpp, S., and Franke, W. W. (1996). Immunological identification and characterization of the desmosomal cadherin Dsg2 in coupled and uncoupled epithelial cells and in human tissues. *Differentiation*, **60**, 99.

29. Shimizu, H., Masunaga, T., Ishiko, A., Kikuchi, A., Hashimoto, T., and Nishikawa, T. (1995). Pemphigus vulgaris and pemphigus foliaceus sera show an inversely graded binding pattern to extracellular regions of desmosomes in different layers of human epidermis. *J. Invest. Dermatol.*, **105**, 153.

30. Wu, H., Wang, Z. H., Yan, A., Lyle, S., Fakharzadeh, S., Wahl, J. K., *et al.* (2000). Protection against pemphigus foliaceus by desmoglein 3 in neonates. *N. Engl. J. Med.*, **343**, 31.

31. King, I. A., Sullivan, K. H., Bennett, R., and Buxton, R. S. (1995). The desmocollins of human foreskin epidermis: identification and chromosomal assignment of a third gene and expression patterns of the three isoforms. *J. Invest. Dermatol.*, **105**, 314.

32. Troyanovsky, S. M., Eshkind, L. G., Troyanovsky, R. B., Leube, R. E., and Franke, W. W. (1993). Contributions of cytoplasmic domains of desmosomal cadherins to desmosome assembly and intermediate filament anchorage. *Cell*, **72**, 561.

33. Norvell, S. M. and Green, K. J. (1998). Contributions of extracellular and intracellular domains of full length and chimeric cadherin molecules to junction assembly in epithelial cells. *J. Cell Sci.*, **111**, 1305.

34. Korman, N. J., Eyre, R. W., Klaus-Kovtun, V., and Stanley, J. R. (1989). Demonstration of an adhering-junction molecule (plakoglobin) in the autoantigens of pemphigus foliaceus and pemphigus vulgaris. *N. Engl. J. Med.*, **321**, 631.

35. Kowalczyk, A. P., Palka, H. L., Luu, H. H., Nilles, L. A., Anderson, J. E., Wheelock, M. J., *et al.* (1994). Posttranslational regulation of plakoglobin expression: Influence of the desmosomal cadherins on plakoglobin metabolic stability. *J. Biol. Chem.*, **269**, 31214.

36. Troyanovsky, S. M., Troyanovsky, R. B., Eshkind, L. G., Krutovskikh, V. A., Leube, R. E., and Franke, W. W. (1994). Identification of the plakoglobin-binding domain in desmoglein and its role in plaque assembly and intermediate filament anchorage. *J. Cell Biol.*, **127**, 151.

37. Mathur, M., Goodwin, L., and Cowin, P. (1994). Interactions of the cytoplasmic domain of the desmosomal cadherin Dsg1 with plakoglobin. *J. Biol. Chem.*, **269**, 14075.

38. Rickman, L., Simrak, D., Stevens, H. P., Hunt, D. M., King, I. A., Bryant, S. P., *et al.* (1999). N-terminal deletion in a desmosomal cadherin causes the autosomal dominant skin disease striate palmoplantar keratoderma. *Hum. Mol. Genet.*, **8**, 971.

39. Hatzfeld, M. (1999). The armadillo family of structural proteins. *Int. Rev. Cytol.*, **186**, 179.

40. Cowin, P. (1994). Plakoglobin. In *Molecular biology of desmosomes and hemidesmosomes* (ed. J. E. Collins and D. R. Garrod), p. 1. R. G. Landes Co., Austin.

41. Cowin, P., Kapprell, H.-P., Franke, W. W., Tamkun, J., and Hynes, R. O. (1986). Plakoglobin: a protein common to different kinds of intercellular adhering junctions. *Cell*, **46**, 1063.

42. Riggleman, B., Wieschaus, E., and Schedl, P. (1989). Molecular analysis of the armadillo locus: uniformly distributed transcripts and a protein with novel internal repeats are associated with a *Drosophila* segment polarity gene. *Genes Dev.*, **3**, 96.

43. Huber, A. H., Nelson, W. J., and Weis, W. I. (1997). Three-dimensional structure of the armadillo repeat region of β-catenin. *Cell*, **90**, 871.

44. Chitaev, N. A., Leube, R. E., Troyanovsky, R. B., Eshkind, L. G., Franke, W. W., and Troyanovsky, S. M. (1996). The binding of plakoglobin to desmosomal cadherins: patterns of binding sites and topogenic potential. *J. Cell Biol.*, **133**, 359.

45. Palka, H. L. and Green, K. J. (1997). Roles of plakoglobin end domains in desmosome assembly. *J. Cell Sci.*, **110**, 2359.

46. Aberle, H., Butz, S., Stappert, J., Weissig, H., Kemler, R., and Hoschuetzky, H. (1994). Assembly of the cadherin-catenin complex *in vitro* with recombinant proteins. *J. Cell Sci.*, **107**, 3655.

47. Jou, T. S., Stewart, D. B., Stappert, J., Nelson, W. J., and Marrs, J. A. (1995). Genetic and biochemical dissection of protein linkages in the cadherin-catenin complex. *Proc. Natl. Acad. Sci. USA*, **92**, 5067.

48. Sacco, P. A., McGranahan, T. M., Wheelock, M. J., and Johnson, K. R. (1995). Identification of plakoglobin domains required for association with N-cadherin and alpha-catenin. *J. Biol. Chem.*, **270**, 20201.

49. Aberle, H., Schwartz, H., Hoschuetzky, H., and Kemler, R. (1996). Single amino acid substitutions in proteins of the armadillo gene family abolish their binding to α-catenin. *J. Biol. Chem.*, **271**, 1520.

50. Witcher, L. L., Collins, R., Puttagunta, S., Mechanic, S. E., Munson, M., Gumbiner, B., *et al.* (1996). Desmosomal cadherin binding domains of plakoglobin. *J. Biol. Chem.*, **271**, 10904.

51. Wahl, J. K., Sacco, P. A., McGranahan-Sadler, T. M., Sauppe, L. M., Wheelock, M. J., and Johnson, K. R. (1996). Plakoglobin domains that define its association with the desmosomal cadherins and the classical cadherins: identification of unique and shared domains. *J. Cell Sci.*, **109**, 1143.

52. Plott, R. T., Amagai, M., Udey, M. C., and Stanley, J. R. (1994). Pemphigus vulgaris antigen lacks biochemical properties characteristic of classical cadherins. *J. Invest. Dermatol.*, **103**, 168.

53. Kowalczyk, A. P., Bornslaeger, E. A., Borgwardt, J. E., Palka, H. L., Dhaliwal, A. S., Corcoran, C. M., *et al.* (1997). The amino-terminal domain of desmoplakin binds to plakoglobin and clusters desmosomal cadherin-plakoglobin complexes. *J. Cell Biol.*, **139**, 773.

54. Chitaev, N. A., Averbakh, A. Z., Troyanovsky, R. B., and Troyanovsky, S. M. (1998). Molecular organization of the desmoglein-plakoglobin complex. *J. Cell Sci.*, **111**, 1941.

55. Smith, E. A. and Fuchs, E. (1998). Defining the interactions between intermediate filaments and desmosomes. *J. Cell Biol.*, **141**, 1229.

56. Ruhrberg, C. and Watt, F. M. (1997). The plakin family: versatile organisers of cytoskeletal architecture. *Curr. Opin. Genet. Dev.*, **7**, 392.

57. Ruhrberg, C., Hajibagheri, M. A. N., Parry, D. A. D., and Watt, F. M. (1997). Periplakin, a novel component of cornified envelopes and desmosomes that belongs to the plakin family and forms complexes with envoplakin. *J. Cell Biol.*, **139**, 1835.

58. Ruhrberg, C., Hajibagheri, M. A. N., Simon, M., Dooley, T. P., and Watt, F. M. (1996). Envoplakin, a novel precursor of the cornified envelope that has homology to desmoplakin. *J. Cell Biol.*, **134**, 715.

59. Steinert, P. and Marekov, L. (1999). Initiation of assembly of the cell envelope barrier structure of stratified squamous epithelia. *Mol. Biol. Cell*, **12**, 4247.

60. Herrmann, H. and Wiche, G. (1987). Plectin and IFAP-300K are homologous proteins binding to microtubule-associated proteins 1 and 2 and to the 240 kilodalton subunit of spectrin. *J. Biol. Chem.*, **262**, 1320.

61. Foisner, R., Leichtfried, F. E., Herrmann, H., Small, J. V., Lawson, D., and Wiche, G. (1988). Cytoskeleton-associated plectin: *in situ* localization, *in vitro* reconstitution, and binding to immobilized intermediate filament proteins. *J. Cell Biol.*, **106**, 723.

62. Skalli, O., Jones, J., and Goldman, R. D. (1992). IFAP 300: a candidate linker of IF to desmosomes. *Mol. Biol. Cell*, **3**, 351a.

63. Virata, M. L. A., Wagner, R. M., Parry, D. A. D., and Green, K. J. (1992). Molecular structure of the human desmoplakin I and II amino terminus. *Proc. Natl. Acad. Sci. USA*, **89**, 544.

64. Green, K. J., Guy, S., Cserhalmi-Friedman, P. B., McClean, I., Christiano, A. M., and Wagner, R. M. (1999). Analysis of the desmoplakin gene reveals striking conservation with other members of the plakin family of cytolinkers. *Exp. Dermatol.*, **8**, 462.

65. Angst, B. D., Nilles, L. A., and Green, K. J. (1990). Desmoplakin II expression is not restricted to stratified epithelia. *J. Cell Sci.*, **97**, 247.

66. Stappenbeck, T. S., Bornslaeger, E. A., Corcoran, C. M., Luu, H. H., Virata, M. L. A., and Green, K. J. (1993). Functional analysis of desmoplakin domains: specification of the interaction with keratin versus vimentin intermediate filament networks. *J. Cell Biol.*, **123**, 691.

67. Green, K. J., Parry, D. A. D., Steinert, P. M., Virata, M. L. A., Wagner, R. M., Angst, B. D., *et al.* (1990). Structure of the human desmoplakins: implications for function in the desmosomal plaque. *J. Biol. Chem.*, **265**, 2603.

68. O'Keefe, E. J., Erickson, H. P., and Bennett, V. (1989). Desmoplakin I and desmoplakin II: Purification and characterization. *J. Biol. Chem.*, **264**, 8310.

69. Stappenbeck, T. S. and Green, K. J. (1992). The desmoplakin carboxyl terminus coaligns with and specifically disrupts intermediate filament networks when expressed in cultured cells. *J. Cell Biol.*, **116**, 1197.

70. Kouklis, P. D., Hutton, E., and Fuchs, E. (1994). Making a connection: direct binding between keratin intermediate filaments and desmosomal proteins. *J. Cell Biol.*, **127**, 1049.

71. Meng, J.-J., Bornslaeger, E. A., Green, K. J., Steinert, P. M., and Ip, W. (1997). Two hybrid analysis reveals fundamental differences in direct interactions between desmoplakin and cell type specific intermediate filaments. *J. Biol. Chem.*, **272**, 21495.

72. Bornslaeger, E. B., Corcoran, C. M., Stappenbeck, T. S., and Green, K. J. (1996). Breaking the connection: displacement of the desmosomal plaque protein desmoplakin from cell–cell interfaces disrupts anchorage of intermediate filament bundles and alters intercellular junction assembly. *J. Cell Biol.*, **134**, 985.

73. Gallicano, G. I., Kouklis, P., Bauer, C., Yin, M., Vasioukhin, V., Degenstein, L., *et al.* (1998). Desmoplakin is required early in development for assembly of desmosomes and cytoskeletal linkage. *J. Cell Biol.*, **143**, 2009.

74. Armstrong, D. K. B., McKenna, K. E., Purkis, P. E., Green, K. J., Eady, R. A. J., Leigh, I. M., *et al.* (1999). Haploinsufficiency of desmoplakin causes a striate subtype of palmoplantar keratoderma. *Hum. Mol. Genet.*, **8**, 143.

75. Norgett, E. E., Hatsell, S. J., Carvajal-Huerta, L., Cabezas, J.-C. R., Common, J., Purkis, P. E., *et al.* (2000). Recessive mutation in desmoplakin disrupts desmoplakin-intermediate filament interactions and causes dilated cardiomyopathy, woolly hair and keratoderma. *Hum. Mol. Genet.*, **9**, 2761.

76. Kapprell, H.-P., Owaribe, K., and Franke, W. W. (1988). Identification of a basic protein of Mr 75,000 as an accessory desmosomal plaque protein in stratified and complex epithelia. *J. Cell Biol.*, **106**, 1679.

77. Hatzfeld, M., Kristjansson, G. I., Plessmann, U., and Weber, K. (1994). Band 6 protein, a major constituent of desmosomes from stratified epithelia, is a novel member of the armadillo multigene family. *J. Cell Sci.*, **107**, 2259.

78. Heid, H. W., Schmidt, A., Zimbelmann, R., Schafer, S., Winter-Simanowski, S., Stumpp, S., *et al.* (1994). Cell type-specific desmosomal plaque proteins of the plakoglobin family: plakophilin 1 (band 6 protein). *Differentiation*, **58**, 113.

79. Mertens, C., Kuhn, C., and Franke, W. W. (1996). Plakophilins 2a and 2b: constitutive proteins of dual location in the karyoplasm and the desmosomal plaque. *J. Cell Biol.*, **135**, 1009.

80. Schmidt, A., Langbein, L., Pratzel, S., Rode, M., Rackwitz, H.-R., and Franke, W. W. (1999). Plakophilin 3-a novel cell-type-specific desmosomal plaque protein. *Differentiation*, **64**, 291.

81. Bonne, S., Hengel, J. V., Nollet, F., Kools, P., and Roy, F. V. (1999). Plakophilin-3, a novel armadillo-like protein present in nuclei and desmosomes of epithelial cells. *J. Cell Sci.*, **112**, 2265.

82. Schmidt, A., Langbein, L., Rode, M., Pratzel, S., Zimbelmann, R., and Franke, W. W. (1997). Plakophilins 1a and 1b: widespread nuclear proteins recruited in specific epithelial cells as desmosomal plaque components. *Cell Tissue Res.*, **290**, 481.

83. Moll, I., Kurzen, H., Langbein, L., and Franke, W. W. (1997). The distribution of the desmosomal protein, plakophilin 1, in human skin and skin tumors. *J. Invest. Dermatol.*, **108**, 139.

84. Mertens, C., Kuhn, C., Moll, R., Schwetlick, I., and Franke, W. W. (1999). Desmosomal plakophilin 2 as a differentiation marker in normal and malignant tissues. *Differentiation*, **64**, 277.

85. Hofmann, I., Mertens, C., Brettel, M., Nimmrich, V., Schnolzer, M., and Herrmann, H. (2000). Interaction of plakophilins with desmoplakin and intermediate filament proteins: an *in vitro* analysis. *J. Cell Sci.*, **113**, 2471.

86. Kowalczyk, A. P., Hatzfeld, M., Bornslaeger, E. A., Kopp, D. S., Borgwardt, J. E., Corcoran, C. M., *et al.* (1999). The head domain of plakophilin-1 binds to desmoplakin and enhances its recruitment to desmosomes: implications for cutaneous disease. *J. Biol. Chem.*, **274**, 18145.

87. Bornslaeger, E. A., Godsel, L. M., Corcoran, C. M., Park, J. K., Hatzfeld, M., Kowalczyk, A. P., *et al.* (2001). Plakophilin 1 interferes with plakoglobin binding to desmoplakin, yet together with plakoglobin promotes clustering of desmosomal plaque complexes at cell–cell borders. *J. Cell Sci.*, **114**, 727.

88. Hatzfeld, M., Haffner, C., Schulze, K., and Vinzens, U. (2000). The function of plakophilin 1 in desmosome assembly and actin filament organization. *J. Cell Biol.*, **149**, 209.

89. McGrath, J. A., McMillan, J. R., Shemanko, C. S., Runswick, S. K., Leigh, I. M., Lane, E. B., *et al.* (1997). Mutations in the plakophilin 1 gene result in ectodermal dysplasia/skin fragility syndrome. *Nature Genet.*, **17**, 240.

90. Ouyang, P. and Sugrue, S. P. (1992). Identification of an epithelial protein related to the desmosome and intermediate filament network. *J. Cell Biol.*, **118**, 1477.

91. Tsukita, S. and Tsukita, S. (1985). Desmocalmin: a calmodulin-binding high molecular weight protein isolated from desmosomes. *J. Cell Biol.*, **101**, 2070.

92. Shi, J. and Sugrue, S. P. (2000). Dissection of protein linkage between keratins and pinin, a protein with dual location at desmosome-intermediate filament complex and in the nucleus. *J. Biol. Chem.*, **275**, 14910.

93. North, A. J., Bardsley, W. G., Hyam, J., Bornslaeger, E. A., Cordingley, H. C., Trinnaman, B., *et al.* (1999). Molecular map of the desmosomal plaque. *J. Cell Sci.*, **112**, 4325.

94. Nagafuchi, A., Shirayoshi, Y., Okasaki, K., Yamada, K., and Takeichi, M. (1987). Transformation of cell adhesion properties by exogenously introduced E-cadherin cDNA. *Nature*, **329**, 341.

95. Amagai, M., Karpati, S., Klaus-Kovtun, V., Udey, M. C., and Stanley, J. R. (1994). The extracellular domain of pemphigus vulgaris antigen (desmoglein 3) mediates weak homophilic adhesion. *J. Invest. Dermatol.*, **102**, 402.

96. Chidgey, M. A. J., Clarke, J. P., and Garrod, D. R. (1996). Expression of full-length desmosomal glycoproteins (desmocollins) is not sufficient to confer strong adhesion on transfected L929 cells. *J. Invest. Dermatol.*, **106**, 689.

97. Kowalczyk, A. P., Borgwardt, J. E., and Green, K. J. (1996). Analysis of desmosomal cadherin-adhesive function and stoichiometry of desmosomal cadherin-plakoglobin complexes. *J. Invest. Dermatol.*, **107**, 293.

98. Marcozzi, C., Burdett, I. D. J., Buxton, R. S., and Magee, A. I. (1998). Coexpression of both types of desmosomal cadherin and plakoglobin confers strong intercellular adhesion. *J. Cell Sci.*, **111**, 495.

99. Tselepis, C., Chidgey, M., North, A., and Garrod, D. (1998). Desmosomal adhesion inhibits invasive behavior. *Proc. Natl. Acad. Sci. USA*, **95**, 8064.

100. Chitaev, N. A. and Troyanovsky, S. M. (1997). Direct Ca^{2+}-dependent heterophilic interaction between desmosomal cadherins, desmoglein and desmocollin, contributes to cell–cell adhesion. *J. Cell Biol.*, **138**, 193.

101. Troyanovsky, R. B., Klingelhofer, J., and Troyanovsky, S. (1999). Removal of calcium ions triggers a novel type of intercadherin interaction. *J. Cell Sci.*, **112**, 4379.

102. Stanley, J. R., Koulu, L., and Thivolet, C. (1984). Distinction between epidermal antigens binding pemphigus vulgaris and pemphigus foliaceus autoantibodies. *J. Clin. Invest.*, **74**, 313.

103. Eyre, R. W. and Stanley, J. R. (1987). Human autoantibodies against a desmosomal protein complex with a calcium-sensitive epitope are characteristic of pemphigus foliaceus patients. *J. Exp. Med.*, **165**, 1719.

104. Stanley, J. R., Yaar, M., Nelson, P. H., and Katz, S. I. (1982). Pemphigus antibodies identify a cell surface glycoprotein synthesized by human and mouse keratinocytes. *J. Clin. Invest.*, **70**, 281.

105. Amagai, M. (1999). Autoimmunity against desmosomal cadherins in pemphigus. *J. Dermatol. Sci.*, **20**, 92.

106. Amagai, M., Hashimoto, T., Shimizu, N., and Nishikawa, T. (1994). Absorption of pathogenic autoantibodies by the extracellular domain of pemphigus vulgaris antigen (Dsg3) produced by baculovirus. *J. Clin. Invest.*, **94**, 59.

107. Amagai, M., Klaus-Kovtun, V., and Stanley, J. R. (1991). Autoantibodies against a novel epithelial cadherin in pemphigus vulgaris, a disease of cell adhesion. *Cell*, **67**, 869.

108. Mahoney, M. G., Wang, Z., Rothenberger, K., Koch, P. J., Amagai, M., and Stanley, J. R. (1999). Explanations for the clinical and microscopic localization of lesions in pemphigus foliaceus and vulgaris. *J. Clin. Invest.*, **103**, 461.

109. Mahoney, M. G., Aho, S., Uitto, J., and Stanley, J. R. (1998). The members of the plakin family of proteins recognized by paraneoplastic pemphigus antibodies include periplakin. *J. Invest. Dermatol.*, **111**, 308.

110. Amagai, M., Nishikawa, T., Nousari, H. C., Anhalt, G. J., and Hashimoto, T. (1998). Antibodies against desmoglein 3 (pemphigus vulgaris antigen) are present in sera from patients with paraneoplastic pemphigus and cause acantholysis *in vivo* in neonatal mice. *J. Clin. Invest.*, **102**, 775.

111. Amagai, M., Natsuyoshi, N., Wang, Z., Andl, C., and Stanley, J. (2000). Toxin in Bullous Impetigo and Staphylococcal scalded skin syndrome targets desmoglein 1. *Nature Med.*, **6**, 1275.

112. Bierkamp, C., McLaughlin, K. J., Schwarz, H., Huber, O., and Kemler, R. (1996). Embryonic heart and skin defects in mice lacking plakoglobin. *Dev. Biol.*, **180**, 780.

113. Ruiz, P., Brinkmann, V., Ledermann, B., Behrend, M., Grund, C., Thalhammer, C., *et al.* (1996). Targeted mutation of plakoglobin in mice reveals essential functions of desmosomes in the embryonic heart. *J. Cell Biol.*, **135**, 215.

114. Bierkamp, C., Schwarz, H., Huber, O., and Kemler, R. (1998). Desmosomal localization of β-catenin in the skin of plakoglobin null-mutant mice. *Development*, **126**, 371.

115. Kowalczyk, A. P., Navarro, P., Dejana, E., Bornslaeger, E. A., Green, K. J., Kopp, D. S., *et al.* (1998). VE-cadherin and desmoplakin are assembled into dermal microvascular endothelial intercellular junctions: a pivotal role for plakoglobin in the recruitment of desmoplakin to intercellular junctions. *J. Cell Sci.*, **111**, 3045.

116. Charpentier, E., Lavker, R. M., Acquista, E., and Cowin, P. (2000). Plakoglobin supresses epithelial proliferation and hair growth *in vivo*. *J. Cell Biol.*, **149**, 503.

117. Koch, P. J., Mahoney, M. G., Ishikawa, H., Pulkkinen, L., Uitto, J., Shultz, L., *et al.* (1997). Targeted disruption of the pemphigus vulgaris antigen (desmoglein 3) gene in mice causes loss of keratinocyte cell adhesion with a phenotype similar to pemphigus vulgaris. *J. Cell Biol.*, **137**, 1091.

118. Stanley, J. R. (1993). Cell adhesion molecules as targets of autoantibodies in pemphigus and pemphigoid, bullous diseases due to defective epidermal cell adhesion. *Adv. Immunol.*, **51**, 291.

119. Koch, P. J., Mahoney, M. G., Cotsarelis, G., Rothenberger, K., Lavker, R. M., and Stanley, J. R. (1998). Desmoglein 3 anchors telogen hair in the follicle. *J. Cell Sci.*, **111**, 2529.

120. Allen, E., Yu, Q.-C., and Fuchs, E. (1996). Mice expressing a mutant desmosomal cadherin exhibit abnormalities in desmosomes, proliferation, and epidermal differentiation. *J. Cell Biol.*, **133**, 1367.

121. Hennings, H., Michael, D., Cheng, C., Steinert, P., Holbrook, K., and Yuspa, S. H. (1980). Calcium regulation of growth and differentiation of mouse epidermal cells in culture. *Cell*, **19**, 245.

122. Jones, J. C. R. and Goldman, R. D. (1985). Intermediate filaments and the initiation of desmosome assembly. *J. Cell Biol.*, **101**, 506.

123. Gumbiner, B., Stevenson, B., and Grimaldi, A. (1988). The role of the cell adhesion molecule uvomorulin in the formation and maintenance of the epithelial junctional complex. *J. Cell Biol.*, **107**, 1575.

124. Lewis, J. E., III, J. K. W., Sass, K. M., Jensen, P. J., Johnson, K. R., and Wheelock, M. J. (1997). Cross-talk between adherens junctions and desmosomes depends on plakoglobin. *J. Cell Biol.*, **136**, 919.

125. Wahl, J. K., Nieset, J. E., Sacco-Bubulya, P. A., Sadler, T. M., Johnson, K. R., and Wheelock, M. J. (2000). The amino- and carboxyl-terminal tails of β-catenin reduce its affinity for desmoglein 2. *J. Cell Sci.*, **113**, 1737.

126. Hanakawa, Y., Amagai, M., Shirakata, Y., Sayama, K., and Hashimoto, K. (2000). Different effects of dominant negative mutants of desmocollin and desmoglein on the cell–cell adhesion of keratinocytes. *J. Cell Sci.*, **113**, 1803.

127. Pasdar, M. and Nelson, W. J. (1988). Kinetics of desmosome assembly in Madin-Darby canine kidney epithelial cells: temporal and spatial regulation of desmoplakin organization and stabilization upon cell–cell contact I. Biochemical analysis. *J. Cell Biol.*, **106**, 677.

128. Penn, E. J., Burdett, I. D. J., Hobson, C., Magee, A. I., and Rees, D. A. (1987). Structure and assembly of desmosome junctions: biosynthesis and turnover of the major desmosome

components of Madin-Darby canine kidney cells in low calcium medium. *J. Cell Biol.*, **105**, 2327.

129. Demlehner, M. P., Schafer, S., Grund, C., and Franke, W. W. (1995). Continual assembly of half-desmosomal structures in the absence of cell contacts and their frustrated endocytosis: a coordinated Sisyphus cycle. *J. Cell Biol.*, **131**, 745.

130. Pasdar, M. and Nelson, W. J. (1988). Kinetics of desmosome assembly in Madin-Darby canine kidney cells: temporal and spatial regulation of desmoplakin organization and stabilization upon cell–cell contact. II. Morphological analysis. *J. Cell Biol.*, **106**, 687.

131. Troyanovsky, S. M., Troyanovsky, R. B., Eshkind, L. G., Leube, R. E., and Franke, W. W. (1994). Identification of amino acid sequence motifs in desmocollin, a desmosomal glycoprotein, that are required for plakoglobin binding and plaque formation. *Proc. Natl. Acad. Sci. USA*, **91**, 10790.

132. Stappenbeck, T. S., Lamb, J. A., Corcoran, C. M., and Green, K. J. (1994). Phosphorylation of the desmoplakin COOH terminus negatively regulates its interaction with keratin intermediate filament networks. *J. Biol. Chem.*, **269**, 29351.

133. Pasdar, M., Li, Z., and Chan, H. (1995). Desmosome assembly and disassembly are regulated by reversible protein phosphorylation in cultured epithelial cells. *Cell Motil. Cytoskel.*, **30**, 108.

134. van Hengel, J., Gohon, L., Bruyneel, E., Vermeulen, S., Cornelissen, M., Mareel, M., *et al.* (1997). Protein kinase C activation upregulates intercellular adhesion of alpha-catenin-negative human colon cancer cell variants via induction of desmosomes. *J. Cell Biol.*, **137**, 1103.

135. Hoschuetzky, H., Aberle, H., and Kemler, R. (1994). β-catenin mediates the interaction of the cadherin-catenin complex with epidermal growth factor receptor. *J. Cell Biol.*, **127**, 1375.

136. Shibamoto, S., Hayakawa, M., Takeuchi, K., Hori, T., Oku, N., Miyazawa, K., *et al.* (1994). Tyrosine phosphorylation of β-catenin and plakoglobin enhanced by hepatocyte growth factor and epidermal growth factor in human carcinoma cells. *Cell Adhes. Commun.*, **1**, 295.

137. Fuchs, M., Muller, T., Lerch, M. M., and Ullrich, A. (1996). Association of human protein-tyrosine phosphatase κ with members of the armadillo family. *J. Biol. Chem.*, **271**, 16712.

138. Kitajima, Y., Aoyama, Y., and Seishima, M. (1999). Transmembrane signaling for adhesive regulation of desmosomes and hemidesmosomes, and for cell–cell detachment induced by pemphigus IgG in cultured keratinocytes: involvement of protein kinase C. *J. Invest. Dermatol. Sym. Proc.*, **4**, 137.

139. Dierick, H. and Bejsovec, A. (1999). Cellular mechanisms of Wingless/Wnt signal transduction. *Curr. Top. Dev. Biol.*, **43**, 153.

140. Gelderloos, J. A., Witcher, L., Cowin, P., and Klymkowsky, M. W. (1997). Plakoglobin, the other 'Arm' of vertebrates. In *Cytoskeletal-membrane interactions and signal transduction* (ed. P. Cowin and M. Klymkowsky), p. 12. Landes Bioscience, Austin.

141. Karnovsky, A. and Klymkowsky, M. W. (1995). Anterior axis duplication in *Xenopus* induced by the over-expression of the cadherin-binding protein plakoglobin. *Proc. Natl. Acad. Sci. USA*, **92**, 4522.

142. Merriam, J. M., Rubenstein, A. B., and Klymkowsky, M. W. (1997). Cytoplasmically anchored plakoglobin induces a WNT-like phenotype in *Xenopus*. *Dev. Biol.*, **185**, 67.

143. Cox, R. T., Pai, L.-M., Miller, J. M., Orsulic, S., Stein, J., McCormick, C. A., *et al.* (1999). Membrane-tethered *Drosophila* Armadillo cannot transduce Wingless signal on its own. *Development*, **126**, 1327.

144. Klymkowsky, M. W., Williams, B. O., Barish, G. D., Varmus, H. E., and Vourgourakis, Y. E. (1999). Membrane-anchored plakoglobins have multiple mechanisms of action in Wnt signaling. *Mol. Biol. Cell*, **10**, 3151.

145. Kodama, S., Ikeda, S., Asahara, T., Kishida, M., and Kikuchi, A. (1999). Axin directly interacts with plakoglobin and regulates its stability. *J. Biol. Chem.*, **274**, 27682.

146. Zhurinsky, J., Shtutman, M., and Ben-Ze'ev, A. (2000). Plakoglobin and β-catenin: protein interactions, regulation and biological roles. *J. Cell Sci.*, **113**, 3127.

147. Simcha, I., Shtutman, M., Salomon, D., Zhurinsky, J., Sadot, E., Geiger, B., *et al.* (1998). Differential nuclear translocation and transactivation potential of β-catenin and plakoglobin. *J. Cell Biol.*, **141**, 1433.

148. Hakimelahi, S., Parker, H., Gilchrist, A., Barry, M., Li, Z., Bleackley, R., *et al.* (2000). Plakoglobin regulates the expression of the anti-apoptotic protein BCL-2. *J. Biol. Chem.*, **275**, 10905.

149. He, T. C., Sparks, A. B., Rago, C., Hermeking, H., Zawel, L., Costa, L. T. D., *et al.* (1998). Identification of c-Myc as a target of the APC pathway. *Science*, **281**, 1509.

150. Tetsu, O. and McCormick, F. (1999). β-catenin regulates expression of cyclin D1 in colon carcinoma cells. *Nature*, **398**, 422.

151. Shtutman, M., Zhurinsky, J., Simcha, I., Albanese, C., D'Amico, M., Pestell, R., *et al.* (1999). The cyclin D1 gene is a target of the β-catenin/LEF-1 pathway. *Proc. Natl. Acad. Sci. USA*, **96**, 5522.

152. Kolligs, F. T., Kolligs, B., Hajra, K. M., Hu, G., Tani, M., Cho, K. R., *et al.* (2000). Gamma-catenin is regulated by the APC tumor suppressor and its oncogenic activity is distinct from that of beta-catenin. *Genes Dev.*, **14**, 1319.

153. Jones, J. C. R., Hopkinson, S. B., and Goldfinger, L. E. (1998). Structure and assembly of hemidesmosomes. *BioEssays*, **20**, 488.

154. Burridge, K., Fath, K., Kelly, T., Nuckolls, G., and Turner, C. (1988). Focal adhesions: transmembrane junctions between the extracellular matrix and the cytoskeleton. *Annu. Rev. Cell Biol.*, **4**, 487.

155. Jones, J. C. R. and Green, K. J. (1991). Intermediate filament-plasma membrane interactions. *Curr. Opin. Cell Biol.*, **3**, 127.

156. Niessen, C. M., Hogervorst, F., Jaspars, L. H., De Melker, A. A., Delwel, G. O., Hulsman, E. H. M., *et al.* (1994). The α6β4 integrin is a receptor of both laminin and kalinin. *Exp. Cell Res.*, **211**, 360.

157. Clark, E. A. and Brugge, J. S. (1995). Integrins and signal transduction pathways: the road taken. *Science*, **268**, 233.

158. Mainiero, F., Pepe, A., Wary, K. K., Spinardi, L., Mohammadi, M., Schlessinger, J., *et al.* (1995). Signal transduction by the α6β4 integrin: distinct β4 subunit sites mediate recruitment of Shc/Grb2 and association with the cytoskeleton of hemidesmosomes. *EMBO J.*, **14**, 4470.

159. Baker, S. E., DiPasquale, A., Stock, E. L., Plopper, G., Quaranta, V., Fitchmun, M., *et al.* (1996). Morphogenetic effects of soluble laminin-5 on cultured epithelial cells and tissue explants. *Exp. Cell Res.*, **228**, 262.

160. Giancotti, F. G. (1996). Signal transduction by the α6β4 integrin: charting the path between laminin binding and nuclear events. *J. Cell Sci.*, **109**, 1165.

161. Hogervorst, F., Kuikman, I., van Kessel, A., and Sonnenberg, A. (1991). Molecular cloning of the human alpha 6 integrin subunit. Alternative splicing of alpha 6 mRNA and chromosomal localization of the alpha 6 and beta 4 genes. *Eur. J. Biochem.*, **199**, 425.

162. Tamura, R. N., Cooper, H. M., Collo, G., and Quaranta, V. (1991). Cell type-specific integrin variants with alternative alpha chain cytoplasmic domains. *Proc. Natl. Acad. Sci. USA*, **88**, 10183.

163. Riddelle, K. S., Green, K. J., and Jones, J. C. R. (1991). Formation of hemidesmosomes *in vitro* by a transformed rat bladder cell line. *J. Cell Biol.*, **112**, 159.

164. Hogervorst, F., Admiraal, L. G., Niessen, C., Kuikman, I., Janssen, H., Daams, H., *et al.* (1993). Biochemical characterization and tissue distribution of the A and B variants of the integrin alpha 6 subunit. *J. Cell Biol.*, **121**, 179.

165. Thorsteinsdottir, S., Roelen, B. A. J., Freund, E., Gaspar, A. C., Sonnenberg, A., and Mummery, C. L. (1995). Expression patterns of laminin receptor splice variants alpha 6A beta 1 and alpha 6B beta 1 suggest different roles in mouse development. *Dev. Dyn.*, **204**, 240.

166. Georges-Labouesse, E., Messaddeq, N., Yehia, G., Cadalbert, L., Dierich, A., and LeMeur, M. (1996). Absence of integrin alpha 6 leads to epidermolysis bullosa and neonatal death in mice. *Nature Genet.*, **13**, 370.

167. Gimond, C., Baudoin, C., van der Neut, R., Kramer, D., Calafat, J., and Sonnenberg, A. (1998). Cre-loxP- mediated inactivation of the alpha6A integrin splice variant *in vivo*: evidence for a specific functional role of alpha6A in lymphocyte migration but not in heart development. *J. Cell Biol.*, **143**, 253.

168. Giancotti, F. G., Stepp, M. A., Suzuki, S., Engvall, E., and Ruoslahti, E. (1992). Proteolytic processing of endogenous and recombinant beta 4 integrin subunit. *J. Cell Biol.*, **118**, 951.

169. Kajiji, S., Tamura, R. N., and Quaranta, V. (1989). A novel integrin (aEB4) from human epithelial cells suggests a fourth family of integrin adhesion receptors. *EMBO J.*, **8**, 673.

170. Klein, S., Giancotti, F. G., Presta, M., Albelda, S. M., Buck, C. A., and Rifkin, D. B. (1993). Basic fibroblast growth factor modulates integrin expression in microvascular endothelial cells. *Mol. Biol. Cell*, **4**, 973.

171. Lee, E. C., Lotz, M. M., Steele, G. D., and Mercurio, A. M. (1992). The integrin alpha 6 beta 4 is a laminin receptor. *J. Cell Biol.*, **117**, 671.

172. Carter, W. G., Kaur, P., Gill, S. G., Gahr, P. J., and Wayner, E. A. (1990). Distinct functions for integrins α3β1 in focal adhesions and α6β4/bullous pemphigoid antigen in a new stable anchoring contact (SAC) of keratinocytes: relation to hemidesmosomes. *J. Cell Biol.*, **111**, 3141.

173. Jones, J. C. R., Kurpakus, M. A., Cooper, H. M., and Quaranta, V. (1991). A function for the integrin α6β4 in the hemidesmosome. *Cell Regul.*, **2**, 427.

174. Stepp, M. A., Spurr-Michaud, S., Tisdale, A., Elwell, J., and Gipson, I. K. (1990). α6β4 integrin heterodimer is a component of hemidesmosomes. *Proc. Natl. Acad. Sci. USA*, **87**, 8970.

175. Sonnenberg, A., Calafat, J., Janssen, H., Daams, H., van der Raaij-Helmer, L. M. H., Falciuoni, R., *et al.* (1991). Integrin α6/β4 complex is located in hemidesmosomes, suggesting a major role in epidermal cell-basement membrane adhesion. *J. Cell Biol.*, **113**, 907.

176. Spinardi, L., Einheber, S., Cullen, T., Milner, T. A., and Giancotti, F. G. (1995). A recombinant tail-less integrin β4 subunit disrupts hemidesmosomes, but does not suppress α6β4-mediated cell adhesion to laminins. *J. Cell Biol.*, **129**, 473.

177. Klatte, D. H., Kurpakus, M. A., Grelling, K. A., and Jones, J. C. R. (1989). Immunochemical characterization of three components of the hemidesmosome and their expression in cultured epithelial cells. *J. Cell Biol.*, **109**, 3377.

178. Guidice, G. J., Emery, D. J., Zelickson, B. D., Anhalt, G. J., Liu, Z., and Diaz, L. A. (1993). Bullous pemphigoid and herpes gestationis autoantibodies recognize a common non-collagenous site on the BP180 domain. *J. Immunol.*, **151**, 5742.

179. Hopkinson, S. B., Riddelle, K. S., and Jones, J. C. R. (1992). The cytoplasmic domain of the 180kD bullous pemphigoid antigen, a hemidesmosomal component: molecular and cell biologic characterization. *J. Invest. Dermatol.*, **99**, 264.

180. Li, K., Giudice, G. J., Tamai, K., Do, H. C., Sawamura, D., Diaz, L. A., *et al.* (1992). Cloning of partial cDNA for mouse 180-kDa bullous pemphigoid antigen (BPAG2), a highly conserved collagenous protein of the cutaneous basement membrane zone. *J. Invest. Dermatol.*, **99**, 258.

181. Hopkinson, S. B. and Jones, J. C. R. (1996). Collagen XVII and collagen related molecules: linked by more than a common motif. *Semin. Cell Dev. Biol.*, **7**, 659.

182. Hirako, Y., Usukura, J., Nishizawa, Y., and Owaribe, K. (1996). Demonstration of the molecular shape of BP180, a 180-kDa bullous pemphigoid antigen and its potential for trimer formation. *J. Biol. Chem.*, **271**, 13739.

183. Masunaga, T., Shimizu, H., Yee, C., Borradori, L., Lazarova, Z., Nishikawa, T., *et al.* (1997). The extracellular domain of BPAG2 localizes to anchoring filaments and its carboxyl terminus extends to the lamina densa of normal human epidermal basement membrane. *J. Invest. Dermatol.*, **109**, 200.

184. Borradori, L., Koch, P. J., Niessen, C. M., Erkeland, S., van Leusden, M. R., and Sonnenberg, A. (1997). The localization of bullous pemphigoid antigen 180 (BP180) in hemidesmosomes is mediated by its cytoplasmic domain and seems to be regulated by the β4 integrin subunit. *J. Cell Biol.*, **136**, 1333.

185. Aho, S. and Uitto, J. (1998). Direct interaction between the intracellular domains of bullous pemphigoid antigen 2 (BP180) and beta 4 integrin, hemidesmosomal components of basal keratinocytes. *Biochem. Biophys. Res. Commun.*, **243**, 694.

186. Hopkinson, S. B., Baker, S. E., and Jones, J. C. R. (1995). Molecular genetic studies of a human epidermal autoantigen (the 180-kD bullous pemphigoid antigen/BP180): identification of functionally important sequences within the BP180 molecule and evidence for an interaction between BP180 and alpha 6 integrin. *J. Cell Biol.*, **130**, 117.

187. Hopkinson, S. B., Findlay, K., and Jones, J. C. R. (1998). Interaction of BP180 (type XVII collagen) and α6 integrin subunit is necessary for stabilization of hemidesmosome structure. *J. Invest. Dermatol.*, **111**, 1015.

188. Stanley, J. R. (1995). Autoantibodies against adhesion molecules and structures in blistering skin diseases. *J. Exp. Med.*, **181**, 1.

189. Hirako, Y., Usukara, J., Uematsu, J., Hashimoto, T., Kitajima, Y., and Owaribe, K. (1998). Cleavage of BP180, a 180kD bullous pemphigoid antigen, yields a 120kD collagenous extracellular polypeptide. *J. Biol. Chem.*, **27**, 9711.

190. Yang, Y., Dowling, J., Yu, Q.-C., Kouklis, P., Cleveland, D. W., and Fuchs, E. (1996). An essential cytoskeletal linker protein connecting actin microfilaments to intermediate filaments. *Cell*, **86**, 655.

191. Guo, L., Degenstein, L., Dowling, J., Yu, Q. C., Wollmann, R., Perman, B., *et al.* (1995). Gene targeting of BPAG1: abnormalities in mechanical strength and cell migration in stratified epithelia and neurologic degeneration. *Cell*, **81**, 233.

192. Hopkinson, S. B. and Jones, J. C. R. (2000). The N-terminus of the transmembrane protein BP180 interacts with the N-terminal domain of BP230, thereby mediating keratin cytoskeleton anchorage to the cell surface at the site of the hemidesmosome. *Mol. Biol. Cell*, **11**, 277.

193. Wiche, G. (1998). Role of plectin in cytoskeleton organization and dynamics. *J. Cell Sci.*, **111**, 2477.

194. Wiche, G., Krepler, R., Artlieb, U., Pytela, R., and Aberer, W. (1984). Identification of plectin in different human cell types and immunolocalization at epithelial basal cell surface membranes. *Exp. Cell Res.*, **155**, 43.

195. Skalli, O., Jones, J. C. R., Gagescu, R., and Goldman, R. D. (1994). IFAP300 is common to desmosomes and hemidesmosomes and is a possible linker of intermediate filaments to these junctions. *J. Cell Biol.*, **125**, 159.

196. Chavanas, S., Pulkkinen, L., Gache, Y., Smith, F. J., McLean, W. H., Uitto, J., *et al.* (1996). A homozygous nonsense mutation in the PLEC1 gene in patients with epidermolysis bullosa simplex with muscular dystrophy. *J. Clin. Invest.*, **98**, 2196.

197. Gache, Y., Chavanas, S., P., L. J., Wiche, G., Owaribe, K., Meneguzzi, G., and Ortonne, J. P. (1996). Defective expresion of plectin/HD1 in epidermolysis bullosa simplex with muscular dystrophy. *J. Clin. Invest.*, **97**, 2289.

198. McLean, W. H. I., Pulkkinen, L., Smith, F. J. D., Rugg, E. L., Lane, E. B., Bullrich, F., *et al.* (1996). Loss of plectin causes epidermolysis bullosa with muscular dystrophy: cDNA cloning and genomic organization. *Genes Dev.*, **10**, 1724.

199. Pulkkinen, L., Smith, F. J., Shimizu, H., Murata, S., Yaoita, H., Hachisuka, H., *et al.* (1996). Homozygous deletion mutations in the plectin gene (PLEC1) in patients with epidermolysis bullosa simplex associated with late-onset muscular dystrophy. *Hum. Mol. Genet.*, **5**, 1539.

200. Smith, F. J. D., Eady, R. A. J., Leigh, I. M., McMillan, J. R., Rugg, E. L., Kelsell, D. P., *et al.* (1996). Plectin deficiency results in muscular dystrophy with epdermolysis bullosa. *Nature Genet.*, **13**, 450.

201. Fuchs, E. (1996). Of Mice and Men: Genetic disorders of the cytoskeleton. *Mol. Biol. Cell*, **8**, 189.

202. Geerts, D., Fontao, L., Nievers, M., Schaapveld, R., Purkis, P., Wheeler, G., *et al.* (1999). Binding of integrin alpha6beta4 to plectin prevents plectin association with F-actin but does not interface with intermediate filament binding. *J. Cell Biol.*, **147**, 417.

203. Nikolic, B., MacNulty, E., Mir, B., and Wiche, G. (1996). Basic amino acid residue cluster within nuclear targeting sequence motif is essential for cytoplasmic plectin-vimentin network junctions. *J. Cell Biol.*, **134**, 1455.

204. Svitkina, T. M., Verkhovsky, A. B., and Borisy, G. G. (1996). Plectin sidearms mediate interaction of intermediate filaments with microtubules and other components of the cytoskeleton. *J. Cell Biol.*, **135**, 991.

205. Rezniczek, G. A., de Pereda, J. M., Reiper, S., and Wiche, G. (1998). Linking integrin alpha6beta4-based cell adhesion to the intermediate filament cytoskeleton: direct interaction between the beta4 subunit and plectin at multiple molecular sites. *J. Cell Biol.*, **141**, 209.

206. Nievers, M. G., Kuikman, I., Geerts, D., Leigh, I. M., and Sonnenberg, A. (2000). Formation of hemidesmosome-like structures in the absence of ligand binding by the α6β4 integrin requires binding of HD1/plectin to the cytoplasmic domain of the β4 integrin subunit. *J. Cell Biol.*, **113**, 963.

207. Sterk, L. M. T., Geuijen, C. A. W., Oomen, L. C. J. M., Calafat, J., Janssen, H., and Sonnenberg, A. (2000). The tetraspan molecule, CD151, a novel constituent of hemidesmosomes, associates with the integrin α6β4 and may regulate the spatial organization of hemidesmosomes. *J. Cell Biol.*, **149**, 969.

208. Boudreau, N. and Bissell, M. J. (1998). Extracellular matrix signalling: integration of form and function in normal and malignant cells. *Curr. Opin. Cell Biol.*, **10**, 640.

209. Aumailley, M. and Smyth, N. (1998). The role of laminins in basement membrane function. *J. Anat.*, **193**, 1.

210. Miner, J. H., Patton, B. L., Lentz, S. I., Gilbert, D. J., Snider, W. D., Jenkins, N. A., *et al.* (1997). The laminin α chains: expression, developmental transitions, and chromosomal locations of α1–5, identification of heterotrimeric laminins 8–11, and cloning of a novel α3 isoform. *J. Cell Biol.*, **137**, 685.

211. Ryan, M. C., Tizard, R., VanDevanter, D. R., and Carter, W. C. (1994). Cloning of the LamA3 gene encoding the α3 chain of the adhesive ligand epiligrin. *J. Biol. Chem.*, **269**, 22779.

212. Baker, S. E., Hopkinson, S. B., Fitchmun, M., Andreason, G. L., Frasier, F., Plopper, G., *et al.* (1996). Laminin-5 and hemidesmosomes: role of the alpha 3 chain subunit in hemidesmosome stability and assembly. *J. Cell Sci.*, **109**, 2509.

213. Rousselle, P., Lunstrum, G. P., Keene, D. R., and Burgeson, R. E. (1991). Kalinin: an epithelium-specific basement membrane adhesion molecule that is a component of anchoring filaments. *J. Cell Biol.*, **114**, 567.

214. Domloge-Hultsch, N., Gammon, W. R., Briggaman, R. A., Gil, S. G., Carter, W. G., and Yancey, K. B. (1992). Epiligrin, the major human keratinocyte integrin ligand, is a target in both an acquired autoimmune and an inherited subepidermal blistering skin disease. *J. Clin. Invest.*, **90**, 1628.

215. Pulkkinen, L., Christiano, A. M., Gerecke, D., Wagman, D. W., Burgeson, R. E., Pittelkow, M. R., *et al.* (1994). A homozygous nonsense mutation in the β3 chain gene of laminin 5 (LAMB3) in Herlitz junctional epidermolysis bullosa. *Genomics*, **24**, 357.

216. Pulkkinen, L., Christiano, A. M., Airenne, T., Haakana, H., Tryggvason, K., and Uitto, J. (1994). Mutations in the γ2 chain gene (LAMC2) of kalinin/laminin 5 in the junctional forms of epidermolysis bullosa. *Nature Genet.*, **6**, 293.

217. Kirtschig, G., Marinkovich, M. P., Burgeson, R. E., and Yancey, K. B. (1995). Anti-basement membrane autoantibodies in patients with anti-epiligrin cicatricial pemphigoid bind the alpha subunit of laminin 5. *J. Invest. Dermatol.*, **105**, 543.

218. Kivirikko, S., McGrath, J. A., Baudoin, C., Aberdam, D., Ciatti, S., Dunnill, M. G., *et al.* (1995). A homozygous nonsense mutation in the α3 chain gene of laminin 5 (LAMA3) in lethal (Herlitz) junctional epidermolysis bullosa. *Hum. Mol. Genet.*, **4**, 959.

219. McGrath, J. A., Pulkkinen, L., Christiano, A. M., Leigh, I. M., Eady, R. A. J., and Uitto, J. (1995). Altered laminin 5 expression due to mutations in the gene encoding the beta 3 Chain (LAMB3) in generalized atrophic benign epidermolysis bullosa. *J. Invest. Dermatol.*, **104**, 467.

220. Epstein, E. H. J. (1996). The genetics of human skin diseases. *Curr. Opin. Genet. Dev.*, **6**, 295.

221. Sakai, L. Y., Keene, D. R., Morris, N. P., and Burgeson, R. E. (1986). Type VII collagen is a major structural component of anchoring fibrils. *J. Cell Biol.*, **103**, 1577.

222. Burgeson, R. E. (1993). Type VII collagen, anchoring fibrils, and epidermolysis bullosa. *J. Invest. Dermatol.*, **101**, 252.

223. Chen, M., Marinkovich, M. P., Veis, A., Cai, X., Rao, C. N., O'Toole, E. A., *et al.* (1997). Interactions of the amino-terminal noncollagenous (NC1) domain of type VII collagen with extracellular matrix components. A potential role in epidermal-dermal adherence in human skin. *J. Biol. Chem.*, **272**, 14516.

224. Rousselle, P., Keene, D. R., Ruggiero, F., Champliaud, M. F., Rest, M., and Burgeson, R. E. (1997). Laminin 5 binds the NC-1 domain of type VII collagen. *J. Cell Biol.*, **138**, 719.

225. Lunstrum, G. P., Sakai, L. Y., Keene, D. R., Morris, N. P., and Burgeson, R. E. (1986). Large complex globular domains of type VII procollagen contribute to the structure of anchoring fibrils. *J. Biol. Chem.*, **261**, 9042.

226. Burgeson, R. E. and Christiano, A. M. (1997). The dermal-epidermal junction. *Curr. Opin. Cell Biol.*, **9**, 651.

227. Woodley, D. T., Burgeson, R. E., Lunstrum, G., Bruckner-Tuderman, L., Reese, M. J., and Briggaman, R. A. (1988). Epidermolysis bullosa acquisita antigen is the globular carboxyl terminus type VII procollagen. *J. Clin. Invest.*, **81**, 683.

228. Uitto, J., Pulkkinen, L., and Christiano, A. M. (1994). Molecular basis of the dystrophic and junctional forms of epidermolysis bullosa: mutations in the type VII collagen and kalinin (laminin 5) genes. *J. Invest. Dermatol.*, **103**, 39S.

229. Kurpakus, M. A., Quaranta, V., and Jones, J. C. R. (1991). Surface relocation of alpha6 beta4 integrins and assembly of hemidesmosomes in an *in vitro* model of wound healing. *J. Cell Biol.*, **115**, 1737.

230. Stahl, S., Weitzman, S., and Jones, J. C. R. (1997). The role of laminin-5 and its receptors in mammary epithelial cell branching morphogenesis. *J. Cell Sci.*, **110**, 55.

231. Riddelle, K. S., Hopkinson, S. B., and Jones, J. C. R. (1992). Hemidesmosomes in the epithelial cell line 804G: their fate during wound closure, mitosis and drug induced reorganization of the cytoskeleton. *J. Cell Sci.*, **103**, 475.

232. Hall, P. A. and Watt, F. M. (1989). Stem cells: the generation and maintenance of cellular diversity. *Development*, **106**, 619.

233. Dans, M., Gagnoux-Palacios, L., Blaikie, P., Klein, S., Mariotti, A., and Giancotti, F. (2001). Tyrosine phosphorylation of the beta 4 integrin cytoplasmic domain mediates Shc signaling to extracellular signal-regulated kinase and antagonizes formation of hemidesmosomes. *J. Biol. Chem.*, **276**, 1494.

234. Niessen, C. M., Hulsman, E. H. M., Rots, E. S., Sanchez-Aparicio, P., and Sonnenberg, A. (1997). Integrin α6β4 forms a complex with the cytoskeletal protein HD1 and induces its redistribution in transfected COS-7 cells. *Mol. Cell. Biol.*, **8**, 555.

235. Mainiero, F., Murgia, C., Wary, K. K., Curatola, A. M., Pepe, A., Blumenberg, M., *et al.* (1997). The coupling of alpha6beta4 integrin to Ras-MAP kinase pathways mediated by Shc controls keratinocyte proliferation. *EMBO J.*, **16**, 2365.

236. Murgia, C., Blaikie, P., Kim, N., Dans, M., Petrie, H. T., and Giancotti, F. G. (1998). Cell cycle and adhesion defects in mice carrying a targeted deletion of the integrin β4 cytoplasmic domain. *EMBO J.*, **17**, 3940.

237. Clarke, A. S., Lotz, M. M., Chao, C., and Mercurio, A. M. (1995). Activation of the p21 pathway of growth arrest and apoptosis by the β4 integrin cytoplasmic domain. *J. Biol. Chem.*, **270**, 22673.

238. Ruoslahti, E. and Reed, J. C. (1994). Anchorage dependence, integrins, and apoptosis. *Cell*, **77**, 477.

239. Frisch, S. M. and Francis, H. (1994). Disruption of epithelial cell–matrix interactions induces apoptosis. *J. Cell Biol.*, **124**, 619.

240. Zhang, Z., Vuori, K., Reed, J. C., and Ruoslahti, E. (1995). The alpha 5 beta 1 integrin supports survival of cells on fibronectin and up-regulates Bcl-2 expression. *Proc. Natl. Acad. Sci. USA*, **92**, 6161.

241. Bachelder, R. E., Marchetti, A., Falcioni, R., Soddu, S., and Mercurio, A. M. (1999). Activation of p53 function in carcinoma cells by the alpha6beta4 integrin. *J. Biol. Chem.*, **274**, 20733.

242. Bachelder, R. E., Ribick, M. J., Marchetti, A., Falcioni, R., Soddu, S., Davis, K. R., *et al.* (1999). p53 inhibits alpha6beta4 integrin survival signaling by promoting the caspase 3-dependent cleavage of AKT/PKB. *J. Cell Biol.*, **147**, 1063.

243. Rabinovitz, I., Toker, A., and Mercurio, A. M. (1999). Protein kinase C-dependent mobilization of the alpha6 beta4 integrin from hemidesmosomes and its association with

actin-rich cell protrusions drive the chemotactic migration of carcinoma cells. *J. Cell Biol.,* **146**, 1147.

244. DiPersio, C., Neut, R. V. D., Georges-Labouesse, E., Kreidberg, J., Sonnenberg, A., and Hynes, R. (2000). α3β1 and α6β4 integrin receptors for laminin-5 are not essential for epidermal morphogenesis and homeostasis during skin development. *J. Cell Sci.,* **113**, 3051.

245. Xia, Y., Gill, S. G., and Carter, W. G. (1996). Anchorage mediated by integrin α6β4 to laminin 5 (epiligrin) regulates tyrosine phosphorylation of a membrane associated 80-kD protein. *J. Cell Biol.,* **132**, 727.

246. Langhofer, M., Hopkinson, S. B., and Jones, J. C. R. (1993). The matrix secreted by 804G cells contains laminin-related components that participate in hemidesmosome assembly *in vitro*. *J. Cell Sci.,* **105**, 753.

247. Schaapveld, R. Q. J., Borradori, L., Geerts, D., van Leusden, M. R., Kuikman, I., Nievers, M. G., *et al.* (1998). Hemidesmosome formation is initiated by the beta4 integrin subunit, requires complex formation of beta4 and HD1/plectin, and involves a direct interaction between beta4 and the bullous pemphigoid antigen 180. *J. Cell Biol.,* **142**, 271.

248. Hormia, M., Falk Marzillier, J., Plopper, G., Tamura, R. N., Jones, J. C. R., and Quaranta, V. (1995). Rapid spreading and mature hemidesmosome formation in HaCaT keratinocytes induced by incubation with soluble laminin-5r. *J. Invest. Dermatol.,* **105**, 557.

249. Goldfinger, L. E., Stack, M. S., and Jones, J. C. R. (1998). Processing of laminin-5 and its functional consequences: role of plasmin and tissue-type plasminogen activator. *J. Cell Biol.,* **141**, 255.

250. Goldfinger, L. E., Jiang, L., Hopkinson, S. B., Stack, M. S., and Jones, J. C. R. (2000). Spatial regulation and activity modulation of plasmin by high affinity binding to the G domain of the α3 subunit of laminin-5. *J. Biol. Chem.,* **275**, 34887.

251. Carter, W. G., Ryan, M. C., and Gahr, P. J. (1991). Epiligrin, a new cell adhesion ligand for integrin α3β1 in epithelial basement membranes. *Cell,* **65**, 599.

252. Delwel, G. O., de Melker, A. A., Hogervorst, F., Jaspars, L. H., Fles, D. L. A., Kuikman, I., *et al.* (1994). Distinct and overlapping ligand specificities of the α3Ab1 and α6Ab1 Integrins: recognition of laminin isoforms. *Mol. Biol. Cell,* **5**, 203.

253. DiPersio, C. M., Hodivala-Dilke, K. M., Jaenisch, R., Kreidberg, J. A., and Hynes, R. O. (1997). α3β1 Integrin is required for normal development of the epidermal basement membrane. *J. Cell Biol.,* **137**, 729.

254. Ragahavan, S., Bauer, C., Mundschau, G., Li, Q., and Fuchs, E. (2000). Conditional ablation of β1 integrin in skin: severe defects in epidermal proliferation, basement membrane formation, and hair follicle invagination. *J. Cell Biol.,* **150**, 1149.

255. Goldfinger, L. E., Hopkinson, S. B., deHart, G. W., Collawn, S., Couchman, J. R., and Jones, J. C. R. (1999). The α3 laminin subunit, α6β4 and α3β1 integrin coordinately regulate wound healing in cultured epithelial cells and in the skin. *J. Cell Sci.,* **112**, 2615.

256. Hopkinson, S. B. and Jones, J. C. R. (1994). Identification of a second protein product of the gene encoding a human epidermal autoantigen. *Biochem. J.,* **300**, 851.

257. Brown, A., Bernier, G., Mathieu, M., Rossant, J., and Kothary, R. (1995). The mouse dystonia musculorum gene is a neural isoform of bullous pemphigoid antigen 1. *Nature Genet.,* **10**, 301.

258. Andra, K., Lassmann, H., Bittner, R., Shorny, S., Fassler, R., Propst, F., *et al.* (1997). Targeted inactivation of plectin reveals essential function in maintaining the integrity of skin, muscle, and heart cytoarchitecture. *Genes Dev.,* **11**, 3143.

259. van der Neut, R., Krimpenfort, P., Calafat, J., Niessen, C. M., and Sonnenberg, A. (1996). Epithelial detachment due to absence of hemidesmosomes in integrin β4 null mice. *Nature Med.*, **13**, 366.

260. Dowling, J., Yu, Q.-C., and Fuchs, E. (1996). β4 integrin is required for hemidesmosome formation, cell adhesion and cell survival. *J. Cell Biol.*, **134**, 559.

261. Ryan, M. C., Lee, K., Miyashita, Y., and Carter, W. G. (1999). Targeted disruption of the LAMA3 gene in mice reveals abnormalities in survival and late stage differentiation of epithelial cells. *J. Cell Biol.*, **145**, 1309.

11 | The tight junction

SHOICHIRO TSUKITA, MIKIO FURUSE, and MASAHIKO ITOH

1. Overview: structure and functions of tight junctions

The tight junction (TJ) is a type of intercellular junction, also called the *zonula occludens*. Similarly to other types of intercellular junctions, the TJ was first identified and defined morphologically at the electron microscopic level (1). On ultrathin section electron microscopy, TJs appear as a series of discrete sites of apparent fusions ('kissing points'), involving the outer leaflets of the plasma membranes of adjacent cells (Fig. 1a) (1). On freeze-fracture electron microscopy, TJs appear as a set of continuous, anastomosing intramembranous particle strands or fibrils (TJ strands or fibrils) in the P-face (the outwardly facing cytoplasmic leaflet) with comple-

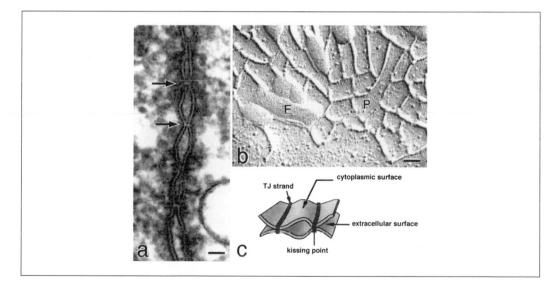

Fig. 1 The structure of tight junctions (TJs). (a) An ultrathin sectional image of TJs in epithelial cells of the epididymis. TJs appear as a series of apparent fusions (kissing points) (*arrows*), involving the outer leaflets of the plasma membranes of adjacent cells. (b) A freeze-fracture image of TJs in intestinal epithelial cells. TJs appear as a set of continuous, anastomosing intramembranous particle strands (TJ strands) in the P-face (P) with complementary grooves in the E-face (E). (c) Schematic drawing of tight junctions. Note that each TJ strand is associated laterally with another TJ strand in apposing membranes of adjacent cells to form a kissing point, where the intercellular space is completely obliterated. Bars, 50 nm.

mentary grooves in the E-face (the inwardly facing extracytoplasmic leaflet) (Fig. 1b) (2, 3). These observations led to our current understanding of the three-dimensional structure of TJs shown in Fig. 1c. Each TJ strand is associated laterally with another TJ strand in apposing membranes of adjacent cells to form a kissing point, where the intercellular space is completely obliterated.

TJs occur mainly in epithelial and endothelial cells, and the appearance of the TJ network revealed by freeze-fracture electron microscopy varies considerably depending on tissue; the number, density, and ramification frequency of TJ strands as well as the morphology of the TJ strand/groove itself show fairly high degrees of diversity (2–4). For example, between adjacent intestinal epithelial cells there is a well-developed complicated network of continuous TJ strands, whereas only one or two discontinuous TJ strands occur between adjacent endothelial cells of capillary blood vessels (5). For the precise definition of TJs, identification of the major components of TJ strands is necessary.

In simple epithelial cells, TJs constitute the so-called 'junctional complex' together with adherens junctions (AJs) (see Chapter 8) and desmosomes (DSs) (see Chapter 10) at the most apical portion of lateral membranes (1), but functionally TJs are thought to be different from AJs and DSs. To date, two possible functions have been proposed for TJs (6–9). The first is the 'barrier function'. The establishment of compositionally distinct fluid compartments by various types of cellular sheets is crucial for the development and function of most organs in multicellular organisms. Since these cell sheets are composed of two-dimensionally arranged epithelial and endothelial cells, some mechanism is required to seal cells to create a primary barrier to the diffusion of solutes through the paracellular pathway. This mechanism is thus essential for multicellular systems, and TJs are now believed to be directly involved in this barrier mechanism.

The second function proposed for TJs is the 'fence function' (10). As mentioned above, epithelial/endothelial cellular sheets form barriers between two distinct fluid compartments, the luminal and serosal compartments, but to maintain the environment of each compartment, various materials are dynamically and selectively transported across these cellular sheets through the transcellular pathway. For this vectorial transport of materials across cellular sheets, their plasma membranes are functionally divided into apical and basolateral domains that face the luminal and serosal compartments, respectively. However, since integral membrane proteins as well as lipids can diffuse freely within the plane of the lipid bilayer of plasma membranes, some diffusion barrier is required at the border between apical and basolateral membrane domains. Apical and basolateral membrane domains differ in their compositions of integral membrane proteins as well as lipids. Since TJs appear like a fence within plasma membranes of the most apical part of lateral membranes on freeze-fracture electron microscopy (2, 3), they have been suggested to be the morphological counterpart for the localized diffusion barrier.

In this chapter, we will first summarize current understanding of the molecular architecture of TJs, and then discuss functional aspects of TJs as well as their mechanism of regulation.

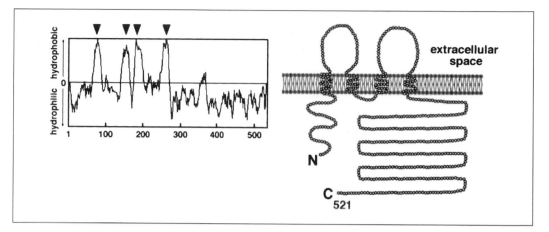

Fig. 2 The structure of occludin. Hydrophilicity plot for mouse occludin prepared using the Kyte and Doolittle program suggests four transmembrane domains (*arrowheads*) in each occludin molecule. As shown in the folding model, occludin is thought to be comprised of four transmembrane domains, a long COOH-terminal cytoplasmic domain, a short NH$_2$-terminal cytoplasmic domain, two extracellular loops, and one intracellular turn.

2. Integral membrane proteins constituting tight junction strands

In the past 20–30 years, studies have been performed to identify integral membrane proteins constituting TJ strands. For example, mAbs were screened for their ability to affect the barrier functions of epithelial cellular sheets (11), but until recently very little was known regarding the molecular architecture of TJ strands. In 1993, occludin was first reported as a TJ-specific integral membrane protein (12), and claudins were identified five years later (13).

2.1 Occludin

The term occludin is derived from the Latin word *occlude*. This ~ 60 kDa integral membrane protein was first identified in chicken (12) and then also in mammals (14). Occludin is comprised of four transmembrane domains, a long COOH-terminal cytoplasmic domain, a short NH$_2$-terminal cytoplasmic domain, two extracellular loops, and one intracellular turn (Fig. 2). One of the most characteristic aspects of its sequence is the high content of tyrosine and glycine residues in the first extracellular loop (~ 60%). The COOH-terminal 150 amino acids are predicted to form a typical α-helical coiled-coil structure. These structural characteristics of occludin are well conserved phylogenetically, although the amino acid sequences of chicken and mammalian occludins are quite divergent (~ 50% identity) (14). No occludin isoforms, i.e. occludin-related genes, have yet been identified in any species, but two forms of occludin were recently found to be generated by alternative splicing (15).

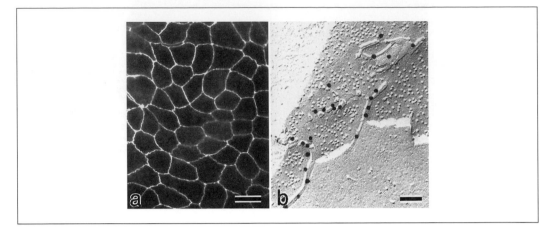

Fig. 3 Subcellular distribution of occludin. (a) Immunofluorescence image of cultured epithelial cells, MDCK, with anti-occludin mAb. Occludin is exclusively concentrated at cell–cell borders. (b) Immunoelectron microscopic image of freeze-fracture replicas of chicken liver with anti-occludin mAb. TJ strands are specifically labelled. Bars (a) 30 μm; (b) 50 nm.

In epithelial cells, occludin has been shown to be exclusively localized at the kissing points of TJs by ultrathin section electron microscopy (12), and on TJ strands *per se* by immunoreplica electron microscopy (Fig. 3) (16, 17). Taking the intense labelling of TJ strands on replicas with anti-occludin antibodies into consideration, occludin appears to be at least one of the major constituents of TJ strands *in situ*.

Occludin has been reported to be a functional component of TJs. Overexpression of full-length occludin in cultured MDCK cells elevates their *trans*-epithelial resistance (TER) (18), and introduction of COOH-terminally truncated occludin into MDCK cells (19) or *Xenopus laevis* embryo cells (20) results in the increased para-cellular leakage of small molecular mass tracers. The TER of cultured *Xenopus* epithelial cells is down-regulated by addition to the culture medium of a synthetic peptide corresponding to the second extracellular loop of chicken occludin (21). TJ fence function against fluorescently labelled lipid probes is also affected when COOH-terminally truncated occludin is introduced into MDCK cells (19).

The structure and functions of TJs, however, cannot be explained by occludin alone. Although TJ strands in most cells contain occludin, occludin is undetectable in TJ strands in most endothelial cells of non-neuronal tissues (22) as well as in Sertoli cells in human testis (23). Recently, both alleles of the occludin gene have been disrupted in embryonic stem (ES) cells. Interestingly, when the occludin-deficient ES cells were differentiated into epithelial cells by embryoid body formation, well-developed TJ strands were formed between adjacent epithelial cells (24). This finding conclusively indicated that occludin is not necessarily required for TJ formation itself, and that there are as yet unidentified TJ integral membrane protein(s) which can form strand structures without occludin. Furthermore, the above findings suggesting the functional importance of occludin in TJs should be re-evaluated.

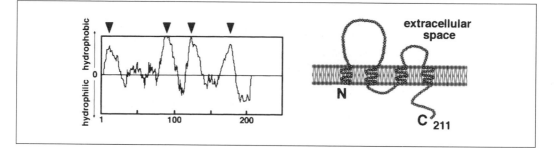

Fig. 4 The structure of claudin-1. Hydrophilicity plot for mouse claudin-1 prepared using the Kyte and Doolittle program suggests four transmembrane domains (*arrowheads*) in each occludin molecule. As shown in the folding model, claudin-1 is thought to be comprised of four transmembrane domains, a short COOH-terminal cytoplasmic domain, two extracellular loops, and one intracellular turn.

2.2 Claudins

The term claudin is derived from the Latin word *claudere* (to close). Two related ~ 23 kDa integral membrane proteins, claudin-1 and -2 (38% identical at the amino acid sequence level), were identified as components of isolated junctional fractions from the liver that were co-partitioned with occludin (13). Both claudin-1 and -2 also possess four transmembrane domains (Fig. 4), but do not show any sequence similarity to occludin. The cytoplasmic domain and the second extracellular loop of claudins are significantly shorter than those of occludin.

FLAG-tagged claudin-1 or -2 expressed in cultured MDCK cells are correctly targeted to and incorporated into pre-existing TJ strands (13). Furthermore, when claudin-1 or -2 was singly transfected into mouse L fibroblasts, these claudins conferred Ca^{2+}-independent cell adhesion activity on L fibroblasts (25). In these transfectants, claudins were concentrated at cell–cell contact planes, not diffusely but in an elaborate network pattern (Fig. 5a, b) (26). At the electron microscopic level, these planes are characterized by a well-developed network of TJ strand-like structures which are indistinguishable from TJ strands *in situ* (Fig. 5c). As these reconstituted TJ strands are exclusively labelled by anti-FLAG mAb on freeze-fracture replicas, it can be concluded that introduced claudins are polymerized into TJ strands within plasma membranes.

Several sequences similar to claudin-1 and -2 were found in the databases, indicating the existence of a new gene family that can be called the 'claudin' family (27). Based on sequence similarity, 19 members of this gene family have been identified to date in mice as well as human, and designated as claudin-1 to -19 (Fig. 6) (28–30). Among these, several members (claudin-3, -4, -5, -10, and -11) have been reported previously (RVP1, CPE-R, TMVCF/mBEC-1, OSPL, and OSP, respectively), although their physiological functions have not been elucidated. All of these claudin family members showed similar patterns on hydrophilicity plots, which predicted four transmembrane domains in each molecule. Similar to claudin-1 and -2, HA-

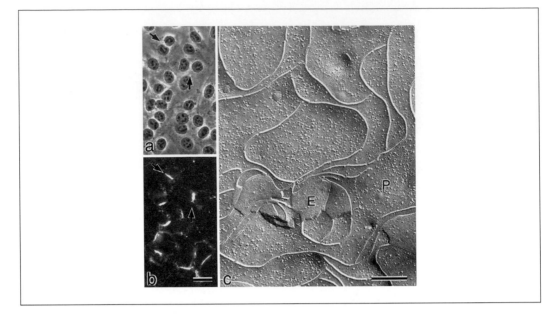

Fig. 5 Reconstitution of TJ strands in L fibroblastic cells. (a, b) Concentration of claudin-1 at cell–cell contact planes (*arrows*) of stable L transfectants expressing FLAG-tagged claudin-1. (a) Phase contrast and (b) immuno-fluorescence microscopic images. Cells were stained with anti-FLAG mAb. (c) Freeze-fracture image of cell contact planes of stable L transfectants expressing FLAG-claudin-1. TJ strands and corresponding grooves are observed in P- (P) and E-faces (E), respectively. These structures are indistinguishable from TJ strands *in situ*. Bars (a, b) 20 μm; (c) 200 nm.

tagged claudin-3 to -8 introduced into cultured MDCK cells are concentrated at pre-existing TJs (27). Claudin-9 to -19 (except for claudin-11) have not been well characterized, but preliminary data favour the notion that these uncharacterized claudins are also directly involved in the formation of TJ strands *in situ*. Although the antigenicity of claudins is generally fairly low, specific antibodies for some claudin members have been obtained (Fig. 7a). These antibodies exclusively label the TJ strand itself *in situ* on freeze-fracture replicas (Fig. 7b) (27, 32).

The expression pattern of each claudin species varies considerably among tissues (13, 27). For example, claudin-3 mRNA is detected in large amounts in lung and liver, and in small amounts in kidney and testis. Claudin-4, -7, and -8 are primarily expressed in lung and kidney. Claudin-6 expression is detected in large amounts in embryos, but not in adult tissues. On the other hand, some types of cells express their own specific species of claudins. For example, claudin-5 is expressed exclusively in endothelial cells of blood vessels (31). The expression of claudin-11 is restricted to the brain and testis (32). This claudin species constitutes TJ strands between lamellae of myelin sheaths of oligodendrocytes in the brain, and those between adjacent Sertoli cells in the testis. Recently, the claudin-11-deficient mice were successfully generated, and these lacked TJ strands both in myelin sheaths and Sertoli cells, indicating clearly that in these types of cells claudin-11 primarily constitutes TJ strands (33).

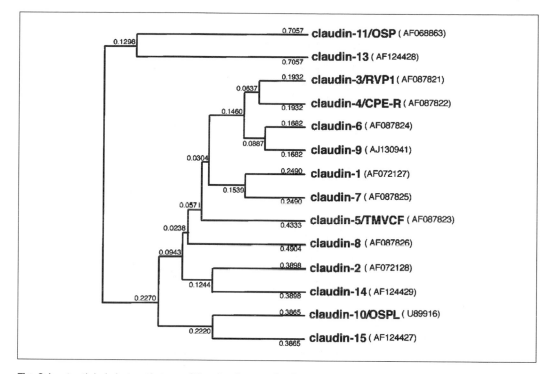

Fig. 6 A potential phylogenetic tree of the claudin gene family (claudin-1 to -15). The tree was constructed by the unweighed pair group method using the calculated genetic distances (presented numerically) between pairs of members. Accession numbers in GenBank/EMBL/DDBJ data base are presented in parentheses.

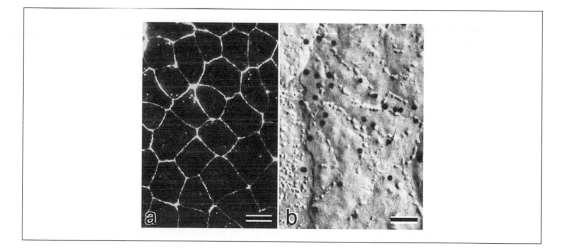

Fig. 7 Subcellular distribution of claudin-3. (a) Immunofluorescence image of cultured mouse Eph4 epithelial cells with anti-claudin-3 pAb. Claudin-3 is exclusively concentrated at cell–cell borders. (b) Immunoelectron microscopic image of freeze-fracture replicas of mouse kidney epithelial cells with anti-claudin-3 pAb. TJ strands are specifically labelled. Bars (a) 30 μm; (b) 60 nm.

Recently, to evaluate the direct involvement of claudins in the barrier function of TJs, the bacterial peptide toxin, *Clostridium perfringens* enterotoxin (CPE), was utilized (34). As described above, claudin-4 was initially identified as a CPE receptor (CPE-R), and the COOH-terminal half of this peptide (C-CPE) was reported to specifically bind to claudin-4/CPE-R (35, 36). When this C-CPE was applied to cultured epithelial cells, MDCK cells that express mainly claudin-1 and -4, claudin-4 was specifically removed from TJs with a concomitant decrease in TER as well as in the number of TJ strands. These findings conclusively indicated that claudins are directly involved in the barrier functions of TJs.

2.3 JAM

JAM (a junction-associated membrane protein), which is a ~ 40 kDa integral membrane protein with a single transmembrane domain, belongs structurally to the immunoglobulin superfamily (37). At the immunofluorescence microscopic level, this molecule is localized at TJs, but it shows no ability to reconstitute TJ strands in L fibroblasts (Itoh, M. and Tsukita, S., unpublished data).

3. Peripheral membrane proteins underlying tight junctions

Analyses of TJ-specific peripheral membrane proteins preceded those of integral membrane proteins. The following six peripheral membrane proteins have been shown to be specifically concentrated at the cytoplasmic surface of TJs, three of which belong to the so-called MAGUK family. Signalling molecules concentrated at TJs will be discussed later.

3.1 MAGUKs

The MAGUKs (membrane-associated guanylate kinase homologues) contain several internal repeats, one SH3 domain, and one guanylate kinase-like (GUK) domain, although they show no guanylate kinase activity (38, 39). The internal repeat is called the PDZ domain named after three members of this family, PSD-95, *dlg*, and ZO-1 (40). These molecules are localized just beneath the plasma membrane, and are generally involved in generation and maintenance of specialized membrane domains such as epithelial intercellular junctions and synaptic junctions. To date, ten and four MAGUKs have been identified in vertebrates and invertebrates, respectively, but the list of MAGUK family members is rapidly growing (Fig. 8).

Our knowledge of the functions of SH3 and GUK domains in MAGUKs is still limited, but PDZ domains have been well characterized (41). PDZ domains are found not only in MAGUKs but also in various other proteins underlying plasma membranes. PDZ domains were initially thought to bind specifically to the COOH-terminal -E-S/T-D-V motif in the cytoplasmic tails of integral membrane proteins,

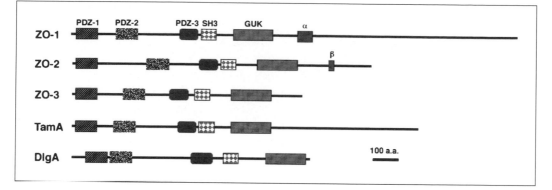

Fig. 8 The structure of MAGUKs (ZO-1, ZO-2, and ZO-3) localized in tight junctions and their related proteins of *Drosophila* (TamA, DlgA). These proteins contain three PDZ domains, one SH3 domain, and one guanylate kinase-like domain (GUK). In ZO-1 and ZO-2, isoforms are generated by alternative splicing at α and β, respectively.

which allows multi-PDZ domain proteins such as MAGUKs to crosslink distinct types of integral membrane proteins to establish specialized membrane domains. For example, Shaker K^+ channel and NMDA R2 subunit, which end in -ETDV and -ESDV, respectively, are crosslinked by PSD-95 in postsynaptic membranes (42, 43).

To date, three MAGUKs, ZO-1, ZO-2, and ZO-3, have been shown to be localized at the cytoplasmic surface of TJs.

3.1.1 ZO-1

ZO (*Zonula occludens*)-1 with a molecular mass of 220 kDa was first identified by monoclonal antibody production against the membrane fraction from mouse liver (44), and it is localized in the immediate vicinity of the plasma membrane of TJs in epithelial and endothelial cells (45). ZO-1 contains three PDZ domains, one SH3 domain, and one GUK domain in its NH_2-terminal half, and a proline-rich domain in its COOH-terminal half (46, 47). There are several isotypes of ZO-1 generated by alternative splicing.

ZO-1 is exclusively concentrated at TJs in simple epithelial cells (Fig. 9a), but is expressed even in other types of cells lacking TJs. ZO-1 occurs in the small junctions of the slit diaphragm in podocytes of kidney glomeruli (48). Furthermore, ZO-1 is precisely co-localized at spot-like AJs with cadherins in fibroblasts and cardiac muscle cells (Fig. 9b) (47, 49). In good agreement with the immunolocalization results, the NH_2-terminal half of ZO-1 directly binds to the cytoplasmic domain of occludin and claudins (50–52) as well as the COOH-terminal region of α-catenin (51, 53) both *in vitro* and *in vivo*. Furthermore, the COOH-terminal half of ZO-1 shows affinity to actin filaments (51, 54). Thus, ZO-1 functions as a crosslinker between occludin/claudins and actin filaments at TJs and between cadherin/catenin complex and actin filaments at spot-like AJs.

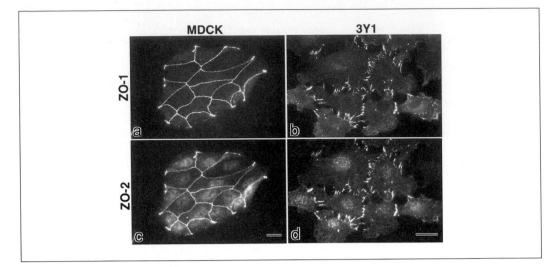

Fig. 9 Subcellular distribution of ZO-1 and ZO-2. Cultured MDCK epithelial cells (a, c) and 3Y1 fibroblasts (b, d) were doubly stained with anti-ZO-1 mAb (a, b) and anti-ZO-2 pAb (c, d). In epithelial cells, both ZO-1 and ZO-2 are co-localized at TJs, whereas in fibroblasts lacking TJs they are concentrated at spot-like adherens junctions. Bars 10 μm.

As ZO-1 shows sequence similarity to lethal *discs large-1* (*dlg*), one of the tumour suppressor genes in *Drosophila melanogaster*, it was suggested to be involved in the regulation of cell proliferation (46, 55). However, to date, there have been no experimental observations to support this notion. Furthermore, *Drosophila tam*, which is more closely related to mammalian ZO-1 than *Drosophila dlg* (56), and human *dlg* (*hdlg*) and PSD-95, which are more similar to *Drosophila dlg* than ZO-1, were identified (57). Thus, the relationship between the *dlg* tumour suppressor and ZO-1 may be more distant than originally suspected. The simple scheme in which TJ-specific MAGUKs are involved in cellular proliferation should therefore be re-evaluated.

3.1.2 ZO-2

ZO-2 with a molecular mass of 160 kDa was identified as a ZO-1 binding protein by immunoprecipitation experiments (58). ZO-2 is concentrated in the immediate vicinity of the plasma membrane of TJs in epithelial and endothelial cells, and contains three PDZ domains, one SH3 domain, and one GUK domain followed by a short COOH-terminal region (59). The second PDZ domain is responsible for the ZO-1/ZO-2 interaction (60). Two isotypes are generated by alternative splicing.

ZO-2 is very similar to ZO-1 in its subcellular distribution (Fig. 9c, d) (60), although it was not detected in the slit diaphragm in podocytes of kidney glomeruli (59). ZO-2 is also concentrated at cadherin-based spot-like AJs together with ZO-1 in fibroblasts and cardiac muscle cells, whereas in epithelial cells it is exclusively localized at TJs, not at AJs. Furthermore, similarly to ZO-1, the NH_2-terminal *dlg*-like domain of ZO-2 binds directly to occludin/claudins as well as α-catenin, and its COOH-terminal region associates with actin filaments (52, 60).

3.1.3 ZO-3

ZO-3 was identified as a phosphorylated 130 kDa protein in the immunoprecipitated ZO-1/ZO-2 complex (61). This protein also contains three PDZ domains, one SH3 domain, and one GUK domain followed by a short COOH-terminal domain (62). Unlike ZO-1 and ZO-2, this molecule contains a proline-rich domain between the second and third PDZ domains.

ZO-3 has not been characterized in detail. When ZO-3 was transfected to cultured epithelial cells, it was correctly targeted to pre-existing TJs. ZO-3, unlike ZO-1 or ZO-2, is not detected in fibroblasts or cardiac muscle cells (52). ZO-3 binds to the cytoplasmic domain of occludin/claudins, and interacts with ZO-1 but not with ZO-2.

3.2 Non-MAGUKs

In addition to ZO-1, ZO-2, and ZO-3, three peripheral membrane proteins, which do not belong to the MAGUK family, have been shown to be concentrated at the cytoplasmic surface of TJs. However, as compared to ZO-1, our knowledge regarding these proteins is still limited.

(a) *Cingulin.* Cingulin was identified by monoclonal antibody production using avian intestinal brush borders as antigens (63). Two forms with molecular masses of 108 kDa and 140 kDa are localized at TJs of both epithelial and endothelial cells. Purified cingulin is heat-stable, and has a rod-like appearance when observed by low-angle rotary shadowing. Its full-length cDNA has been cloned recently, and sequence analyses suggested that cingulin exists as a coiled-coil dimer (64). Cingulin was reported to directly bind to ZO-1, ZO-2, ZO-3, and myosin (64).

(b) *7H6 antigen.* Monoclonal antibodies were raised against the bile canaliculi fractions obtained from rat liver, and one mAb (7H6) recognized an antigen of ~ 155 kDa, which is localized at TJs of epithelial cells (65). 7H6 antigen has been proposed to be involved in the regulation of TJ permeability, but as yet there is no direct evidence to support this suggestion.

(c) *Symplekin.* Symplekin with a molecular mass of 130 kDa was identified by monoclonal antibody production (66), and does not show any significant sequence similarity to previously identified proteins. Both symplekin mRNA and protein occur in a wide range of cell types that do not form TJs. Symplekin is localized in all of these diverse cell types in the nucleoplasm, and only in those cells forming TJs such as epithelial cells is it recruited, partially but specifically, to TJs.

4. A possible model for the molecular architecture of tight junctions

Various models for the molecular architecture of TJs have been proposed to explain their structure and functions, but these can be broadly classified as 'protein' or 'lipid' models (Fig. 10a). In the protein model, TJ strands represent units of integral membrane proteins aggregated linearly. In contrast, in the lipid model, TJ strands are predominantly lipidic in nature, i.e. inverted cylindrical lipid micelles. As shown in Fig. 10a, in the lipid model, TJ strands should be readily solubilized by detergents, and the outer leaflet of apical (and also basolateral) plasma membranes should be continuous between adjacent epithelial cells. However, TJ strands are resistant to detergent extraction (67). Furthermore, when a fluorescently labelled lipid probe is introduced into the outer leaflet of apical membranes of a single cell in the confluent epithelial cell culture sheet, the probe does not diffuse into the apical membrane of adjacent unlabelled cells. Thus, strands do not appear to be composed solely of lipids. The identification of occludin and claudins has provided evidence for the view that the strands are composed of proteins.

4.1 TJ strands as linear co-polymers of occludin and claudins

Several lines of evidence have contributed to our current understanding of how occludin and claudins are involved in the formation of TJ strands:

(a) Anti-occludin (68) as well as anti-claudin antibodies (27) exclusively label TJ strands in various organs, indicating that both occludin and claudins are incorporated into TJ strands *in situ*.

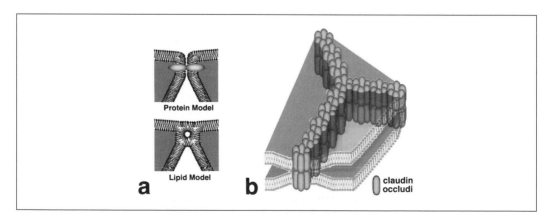

Fig. 10 Structural model for TJ strands. (a) Schematic drawing of protein and lipid models. (b) Co-polymer model. In this model, occludin and claudins bearing four transmembrane domains constitute homo- or hetero-hexamers, and these hexamers are aligned in a liner fashion to form individual TJ strands. Each TJ strand is associated laterally with another TJ strand in apposing membranes of adjacent cells to form kissing points. At present, the arrangement of occludin and claudins in TJ strands is purely speculative (see details in the text).

(b) When frozen sections of various tissues are immunofluorescently stained with anti-occludin antibodies, the staining intensity of occludin correlates well with the number of TJ strands in epithelial/endothelial cells (68), suggesting that the density of occludin molecules in TJ strands remains almost constant. In contrast, the content of each claudin species varies considerably depending on the tissue examined (13, 27).

(c) Occludin is not necessarily required for TJ strand formation because a well-developed network of TJ strands occurs in cells lacking occludin expression such as epithelial cells differentiated from occludin-deficient ES cells (24) and human Sertoli cells (23).

(d) When occludin is transfected into L fibroblasts, only a small number of short strands are induced, which are exclusively labelled with anti-occludin antibodies. In contrast, claudin-expressing L fibroblast transfectants bear a well-developed network of TJ strands. Furthermore, when occludin and claudin are co-transfected into L fibroblasts, occludin is recruited to and incorporated into claudin-based TJ strands (26).

These observations led to the conclusion that several species of claudins are polymerized to form the major 'backbone' of TJ strands, and that occludin is co-polymerized into these strands *in situ* (28). It remains unclear how claudins and occludin are arranged in individual TJ strands, but any model must explain the thickness of TJ strands (~ 10 nm). This thickness is similar to the diameter of gap junction channels (connexons) which consist of six connexin molecules each of which also bears four transmembrane domains (69). Therefore, since occludin and claudins also exhibit four transmembrane domains, it is possible that they form homo- or hetero-hexamers ~ 10 nm in diameter and these hexamers are aligned in a linear fashion to form TJ strands (Fig. 10b). Of course, these are pure speculations at present, and the arrangements of occludin and claudins in individual TJ strands should be determined probably by cryo-electron microscopy.

This 'co-polymer' model raises many new questions including how and where occludin and claudins are polymerized into strand structures. When occludin and/or claudins are introduced into L fibroblasts, strands are not observed on the free surface of the plasma membrane, but instead are restricted to cell–cell adhesion sites to form strands paired laterally with those in apposing membranes (26). Therefore, newly synthesized occludin and claudins are thought to appear on the cell surface as monomers or oligomers, and only when they adhere with occludin or claudins on the cell surface of adjacent cells are they added to the ends of paired strands pre-existing at cell–cell adhesion sites, where they cause elongation of these paired strands.

Another question is whether claudins on apposing membranes can interact (or adhere) with each other in a heterophilic manner. Recent analyses using L transfectants revealed that claudin-3 can adhere with claudin-1 and -2 in a heterophilic manner, but that claudin-1 cannot adhere with claudin-2 (29, 70). Therefore, the molecular interaction of claudins in paired TJ strands appears to be fairly complex (29).

4.2 Interaction between TJ strands and underlying cytoskeleton

As described above, several peripheral membrane proteins are localized specifically to the cytoplasmic surface of TJs. Furthermore, actin filaments are associated with the cytoplasmic surface of TJs (71), and a fairly well developed plaque structure is observed just beneath TJs on ultrathin section electron microscopy (66, 71). Now that occludin and claudins have been identified as major components of TJ strands, this raises the question of the nature of the molecular linkage from these integral membrane proteins to peripheral membrane proteins as well as actin filaments.

ZO-1 associates directly with the COOH-terminal 150 aa (α-helical coiled-coil portion) of occludin reportedly through its GUK domain (50, 54). ZO-2 is also associated with the same portion of occludin (60). These findings suggest that occludin recruits ZO-1 and ZO-2 to the cytoplasmic surface of TJs. However, even in epithelial-like cells (visceral endoderm cells), differentiated from occludin-deficient ES cells, ZO-1 as well as ZO-2 are still recruited to TJs (24), suggesting another membrane binding partner for ZO-1/ZO-2 in TJ strands.

It should be noted that, with the exception of claudin-11, all of the claudin family members end in -Y-V at their COOH-termini (claudin-11 ends in -H-V) (13, 27, 30, 32). As discussed above, the COOH-terminal -E-S/T-D-V motif in the cytoplasmic tails of integral membrane proteins is frequently recognized by PDZ domains. However, some PDZ domains have binding specificities for COOH-terminal sequences that are distinct from -E-S/T-D-V. Interestingly, the PDZ domain of LIN-2/CASK binds to the COOH-terminal tails of neurexin, which ends in -E-Y-Y-V (72). Recently, *in vitro* as well as *in vivo* analyses clearly revealed that the COOH-termini of claudins directly binds to PDZ1 domains of ZO-1/ZO-2/ZO-3 (52).

Since the COOH-terminal portions of ZO-1 and ZO-2 also bind directly to actin filaments, the molecular linkage from occludin/claudins to actin filaments can be speculated as described in Fig. 11. It remains to be determined how other TJ-specific peripheral membrane proteins such as symplekin, cingulin, and 7H6 antigen as well as TJ-associated signalling molecules (described below) are integrated into this scheme.

5. Formation and destruction of tight junctions in epithelial cells

In simple epithelial cells, TJs and AJs are known to show an intimate relationship not only spatially but also functionally in epithelial cells *in vivo*. For example, when AJs are destroyed by treating cultured epithelial cells (11) or 8-cell embryos (73) with EGTA or anti-cadherin mAb, the formation of TJs is suppressed, indicating that the mechanical adhesion activity of cadherin-based AJs is required for the formation and maintenance of TJs. On the other hand, under some experimental conditions, TJ strands themselves can be formed in the absence of cadherin-based cell adhesion activity; in depolarized

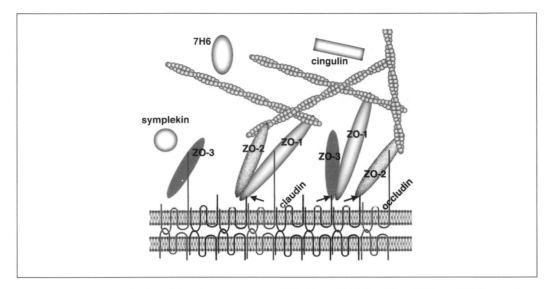

Fig. 11 Molecular organization of the plaque structure of TJs. Three MAGUKs, ZO-1, ZO-2, and ZO-3, constitute scaffolds of the plaque. The information of binding characters of these molecules are summarized in this schematic drawing. These molecules crosslink TJ strands with actin filaments as ZO-1/ZO-2 heterodimers, ZO-1/ZO-3 heterodimers, or monomers. The COOH-terminal regions of ZO-1 and ZO-2 directly bind to actin filaments, although it is not clear about the actin binding activity of ZO-3. The second PDZ domain of ZO-1 directly binds to that of ZO-2 and that of ZO-3. ZO-1, ZO-2, and ZO-3 bind to the COOH-terminal cytoplasmic domain of occludin. The domain responsible for this binding is reported to be narrowed down to the GUK domain on ZO-1 molecules. PDZ1 domains of ZO-1, ZO-2, and/or ZO-3 bind to the COOH-terminal end (-YV) of claudins (*arrows*). It remains unclear how cingulin, 7H6 antigen, and symplekin are associated with the TJ plaque.

cultured epithelial cells such as MDCK cells in low-calcium medium (61, 74) or α-catenin-deficient colon carcinoma cells (75), spot-like (but not belt-like) TJs are observed. Interestingly, when the cadherin activity of α-catenin-deficient colon carcinoma cells is restored by introducing α-catenin cDNA, cells are re-polarized with concomitant formation of belt-like TJs at the most apical part of their lateral membrane (75). Furthermore, as described above, when claudin is expressed in L fibroblasts which lack the expression of cadherins, a well-developed network of TJ strands is induced (26). However, these induced TJs in L transfectants do not surround individual cells continuously. These findings suggest that TJ strand polymerization itself is not dependent on cadherin-based AJs, but that AJs and/or the underlying actin-based cytoskeleton are important for the formation of belt-like TJs.

Information regarding how TJs are formed and sorted from AJs during the polarization process of epithelial cells is still limited. In this respect, the behaviour of E-cadherin, ZO-1, and occludin was recently examined during epithelial cellular polarization by wounding cultured epithelial cells (Fig. 12) (76, 77). At the initial stage of wound healing, very thin cellular protrusions containing actin filaments began to emerge from the front row of the wound. Small cell–cell contact sites are first formed at the tips of these cellular protrusions, where E-cadherin and ZO-1 are simultaneously recruited. This primordial form of spot-like AJs shows no concen-

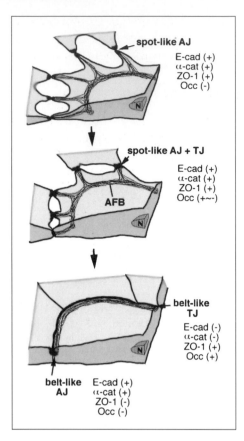

Fig. 12 The behaviour of E-cadherin, α-catenin, occludin, and ZO-1 during epithelial cellular polarization. ZO-1 appears to be transferred from cadherin/catenin (spot-like AJ) to occludin-based TJs. The behaviour of claudins have not yet been examined (see details in the text).

tration of occludin. As cellular polarization proceeds, the E-cadherin/ZO-1-positive spot-like AJs are gradually fused side by side with concomitant shortening of cellular protrusions. As this fusion proceeds, occludin begins to appear in fused junctions, and these E-cadherin/occludin/ZO-1 positive spots are gradually sorted into belt-like TJs containing occludin/ZO-1 and AJs containing E-cadherin. The behaviour of claudins during epithelial cellular polarization has not been examined.

As described above, although ZO-1/ZO-2 were first identified as components of TJ plaques, they are also expressed and concentrated at AJs in non-epithelial cells lacking TJs, such as cardiac muscle cells and fibroblasts. In fibroblasts, ZO-1/ZO-2 are precisely co-localized with cadherins, probably through direct binding to α-catenin, at cell–cell contact sites to form spot-like AJs (49, 51, 60). Thus, epithelial cells at the initial phase of polarization are very similar to fibroblasts in terms of the existence of spot-like AJs. Since in fully polarized epithelial cells, such as intestinal epithelial cells, ZO-1/ZO-2 are exclusively localized at TJs through their direct association with occludin (or possibly also with claudins) (50, 54), during the polarization process ZO-1/ZO-2 must be transferred from cadherin/catenin to TJs. Although the molecular mechanism behind this peculiar behaviour of ZO-1/ZO-2 remains unclear, these findings suggest that ZO-1 as well as ZO-2 play a central role in the formation of TJs and their sorting out from AJs. For further understanding, the behaviour of claudins should also be examined during the polarization process.

6. Regulation of tight junction functions

The tightness of TJs, which is quantitatively measured as TER, varies significantly depending on the type of epithelial/endothelial cells. In general, TER in various epithelial/endothelial cells is correlated well with the number of parallel strands in belt-like TJs (78, 79). The 'leaky' TJs in epithelial cells of kidney proximal tubules are composed of one to two strands, while the 'tight' TJs in those of distal tubules have more strands. More striking examples are found in endothelial cells in brain blood vessels (80) and Sertoli cells in the testis (81), both of which are characterized by numerous well-developed TJ strands. They are directly involved in the so-called 'blood–brain' and 'blood–testis' barriers, respectively, which protect neurons and spermatogenic cells from the external environment.

However, there are some clear exceptions to this strand number hypothesis. MDCK clones, type I and II, which show high and low TER (thousands-fold difference), respectively, have the same number of strands in their TJs (82). Another exception is found in cultured brain endothelial cells. These cells show loose TJ barrier function, but when treated with astrocyte-conditioned medium to elevate their intracellular cAMP level, their barrier function is markedly increased with no changes in the number of TJ strands (80).

The extent of the association of TJ strand particles with the P-face in freeze-fracture replica images is markedly different between MDCK I and II (83) and between non-treated and conditioned medium-treated brain endothelial cells (84); in MDCK I cells and conditioned medium-treated brain endothelial cells (with 'tight' TJs), the strand particles are largely associated with the P-face, leaving continuous grooves occupied by a small number of particles in the E-face (tentatively called 'P-face-associated TJs'). In contrast, in MDCK II cells and non-treated endothelial cells (with 'leaky' TJs), the strands on the P-face are fairly discontinuous, and in the E-face particles form chains that occupy the grooves ('E-face-associated TJs'). Therefore, in addition to the strand number hypothesis, the extent of TJ strand particle association with the P-face appears to be correlated well with the tightness of TJs in epithelial as well as endothelial cells (84). Interestingly, in L cell transfectants expressing claudins, the extent of reconstituted TJ strand particle association with the P-face varies significantly depending on the expressed claudin species. For example, TJ strands reconstituted from claudin-1 particles are largely P-face-associated, whereas those from claudin-2 are E-face-associated (26). These findings strongly suggest that claudin-1-based TJ strands are tighter than claudin-2-based strands, and that the tightness of individual TJ strands *in situ* is determined by the combination of the members of claudin family in each TJ strand. This hypothesis will be evaluated by introducing the members of claudin family into cultured epithelial cells singly or in combination.

The tightness of TJs is dynamically regulated both under physiological and pathological conditions. For example, the permeability of intestinal epithelial cells is controlled during absorption of nutrients; glucose solution significantly reduces the barrier function of TJs in intestinal epithelial cells (85). During inflammatory and immune responses, TJs in endothelial cells (and also in epithelial cells) is repeatedly destroyed and formed during leukocyte transmigration across the cellular sheets (86), and cytokines such as interferon-γ and TNF-α derived from mononuclear cells affect the TJ barrier function of epithelial cells (87). Furthermore, in some pathological conditions such as Crohn's disease, the barrier function of TJ is down-regulated in patients as well as their relatives, suggesting that this defect allows the access of inflammatory agents through the paracellular pathway into the mucosa resulting in aggravation of bowel inflammation (88). As discussed in the next section, the molecular mechanisms behind TJ regulation under physiological as well as pathological conditions is an area of current active investigation, but information is still limited.

Several bacterial toxins affect TJs. Toxins A and B from *Clostridium difficile* inactivate Rho, Rac, and Cdc42 by glycosylation (89), which affects actin-based cytoskeleton and the TJ barrier function (90). A toxin called ZO toxin (ZOT) derived from *Vibrio cholera* also increases the paracellular permeability of intestinal epithelial cells, inducing diarrhoea, probably by affecting the actin-based cytoskeleton and junctional integrity (91). Finally, as discussed above, it should be noted that claudin-4 had been identified as a receptor for *Clostridium perfringens* enterotoxin (CPE) (35). The involvement of claudin family members in CPE toxicity is intriguing.

7. Signalling events at the cytoplasmic surface of tight junctions

To identify intracellular signalling pathways involved in the regulation of TJ functions, the effects of various inhibitors and activators of signalling molecules on the formation and functions of TJs have been extensively analysed. Studies along this line has indicated that heterotrimeric G proteins down-regulate the TJ barrier function, whereas protein kinase C (PKC), phospholipase C (PLC), and calmodulin act as positive regulators for the TJ barrier function (92). Furthermore, small G proteins such as Rho and Rac have also been shown to be involved in TJ barrier function (93, 94).

However, it remains unclear how these signalling molecules regulate the TJ functions in molecular terms. As discussed above, TJ formation/functions are dependent on cadherin-based AJs, and both TJs and AJs are tightly associated with the underlying actin-based cytoskeleton. Therefore, TJ functions are affected by modulating not only TJ components *per se*, but also AJ components and the actin-based cytoskeleton, and it is often difficult to determine which of these are directly modulated by signalling molecules. Likewise, although several TJ-specific peripheral membrane proteins such as ZO-1, ZO-2, ZO-3, and cingulin are serine/threonine phosphorylated (61, 95), a positive correlation has not been observed between TJ assembly and their phosphorylation levels. Occludin is heavily serine/threonine phosphorylated with a time course similar to that of TJ formation after the Ca switch in cultured epithelial cells, and the highly phosphorylated occludin is selectively concentrated at TJs (96). These findings led to the speculation that phosphorylation of occludin, probably by PKC, is required for TJ assembly. However, as occludin-deficient epithelial cells show no abnormalities in TJ formation (24), this does not appear to be the case.

When signalling molecules are concentrated at the cytoplasmic surface of TJs, it is likely that these molecules play some roles in signalling at TJs, although it is not clear whether these molecules are involved in inside-out or outside-in signalling at TJs. Several intriguing signalling molecules have been reported to be localized at TJs (30). For example, two types of heterotrimeric G proteins, $G\alpha 0$ and $G\alpha i2$, are distributed in the region containing TJs (97, 98).

Atypical PKCs, PKC ζ and λ, and their specific binding proteins (ASIP; atypical PKC isotype-specific interacting protein) are also concentrated at the cytoplasmic surface of TJs (99, 100). Interestingly, ASIP is a mammalian homologue of *Caenorhabditis elegans* polarity protein PAR-3. In the PAR-3 mutant of *C. elegans*, the cell division axis in early embryos is affected (101), and the same phenotypic abnormality is detected in mutants lacking the *C. elegans* homologue of PKC λ (102, 103). Thus, it is likely that the PKC ζ/ASIP and/or PKC λ/ASIP complex play an important role(s) in the establishment of cell polarity, not only in *C. elegans* embryos but also mammalian epithelial cells. Further analyses of how this complex regulates epithelial polarity will provide new insight into the functions of TJs.

AF-6, a putative target for *ras*, was reported to be concentrated at TJs (104). However, data regarding the localization at TJs are not conclusive, and an alternative variant called afadin was shown to be specifically localized at AJs at the electron microscopic level (105). Furthermore, it remains unclear whether AF-6 is actually a target for *ras*, thus a link between *ras* signalling and TJ function is speculative at present.

Several proteins thought to be involved in vesicular transport or fusion were also reported to be concentrated at the cytoplasmic surface of TJs. Rab3B (106) and Rab13 (107) are specifically localized at TJs in epithelial cells. However, there have been no reports on the effects of dominant-negative constructs for these Rabs on barrier/ fence functions of TJs or on cellular polarity. Recently, mammalian homologues of yeast *sec6* and *sec8* products were reported to be localized at TJs in epithelial cells (108). *Sec6/8* products are involved in vesicular targeting required for polarized budding in yeasts (109). These findings led to the intriguing hypothesis that TJs function as a site for vesicular targeting and fusion to establish epithelial polarization (110).

8. Conclusions and outlook

The study of TJs has a long history, and there are accumulated large amounts of data regarding the structure and functions of TJs at the cellular as well as tissue levels. Until recently, however, the lack of information on the TJ-specific integral membrane proteins has hampered direct assessment of TJ functions at the molecular level. Now that occludin and claudins have been identified, and TJ strands can be reconstituted from these molecules, the structure and functions of TJs should be determined in detail in molecular terms within the next decade.

The molecular mechanisms of the vectorial transport across the epithelial/ endothelial cell sheets through the 'transcellular' pathway has been examined in detail by analysing various channels and pumps. The 'paracellular' pathway also plays an important role in vectorial transport, and this pathway shows some selectivity for transported materials such as ions: Detailed electrophysiological analyses suggested the existence of aqueous pores within the paired TJ strands (7, 111–113). Interestingly, recent analysis of hereditary hypomagnesemia yielded a new insight that claudins constitute not only the TJ strand itself, but also aqueous pores within the strand (114). The elucidation of how occludin/claudin-based TJ strands can function as barriers which allow selective leakage will be very important for future studies (29).

There is no doubt that TJs are involved in the epithelial/endothelial barrier, but the evidence for their involvement in the fence function is fragmentary. Thus, further experiments are required to determine at the molecular level to what extent and how TJs are required to establish apical and basolateral membrane domains. The specific and shared functions of occludin and each species of claudin should also be further analysed in connection with the barrier and fence functions of TJs.

The mechanism of regulation of TJ functions is another important issue for future studies. At present, no information is available regarding the transcriptional regulation of each species of claudin as well as post-translational regulation by various signalling pathways. Of course, detailed analyses of the interaction and mechanism of regulation between claudins/occludin and underlying cytoskeletal proteins are also required. Research in the area of tight junctions continues to advance rapidly (115). Elucidation of the mechanisms of regulation of TJ functions will lead not only to better understanding of various physiological and pathological events in molecular terms but also to development of a way to modulate TJ functions, which is important for basic developmental/cellular biology as well as clinical medicine.

References

1. Farquhar, M. G. and Palade, G. E. (1963). Junctional complexes in various epithelia. *J. Cell Biol.*, **17**, 375.
2. Staehelin, L. A. (1973). Further observations on the fine structure of freeze-cleaved tight junctions. *J. Cell Sci.*, **13**, 763.
3. Staehelin, L. A. (1974). Structure and function of intercellular junctions. *Int. Rev. Cytol.*, **39**, 191.
4. Hull, B. E. and Staehelin, L. A. (1976). Functional significance of the variations in the geometrical organization of tight junction networks. *J. Cell Biol.*, **68**, 688.
5. Simionescu, M., Simionescu, N., and Palade, G. E. (1976). Segmental differentiations of cell junctions in the vascular endothelium. Arteries and vein. *J. Cell Biol.*, **68**, 705.

6. Schneeberger, E. E. and Lynch, R. D. (1992). Structure, function, and regulation of cellular tight junctions. *Am. J. Physiol.*, **262**, L647.

7. Gumbiner, B. (1993). Breaking through the tight junction barrier. *J. Cell Biol.*, **123**, 1631.

8. Anderson, J. M. and Van Itallie, C. M. (1995). Tight junctions and the molecular basis for regulation of paracellular permeability. *Am. J. Physiol.*, **269**, G467.

9. Yap, A. S., Mullin, J. M., and Stevenson, B. R. (1999). Molecular analysis of tight junction physiology: Insights and paradoxes. *J. Membrane Biol.*, **163**, 159.

10. Rodriguez-Boulan, E. and Nelson, W. J. (1989). Morphogenesis of the polarized epithelial cell phenotype. *Science*, **245**, 718.

11. Gumbiner, B. and Simons, K. (1986). A functional assay for proteins involved in establishing an epithelial occluding barrier: identification of a uvomorulin-like polypeptide. *J. Cell Biol.*, **102**, 457.

12. Furuse, M., Hirase, T., Itoh, M., Nagafuchi, A., Yonemura, S., Tsukita, Sa., *et al.* (1993). Occludin: a novel integral membrane protein localizing at tight junctions. *J. Cell Biol.*, **123**, 1777.

13. Furuse, M., Fujita, K., Hiiragi, T., Fujimoto, K., and Tsukita, Sh. (1998). Claudin-1 and -2: Novel integral membrane proteins localizing at tight junctions with no sequence similarity to occludin. *J. Cell Biol.*, **141**, 1539.

14. Ando-Akatsuka, Y., Saitou, M., Hirase, T., Kishi, M., Sakakibara, A., Itoh, M., *et al.* (1996). Interspecies diversity of the occludin sequence: cDNA cloning of human, mouse, dog, and rat-kangaroo homologues. *J. Cell Biol.*, **133**, 43.

15. Goodenough, D. A. (1999). Plugging the leaks. *Proc. Natl. Acad. Sci. USA*, **96**, 319.

16. Fujimoto, K. (1995). Freeze-fracture replica electron microscopy combined with SDS digestion for cytochemical labeling of integral membrane proteins. Application to the immunogold labeling of intercellular junctional complexes. *J. Cell Sci.*, **108**, 3443.

17. Furuse, M., Fujimoto, K., Sato, N., Hirase, T., Tsukita, Sa., and Tsukita, Sh. (1996). Overexpression of occludin, a tight junction-associated integral membrane protein, induces the formation of intracellular multilamellar bodies bearing tight junction-like structures. *J. Cell Sci.*, **109**, 429.

18. McCarthy, K. M., Skare, I. B., Stankewich, M. C., Furuse, M., Tsukita, Sh., Rogers, R. A., *et al.* (1996). Occludin is a functional component of the tight junction. *J. Cell Sci.*, **109**, 2287.

19. Balda, M. S., Whitney, J. A., Flores, C., González, S., Cereijido, M., and Matter, K. (1996). Functional dissociation of paracellular permeability and transepithelial electrical resistance and disruption of the apical-basolateral intramembrane diffusion barrier by expression of a mutant tight junction membrane protein. *J. Cell Biol.*, **134**, 1031.

20. Chen, Y.-H., Merzdorf, C., Paul, D. L., and Goodenough, D. A. (1997). COOH terminus of occludin is required for tight junction barrier function in early *Xenopus* embryos. *J. Cell Biol.*, **138**, 891.

21. Wong, V. and Gumbiner, B. M. (1997). A synthetic peptide corresponding to the extracellular domain of occludin perturbs the tight junction permeability barrier. *J. Cell Biol.*, **136**, 399.

22. Hirase, T., Staddon, J. M., Saitou, M., Ando-Akatsuka, Y., Itoh, M., Furuse, M., *et al.* (1997). Occludin as a possible determinant of tight junction permeability in endothelial cells. *J. Cell Sci.*, **110**, 1603.

23. Moroi, S., Saitou, M., Fujimoto, K., Sakakibara, A., Furuse, M., Yoshida, O., *et al.* (1998). Occludin is concentrated at tight junctions of mouse/rat but not human/guinea pig Sertoli cells in testes. *Am. J. Physiol.*, **274**, C1708.

24. Saitou, M., Fujimoto, K., Doi, Y., Itoh, M., Fujimoto, T., Furuse, M., *et al.* (1998). Occludin-deficient embryonic stem cells can differentiate into polarized epithelial cells bearing tight junctions. *J. Cell Biol.*, **141**, 397.

25. Kubota, K., Furuse, M., Sasaki, H., Sonoda, N., Fujita, K., Nagafuchi, A., *et al.* (1999). Ca^{2+}-independent cell adhesion activity of claudins, integral membrane proteins of tight junctions. *Curr. Biol.*, **9**, 1035.

26. Furuse, M., Sasaki, H., Fujimoto, K., and Tsukita, Sh. (1998). A single gene product, claudin-1 or -2, reconstitutes tight junction strands and recruits occludin in fibroblasts. *J. Cell Biol.*, **143**, 391.

27. Morita, K., Furuse, M., Fujimoto, K., and Tsukita, Sh. (1999). Claudin multigene family encoding four-transmembrane domain protein components of tight junction strands. *Proc. Natl. Acad. Sci. USA*, **96**, 511.

28. Tsukita, S. and Furuse, M. (1999). Occludin and claudins in tight junction strands: Leading or supporting players? *Trends Cell Biol.*, **9**, 268.

29. Tsukita, S. and Furuse, M. (2000). Pores in the wall: Claudins constitute tight junction strands containing aqueous pores. *J. Cell Biol.*, **49**, 13.

30. Tsukita, S., Furuse, M., and Itoh, M. (1999). Structural and signaling molecules come together at tight junctions. *Curr. Opin. Cell Biol.*, **11**, 628.

31. Morita, K., Sasaki, H., Furuse, M., and Tsukita, Sh. (1999). Endothelial claudin: Claudin-5/TMVCF constitutes tight junction strands in endothelial cells. *J. Cell Biol.*, **147**, 185.

32. Morita, K., Sasaki, H., Fujimoto, K., Furuse, M., and Tsukita, Sh. (1999). Claudin-11/OSP-based tight junctions of myelin sheaths in brain and Sertoli cells in testis. *J. Cell Biol.*, **145**, 579.

33. Gow, A., Southwood, C. M., Li, J. S., Pariali, M., Riordan, G. P., Brodie, S. E., *et al.* (1999). CNS myelin and Sertoli cell tight junction strands are absent in OSP/claudin-11 null mice. *Cell*, **99**, 649.

34. Sonoda, N., Furuse, M., Sasaki, H., Yonemura, S., Katahira, J., Horiguchi, Y., *et al.* (1999). Clostridium perfringens enterotoxin fragment removes specific claudins from tight junction strands: Evidence for direct involvement of claudins in tight junction barrier. *J. Cell Biol.*, **147**, 195.

35. Katahira, J., Inoue, N., Horiguchi, Y., Matsuda, M., and Sugimoto, N. (1997). Molecular cloning and functional characterization of the receptor for Clostridium perfringens enterotoxin. *J. Cell Biol.*, **136**, 1239.

36. Katahira, J., Sugiyama, H., Inoue, N., Horiguchi, Y., Matsuda, M., and Sugimoto, N. (1997). Clostridium perfringens enterotoxin utilizes two structurally related membrane proteins as functional receptors *in vivo*. *J. Biol. Chem.*, **272**, 26652.

37. Martin-Padura, I., Lostaglio, S., Schneemann, M., Williams, L., Romano, M., Fruscella, P., *et al.* (1998). Junctional adhesion molecule, a novel member of the immunoglobulin superfamily that distributes at intercellular junctions and modulates monocyte transmigration. *J. Cell Biol.*, **142**, 117.

38. Kim, S. K. (1995). Tight junctions, membrane-associated guanylate kinases and cell signaling. *Curr. Opin. Cell Biol.*, **7**, 641.

39. Anderson, J. M., Fanning, A. S., Lapierre, L., and Van Itallie, C. M. (1995). Zonula occludens (ZO)-1 and ZO-2: membrane-associated guanylate kinase homologues (MAGuKs) of the tight junction. *Biochem. Soc. Trans.*, **23**, 470.

40. Woods, D. F. and Bryant, P. J. (1993). ZO-1, DlgA and PSD-95/SAP90: homologous proteins in tight, septate and synaptic cell junctions. *Mech. Dev.*, **44**, 85.

41. Gomperts, S. N. (1996). Clustering membrane proteins: It's all coming together with the PSD-95/SAP90 protein family. *Cell*, **84**, 659.

42. Kim, E., Niethammer, M., Rothschild, A., Jan, Y. N., and Sheng, M. (1995). Clustering of shaker-type K^+ channels by interaction with a family of membrane-associated guanylate kinases. *Nature*, **378**, 85.

43. Kornau, H.-C., Schenker, L. T., Kennedy, M. B., and Seeburg, P. H. (1995). Domain interaction between NMDA receptor subunits and the postsynaptic density protein PSD-95. *Science*, **269**, 1737.

44. Stevenson, B. R., Siliciano, J. D., Mooseker, M. S., and Goodenough, D. A. (1986). Identification of ZO-1: a high molecular weight polypeptide associated with the tight junction (zonula occludens) in a variety of epithelia. *J. Cell Biol.*, **103**, 755.

45. Anderson, J. M., Stevenson, B. R., Jesaitis, L. A., Goodenough, D. A., and Mooseker, M. S. (1988). Characterization of ZO-1, a protein component of the tight junction from mouse liver and Madin-Darby canine kidney cells. *J. Cell Biol.*, **106**, 1141.

46. Willott, E., Balda, M. S., Fanning, A. S., Jameson, B., Van Itallie, C., and Anderson, J. M. (1993). The tight junction protein ZO-1 is homologous to the *Drosophila* discs large tumor suppresser protein of septate junctions. *Proc. Natl. Acad. Sci. USA*, **90**, 7834.

47. Itoh, M., Nagafuchi, A., Yonemura, S., Kitani-Yasuda, T., Tsukita, Sa., and Tsukita, Sh. (1993). The 220-kD protein colocalizing with cadherins in non-epithelial cells is identical to ZO-1, a tight junction-associated protein in epithelial cells: cDNA cloning and immunoelectron microscopy. *J. Cell Biol.*, **121**, 491.

48. Schnabel, E., Anderson, J. M., and Farquhar, M. G. (1990). The tight junction protein ZO-1 is concentrated along slit diaphragms of the glomerular epithelium. *J. Cell Biol.*, **111**, 1255.

49. Itoh, M., Yonemura, S., Nagafuchi, A., Tsukita, Sa., and Tsukita, Sh. (1991). A 220-kD undercoat-constitutive protein: Its specific localization at cadherin-based cell–cell adhesion sites. *J. Cell Biol.*, **115**, 1449.

50. Furuse, M., Itoh, M., Hirase, T., Nagafuchi, A., Yonemura, S., Tsukita, Sa., *et al.* (1994). Direct association of occludin with ZO-1 and its possible involvement in the localization of occludin at tight junctions. *J. Cell Biol.*, **127**, 1617.

51. Itoh, M., Nagafuchi, A., Moroi, S., and Tsukita, Sh. (1997). Involvement of ZO-1 in cadherin-based cell adhesion through its direct binding to α catenin and actin filaments. *J. Cell Biol.*, **138**, 181.

52. Itoh, M., Furuse, M., Morita, K., Kubota, K., and Tsukita, Sh. (1999). Direct binding of three tight junction-associated MAGUKs, ZO-1, ZO-2 and ZO-3, with the COOH-termini of claudins. *J. Cell Biol.*, **147**, 1351.

53. Imamura, Y., Itoh, M., Maeno, Y., Tsukita, Sh., and Nagafuchi, A. (1999). Functional domains of α-catenin required for the strong state of cadherin-based cell adhesion. *J. Cell Biol.*, **144**, 1311.

54. Fanning, A. S., Jameson, B. J., Jesaitis, L. A., and Anderson, J. M. (1998). The tight junction protein ZO-1 establishes a link between the transmembrane protein occludin and the actin cytoskeleton. *J. Biol. Chem.*, **273**, 29745.

55. Tsukita, Sh., Itoh, M., Nagahchi, A., Yonemura, S., and Tsukita, Sa. (1993). Submembranous junctional plaque proteins include potential tumor suppressor molecules. *J. Cell Biol.*, **123**, 1049.

56. Takahisa, M., Togashi, S., Suzuki, T., Kobayashi, M., Murayama, A., Kondo, K., *et al.* (1996). The *Drosophila* tamou gene, a component of the activating pathway of extramacrochaetae expression, encodes a protein homologous to mammalian cell–cell junction-associated protein ZO-1. *Genes Dev.*, **10**, 1783.

57. Lue, R. A., Marfatia, S. M., Branton, D., and Chishti, A. H. (1994). Cloning and characterization of hdlg: the human homologue of the *Drosophila* discs large tumor suppressor binds to protein 4.1. *Proc. Natl. Acad. Sci. USA*, **91**, 9818.

58. Gumbiner, B., Lowenkopf, T., and Apatira, D. (1991). Identification of a 160kDa polypeptide that binds to the tight junction protein ZO-1. *Proc. Natl. Acad. Sci. USA*, **88**, 3460.

59. Jesaitis, L. A. and Goodenough, D. A. (1994). Molecular characterization and tissue distribution of ZO-2, a tight junction protein homologous to ZO-1 and the *Drosophila* discs-large tumor suppresser protein. *J. Cell Biol.*, **124**, 949.

60. Itoh, M., Morita, K., and Tsukita, Sh. (1999). Characterization of ZO-2 as a MAGUK family member associated with tight as well as adherens junctions with a binding affinity to occludin and α catenin. *J. Biol. Chem.*, **274**, 5981.

61. Balda, M. S., Gonzalez-Mariscal, L., Matter, K., Cereijido, M., and Anderson, J. M. (1993). Assembly of the tight junction: the role of diacylglycerol. *J. Cell Biol.*, **123**, 293.

62. Haskins, J., Gu, L., Wittchen, E. S., Hibbard, J., and Stevenson, B. R. (1998). ZO-3, a novel member of the MAGUK protein family found at the tight junction, interacts with ZO-1 and occludin. *J. Cell Biol.*, **141**, 199.

63. Citi, S., Sabanay, H., Jakes, R., Geiger, B., and Kendrick-Jones, J. (1988). Cingulin, a new peripheral component of tight junctions. *Nature*, **333**, 272.

64. Cordenousi, M., DÔAtri, F., Hammar, E., Parry, D. A. D., Kendrick-Jones, J., Shore, D., *et al.* (1999). Cingulin contains globular and coiled-coil domains and interacts with ZO-1, ZO-2, ZO-3 and myosin. *J. Cell Biol.*, **147**, 1569.

65. Zhong, Y., Saitoh, T., Minase, T., Sawada, N., Enomoto, K., and Mori, M. (1993). Monoclonal antibody 7H6 reacts with a novel tight junction-associated protein distinct from ZO-1, cingulin, and ZO-2. *J. Cell Biol.*, **120**, 477.

66. Keon, B. H., Schäfer, S., Kuhn, C., Grund, C., and Franke, W. W. (1996). Symplekin, a novel type of tight junction plaque protein. *J. Cell Biol.*, **134**, 1003.

67. Stevenson, B. R. and Goodenough, D. A. (1984). Zonula occludentes in junctional complex-enriched fractions from mouse liver: preliminary morphological and biochemical characterization. *J. Cell Biol.*, **98**, 1209.

68. Saitou, M., Ando-Akatsuka, Y., Itoh, M., Furuse, M., Inazawa, J., Fujimoto, K., *et al.* (1997). Mammalian occludin in epithelial cells: its expression and subcellular distribution. *Eur. J. Cell Biol.*, **73**, 222.

69. Kumar, N. M. and Gilula, N. B. (1996). The gap junction communication channel. *Cell*, **84**, 381.

70. Furuse, M., Sasaki, H., and Tsukita, Sh. (1999). Manner of interaction of heterogeneous claudin species within and between tight junction strands. *J. Cell Biol.*, **147**, 891.

71. Madara, J. L. (1987). Intestinal absorptive cell tight junctions are linked to cytoskeleton. *Am. J. Physiol.*, **253**, C171.

72. Hata, Y., Butz, S., and Sudhof, T. C. (1996). CASK: a novel dlg/PSD95 homolog with an N-terminal calmodulin-dependent protein kinase domain identified by interaction with neurexins. *J. Neurosci.*, **16**, 2488.

73. Fleming, T. P., MacConnell, J., Johnson, M. H., and Stevenson, B. R. (1989). Development of tight junctions de novo in the mouse early embryo: control of assembly of the tight junction-specific protein ZO-1. *J. Cell Biol.*, **108**, 1407.

74. Gonzarez-mariscal, L., Chavez de Ramirez, B., and Cereijido, M. (1985). Tight junction formation in cultured epithelial cells (MDCK). *J. Membrane Biol.*, **86**, 113.

75. Watabe-Uchida, M., Uchida, N., Imamura, Y., Nagafuchi, A., Fujumoto, K., Uemura, T., *et al.* (1998). α catenin-vinculin interaction functions to organize the apical junctional complex in epithelial cells. *J. Cell Biol.*, **142**, 847.

76. Yonemura, S., Itoh, M., Nagafuchi, A., and Tsukita, Sh. (1995). Cell-to-cell adherens junction formation and actin filament organization: Similarities and differences between non-polarized fibroblasts and polarized epithelial cells. *J. Cell Sci.*, **108**, 127.

77. Ando-Akatsuka, Y., Yonemura, S., Itoh, M., Furuse, M., and Tsukita, Sh. (1999). Differential behavior of E-cadherin and occludin in their colocalization with ZO-1 during the establishment of epithelial polarity. *J. Cell. Physiol.*, **179**, 115.

78. Claude, P. and Goodenough, D. A. (1973). Fracture faces of zonuolae occludentes from 'tight' and 'leaky' epithelia. *J. Cell Biol.*, **58**, 390.

79. Madara, J. L. and Pappenheimer, J. R. (1987). Occluding junction structure-function relationship in a cultured epithelial monolayer. *J. Cell Biol.*, **101**, 2124.

80. Rubin, L. L. (1991). The blood-brain barrier in and out of cell culture. *Curr. Opin. Neurobiol.*, **1**, 360.

81. Dym, M. and Fawcett, D. W. (1970). The blood-testis barrier in the rat and the physiological compartmentation of the seminiferous epithelium. *Biol. Reprod.*, **3**, 308.

82. Stevenson, B., Anderson, J. M., Goodenough, D. A., and Mooseker, M. S. (1988). Tight junction structure and ZO-1 content are identical in two strains of Madin-Darby Canine Kidney cells which differ in transepithelial resistance. *J. Cell Biol.*, **107**, 2401.

83. Zampighi, G., Bacallao, R., Mandel, L., and Cereijido, M. (1991). Structural differences of the tight junction between low and high resistance MDCK monolayers. *J. Cell Biol.*, **115**, 479.

84. Wolburg, H., Neuhaus, J., Kniesel, U., Krauss, B., Schmid, E. M., Ocalan, M., *et al.* (1994). Modulation of tight junction structure in blood-brain barrier endothelial cells. Effects of tissue culture, second messengers and cocultured astrocytes. *J. Cell Sci.*, **107**, 1347.

85. Ballard, S. T., Hunter, J. H., and Taylor, A. E. (1995). Regulation of tight-junction permeability during nutrient absorption across the intestinal epithelium. *Annu. Rev. Nutr.*, **15**, 35.

86. Milks, L. C., Conyers, G. P., and Cramer, E. B. (1986). The effect of neutrophil migration on epithelial permeability. *J. Cell Biol.*, **103**, 2729.

87. Schneeberger, E. E. and Lynch, R. D. (1994). Tight junctions: Their modulation under physiological and pathological states. In *Molecular mechanisms of epithelial cell junctions: From development to disease* (ed. S. Citi), p. 123. R. G. Landes Company, Boca Raton.

88. Hollander, D. (1988). Crohn's disease-permeability disorder of the tight junction. *Gut*, **29**, 1621.

89. Aktories, K. (1979). Bacterial toxins that target rho proteins. *J. Cell Biol.*, **99**, 827.

90. Madara, J. L. (1998). Regulation of the movement of solutes across tight junctions. *Annu. Rev. Physiol.*, **60**, 143.

91. Fasano, A., Fiorentini, C., Donelli, G., Uzzau, S., Kaper, J. B., Margaretten, K., *et al.* (1995). Zonula occludens toxin modulates tight junctions through protein kinase C-dependent actin reorganization, *in vitro*. *J. Clin. Invest.*, **96**, 710.

92. Balda, M. S., Gonzalez-Mariscal, L., Contreras, R. G., Macias-Silva, M., Torres-Marquez, M. E., Garcia-Sainz, J. A., *et al.* (1991). Assembly and sealing of tight junctions: possible participation of G-proteins, phospholipase C, protein kinase C and calmodulin. *J. Membrane Biol.*, **122**, 193.

93. Jou, T. S., Schneeberger, E. E., and Nelson, W. J. (1998). Structural and functional regulation of tight junctions by RhoA and Rac1 small GTPases. *J. Cell Biol.*, **142**, 101.

94. Jou, T. S. and Nelson, W. J. (1998). Effects of regulated expression of mutant RhoA and Rac1 small GTPases on the development of epithelial (MDCK) cell polarity. *J. Cell Biol.*, **142**, 85.

95. Citi, S. and Denisenko, N. (1995). Phosphorylation of the tight junction protein cingulin and the effects of protein kinase inhibitors and activators in MDCK epithelial cells. *J. Cell Sci.*, **108**, 2917.

96. Sakakibara, A., Furuse, M., Saitou, M., Ando-Akatsuka, Y., and Tsukita, Sh. (1997). Possible involvement of phosphorylation of occludin in tight junction formation. *J. Cell Biol.*, **137**, 1393.

97. Denker, B. M., Saha, C., Khawaja, S., and Nigam, S. K. (1996). Involvement of a heterotrimeric G protein α subunit in tight junction biogenesis. *J. Biol. Chem.*, **271**, 25750.

98. Saha, C., Nigam, S. K., and Denker, B. M. (1998). Involvement of Gαi2 in the maintenance and biogenesis of epithelial cell tight junctions. *J. Biol. Chem.*, **273**, 21629.

99. Stuart, R. O. and Nigam, S. K. (1995). Regulated assembly of tight junctions by protein kinase C. *Proc. Natl. Acad. Sci. USA*, **92**, 6072.

100. Izumi, Y., Hirose, T., Tamai, Y., Hirai, S., Nagashima, Y., Fujimoto, T., *et al.* (1998). An atypical PKC directly associates and colocalizes at the epithelial tight junction with ASIP, a mammalian homologue of *Caenorhabditis elegans* polarity protein PAR-3. *J. Cell Biol.*, **143**, 95.

101. Kemphues, K. J., Priess, J. R., Morton, D. G., and Cheng, N. S. (1988). Identification of genes required for cytoplasmic localization in early *C. elegans* embryos. *Cell*, **52**, 311.

102. Tabuse, Y., Izumi, Y., Piano, F., Kemphues, K. J., Miwa, J., and Ohno, S. (1998). Atypical protein kinase C cooperates with PAR-3 to establish embryonic polarity in *Caenorhabditis elegans*. *Development*, **125**, 3607.

103. Wu, S. L., Staudinger, J., Olson, E. N., and Rubin, C. S. (1998). Structure, expression, and properties of an atypical protein kinase C (PKC3) from *Caenorhabditis elegans*. PKC3 is required for the normal progression of embryogenesis and viability of the organism. *J. Biol. Chem.*, **273**, 1130.

104. Yamamoto, T., Harada, N., Kano, K., Taya, S., Canaani, E., Matsuura, Y., *et al.* (1997). The Ras target AF-6 interacts with ZO-1 and serves as a peripheral component of tight junctions in epithelial cells. *J. Cell Biol.*, **139**, 785.

105. Mandai, K., Nakanishi, H., Satoh, A., Obaishi, H., Wada, M., Nishioka, H., *et al.* (1997). Afadin: A novel actin filament-binding protein with one PDZ domain localized at cadherin-based cell-to-cell adherens junction. *J. Cell Biol.*, **139**, 517.

106. Weber, E., Berta, G., Tousson, A., St John, P., Green, M. W., Gopalokrishnan, U., *et al.* (1994). Expression and polarized targeting of a rab3 isoform in epithelial cells. *J. Cell Biol.*, **125**, 583.

107. Zahraoui, A., Joberty, G., Arpin, M., Fontaine, J. J., Hellio, R., Tavitian, A., *et al.* (1994). A small rab GTPase is distributed in cytoplasmic vesicles in non polarized cells but colocalizes with the tight junction marker ZO-1 in polarized epithelial cells. *J. Cell Biol.*, **124**, 101.

108. Grindstaff, K. K., Yeaman, C., Anandasabapathy, N., Hsu, S. C., Rodriguez-Boulan, E., Scheller, R. H., *et al.* (1998). Sec6/8 complex is recruited to cell-cell contacts and specifies transport vesicle delivery to the basal-lateral membrane in epithelial cells. *Cell*, **93**, 731.

109. TerBush, D. R., Maurice, T., Roth, D., and Novick, P. (1996). The Exocyst is a multiprotein complex required for exocytosis in *Saccharomyces cerevisiae*. *EMBO J.*, **15**, 6483.

110. Hsu, S.-C., Hazuka, C. D., Foletti, D. L., and Scheller, R. H. (1999). Targeting vesicles to specific sites on the plasma membrane: the role of the sec6/8 complex. *Trends Cell Biol.*, **9**, 150.

111. Diamond, J. (1977). The epithelial junction: bridge, gate, and fence. *Physiologist*, **20**, 10.
112. Claude, P. (1978). Morphological factors influencing transepithelial permeability: a model for the resistance of the zonula occludens. *J. Membrane Biol.*, **10**, 219.
113. Reuss, L. (1992). Tight junction permeability to ions and water. In *Tight junctions* (ed. M. Cereijido), p. 49. CRC Press, London.
114. Simon, D. B., Lu, Y., Choate, K. A., Velazquez, H., Al-Sabban, E., *et al.* (1999). Paracellin-1, a renal tight junction protein required for paracellular Mg^{2+} resorption. *Science*, **285**, 103.
115. Tsukita, S., Furuse, M., and Itoh, M. (2001). Multifunctional strands in tight junctions. *Nature Rev. Mol. Cell Biol.*, **2**, 285.

Index